1994

Ant – Plant Interactions

A volume arising from the symposium held at
Oxford in July 1989, by the Linnean Society of
London and the Plant Sciences and Zoology
Departments of the University of Oxford

Ant – Plant Interactions

Edited by

CAMILLA R. HUXLEY

Department of Plant Sciences, University of Oxford

and

DAVID F. CUTLER

Royal Botanic Gardens, Kew, Richmond, Surrey

Oxford   New York   Tokyo
OXFORD UNIVERSITY PRESS
1991

Oxford University Press, Walton Street, Oxford OX2 6DP

Oxford New York Toronto
Delhi Bombay Calcutta Madras Karachi
Petaling Jaya Singapore Hong Kong Tokyo
Nairobi Dar es Salaam Cape Town
Melbourne Auckland

and associated companies in
Berlin Ibadan

Oxford is a trade mark of Oxford University Press

Published in the United States
by Oxford University Press, New York

© Oxford University Press 1991

A catalogue record for this book is
available from the British Library

Library of Congress Cataloging in Publication Data
Ant–plant interactions/edited by Camilla R. Huxley and David F. Cutler.
Composed largely of contributions to the International Symposium
on Interactions between Ants and Plants, held in Oxford on 6–8 July
1989 and jointly sponsored by the Linnean Society of London and the
Zoology and Plant Sciences departments of the University of Oxford.
Includes bibliographical references and indexes.
1. Ant—Ecology—Congresses. 2. Ants—Evolution—Congresses.
3. Insect–plant relationships—Congresses. I. Huxley, Camilla R.
II. Cutler, D. F. (David Frederick), 1939– . III. Linnean Society
of London. IV. University of Oxford. Dept. of Zoology.
V. University of Oxford. Dept. of Plant Sciences.
VI. International Symposium on Interactions Between Ants and Plants
(1989: Oxford, England)
QL568.F7A55 1991 595.79′6′0452482—dc20 91-16137
ISBN 0-19-854639-4

Typeset by Joshua Associates Ltd, Oxford
Printed in Great Britain by
Biddles Ltd, Guildford and King's Lynn

This book is dedicated to the memory of
Mary Snow, Oxford botanist, 1902–1978

152,104

Preface

This book is composed largely of contributions to the international symposium on interactions between ants and plants held in Oxford on 6–8 July 1989 and jointly sponsored by the Linnean Society of London and the Zoology and Plant Sciences Departments of the University of Oxford. We are most grateful to these institutions and also to the Royal Society and the Tansley Fund for their financial support. Our warmest thanks go in particular to Professor Sir Richard Southwood, Dr B. E. Juniper, and Professor R. M. May, for their personal support.

The papers presented here are grouped into sections according to the nature of the interaction. This inevitably results in an artificial and sometimes awkward arrangement. Moreover, topics are unevenly covered: areas in which there is currently a large body of significant work are well represented while other fields, such as seed harvesting and plant structural adaptation, are relatively neglected. There is some overlap between chapters, but this has been retained only where necessary to preserve the flow of the text. The references are grouped at the end of each part in order to provide a good bibliography on the topic. An exception to this rule is Part 6, where the references remain at the end of each chapter because the subject matter of the four chapters concerned is so diverse. We would like to thank David Nash for his work on the figures.

Oxford   C.R.H.
Kew   D.F.C.
August 1990

Contents

PART 3   EXTRAFLORAL
NECTARY-MEDIATED INTERACTIONS

PART 4   SYMBIOSIS BETWEEN PLANTS AND ANTS

PART 5   POLLINATION, ANT EXCLUSION, AND DISPERSAL

PART 6   ANTS, VEGETATION ECOLOGY, AND THE FUTURE OF ANT–PLANT RESEARCH

Contributors

The numbers in square brackets refer to the author's chapters.

John F. Addicott Department of Zoology, University of Alberta, Edmonton, Alberta T6G 2E9, Canada [8]

Alan N. Andersen Division of Wildlife and Ecology, CSIRO Tropical Ecosystems Research Centre, PMB 44 Winnellie, NT 0821, Australia [33, 36]

M. Baylis Tsetse Research Laboratory, Department of Veterinary Sciences, Langford House, Langford, Bristol BS18 7DU, UK [11]

Andrew J. Beattie School of Biological Sciences, Macquarie University, Sydney, NSW 2109, Australia [27, 37]

David H. Benzing Department of Biology, Oberlin College, Oberlin, Ohio 44074, USA [23]

W. J. Bond Botany Department, University of Cape Town, Private Bag, Rondebosch 7700, South Africa [30]

Carlos R. F. Brandão Museu de Zoologia, Universidade de São Paulo, O1051 São Paulo SP, Brazil [14]

Catherine M. Bristow Department of Entomology, Michigan State University, East Lansing, MI 48824, USA [9]

E. R. Carper Department of Biology, Princeton University, Princeton, NJ 08544-1003, USA [11]

Silvia Claver CRICYT, Parque San Martin, Mendoza, Argentina [5]

Stephen G. Compton Department of Zoology and Entomology, Rhodes University, PO Box 94, Grahamstown 6140, South Africa [10, 16]

J. Hall Cushman School of Biological Sciences, Macquarie University, Sydney, NSW 2109, Australia [8]

Diane W. Davidson Department of Biology, University of Utah, Salt Lake City, Utah 84112, USA [20]

P. J. DeVries Department of Zoology, University of Texas at Austin, Austin, TX 78712, USA [12]

D. H. Feener, Jr. Department of Biology, University of Utah, Salt Lake City, Utah 84112, USA [13]

Brigitte Fiala Zoologisches Institut III, Universität Würzburg, Röntgenting 10, D-8700, Würzburg, Germany [18]

Brian L. Fisher Department of Biology, University of Utah, Salt Lake City, Utah 84112, USA [20]

H. G. Fowler Instituto Biociencias, Universidade Estadual Paulista, Rio Claro, Brazil [5]

Steven N. Handel Department of Biological Sciences, Rutgers University, Piscataway, NJ 08855, USA [27]

Ray Harley Royal Botanic Gardens, Kew, Richmond, Surrey TW9 3AB, UK [28]

Seigo Higashi Graduate School of Environmental Science, Hokkaido University, Sapporo 060, Japan [32]

Carol C. Horvitz Department of Biology, University of Miami, Coral Gables, FL 33124, USA [31]

Jerome J. Howard Department of Entomology, University of Arizona, Tucson, AZ 85721, USA [4]

L. Hughes School of Biological Sciences, Macquarie University, Sydney, NSW 2109, Australia [29]

Camilla R. Huxley Department of Plant Sciences, University of Oxford, South Parks Road, Oxford OX1 3RB, UK [1]

Fuminori Ito Graduate School of Environmental Science, Hokkaido University, Sapporo 060, Japan [32]

Matthew Jebb Christensen Research Institute, PO Box 305 Madang, Papua New Guinea [24]

Pierre Jolivet 67 Boulevard Sould, 75012 Paris, France [26]

Timothy J. King 93 Kingston Road, Oxford OX2 6RL, UK [34]

Suzanne Koptur Department of Biological Sciences, Florida International University, Miami, FL 33199, and Fairchild Tropical Garden, Miami, FL 33156, USA [15]

John H. Lawton Centre for Population Biology, Imperial College, Silwood Park, Ascot, Berks SL5 7PX, UK [16]

D. K. Letourneau College Eight, University of California, Santa Cruz, CA 95064, USA [25]

John T. Longino Allyn Museum of Entomology, 3621 Bay Shore Road, Sarasota, FL 34234, USA [19]

D. A. Mackay School of Biological Sciences, Flinders University, Bedford Park, SA 5042, Australia [17]

Doyle McKey Department of Biology, University of Miami, PO Box 249118, Coral Gables, FL 33124-0421, USA [21]

Ulrich Maschwitz Fachbereich Biologie-Zoologie, J. W. Goethe-Universität, Siesmayerstrasse 70, D-6000 Frankfurt-am-Main, Germany [18]

D. R. Nash Centre for Population Biology, Imperial College, Silwood Park, Ascot, Berks SL5 7PX, UK [11]

Paulo S. Oliveira Departamento de Zoologia, Universidade Estadual de Campinas CP 6109, 13081 Campinas SP, Brazil [14]

Rod Peakall School of Biological Sciences, Macquarie University, Sydney, NSW 2109, Australia [27]

N. E. Pierce Department of Biology, Princeton University, Princeton, NJ 08544, USA [11]

R. J. Powell Paignton Zoological and Botanical Gardens, Totnes Road, Paignton, Devon TQ4 7EU, UK [3]

Vanessa K. Rashbrook School of Biological Sciences, University of Sydney, Sydney, NSW 2006, Australia [16]

B. L. Rice School of Biological Sciences, Macquarie University, Sydney, NSW 2109, Australia [29]

H. G. Robertson The South African Museum, PO Box 61, Cape Town 8000, South Africa [10]

Rainer Rosengren Department of Zoology, University of Helsinki, N. Järnvägsgatan 13, Helsinki 10, Finland [7]

Eugene W. Schupp Savannah River Ecology Laboratory, Drawer E. Aiken, SC 29802, USA [13]

W. D. Stock Botany Department, University of Cape Town, Private Bag, Rondebosch 7700, South Africa [30]

D. J. Stradling Department of Biological Sciences, Hatherly Laboratories, Prince of Wales Road, Exeter EX4 4PS, UK [2, 3]

Liselotte Sundström Department of Zoology, University of Helsinki, N. Järnvägsgatan 13, Helsinki 10, Finland [7]

Tho Yow Pong Forest Research Institute Malaysia, 52109 Kepong, Malaysia [18]

John E. Tobin Harvard University, Museum of Comparative Zoology Laboratories, 26 Oxford Street, Cambridge, MA 02138, USA [35]

Philip S. Ward Department of Entomology, University of California, Davis, CA 95616, USA [22]

M. Westoby School of Biological Sciences, Macquarie University, Sydney, NSW 2109, Australia [29]

M. A. Whalen School of Biological Sciences, Flinders University, Bedford Park, SA 5042, Australia [17]

J. B. Whittaker Division of Biological Sciences, Institute of Environmental and Biological Sciences, Lancaster University, Lancaster LA1 4YQ, UK [6]

Stanley R. J. Woodell Department of Plant Sciences, University of Oxford, South Parks Road, Oxford OX1 3RB, UK [34]

R. Yeaton Botany Department, University of Natal, Pietermaritzburg, South Africa [30]

1

Ants and plants: a diversity of interactions

Camilla R. Huxley

Partners and antagonists

Interactions between ants and plants are extraordinarily diverse. Ants may act on plants as herbivores, defensive agents, seed dispersers, pollinators, or providers of macronutrients (nitrogen, phosphorus, potassium). Conversely, plants may supply a range of nutrients, or nest sites, or attract ants with chemicals which are secondary to the main metabolic processes. Many plants also exclude ants as pollinators, and certain carnivorous plants specialize in trapping ants. Among this wide range of interactions, the extent of benefit or loss to each participant depends critically on the local conditions, in particular the biotic environment. The defensive effect of ants, for instance, depends directly on the susceptibility of the pests of a plant to attack by ants. On the other hand ants often promote certain herbivores such as sap-sucking Homoptera and lycaenoid butterfly larvae.

An outstanding feature of the interactions between ants and plants is the high frequency with which these interactions are mutualistic, i.e. of benefit to both parties. A vast majority of insects antagonistically consume plants; but among ants only the seed harvesters and the neotropical leaf-cutters are extensive and direct herbivores. It has long been known that ants, like other animals, act mutualistically as dispersal agents of plant seeds. Certain plants, myrmecochores, are specifically adapted for ant dispersal having oil-rich elaiosomes (seed appendages which are attractive to ants). Many unspecialized seeds, and especially fruits of grasses, are collected by seed harvester ants; but the success of seed dispersal relative to consumption by different ants is unknown, and frequently may be low, depending on climatic and biotic conditions.

The role of ants as pollinators is also equivocal with many flowers having ant-exclusion devices, and only a very few having been shown conclusively to be ant pollinated. It appears that antibiotic secretions of ants inhibit pollen germination. Besides the mutualisms of dispersal and pollination, ants enter extensively into two other mutualisms with plants—defending plants against herbivore attack, and providing a source of macronutrients. No other animals regularly enter into such mutualisms. Incidental defence of plants by predacious ants is extensive and is sometimes augmented by the presence of

Homoptera which attract ants to the plant. Many plants produce extrafloral nectar which attracts ants. The presence and aggressive behaviour of visiting ants has been shown to reduce herbivory levels on plants with extrafloral nectaries. The plants which obtain nutrients from ants are mostly epiphytes which have roots permeating the carton nests of certain arboreal ant species. However, both these defence and nutrient mutualisms are open to parasitization by ants and plants which take the rewards without conferring any benefit.

Not only are many interactions between ants and plants mutualistic, some hundreds of symbiotic associations are known, with the ants actually nesting inside the plants (myrmecophytes) (for a recent listing see Wilson and Hölldobler 1990). The preformed nest sites (domatia) may be derived from naturally hollow structures such as stems, which additionally have a weak spot where ants can gain access; food bodies and/or nectar are also produced in some genera. Other domatia are formed by invagination of stems, or by leaf sacs; in these species food bodies are usually absent and the ants tend Homoptera inside the domatia. The evolution of domatia raises the question of how such specializations first arose and how they are selected for. Some genera of symbiotic ants and plants are highly diverse and species specific in their symbioses, although mutualism does not tend to increase species-specificity in the way in which parasitism does (Howe 1984; Law and Koptur 1986). Phylogenetic analysis is used in this book for the first time to begin to unravel the history of two of these interactions.

Symbiosis between ants and plants is outstanding as the only extensive suite of symbioses between macroscopic animals and higher plants. Symbiosis between higher and lower organisms is, on the other hand, frequent; most herbivores depend on bacteria, or protista, to assist in detoxification and digestion of plant material. Many plants are symbiotic either with nitrogen-fixing prokaryotes or with fungi that accumulate macronutrients. Ants also enter a symbiosis with lower organisms in the fungus gardens tended by leaf-cutting ants. Here again it is the biochemical capacity of the fungi, as detoxifying and digesting agents, which is of importance. The symbiosis between ants and higher plants, however, usually arises from the nutritive and structural capacity of the plants, and the unspecialized defensive action of ants. It is notable that at least two other groups of organisms interact with ants in a rather similar way. Sap-sucking Homoptera and lycaenoid butterfly larvae form mutualisms in which they provide nutrients for tending ants, and in turn are protected from predators and parasitoids. It is striking that both ant-associated plants and insects sometimes use chemicals to appease or deceive the ants rather than providing a real, beneficial reward.

The study of mutualisms was unfashionable early this century, perhaps partly for political reasons (Boucher 1985). The pioneering work of D. H. Janzen (1966, 1969, 1972, 1974) on ant–plant mutualisms has stimulated much work on these systems, and on mutualisms generally. It is perhaps significant that interest in mutualisms is reflected in the gender of the investigators; a relatively high proportion (just over a fifth) of the senior authors in this book are female. The study of mutualisms has also been favoured by recent advances in the analysis and modelling of systems which depend on many factors, being conditional on a diversity of abiotic and biotic circumstances (Letourneau 1988). It is not unusual for the size, and even the sign, of the effect of one organism on another to be highly dependent on other factors. With the study of chaos now beginning to be applied in ecology we shall also have to consider the implications of cyclical chaotic behaviour on the complex systems described here.

One of the questions not raised in this book is, why ants? In dispersal and pollination mutualisms, the animal provides the mobility which plants lack. However, for plant defence and nutrient provision, it is the eusociality of ants and their large, often permanent colonies that are important (Huxley 1986), and the walking habit of ants may also contribute to their effectiveness in these roles. However, the eusociality of ants may have had various effects. The workers are physically little specialized, but behaviourally very flexible relative to other social Hymenoptera. More interesting is the possibility that the eusocial habit reduces the directness of selective pressure on the behaviour of individual workers. This is illustrated by the leaf-cutter ants which appear to have failed to evolve not to cut fungicidal plants (Powell and Stradling, Chapter 3). Hence, perhaps, the observation that in ants there is a high frequency of loitering among workers and parasitism between species, as well as mutualism with both insects and plants.

Overview of the chapters

Part 1

In Part 1 the interactions of leaf-cutter ants with plants are described. In Chapter 2 Stradling has provided an introduction to the biology of these neotropical ants which bring vegetable matter into their nests where it is detoxified and digested by fungi. This symbiosis has led to the leaf-cutter ants being remarkably polyphagous. Chapter 3 by Powell and Stradling is a detailed investigation of the capacity of the fungi of leaf-cutter ants to survive and utilize various secondary plant compounds. An extensive comparison is made of the response of both ants and fungi to a range of plant species. Ants often collect quite inappropriate material; A. M. Sugden (personal communication) reported leaf-cutter ants cutting and bringing back discs from plastic labels; these discs were discarded around the nest entrance. Cutting plastic discs cannot be explained as providing a sap source for the foragers to drink. These results suggest that natural selection, acting via symbionts in large colonies, is unable to provide an effective feedback on ant choice. It appears that the solution adopted by leaf-cutters is to maintain a high diversity of substrate and thus dilute the effects of toxic allelochemicals.

Another aspect of resource use by leaf-cutter ants is discussed in Chapter 4 by Howard who analyses the relative costs of foraging on trees at different distances from the nest. The somewhat counter-intuitive results lead Howard to suggest various hypotheses to account for the behaviour. However, it is difficult to propose tests which would distinguish between the hypotheses.

In Chapter 5, the last on leaf-cutter ants, Fowler and Claver describe the richness of species assemblies with respect to geographical area and openness of habitat. An unexpected contrast is found between the open habitats of subtropical South America and the perhumid tropics. The correlation between local species richness and regional diversity has implications for the saturation of local communities.

References. At the end of Part 1 will be found the entire set of references for the chapters in this section. The aim here is to provide a good bibliography of the topic, and Parts 2–5 follow suit. In Part 6 the references are given at the end of each chapter, as a unified bibliography for these four diverse topics would be of little use.

Part 2

Interactions in which unspecialized plants benefit (or otherwise) from the predacious and honeydew-collecting activities of ants are discussed in this section. Whittaker, in Chapter 6, describes how the predacious behaviour of wood ants, *Formica*, affects trees growing close to their large, long-lasting nest mounds. Predation by *Formica* is particularly effective during epidemics of susceptible herbivores, such as lepidopteran larvae. However, the level of ant activity on a tree, and hence predation, is highly dependent on the honeydew production by Homoptera. Rosengren and Sundström (Chapter 7) analyse the interaction between *Formica*, pine aphids, and pine growth. The ants utilize honeydew for carbohydrate requirements and also feed on nymphal aphids, perhaps for the protein requirements of the brood, but also in ways which benefit both the pines and the remaining aphids. Rosengren and Sundström raise the possibility of reciprocal altruism by both aphids and pines. This would be promoted by the high degree of relatedness of generations of ants occupying the same nest, and the not very mobile pine population. However, in many biotic environments the tending of Homoptera by ants is deleterious for plants, and Beccara and Venable (1989) suggest that the presence of extrafloral nectaries may sometimes benefit the plant by attracting the ants away from the Homoptera.

The variation in size, and even change of sign, of the effects of interactions between organisms is the subject of Chapter 8 by Cushman and Addicott. Ant–homopteran interactions vary not only with the species of ant and homopteran, but also from season to season and year to year as the levels of tending and of predators and parasitoids vary. The short- and long-term implications of such varying selection pressure on the evolution of mutualisms have yet to be fully worked out.

Bristow (Chapter 9) discusses another factor which influences the outcome of ant–homopteran interactions; namely the quality of honeydew as perceived by the ants. This quality varies both with host plant species and with the part of the plant where the Homoptera feed. Only some 20 per cent of homopterans are ant-tended; the rest have different means of defence against predators and parasitoids. It is perhaps noteworthy that both Bristow, working on oleander, and Compton and Robertson, on fig trees, found that Homoptera feeding near the flowers attracted more ants than did Homoptera feeding on vegetative parts of the same plants. This concentration of ants in the neighbourhood of the developing seeds might have beneficial effects for the plants. In the case of the Cape fig studied by Compton and Robertson (Chapter 10), the presence of ants around the developing figs had complex results owing to the presence of internally and externally ovipositing parasitic wasps and also to predation by the ants on both parasitic and pollinating fig wasps.

Besides Homoptera another major group of insects are tended by ants, the lycaenoid butterflies in the larval, and sometimes pupal, stages. In Chapter 11 Pierce, Nash, Baylis, and Carper describe experiments to investigate the effects of distance from the ant nest, alternative food sources, and quality of secretion on the level of tending by ants and degrees of benefit to a lycaenid butterfly. Related to the lycaenids are the riodinids which are less often ant-associated, and usually prefer plants with extrafloral nectar. In Chapter 12 DeVries integrates the ecology with work on the homology of the organs connected with ant interaction to shed light on the evolution of interaction with ants. DeVries also draws attention to similarities in function between the secretions of lycaenoids, Homoptera, and extrafloral nectaries.

Part 3

Here extrafloral nectaries, their attraction to ants, and the benefits which these ants may or may not bring are discussed. Extrafloral nectaries are usually functionally defined as nectaries which attract animals that are not involved in pollination. Thus the petiolar nectaries of acacia (Knox *et al.* 1986) which attract pollinating birds would not count as extrafloral, while the secretory discs on developing fruit do count. This terminology may however be disrupted by the suggestion by Dominguez *et al.* (1989) that flower nectar of *Croton* sp. may attract predatory wasps which both pollinate the flowers and reduce herbivory. Extrafloral nectaries are most common in the wet tropics. Schupp and Feener (Chapter 13), working on Barro Colorado Island, Panama, found that about a third of the woody dicotyledons and herbaceous vines had vegetative extrafloral nectaries and/or food bodies which are probably collected by ants. They then analysed the correlation of extrafloral nectaries with the lifeform, habitat, and phylogeny of the species surveyed.

The ant fauna visiting extrafloral nectary plants has been studied by Oliveira and Brandão in the cerrados of Brazil (Chapter 14). A wide range of ants was found visiting extrafloral nectaries on two species of shrub, with different ant species visiting by day and by night. However, when ants were censused by recording attacks on live termite baits, a different ant assembly was found. Koptur (Chapter 15) describes the differential distribution of extrafloral nectaries on reproductive and vegetative parts of herbs and woody plants. She compares predation due to ants and to parasitoids which also visit extrafloral nectaries. A model is used to simulate the effects of parasitoids on trait groups in populations with differing proportions of plants with and without extrafloral nectaries.

The fallibility of ant-guard systems mediated by extrafloral nectaries is illustrated by the work of Rashbrook, Compton, and Lawton (Chapter 16), and of Mackay and Whalen (Chapter 17). Rashbrook *et al.* discuss a series of studies on the effects of ants visiting bracken fern nectaries; or rather the lack of effects, for no significant reduction in herbivory has been found when ants are present. Various hypotheses might account for the presence of active nectaries in bracken. For instance, ants may in fact be maintaining pests at low levels at which they are not significantly affected by the ants; but if ants were excluded from a substantial area for a longer time, pest numbers might build up to levels at which ants do cause significant reduction in herbivory. Alternatively, ants might be limiting in epidemic years; bracken is a perennial plant with a very low rate of new establishment and might therefore be endangered by epidemic outbreaks of pests sufficient to kill local populations. Bracken is also well equipped with chemical defences, so ant-exclusion experiments on an ecological time-scale might not result in extra herbivory. On an evolutionary time-scale, having multiple defences may be important in slowing the rate of adaptation of herbivores to overcome any particular defence. Another explanation is that this may be a 'ghost of mutualism past', in Rosengren and Sund-ström's twist to Connell's phrase, where there is no current selection pressure in favour of nectaries, but this pteridophyte is too conservative to lose them.

Mackay and Whalen's study (Chapter 17) covers part of the interesting series of euphorb trees in the Far East. This is a latitudinal series which includes myrmecophytic trees (with domatia and food bodies) restricted to Malaysia, there are extrafloral nectary species in Malaysia, New Guinea, and Northern Australia, and plants with no association with ants in all these areas and extending further south. In New Guinea, Mackay and Whalen were able to demonstrate reduction of herbivory on extrafloral nectary trees by ants, but comparable experiments in Australia did not show an effect. The background levels of herbivory tended to be lower in Australia suggesting that the decreasing frequency of complexity of association with ants with increasing latitude might reflect lower herbivore pressure. The productivity of the environment is probably also dropping with increase in latitude, making biotic defence less efficient than chemical defence.

Part 4

This section covers myrmecophytes which are all plants of the ever-wet or seasonal tropics. These plants fall into two categories both through life form and the major role of the ants. There are ant-trees which have domatia in the stems, petioles, or leaves, and are primarily protected by their ants from herbivores; and there are ant-house epiphytes which house ants in a variety of stem and leaf structures, and benefit mainly from nutrients collected by the ants.

Many ants inhabiting ant-trees are not only aggressive to other insects but also prune encroaching vegetation, preventing access by potential enemies of the ants and also by walking herbivores. This defence may be effective against leaf-cutter ants, a peculiarly neotropical group (Jolivet 1990). Some ant-trees are relatively unspecialized such as the 'new' myrmecophyte *Tetrathylacium costaricense* (Flacourtiaceae) described by L. E. Tennant (unpublished work), where 62 per cent of trees had ants in the hollow twigs, and 16 different species of ant were found altogether. These trees produce no food bodies or nectar; the ants feed largely on honeydew from Homoptera.

In contrast Fiala Maschwitz, and Tho Yow Pong (Chapter 18) describe the most extensive Malaysian tree myrmecophyte *Macaranga* (Euphorbiaceae). Here, a series of some nine plant species are mainly inhabited by one species of *Crematogaster* ant. This ant is highly aggressive towards insect intruders, although it does not eat them. The ant also prunes vines and shows cleaning behaviour which removes herbivore eggs.

Ant-trees and tree-ants reach their greatest abundance in the neotropics. This is illustrated by Longino's work on five species of *Azteca* which inhabit the stems of four species of *Cecropia* in Costa Rica (Chapter 19). The varied behaviour of these ants suggests that they provide different levels of protection to the plants; some may be effectively parasites whilst others may be in an early stage of adaptation to mutualism. It is interesting to consider whether there might be a tendency for mutualists to become parasitic or at least commensal, taking the rewards but incurring fewer costs; rather like the converse of some diseases which start virulent and become benign.

In Chapter 20 Davidson and Fisher analyse a complex series of *Cecropia* interactions with respect to light regime. In Peruvian Amazonia one of each of three pairs of *Cecropia* species occupies closed, and the other disturbed, forest habitats. The light level in different habitats apparently correlates with the resource provisioning for ants relative to the presence of chemical defences in a complex manner, depending on leaf lifespan. It is interesting that there appears to be greater niche separation for the ants among the shade-tolerant species; one ant is predominant in all three high-light species, while each shade-tolerant tree species is mainly inhabited by a different ant species.

A new and powerful tool for formulating hypotheses about the evolutionary history of animal–plant associations is phylogenetic analysis. A cladistic analysis of the group gives likely phylogenies which are then matched against functional information about the association to give the frequency and relative timing of association events. Ideally cladistic analysis should be applied to both participants in an association and the phylogenetic trees compared. So far however, this has been done for only the ants or the plants separately. McKey's work on *Leonardoxa* (Caesalpiniaceae) shows how the approach could be used to demonstrate co-evolution in the strict sense (Chapter 21). Ward, in Chapter 22, applies the cladistic technique to representative species of the largest group of obligate plant-ants, the Pseudomyrmecinae, which inhabit a wide range of neotropical myrmecophytes including *Acacia*, *Cordia*, *Triplaris*, and *Tachigali*. Evidence emerges for association with myrmecophytes having evolved independently several times in this group, and for ant lines switching from one myrmecophyte to another. Evidence for species-specific pairing and co-evolution are discussed.

The ant-associated epiphytes are outlined in Chapter 23 by Benzing, who brings together information on ant gardens in which an arboreal, carton ant nest is supported by roots of epiphytes, and ant-house epiphytes where the plant forms some kind of domatium within which the ants live. Ant-gardens are a feature of the neotropics in particular, each 'garden' is usually inhabited by two ant species living in parabiosis, sometimes with a third, probably parasitic, species. Ant-garden plants produce seeds with elaiosomes; the ants collect the diaspores and incorporate the seed into the carton-nest walls. The seedlings probably derive nutrients from the carton nest and the nest is structurally supported by the roots. The work of Davidson and Epstein (1989) has shown that chemicals in the elaiosomes may play a complex role in attracting ants and having beneficial fungistatic effects in the garden.

Unlike the ant-trees, any one ant-house epiphyte usually harbours only a part of an ant colony, which is distributed among many epiphytes, often of different genera, in a given host tree. Domatia of ant-epiphytes are formed from leaves, e.g. *Dischidia*, rhizomes, e.g. the fern *Lecanopteris*, or tubers, e.g. the Hydnophytinae, a subtribe of the Psychotrieae, Rubiaceae containing *Hydnophytum* and *Myrmecodia*. In Chapter 24 Jebb outlines the remarkable tuber structures of the Hydnophytinae, and their pattern of ant occupation. Domatia of ant-epiphytes have to carry out two functions: to house the ant colony and to absorb nutrients from the ant waste. These functions may be separated in time, as in *Dischidia* where young leaves house brood while older leaves typically have adventitious roots ramifying among the debris. The two functions are separated in space in the Hydnophytinae, some of the cavity surfaces being smooth and impervious, while others are warted and absorptive. Among the genera of the *Hydnophytinae* different mechanisms maintain the proportion of smooth to warted surfaces during the growth of the plant.

Parasitization emerges as a major problem of ant–plant mutualisms. Ant species which are not such efficient defenders may compete for occupation of the hollow stems of ant-trees. Plants may insert their roots into the nutrient-rich carton of ant-gardens or into the domatia of ant-house epiphytes. Other insects may adapt to feeding on food bodies or nectar, and by repelling or appeasing the ants be able to live alongside them. Letourneau (Chapter 25) describes a remarkable parasitization of the already remarkable symbiosis between *Piper* and *Pheidole* in Central America. In *Piper* the presence of the symbiotic ants is necessary to stimulate the production of food bodies. However, a beetle, normally predacious on ant brood, seems to have gone one step further. Various features of the biology of ant-plants may be interpreted as reducing access for parasites. The most diverse and extensive genus of myrmecophytes, *Cecropia*, is also probably the most parasitized. In Chapter 26 Jolivet describes the remarkable adaptations of *Coelomera* beetles to living with *Azteca* on these ubiquitous plants.

Part 5

This section includes the role of ants both in floral mechanisms and in seed dispersal and predation. Ants are rarely agents of pollination; it is far more frequent to find ant-exclusion devices in and around flowers. Harley (Chapter 28) describes a group of neotropical labiates where the stem and pedicel structure appears to have implications for ant access. In Chapter 27, Peakall, Handel, and Beattie review the reported cases of ant pollination, concluding that only two are substantiated. Beattie (Chapter 37) discusses features which enable these plants to avoid the antibiotic secretions of ants.

Myrmecochory, ant-dispersal of seeds with elaiosomes, is far more frequent in plants of nutrient-poor soils than in plants of nutrient-rich areas. The explanation for this has been elusive, though studies led by Bond in South Africa and by Westoby in Australia appear to be about to reach a similar conclusion, based on the relative cost in nutrients of ant and bird dispersal (Chapters 29 and 30). Bond *et al.* also discuss the effect of wind dispersal which tends to concentrate seeds around existing vegetation while ants disperse seed to different locations.

Habitat difference favouring dispersal by birds or ants is also the subject of Chapter 31 by Horvitz, but in quite different plants, perennial herbaceous Marantaceae of the disturbed and closed Costa Rican rainforest. The size of gap appears to be crucial to the relative advantages of adaption for bird or ant dispersal. Higashi and Ito (Chapter 32) have studied a temperate plant (*Trillium*) which might be evolving from bird to ant dispersal, but in the habitat studied it is suffering excessively from nocturnal ground beetles which eat the elaiosomes. Different aspects of release of seeds might be adaptations which would help circumvent this predator.

Unfortunately no new studies on seed harvesting are presented in this book and the question of the extent to which harvesting constitutes predation or dispersal for different ant–plant pairs in different environments does not seem to have been fully addressed. In Chapter 33 Andersen reviews the extraordinarily high diversity of seed-harvesting ants in Australia, and the prevalence of not only herbs, but also of shrubs and trees, with seeds harvested by ants.

Part 6

The final section of this book concerns the role of ants in different vegetation types, parallels between plants and ants and their communities, and the future of ant–plant research.

Woodell and King (Chapter 34) describe the various effects of mound-building ants on temperate grassland vegetation. As mounds are built, the soil-heaping buries some plants but maintains open regeneration niches suitable for others. Other physical conditions on mounds also affect plant distribution. The presence of ant-mounds maintains a series of species which would otherwise be lost from the sward.

In Chapter 35 Tobin gives a resumé of the preliminary results from a study to determine the total canopy fauna of two forest trees in Peruvian Amazonia. A surprisingly high percentage of arthropod individuals and biomass were ants. This has implications for the role of canopy ants as primary consumers or predators. But it is arguable whether feeding on honeydew, possibly a major food source, should count as primary or secondary consumption.

Andersen (Chapter 36) develops a theme of parallels between plants and ants, and the effects this may have had on their community structure. Similarities between the pattern of resource requirement and growth of ants and plants are explored. In addition, the fixed position of both plants and most ant colonies means that their competitive relationships are similar, and differ from those of mobile organisms. Andersen describes features of ant and plant communities in Australia which suggest parallel processes, e.g. dominance and opportunism. The modular growth of both groups may make it easier for symbiotic partners to grow in step with one another.

Finally Beattie (Chapter 37) reviews the areas where research into ant–plant interactions is needed most pressingly. The demography and genetics of mutualistic organisms are almost untouched, as are the effects of pollinators and dispersers on gene flow.

However, as R. M. May reminded the meeting, it is increasingly and sadly apparent that the complete ecosystems in which many mutualisms arose and evolved are being rapidly diminished. Rosengren and Sundström suggest that the interaction between *Formica*, pine aphids, and Scots pine in Finland has already been modified by air pollution. In the Cape fynbos, Bond describes how an introduced ant is potentially lethal to plant species because of its inappropriate treatment of native seeds. While conservationists might achieve the aim of preserving functional examples of most mutualisms in reserves, it would be a more fundamental aim to retain the conditions where evolutionary change of species can continue. For, in nature, no balance is likely to be permanently stable. However static a system may appear, it is always potentially in a state of flux. For instance, participants in successful mutualisms will eventually suffer pressure from further evolution of parasites and predators. It is this ever more complex and changing pattern of species which will be the most challenging subject of study and the most difficult eco-system, or evosystem, to conserve.

References

Beccarra, J. X. I. and Venable, D. L. (1989). Extrafloral nectaries: a defense against ant-Homoptera mutualism? *Oikos*, **55**, 276–80.

Boucher, D. H. (1985). The idea of mutualism, past and future. In *The biology of mutualism*, (ed. D. H. Boucher), pp. 1—25. Croom Helm, London.

Davidson, D. W. and Epstein, W. W. (1989). Epiphytic associations with ants. In *Vascular plants as epiphytes* (ed. U. Lüttge), pp. 200—33. Springer Verlag, Berlin.

Dominguez, C. A., Dirzo, R., and Bullock, S. H. (1989). On the function of floral nectar in *Croton suberosus* (Euphorbiaceae). *Oikos*, **56**, 109–14.

Howe, H. F. (1984). Constraints on the evolution of mutualisms. *The American Naturalist*, **123**, 764–77.

Huxley, C. R. (1986). Evolution of benevolent ant–plant relationships. In *Insects and the plant surface* (ed. B. E. Juniper and T. R. E. Southwood), pp. 257–82. Arnold, London.

Janzen, D. H. (1966). Coevolution of mutualism between ants and acacias in Central America. *Evolution*, **20**, 249–75.

Janzen, D. H. (1969). Allelopathy by myrmecophytes: the ant as an allelopathic agent of *Cecropia. Ecology*, **50**, 147–53.

Janzen, D. H. (1972). Protection of *Barteria* by *Pachysima* ants (Pseudormyrmecinae) in a Nigerian rainforest. *Ecology*, **53**, 885–92.

Janzen, D. H. (1974). Epiphytic myrmecophytes in Sarawak, Indonesia: mutualism through the feeding of plants by ants. *Biotropica*, **6**, 237–59.

Jolivet, P. (1990). Relative protection of *Cecropia* trees against leaf-cutting ants in tropical America. In *Applied Myrmecology*, (ed. Vander Meer *et al.*), pp. 251–4. West View Press, Boulder, Colorado, USA.

Knox, R. B., Margison, R., Kenrick, J., and Beattie, A. J. (1986). The role of extrafloral nectaries in *Acacia*. In *Insects and the plant surface* (ed. B. E. Juniper and T. R. E. Southwood), pp. 295–307. Arnold, London.

Law, R. and Koptur, S. (1986). On the evolution of non-specific mutualism. *Biological Journal of the Linnean Society*, **27**, 251–67.

Letourneau, D. K. (1988). Conceptual framework of three-trophic-level interactions. In *Novel aspects of insect-plant interactions* (ed. P. Barbosa and D. K. Letourneau). Wiley, New York.

Wilson, E. O. and Hölldobler, B. (1990). *The ants*. Springer Verlag, New York.

Part 1

Antagonistic interactions: the leaf-cutter ants

2

An introduction to the fungus-growing ants, Attini

D. J. Stradling

Although the modern Formicidae are typically opportunist omnivores, the diets of most species, like those of their carnivorous ancestors, still contain a large proportion of nutritionally rich, animal protein. The subfamily Myrmicinae contains some notable exceptions, including the granivorous ants of arid regions for which seeds constitute a major component of the diet. Completely herbivorous ants appear to be confined to the 12 genera of the Myrmicine tribe Attini, all of which are obligate symbionts with fungi, that are cultivated on substrates of vegetal origin. The fungus serves as the sole larval food and probably as a principal part of the adult diet for primitive attines, although workers of those species that cut leaves from living plants also drink sap directly. There are approximately 190 species of attines (Weber 1972) distributed in the Nearctic and Neotropical biogeographical regions between latitudes of approximately 40°N and 40°S.

Wilson (1971) grouped the attine genera into primitive, transitional, and advanced categories on the basis of colony size, worker polymorphism, and the substrate used for fungiculture. The six primitive genera typically have small colonies, with no more than a few hundred monomorphic workers, and cultivate their fungi on faeces of phytophagous insects and dead vegetable matter. The three transitional genera form similar, or slightly larger, colonies and the workers, except in some *Trachymyrmex*, are also monomorphic. The material they use for the fungal substrate however, also contains significant proportions of freshly fallen flowers and fruit.

The advanced group contains three genera, one of which, *Pseudoatta*, is a workerless social parasite. The other two genera, *Acromyrmex* and *Atta*, differ from the other attines in forming large colonies with well-developed, functional worker polymorphism, and a distinctive soldier caste in *Atta* species. Both genera cut fresh leaves, flowers, and fruits from living plants for use as the fungal substrate, a habit which earns them the vernacular name of leaf-cutter ants.

The range of plant species used as the fungal substrate is partly determined by the richness of the habitat. Attines that live in savannahs utilize a relatively narrow range of Poaceae whilst those inhabiting tropical rainforest exploit a

comparatively wide spectrum of tree species. Cherrett (1968, 1972a)
collected data indicating that, in a Guyanese rainforest, an *Atta cephalotes* L.
colony exploited 30–50 per cent of accessible plants in a period of ten weeks.
This represents a substantially greater polyphagy than is usual for phytophagous insects.

Species of *Acromyrmex* cultivate their fungus in a few subterranean nest
chambers and attain colony populations of 10–20×10^3. *Atta* colonies are
usually much larger, sometimes occupying in excess of 1000 nest chambers,
reaching estimated worker populations of 7.0×10^6 in the case of *A. vollenweideri* Forel (Jonkman 1977, 1978). Clearly, such enormous colonies at a
fixed nest site make major demands on the food resources of the immediate
environment and foragers may retrieve plant material from distances of over
100 m from the nest.

The biology of the primitive and transitional attine genera is poorly understood, most work having focused on *Atta* and *Acromyrmex* whose large
colonies and leaf-cutting activities make them agricultural pests. Fowler
(1983b) has pointed out that there are significantly more species of advanced
attines in the subtropics than the tropics. The distribution and complexity of
attine communities in relation to habitat is examined by Fowler and Claver in
Chapter 5.

The symbiotic fungi are cultivated in 'fungus gardens' located in subterranean chambers which attain diameters of 25–30 cm in *Atta*. The ants
benefit their fungal symbionts by removing the physical and associated
chemical barriers that plant leaves possess to prevent the entry of pathogenic
fungi (Cherrett *et al.* 1989). The fungi are obligate symbionts and free-living
stages have never been observed. They are also poor competitors; fungus
gardens removed from the care of the ants are rapidly overrun by contamin-
ant fungi and bacteria. Schildknecht and Koob (1971) provided evidence
that higher attines secrete the antibiotics phenylacetic acid and β-
hydroxydecenoic acid (myrmicacin) from the metapleural glands. These
antibiotics suppress the growth of bacteria and the germination of fungal
spores, respectively, thus giving the ants' symbiont a competitive advantage.
The fungus in turn provides the workers with digestive enzymes which are
concentrated and redistributed throughout the fungus garden (Martin and
Boyd 1974). The complexities of this digestive alliance have been summarized by Stradling (1987). The workers plant tufts of mycelium on newly
prepared substrate and these ramify over the surface producing clusters
(staphylae) of swollen-ended hyphae (gongylidia) which serve as the principal, if not sole, larval food. Analyses by Martin *et al.* (1969) indicated that
this provides a food which is rich in protein and carbohydrate, but lacking in
lipid.

The taxonomic position of the attine fungal symbiont has been reviewed by
Cherrett *et al.* (1989). Controversy surrounds its identity, largely due to
doubts over sporophores. The first detailed morphological descriptions were

those of Möller (1893) who examined isolates from fungus gardens of various attine species. He reported finding sporophores on the surface of leaf-cutting ant nests on several occasions and named these *Rozites gongylophora* Möller (Cortinariaceae), a basidiomycete. Although sporophores have since been reported from both laboratory cultures and the field by other authors, Möller (1893) appears to be the only one to have propagated gongylidia-bearing mycelia from the basidiospores. The variety of names which have been applied to these different sporophores suggests that many of them are contaminants. The use of *Attamyces bromatificus* by Kreisel (1972) to name the vegetative mycelium of the symbiont of *Atta insularis* Guerin does not rely on dubious sporophores and avoids these difficulties. Möller's original opinion that the fungus associated with the higher genera was a basidiomycete is supported by Powell (1984) who demonstrated the presence of dolipore septa.

The Attini are a taxonomically compact group whose phylogeny is obscure and provides no obvious clues to the origin of fungiculture. One plausible theory suggested by Garling (1979) is that the fungus is descended from an ectotrophic mycorrhiza encountered by the ants during excavation of subterranean nest chambers, and which thrived on their refuse.

In the absence of a sporulating stage *Attamyces* is dependent upon its symbiotic ants for dispersal. This occurs annually at the time of the nuptial flight, with each gyne leaving the nest with a sample of mycelium in her infrabuccal pocket and using it as an inoculum in establishing a new colony. *Attamyces* is therefore cloned by the ants and the implications for attine speciation have been discussed by Stradling and Powell (1986).

Initially, the discovery that the attine fungi produce cellulase (Martin and Weber 1969) appeared to confirm that the primary selective advantage of fungiculture to the ants is that of providing access to the vast carbohydrate store in the vegetation (Martin 1974). However, Cherrett (1980) argued that this would be unlikely to be the central role on ergonomic grounds, since most of the colony's carbohydrate comes from plant sap drunk by foragers (Littledyke and Cherrett 1976; Quinlan and Cherrett 1979). Therefore, the importance of the fungus to the ants must lie elsewhere. It is likely that the ability of the fungus to concentrate the nutritionally poor, chemically defended, and indigestible plant tissue and debris into a nutritious larval food may provide the answer. The resource quality of substrates and the economics of foraging in leaf-cutting ants are examined by Howard in Chapter 4.

Among plant-feeding insects, with the possible exception of some locusts, the leaf-cutting ants probably contain the only truly polyphagous species. Most other phytophagous insects are monophagous or oligophagous and can fly over relatively large distances in search of a particular food plant. Since ants are walking insects normally tied to a fixed nest site, monophagous herbivory would impose severe restrictions on food availability in tropical habitats of high floral diversity. There is evidence to suggest that fungiculture confers a major advantage on species of *Atta* which inhabit neotropical rain-forests by allowing them to overcome a wide range of plant, chemical defences and thus to be ecologically polyphagous in the midst of a highly diverse flora.

Since foragers drink sap directly from the leaf material that they cut as a substrate for the fungus, they might be expected to exhibit selectivity for plants whose foliage is directly palatable and which provide a suitable fungal substrate at the same time. This postulate is examined by Powell and Stradling in Chapter 3.

3

The selection and detoxification of plant material by fungus-growing ants

R. J. Powell and D. J. Stradling

Introduction

Most phytophagous insects have digestive systems which are adapted to cope with a limited range of plant chemical defences and are monophagous or oligophagous. However, the ability to fly over relatively large distances facilitates the search for specific food plants. It is advantageous for walking, centre-place foragers like ants to have less specialized diets, and among plant-feeding insects the leaf-cutting ants probably contain the only truly polyphagous species.

Southwood (1971) pointed out that a 'nutritional hurdle' exists for herbivores because they generally contain more protein than the plant tissues upon which they feed. Tannins and other digestibility-reducing substances make the hurdle higher by limiting both the quantity and quality of the protein available (Feeny 1976). For most herbivores, during the course of evolution, a micro-organism has become the 'biochemical broker' in the plant–animal interaction (Southwood 1985). For attines, Cherrett (1980) realized that the fungi overcome the nutritional hurdle because, on average, they contain more protein than is found in plants (17 and 13 per cent respectively). The fungi are the major source of this nutrient for the growing larvae.

From the approximately 250 000 angiosperm species known, about 30 000 different, secondary compounds have been described (Harborne 1977). Many of these are highly toxic to insects and most angiosperms tend to have at least one substance at a concentration which is effective against some insects. Feeny (1975, 1976) proposed that dominant, climax-forest tree species are 'apparent' to herbivores, owing to their large size and longevity. The optimum defence strategy for such plants is to make leaves indigestible early in their development. As leaves age, their nutrient and water content decreases whilst the content of tannins and/or latex, resins, and gums increases (Southwood 1971). Such compounds are generalized (quantitative) in their action and are not easily countered by metabolic adaptations of the herbivores. Less apparent, herbaceous plants tend to invest in specific (qualitative) defences, including alkaloids and terpenoids.

These are effective in relatively small concentrations against non-adapted herbivores, and they may also serve as behavioural deterrents. Such compounds are likely to be susceptible to counter-adaptation however, and adapted herbivores may exploit them as attractants or feeding stimulants.

A large proportion of the plants attacked by leaf-cutting ants are 'apparent' types, notably mature trees in rainforest (Cherrett 1968; Rockwood 1976), which have quantitative defences in most cases. However, the ants tend to prefer the new leaves of these plants (Rockwood and Glander 1979) which are less apparent.

Cherrett (1968, 1972a) presented data indicating that *Atta cephalotes* L. exploited 30–50 per cent of accessible plants in a Guyanese rainforest. By further analysis of these data Stradling (1978) demonstrated that there was significantly less attack on plant species which possessed latex, an effective physical block to forager recruitment, and those containing chemically noxious compounds. Two substances of the latter type, limonene in citrus fruit flavedo (Cherrett 1972b) and a terpenoid in the leaves of *Hymenaea courbaril* L. (Caesalpiniaceae) (Ales *et al.* 1981) are known to deter attines. Rockwood (1976) found that foragers of *Atta colombica* and *A. cephalotes* in Costa Rica cut fewer mature leaves than flushes, and exploited a maximum of 31 per cent of available plants. The nutritional value, moisture content, and secondary compound content of the leaves were implicated as prominent factors in selection by foragers (see also Howard *et al.* 1988).

Tannins are phenolic compounds which present the most important barrier to herbivore feeding (Harborne 1977) and to fungal attack (Levin 1976) in angiosperms. They exhibit wide structural variation but fall into two main groups: hydrolysable and condensed tannins. Hydrolysable tannins (e.g. tannic acid) are mostly water-soluble and occur only in dicotyledons. Condensed tannins are larger molecules, are less water-soluble, and are more resistant to hydrolysis. They are found in 54 per cent of angiosperms (62 per cent of dicotyledons and 29 per cent of monocotyledons) (Swain 1979). Their protective value lies in their ability to reduce the digestibility of plant proteins, polysaccharides, and nucleic acids (Swain 1979) by the formation and strengthening of chemical links between these already large molecules to produce 'complexes' which are inert to hydrolysis by proteases (Goldstein and Swain 1965; Feeny 1969). Tannins may also reduce the palatability of a plant tissue or juice (Rhoades and Cates 1976). Furthermore, digestive enzymes, belonging to both herbivores and fungal pathogens, are denatured by tannins.

Cherrett (1980) proposed that *Attamyces* might serve in the destruction of these digestibility reducing substances, although its ability to degrade tannins and other phenolics had not been reported at that time. However, phenol oxidase activity has been demonstrated in a species of *Lepiota* which is the fungal symbiont of the primitive attine *Myrmicocrypta buenzlii* Borgmeier (Seaman 1984). Here we report the enzymatic capabilities of *Attamyces* with these substances, their effects on its growth, and their fate during the process of digestion in ant-maintained fungus-gardens.

Hubbell *et al.* (1983) suggested that natural selection favours leaf-cutting ants capable of discriminating between plant species which are toxic or non-toxic to their fungus. Our studies on the responses of both partners in the symbiosis towards some of the plants described by previous workers are reported here.

Materials

Six colonies of *Atta cephalotes*, five of *Acromyrmex octospinosus* Reich, and three of *Trachymyrmex urichi* Forel, collected in Trinidad, West Indies, were maintained in the laboratory at 25 °C, > 70 per cent relative humidity (RH), and in a 12:12 photo-period. Isolates of symbiotic fungi from these colonies were cultured on potato dextrose agar (PDA) plates as described by Powell and Stradling (1986). For bioassay, compounds to be tested were either incorporated into the medium before inoculation, or introduced through wells cut in the agar.

Samples of plant leaves were collected at midday from various locations in Trinidad during the period from July to September 1981. The 29 species sampled are shown in Table 3. 1 in the categories used by Stradling (1978) and Rockwood (1976). Mature leaves were separated from new flushes, air-dried, ground using a hammer mill (grid size 1 mm), and stored over silica gel.

Chemical analyses of leaves

Total, soluble phenolic substances were estimated by the colorimetric method of Allen *et al.* (1974). The tannin content was estimated using the relative astringency technique of Bate-Smith (1973) which measures protein precipitation by the tannins, relative to a standard tannic acid solution.

The nitrogen concentration of samples was measured using an automatic nitrogen analyser (Carlo–Erba, model 1400). Protein content was estimated by multiplication of the nitrogen concentration by a factor of 6.25 (Allen *et al.* 1974). 'Available protein' is the ratio of protein content to relative astringency.

The acidity of the leaf sap was measured using a flat-tipped glass probe (Pye Unicam 403–30M8) which permitted direct pH measurement from leaf surfaces, from small volumes of liquid extracted from plant tissues, and from culture media containing plant material.

The results of these analyses are summarized in Table 3. 1. There was a significant correlation between the content of soluble phenolic compound and the relative astringency of mature leaves ($r = 0.645$, $P < 0.001$). Multiple regression analysis indicated that the nitrogen content of leaves accounted for 32.7 per cent of the observed variation in their pH, whilst soluble

Table 3.1. Chemical analyses of mature leaves: mean values are shown for each variable. $\bar{x}$ = mean for each category of species.

Family and species	Abbrev.	pH Leaf	Agar	Protein (% w/v)	Available protein (%)	Soluble phenol (mg/g)	Relative astringency (%)
(1) Laticiferous species (Stradling 1978)							
Guttiferae:							
Vismia guianensis (Aubl.) Choisy	Vg	4.2	4.8	8	18	11	45
Vismia falcata Rusby	Vf	4.4	4.8	11	23	8	48
Clusia tocuchensis Britt.	Clt	3.7	4.6	5	8	20	55
Symphonia globulifera L.f.	Sg	4.7	5.0	10	23	14	42
Rheedia macrophylla Planch. & Triana	Rm	4.5	4.9	8	16	15	50
Moraceae:							
Cecropia peltata L.	Cp	5.5	5.2	19	36	14	53
Ficus sp.	F	5.6	5.1	7	13	20	56
Chlorophora tinctoria (L.) Gaudich.	Cht	8.0	6.9	20	54	3	36
Sapotaceae:							
Pouteria hartii (Hemsl.) Dubard.	Ph	5.2	5.2	10	12	32	86
$\bar{x}$		5.1	5.3	10	20	15	52
(2) Non-laticiferous species having possible attine-deterrent properties (Stradling 1978)							
Flacourtiaceae:							
Ryania speciosa Vahl	Rs	5.1	5.0	12	20	12	59
Anacardiaceae:							
Tapirira guianensis Aubl.	Tg	4.2	4.6	8	14	12	58

Simarubaceae:

Simaba multiflora A. Juss.	Sim	5.5	5.3	10	15	6	67
Simaba cedron Planch.	Sc	6.8	5.5	9	23	3	40
Simaba sp.	S	6.4	5.4	9	21	2	45
x̄		5.6	5.2	10	19	7	54

(3) **Non-laticiferous species having no known attine-deterrent properties (Stradling 1978)**

Combretaceae:							
Terminalia amazonia (J. F. Gmel.) Excell	Ta	4.4	4.5	11	12	45	93
Flacourtiacea:							
Casearia guianensis (Aubl.) Urb.	Cg	4.8	4.8	10	23	12	44
Lauraceae:							
Nectandra martinicensis (Jacq.) Mez.	Nm	4.0	5.5	19	58	7	33
Phoebe elongata (Vahl) Nees	Pe	5.6	5.3	12	29	7	43
Melastomaceae:							
Miconia acinodendron Sweet	Ma	4.0	5.6	11	22	9	51
Rosaceae:							
Licania membranacea Sagot ex Laness.	Lm	4.9	5.1	8	13	27	59
x̄		4.6	5.1	12	26	18	54

(4) **Species with mature leaves palatable to *Atta* spp. (Rockwood 1976)**

Fabaceae:							
Dioclea guianensis Benth.	Dg	5.5	5.6	18	32	8	56
Lysiloma sabicu Benth.	Ls	5.3	5.2	11	27	17	41
Albizzia lebbek (L.) Benth.	Al	5.9	5.3	29	69	4	42
Burseraceae:							
Bursera gummifera L.	Bs	4.4	4.9	11	18	8	63
Tiliaceae:							
Muntingia calabura L.	Mc	5.1	5.3	13	29	21	44

Table 3.1. (*cont.*)

Family and species	Abbrev.	pH		Protein (% w/v)	Available protein (%)	Soluble phenol (mg/g)	Relative astringency (%)
		Leaf	Agar				
Annonaceae:							
Annona trinitensis Saffordd	At	5.2	5.2	19	39	4	50
Sterculiaceae:							
Guazuma ulmifolia (Lam.) Oken	Gu	7.5	5.9	21	62	6	34
Anacardiaceae:							
Spondias mombin L. = *purpurea* L.	Sm	3.7	4.2	18	38	17	47
Boraginaceae:							
Cordia alliodora (Ruiz & Pav.) Cham.	Ca	8.1	6.3	19	65	8	29
$\bar{x}$		5.6	5.4	18	42	10	45
Grand mean		5.3	5.2	13	29	13	51

phenolic compounds and relative astringency together accounted for 29.7 per cent. This results in a significant correlation between pH and available protein (Fig. 3. 1).

Digestive capabilities of *Attamyces*

Phenolic substances

Various phenolic substances (Table 3. 2) were presented to well-established plate cultures of fungal symbionts from each ant species. This was achieved by cutting four 8 mm diameter wells in the agar of each plate 1 cm from the edge of the mycelium, and filling the wells with O.1 per cent aqueous solution of the phenolic compound to be tested, or distilled water as a control. Plates were then incubated at 25 °C in darkness for two hours, after which they were examined for obvious colour reactions relating to the wells.

The pattern of browning (Table 3. 2) suggests that the fungi are only able to break down *ortho*-hydroxyl (*o*-OH) groups, perhaps by means of phenol oxidase.

Lignin

Agar plates containing either Indulin AT (Westvaco Polychemicals) or Peritan Na (sodium lignosulphonate, Norcem Ltd) at a concentration of 0.05 per cent w/v were prepared, inoculated with the three fungal strains, and incubated for 14 days. Lignin dephenolization was estimated by the method of Sundman and Nase (1971).

On Indulin and on Peritan agar, cultures showed zones of dephenolization immediately beneath and around the mycelia, indicating that the fungi were capable of at least partial degradation of lignin.

Tanned protein

Peptone was chosen as a readily available protein source, and was incorporated in the agar medium and flocculated with tannic acid. The method of Feeny (1969) was employed to confirm that 0.05 per cent w/v tannic acid was sufficient to produce an opaque medium, contrasting with that containing untreated peptone agar. Replica plates of both media were inoculated with the three fungal strains and incubated for four weeks. The colony radius, form, and the physical state of the medium were then recorded.

Cultures grew normally on flocculated peptone media, forming mycelia surrounded by a dark brown, translucent zone which contrasted sharply with the opaque medium. Addition of tannic acid to peptone media significantly improved growth ($P < 0.001$) in the case of strains from *Atta cephalotes* and *Acromyrmex octospinosus*, but not from *Trachymyrmex urichi*.

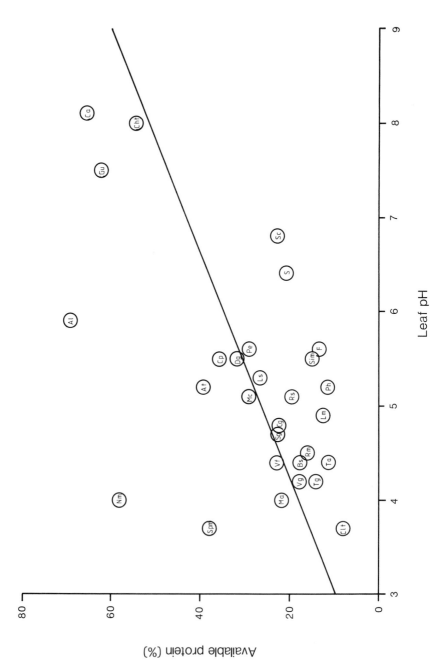

Fig. 3.1. The relationship between leaf-sap pH and available protein for the plant species listed in Table 3.1.

Table 3.2. Phenolic compounds presented to *Attamyces* cultures on potato dextrose agar plates, indicating the possession of ortho-OH groups and the colour reaction in the vicinity of the fungal mycelium. Compounds are listed in ascending order of molecular mass.

Compound	ortho-OH	Browning
Phenol	−	−
Catechol	+	+
Quinol	−	−
Resorcinol	−	−
Pyrogallol	+	+
Protocatechuic acid	+	+
3,4-dihydroxyphenylacetic acid	+	+
3,4-dihydroxyphenylalanine	+	+
p-Phenyldiamine	−	−
Gallic acid	+	+
Catechin	+	+
4-Methylcatechin	+	+
Ellagic acid	+	+
Tannic acid	+	+
Quebracho (condensed) tannin	−	−

Effects of phenolics on fungal growth

Tannins

Replica plates were prepared containing either 0.5 per cent w/v gallic acid agar or 0.5 per cent w/v tannic acid agar (Davidson *et al.* 1938). Further plates were prepared containing three different concentrations of tannic acid (0.01, 0.05, and 0.1 per cent w/v), quebracho (condensed) tannin, and the basal medium alone as a control. The plates were inoculated separately with the three fungal strains and incubated for five days, after which the extent of oxidation of the tannic acid was assessed using an arbitrary scale for the size of oxidation zones and intensity of colour development. The diameters of the fungal colonies were recorded after 14 days.

All three strains grew normally on tannic acid media at the three con-centrations used. Although cultures on media containing 0.05 per cent tannic acid were not significantly different in size from the controls ($P > 0.05$), those on media containing O.1 per cent were significantly smaller ($P < 0.001$) (Table 3. 3). The strain from *A. cephalotes* produced significantly larger

Table 3.3. Colony diameter (cm) of *Attamyces* cultures after 14 days when grown on media containing different concentrations of tannic acid. $\bar{x}$ = mean value.

Ant species	Tannic acid concentration (%)					
	0	0.01	0.05	0.1	$\bar{x}$	n
Atta cephalotes	0.64	0.67	0.67	0.61	0.65	36
Acromyrmex octospinosus	0.63	0.58	0.62	0.54	0.59	36
Trachymyrmex urichi	0.43	0.45	0.43	0.34	0.41	36
$\bar{x}$	0.57	0.57	0.57	0.50	0.55	

colonies ($P < 0.001$) than that from *Ac. octospinosus*, which in turn produced significantly larger colonies than the *T. urichi* strain ($P < 0.001$). Zones of browning around all three strains were equal in size, indicating consistent enzyme diffusion rates, but there was an increase in colour intensity with tannic acid concentration.

No growth was observed on media containing condensed tannin at a concentration of 0.01 per cent w/v or above. Nevertheless, zones of browning were discernible around the points of inoculation indicating that enzymes transferred with the inocula were capable of degrading some components of the condensed tannin, perhaps phenolic impurities.

Leaf material

Samples (18 g) of each powdered leaf were sterilized by exposure to gamma radiation from a cobalt-60 source at a dose of 2×10^7 rads for 48 hours. Each sample was added to 360 cm^3 of molten PDA at 45 °C to form a 5 per cent suspension (w/v). The resulting medium was poured into petri dishes and inoculated when cool. The colony form, and any change in the medium during incubation, were noted. The fungal yield (total fresh mass) was estimated as in Powell and Stradling (1986).

The mean yield for each strain and each of the 29 test media (plus 1 control) are shown in Fig. 3. 2. There were significant differences in yield between the treatments and between strains ($P < 0.001$). Since the three strains grew in closely similar rank order on each type of medium, data were pooled. The resulting composite rank order of plants was divided into three subjective categories (no, or poor growth; moderately good growth; and good growth). The boundaries of these categories were set by one standard deviation from the overall mean fungal fresh mass (Fig. 3. 2).

Addition of leaf material generally resulted in a lower yield than the

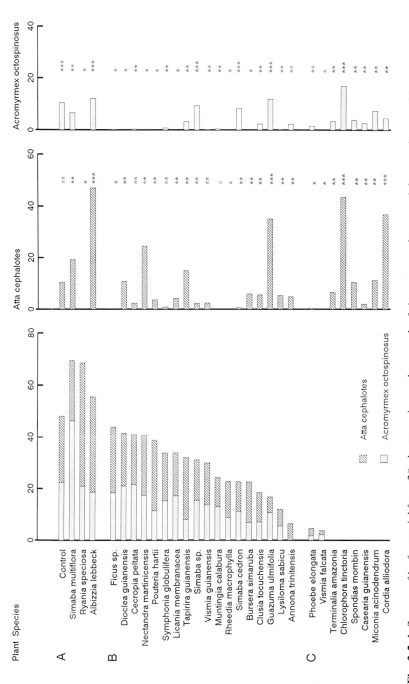

Fig. 3.2. Influence of leaf material from 29 plant species on the growth of *Attamyces* cultures and the recruitment of foragers to sucrose solution in two leaf-cutting ant species. A: Good growth, B: moderately good growth, C: no or poor growth, *** palatable, ** moderately palatable, * marginally palatable or unpalatable.

control. The four plant species *Cordia alliodora*, *Casearia guianensis*, *Chlorophora tinctoria*, and *Terminalia amazonia* completely inhibited growth. Significant increase in the yield only occurred in media prepared from the mature leaves of *Simaba multiflora*, *Albizzia lebbeck*, and *Ryania speciosa*. A brown ring surrounded the fungal mycelia except in media prepared from leaves having low soluble phenolics and relative astringency, such as *Cordia alliodora* and *Chlorophora tinctoria*.

There was no correlation between the available protein content of the mature leaves and the fungal yield, the fungus cultivated by *Atta* giving $r = 0.0203$ ($P > 0.05$). Multiple regression revealed that the pH of the leaf samples (expressed as a quadratic function) was the largest contributing factor to fungal yield (Fig. 3. 3), accounting for 49 and 33 per cent of the variation for *A. cephalotes* and *Ac. octospinosus* respectively.

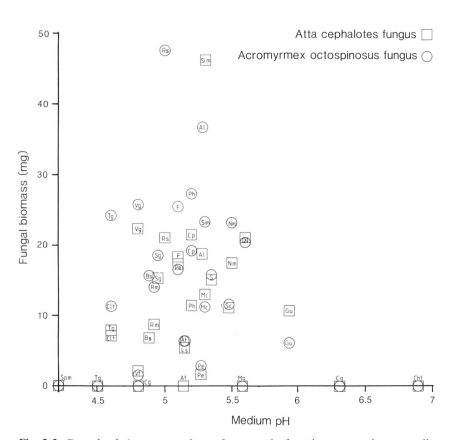

Fig. 3.3. Growth of *Attamyces* cultures from two leaf-cutting ant species on media containing macerated leaves of the plant species listed in Table 3.1 in relation to the pH of the growth medium.

The fate of foliar phenolic compounds in fungus gardens

Qualitative study

Qualitative changes in the phenolic content of leaf fragments during their passage through fungus gardens were studied in one colony each of *A. cephalotes* and *Ac. octospinosus*. The colonies were supplied with fresh leaves of privet (*Ligustrum ovalifolium* Hassk.) for several months, collected daily at noon from the same plants. Estimates of substrate turnover in fungus gardens showed general agreement with Weber's (1972) value of seven weeks. The period over which the colonies were provided with this material was therefore sufficient to ensure that, at the time of sampling, the entire fungus-garden substrate comprised *L. ovalifolium* fragments.

Approximately 10 g (fresh mass) samples of the following five types of material were taken for analysis:

1. newly flushing privet leaves;
2. mature privet leaves;
3. fungus garden with recently added substrate;
4. mature fungus garden; and
5. recently ejected debris.

The samples were prepared in the same way as field-collected leaf material.

Analysis was by means of two-dimensional paper chromatography using the following mobile phases: (a) *n*-butanol, ethanoic acid, and distilled water (B:A:W) (40:10:22) and (b) 2 per cent w/v ethanoic acid solution (MeCOOH). The separated phenolics were located by treating replica chromatograms with the reagents described by Swain (1969), Ribereau-Gayon (1972), and Harborne (1973) and the R_f values and colour reactions were compared with the results of these workers.

A representation of the two-dimensional chromatograms obtained is shown in Fig. 3. 4. Eighteen different spots were observed, each representing an individual phenolic substance. Their R_f values and reactions with various reagents enabled tentative identifications to be made (Table 3. 4). The compounds represented various groups of phenolic substances including simple phenols, phenolic acids, coumarins, flavonoids, and tannins. Most were present as aglycones due to hydrolysis of glycosides during extraction.

The samples of young and old privet leaves each possessed 16 different phenolic compounds. Four of them were not common to both: e and f were only found in old leaves, while l and r were only present in young leaves. These were tentatively identified by their R_f values as glycosides of flavonoids. Both samples possessed condensed tannin and a pink anthocyanidin.

Chromatograms prepared from samples of privet leaf fragments which had been incorporated as fungal substrate showed fewer spots than those prepared from the fresh leaves. As Fig. 3. 4 indicates, fragments ejected from

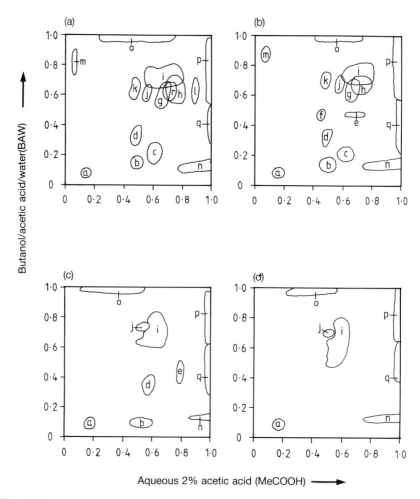

Fig. 3.4. Two-dimensional chromatographs showing the separation of phenolic compounds in samples of privet (*Ligustrum vulgare*) leaves and fungus-garden refuse ejected from attine colonies provided with privet. (a) Newly flushed leaves; (b) mature leaves; (c) refuse from *Acromyrmex octospinosus*; and (d) refuse from *Atta cephalotes*. Scales are R_f units. $a - r$ = phenolic compounds identified in Table 3.4.

Ac. octospinosus colonies possessed only 10 of the 18 possible substances, while those removed from *A. cephalotes* colonies contained only seven. The former possessed all of the substances found in the latter and substances b, d, and e in addition.

   Among the compounds not removed from the fungus-garden material were the condensed tannin and the pink anthocyanidin. Substances absent

Table 3.4. Tentative identification of phenolic substances separated by two-dimensional paper chromatography of privet leaf and fungus garden samples shown in Fig. 3.4.

Substance	B:A:W R_f	MeCOOH R_f	Possible identity
a	0.08	0.14	Condensed tannin
b	0.14	0.50	A trihydroxyphenolic compound
c	0.20	0.62	A substance possessing phenolic groups
d	0.32	0.52	A substance possessing phenolic groups
e	0.45	0.70	A trihydroxyphenolic compound
f	0.48	0.45	Unknown
g	0.60	0.65	A flavanol
h	0.64	0.70	An anthocyanin
i	0.74	0.65	Simple phenol of flavanol-3-glycoside
j	0.65	0.55	Aurone or flavanol
k	0.68	0.45	A flavanol
l	0.64	0.90	A substance possessing phenolic groups
m	0.84	0.10	A flavone
n	0.14	0.95	An anthocyanidin
o	0.98	0.40	Aurone or flavanol
p	0.80	0.98	Aurone, chalkone, or flavanol-3-glycoside
q	0.45	0.98	Aurone or flavanol
r	0.64	0.70	Simple phenolic compound

from the fungus-garden material of both ant species included simple phenolic compounds.

Quantitative study

The concentration of soluble phenolic compounds and their relative astringency (Bate-Smith 1973) was measured in samples representing the same sequential phases of substrate passage through the fungus garden. The results are shown in Fig. 3. 5. Although young and old leaves did not differ significantly in astringency young leaves contained a mean concentration of soluble phenolic substances which was 12 per cent greater than in old leaves ($P < 0.001$).

There were differences in astringency between the fresh leaves and the fungus-garden samples ($P < 0.001$). On being incorporated into a young fungus garden of *A. cephalotes*, the astringency of the privet decreased by 13 per cent and in a fungus garden of *Ac. octospinosus* it decreased by 31 per cent. The young fungus gardens of both species also contained significantly

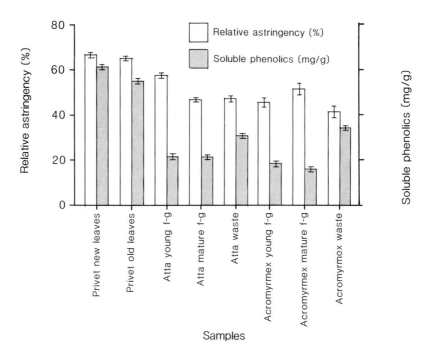

Fig. 3.5. Relative astringency and soluble phenolic content of privet-leaf material before, during, and after passage through fungus gardens of *Atta cephalotes* and *Acromyrmex octospinosus*.

lower phenolic concentrations than the fresh leaves ($P < 0.001$). Those of *A. cephalotes* provided a 63 per cent decrease, and those of *Ac. octospinosus*, 68 per cent.

In *A. cephalotes*, there was a further 16 per cent decrease in astringency during the passage of substrate into the mature zones of the fungus garden. However, significant decrease in phenolics ($P < 0.01$) was only observed in *Ac. octospinosus*. Here, the concentration of phenolics in leaf fragments had decreased by 73 per cent on reaching the mature zones of the fungus garden.

The debris ejected by *Ac. octospinosus* workers possessed significantly lower astringency than the fungus garden ($P < 0.001$) while that of *A. cephalotes* showed no change ($P. > 0.05$). However, for both ant species there was a significantly higher concentration of phenolic substances in the debris ejected than in the fungus gardens ($P < 0.001$). Whereas in *A. cephalotes* debris, the concentration rose to about half that present in fresh leaves, in *Ac. octospinosus* debris it was 59 per cent. In both cases the debris contained the carcasses of dead ants and many dry leaf fragments which had never been incorporated as substrate.

Observations on ant feeding behaviour

Laboratory colonies of *A. cephalotes* and *Ac. octospinosus* were used in feeding preference experiments which followed the design of Littledyke and Cherrett (1975). Prior to testing, the ants were denied vegetation for 24 hours to encourage consistent, rapid foraging responses. Test solutions or suspensions in 100 μl quantities per solution were randomized in a grid pattern on a glass plate, each drop being spread to occupy a circle of approximately 2.5 cm^2. Ants were then allowed access for 15 min. At the end of this time the number of individuals feeding at each drop was recorded. This provided an estimate of the relative arrestiveness of the solutions tested. Trials were repeated five times for each ant colony, the glass being thoroughly washed after each trial to remove any pheromone deposits laid by workers in preceding trials. The positions of the solutions were re-randomized for successive trials.

Effects of tannins

Solutions of 10 per cent sucrose, containing either tannic acid or quebracho tannin, were presented to the ants at concentrations between 0 and 2 per cent w/v.

The addition of tannic acid at concentrations up to 2 per cent, to sucrose solutions, significantly reduced the arrestive qualities of the solutions to both *A. cephalotes* and *Ac. octospinosus* ($F_{2,126} = 833.6$) ($P < 0.001$) (Fig. 3. 6). Foragers of both species (80 per cent of *A. cephalotes* and 86 per cent of *Ac. octospinosus*) preferred the control solutions to those containing tannic acid, although a few individuals of both species drank from solutions containing up to 0.25 per cent tannic acid. Foragers encountering concentrations greater than this exhibited some distinctive behaviour patterns, stepping backwards and examining the liquid with their antennae. At the same time the mouthparts were fully retracted, the mandibles parted, and the gaster raised high above the surface of the substratum. After a few seconds' investigation, these foragers would retreat with gasters raised, either towards the nest entrance or to another drop of sucrose solution.

Many ants that had returned to the nest after an encounter with tannic acid at a repellant concentration, later returned carrying soil particles or dry leaf fragments. This debris was deposited in the repellent liquid while the ants held their gasters high and apparently avoided touching it. Samples which had received this treatment were subsequently ignored by all foragers.

The effects of tannic acid and quebracho tannin concentrations up to 0.25 per cent are shown in Fig. 3. 6. The 0.01 per cent tannic acid was the only concentration producing a significant increase ($P < 0.05$) in the arrestive properties of the sucrose solution with either species.

Quebracho tannin produced a similar response to tannic acid. However, a

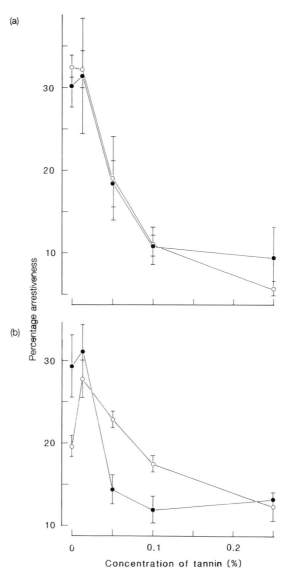

Fig. 3.6. Effects of tannin concentration on the arrestive properties of 10 per cent sucrose solutions to foragers of (a) *Atta cephalotes* and (b) *Acromyrmex octospinosus*. ● = Tannic acid, ○ = quebracho tannin.

striking difference was observed in that both 0.01 and 0.05 per cent quebracho were significantly more arrestive to *Ac. octospinosus* foragers ($P < 0.001$) than the control. *Atta cephalotes* foragers, however, found only the 0.01 per cent quebracho to be as arrestive as the control, the other concentrations being significantly less so ($P < 0.001$).

Leaf material bioassay

The mature leaves of all plant species listed in Table 3. 1 were employed in bioassays. To eliminate variation in forager response to whole leaves due to their physical structure, as described by Barrer and Cherrett (1972) and Waller (1982*b*), leaf samples were presented as 0.5 g of finely divided powder, suspended in 10 per cent sucrose solutions.

This assay exploited the readiness of foragers to drink sucrose solutions (Powell 1984). Although it is unlikely that the chemical composition of dried leaf powder is identical to that of fresh leaves, many of the more stable components such as phenolic compounds and tannins would be expected to remain unchanged.

The results are presented in Fig. 3. 2 where the relative palatability of each plant species to both *A. cephalotes* and *Ac. octospinosus* is expressed as a function of the number of foragers recruited. The resulting composite rank order of plants was divided into three subjective categories shown in Fig. 3. 2 (unpalatable or marginally palatable, moderately palatable, and palatable), the boundaries of which were set by one standard deviation from the overall mean number of foragers recruited.

There were significant differences in the foraging effort expended on plant species ($P < 0.001$) and also in the intensity of foraging between the two species of ants ($P < 0.001$). Figure 3. 2 shows that there were differences in relative palatability between the ant species. Three samples were more palatable to *Ac. octospinosus* than to *A. cephalotes* (*Simaba cedron*, *Simaba* sp., and the control). Conversely, two samples (*Cordia alliodora* and *Nectandra martinicensis*) were more palatable to *A. cephalotes* than to *Ac. octospinosus*. Despite these differences, Spearman Rank Correlation analysis revealed a significant correlation between the preferences of the two ant species ($r = 0.5093$, $P < 0.01$). Seven and five plant samples were completely unpalatable (i.e. recruited no ants) to *Ac. octospinosus* and *A. cephalotes* respectively, with two species being common to both, *Ryania speciosa* and *Rheedia macrophylla*.

Correlation between the available protein content of leaves and the palatability to the ants was highly significant for *A. cephalotes* ($r = 0.803$, $P < 0.001$) but less so for *Ac. octospinosus* ($r = 0.470$, $P < 0.05$). Zero palatability data were not used in these calculations.

The fact that palatability of leaf material to ants and its ability to support optimum fungal growth did not coincide in many cases is evident from Fig. 3. 2. There was no significant correlation between palatability and fungal yield for either ant–fungus partnership. Thus the leaf samples selected by foragers were not necessarily the most suitable for fungiculture. Several species are outstanding in this respect, the most striking examples being *Ryania speciosa*, *Rheedia macrophylla*, and *Ficus* sp. which were unpalatable to both ant species but which supported significant fungal yields. Conversely, *Chlorophora tinctoria*, *Miconia acinodendron*, *Cordia alliodora*, *Spondias mombin*, *Terminalia amazonia*, and *Casearia guianensis* were palatable but supported no, or poor, fungal yields.

Discussion

Fungal digestion and growth

Phenolic compounds in plant leaves have been considered to be a biochemical hurdle for the ant–fungus symbiosis (Quinlan and Cherrett 1979; Cherrett 1980). However, the present findings show that the fungal symbionts of at least three attine genera secrete enzymes capable of breaking down a range of phenolic compounds.

We found evidence of at least two detoxifying enzymes secreted by *Attamyces*. First, tannase which cleared protein flocculated by tannic acid. This enzyme, which is a type of esterase, releases any proteins complexed with the tannin. Secondly, phenolic acids, released by this process, were further attacked by polyphenol oxidase. The conditions found in attine nests, of pH 4.5–5.0 and 25 °C (Powell and Stradling 1986), would therefore appear to be ideal for the action of these enzymes.

Boyd and Martin (1975) were of the opinion that attine fungi have lost their ability to secrete enzymes, and suggested that they are released during passage through the ants' guts and subsequently distributed over the fungus garden in their faeces. The development of diffusion zones on plate cultures containing tannic acid suggests that some direct secretion of these detoxifying enzymes does occur.

Since the leaf-cutting ant fungus *Attamyces bromatificus* is a basidiomycete (Powell 1984) capable of secreting polyphenol oxidase, it might be classified as a white-rot fungus. This was confirmed by its capacity for at least partial dephenolization of lignin. Phenol oxidases are involved in the ultimate oxidation of the constituent phenolic groups of lignin. However, Martin and Weber (1969) found that whilst the cellulose level dropped, no appreciable change in the lignin level occurred during the passage of plant tissue through fungus gardens of *Atta colombica tonsipes* Santschi. It is unlikely that the fungus makes major use of lignin, but destruction of lignin may facilitate the penetration of plant cells by fungal hyphae.

Clearly, an important aspect of the oxidation of phenolic compounds by attine fungi is that of reducing the toxicity of plant-hydrolysable tannins. Our results show that attine fungi can obtain nitrogen from peptone that has been precipitated by tannic acid. Without the detoxifying enzymes, the peptone could not have been hydrolysed by most proteases. The protein-precipitating action of tannins is a plant defence which few other herbivores have managed to breach.

Our observations that the attine fungi were capable of growing normally on media containing up to 0.1 per cent tannic acid, and producing diffusion zones of brown quinones, contrasts with the results of Seaman (1984), who showed that the fungal symbiont of the primitive attine *Myrmicocrypta buenzlii* is prevented from growing by tannic acid at concentrations in excess of 0.05 per cent.

Condensed tannin was at least ten times more inhibitory of fungal growth than hydrolysable tannin. Degradation of condensed tannin requires an exceptional enzyme which itself resists the strong, non-specific tannin–enzyme interactions. The fungus, *Penicillium adametzi*, is one of the few micro-organisms known to produce such a substance (Grant 1976). Condensed tannins are generally considered to be less harmful to insect guts than hydrolysable tannins (Bernays 1978).

The brown colour of the breakdown products in media containing phenolics or macerated plant tissues, is strikingly similar to that of mature fungus gardens. Our data show that there is a decline in the range and concentration of phenolic substances during the passage of leaf fragments through a fungus garden which may explain its gradual change in colour with age. However, by no means all phenolic substances are degraded *en route*. Of those contained in privet leaves, some of the larger molecules survived, particularly a condensed tannin.

Observations on ant feeding behaviour

Like most herbivores, the higher attines select plant tissues with high available protein and thus the fungus is assured of an abundant protein supply to fulfil its role in the provision of essential nutrients for larvae (Cherrett *et al.* 1989). The main limiting factor for fungal growth appears to be substrate acidity. For this the fungus has a narrow optimum range close to pH 5 (Powell and Stradling 1986). Since plant tissue acidity is strongly correlated with phenolic content, by choosing less acidic tissues the ants procure those with more available protein. Indeed the ants show a preference for materials which are less acidic than the optimum for their fungus. Nevertheless, the acidity of most substrate within the fungus garden was found to be close to the optimum, irrespective of the plant tissue source, only becoming slightly more acidic (pH 4.8) in the older regions. This regulation is achieved by the addition of several acidic secretions by the ants during substrate preparation. This implies that they are incapable of the reverse process of rendering more acidic material suitable for fungiculture.

Since young leaves contain less tannin than mature ones, they will be better sources of available protein. Furthermore, young leaves principally contain hydrolysable tannins in contrast to the condensed ones associated with ageing and toughening. Barrer and Cherrett (1972) found that attine foragers discriminated markedly in favour of young leaves and Waller (1982b) showed that this preference was strongly influenced by moisture content. Such a selection of substrate material is to the benefit of *Attamyces* since its enzymes are ineffective with condensed tannins and its growth is inhibited by them. Members of some plant families, such as the Moraceae and Boraginaceae, are completely lacking in condensed tannin (Bate-Smith 1972). This may explain why *Chlorophora tinctoria* and *Cordia alliodora*, which belong to these two families, respectively, possessed high available protein and were the least acidic of all our samples. These were significantly arrestive to foragers, and in the case of *A. cephalotes* more so than the control.

The mechanism by which foragers select plant material is unknown but the sap which they drink from the edge of newly cut leaves must provide them with chemical information. By rejecting those leaves with very acidic sap they will avoid high concentrations of phenolic compounds and tannins and low available protein content.

The present results show that a number of plants attacked contain phenolic compounds such as soluble, hydrolysable tannins even in the new leaves. These are potentially detrimental to digestive processes and Bernays (1978) has shown that they are highly damaging to insect guts. However, foragers that ingest them in the sap from the cut edges of young leaves seem to be unaffected by them. Feeny (1969) showed that in some insects, a gut pH greater than nine was sufficient to cause dissociation of protein–tannin complexes, but attine worker guts have an acid pH (Febvay 1981). The presence of fungal tannase and phenol oxidase in worker guts, which we have demonstrated, may offer an alternative protection to the foragers from the effects of hydrolysable tannins in the plant sap. This would support the concept of acquired detoxification enzymes proposed by M. M. Martin (personal communication).

Our results provide no support for optimal foraging hypotheses. On the contrary, analysis revealed the strong influence of high protein and low phenolic content of leaves upon foraging preferences, and the importance of plant tissue pH for the fungus. Many preferred plant species fitted this model but notable exceptions included those that had an appropriate acidity and available protein but were either avoided by foragers or did not support fungal growth. Such plants may have contained deterrent secondary metabolites other than phenolics and tannins, or were fungistatic/fungicidal.

Howard *et al.* (1988) provided evidence that three or four terpenoids, occurring in the leaves of three neotropical tree species, acted both as deterrents in pick-up tests and inhibited the growth of *Attamyces* on plate cultures. Our data, however, do not support the more general hypothesis that ants only select plants which are not toxic to their fungus. Some plants, such as *Ryania speciosa*, a source of the naturally occurring insecticide ryanodine (Hill and Murtha 1962), were avoided by foragers, but proved to be good fungal substrates. By contrast, some plants acceptable to foragers, and possessing appropriate acidity and available protein, completely inhibited growth of *Attamyces*. These included *Miconia acinodendron* which produces the potent antibiotic, primin, and is active against bacteria and fungi (Goncalves de Lima *et al.* 1970). Other plant species, such as *Casearia guianensis*, *Phoebe elongata*, and *Annona trinitensis* were also inhibitory to fungal growth and require further investigation. Such plant species acceptable to foragers may contain antifungal compounds that the ants cannot detect. Substrate selection by higher attines, although constrained by the presence of substances directly toxic to the foragers, includes the collection of material which supports little or no fungal growth. We therefore suggest that the repeated switching between a wide variety of plant species in nature will dilute and minimize any deleterious effects due to the collection of substrates unsuitable for fungal growth, whilst at the same time obviating any necessity to scout on foot for a restricted number of target species amidst the diversity of a complex habitat.

Extensive poisoning of a fungus garden would only occur if foragers consistently exploited only a narrow range of plant species containing fungistatic/fungicidal compounds. That this could occur is suggested by the work of Mullenax (1979) who provided field *Atta* colonies with jackbean, *Canavalia ensiformis* D.C. (Fabaceae s.s.), which contains the fungicides demethylhomopterocarpin (Lampard 1974) and canavanine (Schlueter and Bordas 1972). Foragers preferred the jackbean to the surrounding vegetation and after several days the colonies became inactive. The ants were unable to respond to the toxicity of the plant to their fungus.

Acknowledgements

We would like to thank Dr Jack Fisher for providing samples of industrial lignins and Dr Elizabeth Bernays for supplying quebracho tannin. We are also grateful to the staff of the Trinidad and Tobago National Herbarium and of the Department of Biological Sciences, University of the West Indies, Trinidad for their assistance with field work. This project was funded by the SERC.

4

Resource quality and cost in the foraging of leaf-cutter ants

Jerome J. Howard

Introduction

Leaf-cutter ant colonies may contain over a million workers and a thousand fungus gardens (Weber 1972), and to support these large colonies the ants continually search out and harvest suitable plants. However, ant colonies face limits to their ability to locate and harvest resources efficiently. At any given time most foraging workers are engaged in harvesting known resources, with only a small fraction searching for new resources (Johnson *et al.* 1987). Ants may attack plants over 100 m from the colony, and a single foraging trip may require several hours (Lewis *et al.* 1974; Hubbell *et al.* 1980). The time and energy required to locate and harvest rare or distant plants of high quality may, at times, make it more efficient for ants to harvest more readily available resources of lower quality.

Analysis of this potential tradeoff between resource quality and cost is central to understanding overall colony foraging strategies. However, differing requirements of the partners in the mutualistic relationship may result in conflicts over resource quality. In addition, foraging costs may be measured in a variety of ways, and it is unclear which, if any, measure is relevant to leaf-cutter ants. In this chapter the resource quality and cost for leaf-cutter ants is discussed and suggestions made as to how this information can be used to test critically hypotheses advanced to explain leaf-cutter ant foraging behaviour.

What is an appropriate measure of resource quality?

One of the central assumptions of optimal foraging theory is that different resources may be ranked according to some measure of quality. It is assumed that, all things being equal, the resources consumed by an animal contribute to fitness as a direct function of their quality (Pyke *et al.* 1977). Thus, the ultimate measure of resource quality would be the relative reproductive success of individuals consuming different resources. For many animals this is difficult or impossible to measure, and other objective measures of resource quality, such as net energy intake per unit time, are usually estimated instead.

At first glance an obvious measure of resource quality for leaf-cutter ants would appear to be the rate of fungus growth achieved on various plant species. However, adult leaf-cutters obtain as much as 95 per cent of their daily energy needs directly from plant sap and juices ingested during leaf cutting and handling (Quinlan and Cherrett 1979). It is possible that the availability of soluble carbohydrates and other small molecules may significantly affect ant preferences even though these substances may have only minor effects on fungus growth. Another serious complication in estimating resource quality for leaf-cutters is that, for other herbivores, plants are frequently complementary rather than substitutable resources due to differences in nutrient availability and the presence of deleterious allelochemicals (Pulliam 1975; Tillman 1980). If plant resources are complementary then plants cannot be ranked by any single measure of quality. These two issues, conflict within the mutualism and the nature of plant resources, must be resolved before any general understanding of resource quality can be reached.

Resource quality as assessed by ants and fungus

Several studies have documented variation in ant utilization of available plants. Colonies foraging in the field sample many plants but cut large amounts of material from only a few species at any given time (Cherrett 1968; Rockwood 1976; Rockwood and Hubbell 1987). Differences in plant species harvested are consistently reproduced in experiments equalizing plant availability and distance from the colony, suggesting that differential cutting is, in fact, due to ant preference for particular plants (Waller 1982a,b; Hubbell and Wiemer 1983; Hubbell et al. 1984; Howard 1987, 1988). It is generally assumed that these preferences are based on significant differences in the ability of plants to support fungus growth; but see the discussion by Powell and Stradling in Chapter 3.

Investigations of ant longevity and fungus growth with different diets have become an important tool in assessing possible intra-mutualism conflicts and the nature of plant resources. At present, there is no information available on the longevity of leaf-cutter ant workers fed saps or juices of different plants. However, ants have been maintained for up to several weeks on totally artificial diets incorporating 5 per cent sucrose or glucose, 1 per cent hydrolysed protein, and 0.1 per cent vitamin mixture (Boyd and Martin 1975; Howard et al. 1988). Survivorship studies using modifications of this basic diet reveal no significant differences in ant longevity with diets incorporating varying concentrations of sugar and hydrolysed protein (J. J. Howard, personal communication). The concentrations tested are comparable to those of soluble sugars and protein found in tropical plants available to leaf-cutters (Howard 1987). It appears that resting ant energy requirements can be met by most plants. Ants engaged in colony activities are likely to need considerably more energy, but studies using totally dry fungus substrate supplemented with liquid ant food have not been attempted.

Other investigations have attempted to assess the growth of the fungus on plants, plant extracts, or chemically defined diets. Ant fungi grow significantly better on leaves pre-treated by ant processing or chemical dewaxing and sterilization (Quinlan and Cherret 1977). Extracts of cultivated plants and artificial substrates, such as beans and cereals, differ greatly in their ability to support fungus growth (Quinlan and Cherrett 1978; Mudd and Bateman 1979), but mixed-substrate diets have not been specifically compared with single-substrate diets. The fact that leaf-cutter colonies have been maintained for years on largely mono-specific diets of lilac (*Syringa*) or privet (*Ligustrum*) suggests that some plant species may be complete resources for ants and fungus. However, temperate plants used to culture leaf-cutter ants may be chemically very different from the neotropical plants that evolved under leaf-cutter attack.

Only one study has compared the effects of dietary constituents on ant longevity and fungus growth (Howard *et al.* 1988). Three out of four terpenoids that significantly repelled leaf-cutter ants in behavioural bioassays inhibited fungus growth or killed cultures outright. Only two of these compounds reduced ant survival at biologically realistic concentrations, suggesting that ants might avoid plants containing substances deleterious to the fungus but harmless to themselves.

The possibility that ants and fungus have conflicting requirements raises questions about the nature of the ants' contribution to this mutualism. At one extreme, ants may in fact forage for resources that support the maximal rate of fungus growth, imbibing liquids only as an adjunct to this process. This is almost certainly the case in lower attine genera, which harvest more detritus than fresh plant material (Weber 1972). At the other extreme, *Atta* and *Acromyrmex* may forage on plants that provide rich supplies of immediately available energy, leaving the fungus to convert structural carbohydrates into usable energy and detoxify any allelochemicals present (Littledyke and Cherrett 1976; Quinlan and Cherrett 1978; Powell and Stradling, Chapter 3, this volume). Maximizing fungus growth is not necessarily the object of resource harvest if ants also forage for immediate energetic gain.

Does cost of resource acquisition affect foraging behaviour?

Interest in quantifying leaf-cutter ant foraging costs is relatively recent, although it is commonly assumed that costs influence ant foraging patterns. Surface activities related to foraging include trail construction, searching for new resources, and harvesting existing resources. Each of these activities contributes to the overall cost of obtaining resources. At present little or nothing is known about the time or energy costs of trail construction and maintenance. Trunk trails may be over 10 cm wide and 100 m long, and are cleared to bare soil (Weber 1972). This suggests that trail construction and maintenance is costly, but it is rewarded by four- to tenfold increases in ant running speed compared to uncleared ground (Rockwood and Hubbell 1987). Once established, trails may channel ant activity into portions of their territory for many months, presumably because the costs of harvesting on existing trails are lower than those involved in constructing trails to new resources.

Searching for new resources is a critical and relatively uninvestigated aspect of leaf-cutter foraging. Ant colonies strike a balance between searching for new resources and harvesting existing resources, and the efficiency of one function can be increased only by diverting workers from the other (Johnson *et al.* 1987). In theory, colonies should allocate enough workers to searching for resources to ensure that the average quality of incoming resources remains constant over time. In practice, limiting the number of searchers might cause high-quality resources to escape detection, forcing ants to cut less preferred plants in order to maintain acceptable rates of fungus substrate harvest.

Costs associated with trail construction and searching are difficult to measure but are likely to influence long-term ant foraging patterns, because they operate primarily through allocation of workers away from harvesting and into different functions. In contrast, costs associated with travelling to and from available plants are easy to measure and are likely to affect short-term decisions about which plants to cut. Leaf-cutter ants generally travel at $1-2$ m min^{-1}, with laden ants travelling more slowly than unladen ants (Lewis *et al.* 1974; Hubbell *et al.* 1980; Rudolph and Loudon 1986; Lighton *et al.* 1987). However, larger ants travel faster than small ones, suggesting that ants of different sizes vary in efficiency of load carriage (Rudolph and Loudon 1986; Lighton *et al.* 1987). So far, information on the time costs of foraging has not been used to determine whether ants would cut less preferred plants that can be harvested more rapidly than more preferred plants.

Only recently have estimates of the energy costs of foraging become available. Lighton *et al.* (1987) developed a model of energetic costs of load carriage in *Atta colombica* which is widely applicable to attines and other ants. Total, gross cost of transport increases with ant size, but the cost per gram transported decreases with body size. Larger ants are thus more efficient at transporting loads of a given size in terms of energy as well as time. However, large ants cut relatively smaller loads (Lutz 1929; Rudolph and Loudon 1986; Feener *et al.* 1988), so the overall energy efficiency of load retrieval of workers may not be a simple function of body size.

Time and energy costs of foraging need not be of equal importance to leaf-cutter ants, and it is of significance and interest to determine whether one might be more limiting than the other. Some insights into this question may be gained from simple calculations based on observations of foraging attines. In the following discussion it is assumed that: (1) each load harvested by foraging ants offers a complete diet (although not necessarily an optimal diet) to both ants and fungus; and (2) gross energy content is an index of plant quality.

Gross rates of energy gain per foraging trip were estimated using several colonies of *Acromyrmex octospinosus* at Phoenix, Arizona, a sample of *Atta colombica* from Barro Colorado Island, Panama, collected in January 1989, and data for *A. colombica* from Barro Colorado Island reported by Lighton *et al.* (1987). The total length of each trail was determined for eight *Ac. versicolor* colonies ($n = 52$) and six *A. colombica* colonies ($n = 41$). For the first two samples workers were timed over a flat section of trail of known length, and then the ant and its load were collected and weighed. Loads were then dried to obtain estimates of dry biomass. The unladen and laden gross cost of transport were calculated for each ant using equations 18 and 19, respectively, given in the paper by Lighton *et al.* (1987). For the third sample, equivalent data were taken from published values or calculated from regressions used in Lighton *et al.* (1987). Dry load mass was estimated assuming 75 per cent moisture content. The usable energy content of loads was calculated assuming 4.5 k cal g^{-1} plant material (4.186 J cal^{-1}) and that the fungus utilizes 45 per cent of the total energy in the leaves (Martin and Weber 1969).

The data reveal large differences in distance travelled, ant and burden size, cost estimate, and rate of energy gain between the three samples (Fig. 4. 1). However, in each sample the estimated net energy obtained (usable energy in a load minus energy costs of load harvest) are very similar for median and maximum foraging distances. The energy costs of travel are quite small relative to the potential usable energy contained in each load harvested, suggesting that foraging at greater distances from the colony entails little loss in net energy obtained. The rate of energy gain is much more sensitive to the time costs of harvesting: increasing foraging distance decreases the rate of energy gain. Estimates from these three samples suggest that increasing foraging from the median to the maximum foraging distance may result in a 2.5- to fourfold decrease in the rate of energy gain.

Clearly, there is reason to believe that ant foraging may be more limited by time than by the energy required to travel long distances. This pattern is consistent with studies of estimated foraging costs for seed-harvesting ants (Fewell 1988). The data presented here give no more than rough estimates, and considerable improvements in cost estimates are to be expected from systematic studies using uniform methods of data collection and calculation. Nevertheless, estimated rates of energy gain on trails of median length vary by less than a factor of two in these samples. This consistency may indicate that leaf-cutter species are governed by essentially similar fungus growth dynamics, requiring similar rates of leaf harvest.

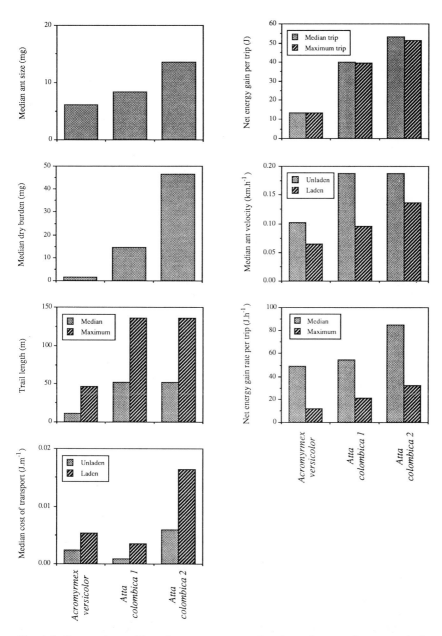

Fig. 4.1. Comparison of foraging parameters and estimated rates of energy gain for three samples of leaf-cutter ant. Ac = *Acromyrmex versicolor*, *n* = 29 workers. At 1 = *Atta colombica* sample 1, *n* = 199 workers. At 2 = *Atta colombica* data from Lighton *et al*. 1987, *n* = 154.

Quality, cost, and hypotheses about leaf-cutter ant foraging

Resolving these issues of resource quality and cost will ultimately lead to greater understanding of leaf-cutter ant foraging patterns. The foraging of leaf-cutters is typified by patterns that challenge our understanding of foraging, and a number of hypotheses which aim to explain these patterns have been advanced.

Three features of leaf-cutter ant foraging in particular have generated much interest. First, despite having strong preferences for a few plant species, leaf-cutters persist in sampling numerous species that consistently remain only moderately to slightly preferred (Rockwood 1976). Secondly, even highly preferred plants are usually abandoned before they are completely defoliated (Cherrett 1968; Fowler and Stiles 1980). Finally, ants travel long distances on trunk trails to reach plant resources, often bypassing conspecific plants of apparently equal quality much closer to the colony (Cherrett 1982).

Three general hypotheses seek to explain these patterns (Table 4. 1). The *optimal foraging* hypothesis (Fowler and Stiles 1980; Rockwood and Hubbell 1987) suggests that observed patterns are the result of foraging to maximize the quality of material harvested per unit of time or energy. According to this hypothesis a distant plant of moderate palatability at the edge of an existing trail may provide a higher rate of return than a nearby plant of high quality but at some distance from the nearest trail. Ants may abandon plants being harvested when a new plant offering a higher rate of return becomes available, or when defoliation-induced changes in chemistry decrease quality (palatability?) relative to other available plants. The *resource conservation* hypothesis (Cherrett 1982) suggests that long-lived leaf-cutter colonies conserve resources over the colony lifespan by limiting damage inflicted on highly preferred or nearby resources. Thus, leaf-cutters harvest some material that is either of higher cost than necessary or of lower quality than the best available at a given time. The *nutrient balance* hypothesis argues that plant resources are complementary, and that ants must seek out a variety of plants in order to provide a suitable mix of nutrients for fungus growth. In this case, the cost of harvesting resources may play little or no part in foraging decisions which involve a specific deficiency that can be satisfied by only one available plant species; the colony must cut the plant at whatever cost or perish. The *allelochemical* hypothesis stems from the need to avoid high levels of any single toxic allelochemical. A mix of species will dilute toxic chemicals present in any one species.

It has proven difficult to evaluate the extent to which ant foraging patterns are uniquely predicted by any one of these hypotheses. For example, both optimal and nutrient-balancing strategies can generate patterns that fit the conservational foraging hypothesis. However, each of these hypotheses makes different predictions about the behaviour of ants in relation to resource quality and cost (Table 4. 1). By following long-term ant foraging

Table 4.1. Predicted relationship of resource quality and cost to foraging behaviours of leaf-cutter ants, from three competing hypotheses to explain leaf-cutter foraging systems. Other explanations may also be possible.

Model	Resource type	Explanation for foraging behaviour		
		Sampling many species	Partial defoliation	Cutting distant plants
Optimal	Substitutible	Highest quality/unit cost sometimes provided by less-preferred plants	Plants abandoned when new ones providing higher quality/unit cost are located	Distant plants may yield higher quality/unit cost than near ones due to trail system
Conservational	Substitutible	Lower quality/higher cost accepted to limit damage to best resources	Limiting damage requires abandonment even if other resources are of lesser quality/unit cost	Higher costs incurred to prevent over exploitation of plants near the colony
Nutrient/ allelochemical balancing	Complementary (no ranking by quality possible)	Many plants required to provide proper nutrient/ chemical balance for fungus growth	Imbalances in chemistry prompt search for complementary plant	Distant plants may be required to complement other plants being harvested

patterns and performing critical experimental manipulations on resource availability, it may ultimately be possible to distinguish between competing explanations of ant foraging behaviour. As leaf-cutters are the most important generalized herbivores in the neotropics, such knowledge will be of critical importance in establishing the role of leaf-cutters in increasingly fragmented tropical ecosystems.

Acknowledgements

I wish to thank J. Cazin, D. Feener, J. Lee, S. Rissing, and D. Wiemer for discussions. B. Terkanian and R. Stanford collected ants in Panama. Financial support was provided by the Maytag Fellowship in Zoology, in the Department of Zoology, Arizona State University, and by a post-doctoral fellowship from the National Science Foundation of the United States.

5

Leaf-cutter ant assemblies: effects of latitude, vegetation, and behaviour

H. G. Fowler and Silvia Claver

Introduction

Although leaf-cutter ants (*Acromyrex* and *Atta*) are commonly considered to be characteristic of tropical forests, leaf-cutters are actually more abundant and diverse in the subtropics of South America (Fowler 1983*a,b*) (Fig. 5. 1). Due to this biogeographical pattern, co-occurrences of taxa are likely, especially in the subtropical regions. Furthermore, not all species of leaf-cutter ant occur in forests, a sizeable proportion of the recognized taxa are open habitat specialists (Table 5. 1). Many of these open habitat specialists are grass-cutters, while the remainder, and the closed habitat taxa, are true leaf-cutters; although many species forage leaf, flower, and fruit fall for the fungal substrate (Fowler and Stiles 1980; Fowler 1985; Fowler *et al.* 1986).

The inverse distribution of species diversity (Fig. 5. 1) results in some assemblies of leaf-cutting ants containing as many as seven species (Fowler 1983*a*, 1984). However, as regional taxonomic richness (gamma diversity) is

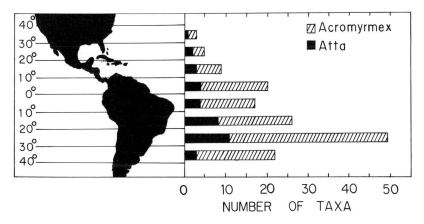

Fig. 5.1. The latitudinal distribution of the taxa of leaf-cutting ants (*Atta* and *Acromyrmex*). All taxonomically valid categories, including subspecies are included.

Table 5.1. Species richness patterns among leaf-cutting ants in the
Neotropics (genera **Atta** and **Acromyrmex**).

Comparison	Number of species of:		Total	G statistic
	Atta	*Acromyrmex*		
Geographic range				
Tropics	8	6	14	
Subtropics	4	14	18	
Both	4	5	9	25.82**
Habitat occupied				
Open	7	11	18	
Closed	7	12	19	
Both	2	2	4	0.26
Foraging Behaviour				
Browser	10	14	24	
Grazer	5	7	12	
Browser/grazer	1	2	3	
Opportunist	0	2	2	2.15

** $= P < 0.001$.

higher in the subtropics (Fig. 5. 1), we must consider the potential inclusion
of species in any assembly analysis (Ricklefs 1987). If we do not, inferences
about spatial patterning due to competition (Fowler 1984; Conner and
Bowers 1987) may be misleading. Taking note of this warning, we have opted
to examine the interaction of regional and local species richness with respect
to latitude and habitat type.

Community interactions of attines have been studied by various people
(Bucher and Montenegro 1974; Fowler and Stiles 1980; Fowler 1983a,
1984; Fowler and Haines 1983) and these studies have focused on patterns
of resource use and nest-spacing strategies for local assemblies. Here we have
attempted to provide information on a regional scale to shed some light on
the evolution and species diversification of these interesting and important
neotropical faunal elements.

To give an adequate picture of the types of species differences found
among taxa of leaf-cutting ants, a dissimilarity matrix is shown for the
Paraguayan fauna (Fig. 5. 2), and taxa can thus be separated by ecology and
behaviour (Fig. 5. 3). However, clusters of species are present due to similarities in ecology, behaviour, and distribution (Fowler 1985). Some taxa show a

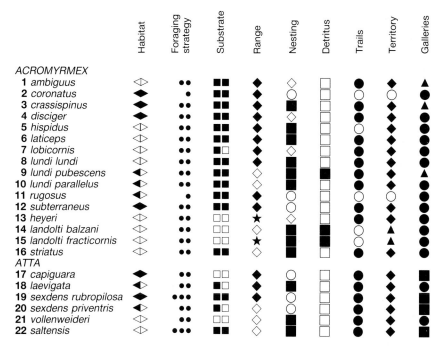

Fig. 5.2. Dissimilarity matrix of the Paraguayan leaf-cutting ants. Dissimilarities are identified by differing symbols. Key: Habitat—open = range lands, closed = woodlands, open/closed = both; Foraging strategy—one circle = forage on fallen flowers and leaves, two circles = actively cuts vegetation, three circles = employs staged foraging when cutting; Substrate—closed = broad-leaf plants, open = grasses, closed/open = both; Range—open = Chaco, closed = eastern Paraguay, star = both; Nesting—square = subterranean nests; circle = dispersed subterranean nests, diamond = superficial or dome nests; Detritus—open = buried in special chambers, closed = carried and dumped on surface; Trails—open = physically defined trails not well developed or absent, closed = well-defined physical trails present; Territory—circle = no well-defined territories; triangle = linear foraging territories; diamond = circular foraging territories; Galleries—circle = foraging galleries absent, square = foraging galleries well-defined and present, triangle = foraging galleries formed by excavating portion of foraging trail and then thatching this with vegetation. Further details in Fowler (1985).

strong positive correlation with local species richness (Table 5. 2) suggesting their inclusion in many local assemblies, while other taxa are apparently more restricted geographically in less variable species-number assemblies. However, it is not economical to obtain such data over a large geographical area, but species presence/absence data as presented here are relatively easy to obtain.

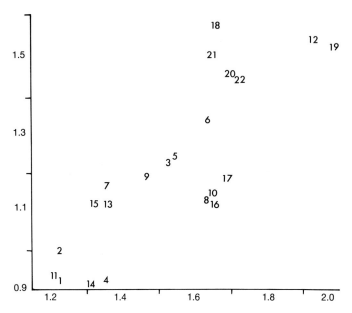

Fig. 5.3. Taxa ordination of Paraguayan leaf-cutting ants based upon dissimilarity matrices (Fig. 5.2) using canonical variate analysis. Taxon numbers are those of Fig. 5.2.

Data base of leaf-cutter ant assemblies

During the last 15 years, we have had the opportunity to sample leaf-cutter ants in many areas from the United States to Argentina. Of these collecting opportunities, some produced quantitative estimates of nest densities while others were restricted to collecting only. Limited bibliographic information was also utilized (e.g. Escalante 1976; Bucher and Montenegro 1974). A total of 550 leaf-cutter assemblies, ranging from tropical wet forests to arid regions (caatinga, cerrado, chaco, patagonia, desert), provided suitable data. We have limited our analysis to species level taxa and defined regional diversity on latitudinal, longitudinal, and phyto-sociological zones. A more detailed analysis will be presented elsewhere. For the present purpose, we have classified habitat types as closed for any woodland, forest, or tree crop, and open for all others, including shrub savannahs. We have defined sub-tropical areas as lying at latitudes greater than 25° north or south, and tropical areas to include all other areas.

Ants were classified into three types according to foraging behaviour:

1. opportunists that rely principally upon leaf, fruit or flower fall, and do not generally cut vegetation (Fowler and Stiles 1980);

Table 5.2. The correlation between colony density and species richness in Paraguayan assemblies of leaf-cutting ants. Kendall's rank correlation and significance level are shown.

Taxon	Correlation	P
Acromyrmex rugosus (Fr. Smith)	0.748	0.0001
Acromyrmex balzani (Emery)	0.668	0.0001
Atta capiguara Goncalves	0.658	0.0001
Atta vollenweideri (Forel)	0.207	0.0001
Atta laevigata (Fr. Smith)	0.186	0.0001
Atta sexdens rubropilosa (Forel)	0.164	0.0005
Acromyrmex subterraneus (Forel)	0.146	0.0022
Acromyrmex crassispinus (Forel)	0.142	0.0029
Acromyrmex heyeri (Forel)	0.132	0.0057
Acromyrmex laticeps (Emery)	0.106	0.0255
Acromyrmex fracticornis (Forel)	0.100	0.0350
Acromyrmex lundi lundi (Guerin)	0.084	0.0786
Acromyrmex hispidus fallax Santschi	0.080	0.0919
Acromyrmex lundi pubescens (Emery)	0.058	0.2275
Acromyrmex coronatus (Fabricius)	0.051	0.2898

2. grazers, that cut grasses or low herbs to low levels in open habitats; and
3. browers, that climb erect vegetation and cut leaf, fruit, or flower fragments.

Some species, such as *Atta laevigata*, could thus be classified as both browsers and grass-cutters, but in such cases we have classified the ants with respect to the species composition of the local assembly. If other grass-cutters were present and the habitat had erect vegetation, such ants were classified as browsers, but if no erect vegetation was present or if other species of grass-cutting *Atta* were not present they were classified as grazers. Species such as those occurring in closed habitats were always classified as browsers. Local species richness was calculated for approximately 5-hectare areas, and thus the values used here are lower than previously reported values (Fowler 1983*b*).

Assembly patterns

Species richness

We separated assemblies into tropical and subtropical by habitat type and species richness (Table 5. 3). No significant differences in the number of

Table 5.3. Species richness of leaf-cutting ant assemblies in open and closed habitats in the tropics and subtropics. H_o is species richness/5ha.

| Species richness (alpha) | Number of assemblies per type | | | | H_o: | |
| | open | | closed | | $O = C$ | $S = T$ |
	Tropics	Subtropics	Tropics	Subtropics		
1	35	73	47	45	n.s.	S > T
2	34	48	51	42	n.s.	n.s.
3	12	15	26	12	n.s.	n.s.
4	2	37	8	13	O < C	S > T
5	0	18	1	6	O < C	S > T
6	0	18	0	7	O < C	S > T
x^2	**	**	**	**		

O = open; C = closed; S = subtropics; T = tropics.
** = $P < 0.001$.

open vs. closed habitat assemblies studied were present ($x^2 = 1.02$; $P > 0.05$), although more assemblies were studied in subtropical areas ($x^2 = 25.31$; $P < 0.05$). Furthermore, sampling effort was not homogeneous across habitat types and subtropical and tropical regions ($x^2 = 39.20$; $P < 0.05$). Nevertheless, due to the number of assemblies studied, this was not considered to be a serious defect of the data (Table 5.3). For any given assembly species richness, no significant differences in sampling were present between habitats; although in many instances subtropical habitats had many more assemblies than did tropical regions for a given species richness. Such a pattern does not depend so much upon sampling intensity, but rather on the fact that assemblies in the subtropics tended to be richer (Fig. 5.1). In fact, mean assembly species richness was higher in the subtropics in both open and closed habitats ($x = 2.7$ and 2.4, respectively) than in their tropical counterparts ($x = 1.8$ and 2.0, respectively). Furthermore, in all cases, the number of assemblies was not uniformly distributed among species richness, but concentrated at lower species richness values, with subtropical regions having the most species-rich assemblies.

Generic structure and foraging behaviour

Analysing assemblies by genus and type of foraging behaviour (Table 5.4), revealed trends in the structuring of local assemblies. Opportunistic species do not occur in the genus *Atta* (Table 5.4). Based upon the species richness of local assemblies, the frequencies of representation by the two genera were

Table 5.4. Patterns of species richness by genus and foraging behaviour of leaf-cutting ant assemblies (Table 5.3), registered by occurrences of species. *Acro* = *Acromyrmex*; H_o is species richness/5ha.

Species richness	Opportunists		Browsers (B)		Grazers (G)		H_o:	
	Atta	*Acro.*	*Atta*	*Acro.*	*Atta*	*Acro*	G:B	Ac:At
1	0	6	12	86	34	62	n.s.	Ac > At
2	0	29	58	130	79	54	G < B	Ac > At
3	0	34	33	56	45	27	n.s.	Ac > At
4	0	43	55	55	46	41	G < B	Ac > At
5	0	8	27	29	26	35	n.s.	Ac > At
6	0	10	31	33	38	38	n.s.	n.s.

not homogeneous, but varied with respect to assembly species richness values (grazers: $G = 159.9$, $P < 0.05$; browsers: $G = 1640.5$, $P < 0.05$). As species richness increased, proportionally more species of *Acromyrmex* were found in assemblies, and within any given species richness value, species of *Acromyrmex* occurred significantly more commonly than species of *Atta*, as would be expected (Fig. 5. 1; Table 5. 1). Browser species are more important than grazers at richness levels of 2 and 4 (Table 5. 3).

Local vs. regional species richness

Comparing local with regional species richness (Fig. 5. 4), no pattern of increasing local species richness was found as regional species richness increased. This relation held for linear ($r^2 = 0.03$) and exponential ($r^2 = 0.06$) models. We may therefore discount regional species richness patterns as being influential in determining the number of species in local assemblies.

Discussion

The general patterns of leaf-cutting ant assemblies shows that taxa are not distributed evenly among assemblies in either species richness, generic composition, or foraging behaviour. Subtropical regions have richer species assemblies, and in particular, open, subtropical habitats. Behaviour and worker polymorphism (Figs 5. 2 and 5. 3) strongly determine the roles of species in assemblies, as well as influencing which species combinations may occur. Thus, although historical factors are important in determining which species are capable of co-occurrence, these determine neither specific combinations of species within assemblies, nor the numbers of their component species (Fowler 1984). Given these patterns, as well as the daily use

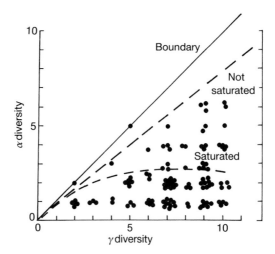

Fig. 5.4. The relation between alpha and gamma diversity of assemblies of leaf-cutter ants, in localities throughout Mexico, Central, and South America. Some values are taken from the literature. Each value represents five local assemblies.

of resources (Fig. 5. 5) (Fowler 1984), regional species richness has little effect upon species assemblies, suggesting that most local assemblies are species saturated.

Leaf-cutting ants have been shown to be dominant key stone elements in open subtropical habitats (Fowler and Haines 1983), they are easily identifiable components of nearly all neotropical habitats, and therefore studies of factors which structure local assemblies are crucial to understanding community organization. Preliminary work has documented separation in resource uses, daily patterns of activity, nest spacing, and foraging strategy as correlates of species co-occurrence in assemblies (Bucher and Montenegro 1974; Fowler 1984, 1985; Fowler *et al.* 1986). Other factors such as rates of colonization and nest movement (Fowler 1981), interactions with vertebrate herbivores (competitive interactions), and predators may be necessary to explain assembly variation more fully.

Acknowledgements

We gratefully acknowledge the assistance provided by the Brazilian Conselho Nacional de Desenvolvimento Cientifico e Tecnológico (CNPq), and the Argentine Consejo Nacional de Investigaciónes Científicas y Técnicas (CONICET) for support in the form of grants and fellowships. Additional support for portions of these studies came from the Ministry of Overseas Development, the US Peace Corps, the United Nations Development Pro-

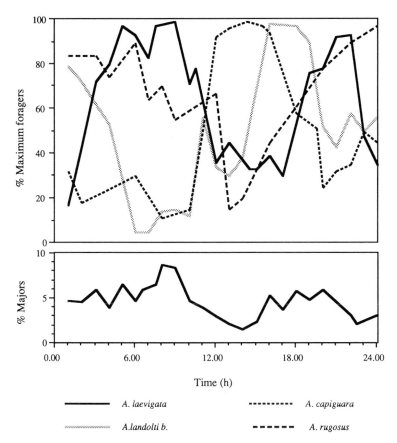

Fig. 5.5. Daily foraging patterns in an assembly of four species of leaf-cutter ants. *Atta laevigata*, which has among the largest major workers of any species, deploys these on foraging trails in large numbers. Other species which do not use this strategy are restricted to using hours of the day when *A. laevigata* is not active. For all other species there is negative correlation of foraging with numbers of active *A. laevigata* major workers.

gramme (UNDP), the Paraguayan Ministry of Agriculture, FUNDUNESP, the World Wildlife Foundation, the Fundação S.O.S. Mata Atlantica, Comisión de Estudios para la Promoción de la Lana Argentina (CEPLA), Rutgers University, and the Special Program for the Improvement and Development of Ecological Research in Argentina (SPAIDERA). Over the years, many people have provided sounding boards for our ideas, but we would especially like to acknowledge Liuz Carlos Forti, Jaques Delabie, Malcolm Cherrett, Ligia F. T. de Romagnano, Jorge Rabinovich, and Enrique Bucher; Nozer Pinto kindly penned the drawings.

References to Part 1

Ales, D. C., Weimer, D. F., and Hubbel, S. P. (1981). A natural repellent of leafcutter ants. *Proceedings of the Iowa Academy of Science*, **88**, 19.

Allen, S. E., Grimshaw, H. M., Parkinson, J. A., and Quarmby, C. (1974). *Chemical analysis of ecological materials*, 365 pp. Blackwell Scientific Publications, Oxford.

Barrer, P. M. and Cherrett, J. M. (1972). Some factors affecting the site and pattern of leaf-cutting activity in the ant *Atta cephalotes* L. *Journal of Entomology (A)*, **47**, 15–27.

Bate-Smith, E. C. (1972). Detection and determination of ellagitannins. *Phytochemistry*, **11**, 1153–6.

Bate-Smith, E. C. (1973). Haemanalysis of tannins: the concept of relative astringency. *Phytochemistry*, **12**, 907–12.

Bernays, E. A. (1978). Tannins: an alternative viewpoint. *Experimental and Applied Entomology*, **24**, 44–53.

Boyd, N. D. and Martin, M. M. (1975). Faecal proteinases of the fungus-growing ant, *Atta texana* (Hym, Formicadae): their fungal origin and ecological significance. *Journal of Insect Physiology*, **21**, 1815–20.

Bucher, E. H. and Montenegro, R. (1974). Habitos forrajeros de cuatro hormigassimpatridas del genero *Acromyrmex* (Hymenoptera, Formicidae). *Ecologia (Argentina)*, **2**, 47–53.

Cherrett, J. M. (1968). The foraging behaviour of *Atta cephalotes* L. (Hymenoptera, Formicidae). 1. Foraging pattern and plant species attacked in tropical rain forest. *Journal of Animal Ecology*, **37**, 387–403.

Cherrett, J. M. (1972*a*). Chemical aspects of plant attack by leaf-cutting ants. In *Phytochemical ecology* (ed. J. B. Harbourne), pp. 13–24. Academic Press, New York.

Cherrett, J. M. (1972*b*). Some factors involved in the selection of vegetable substrate by *Atta cephalotes* (L.) (Hymenoptera: Formicidae) in tropical rain forest. *Journal of Animal Ecology*, **41**, 647–60.

Cherrett, J. M. (1980). Possible reasons for the mutualism between leaf-cutting ants (Hymenoptera, Formicidae) and their fungus. *Biologie–Ecologie Méditérranéenne*, **7**, 113–22.

Cherrett, J. M. (1982). Resource conservation by the leaf-cutting ant *Atta cephalotes* in tropical rain forest. In *Tropical rain forest: ecology and management* (ed. S. L. Sutton, T. C. Whitmore, and A. C. Chadwick), pp. 253—63. Blackwell, London.

Cherrett, J. M., Powell, R. J., and Stradling, D. J. (1989). The mutualism between leaf-cutting ants and their fungus. In *Insect–fungus interactions* (ed. N. Wilding, N. M. Collins, P. M. Hammond, and J. F. Webber), pp. 93—120. Academic Press, London.

Conner, E. F. and Bowers, M. A. (1987). The spatial consequences of interspecific competition. *Annales Zoologici Fennici*, **24**, 213–26.

Davidson, R. W., Campbell, W. A., and Blaisdell, D. J. (1938). Differentiation of wood-decaying fungi by their reactions on gallic and tannic acid medium. *Journal of Agricultural Research*, **57**, 683–95.

Escalante, G. J. A. (1976). Hormigas del valle de K'osnipata (Paucartambo, Cusco). *Revista Peruana de Entomologia Agricola*, **19**, 107–8.

Febvay, G. (1981). Quelques aspects (anatomie et enzymologie) des relations nutritionnelles entre la fourmi attine *Acromyrmex octospinosus* (Hymenoptera, Formicidae) et son champignon symbiotique. Ph.D. thesis, University of Paris.

Feener, D. H., Jr., Lighton, J. R. B., and Bartholomew, G. A. (1988). Curvilinear allometry, energetics and foraging ecology: a comparison of leaf-cutting ants and army ants. *Functional Ecology*, **2**, 509–20.

Feeny, P. (1969). Inhibitory effect of oak leaf tannins on the hydrolysis of proteins by trypsin. *Phytochemistry*, **8**, 2119–26.

Feeny, P. (1975). Biochemical coevolution between plants and their insect herbivores. In *Coevolution of animals and plants* (ed. L. E. Gilbert and P. H. Raven), pp. 3–19. University of Texas Press, Austin.

Feeny, P. (1976). Plant apparency and chemical defense. *Recent Advances in Phytochemistry* **10** 1–40.

Fewell, J. H. (1988). Energetic and time costs of foraging in harvester ants, *Pogonomyrmex occidentalis*. *Behavioral Ecology and Sociobiology*, **22**, 401–8.

Fowler, H. G. (1981). On the emigration of leaf-cutting ant colonies. *Biotropica*, **13**, 316.

Fowler, H. G. (1983*a*). Latitudinal gradients and diversity of the leaf-cutting ants (*Atta* and *Acromyrmex*). *Revista de Biologia Tropical*, **31**, 213–16.

Fowler, H. G. (1983*b*). Distribution patterns of Paraguayan leaf-cutting ants (*Atta* and *Acromyrmex*) (Hymenoptera: Formicidae: Attini). *Studies on the Neotropical Fauna and Environment*, **18**, 121–38.

Fowler, H. G. (1984). A organização das comunidades de formigas cortadeiras. *Anais semanario Regional Ecologia, São Carlos*, **4**, 151–63.

Fowler, H. G. (1985). The leaf-cutting ants (*Atta* and *Acromyrmex*) of Paraguay. *Deutsche entomologische Zeitschrift N.F.*, **32**, 19–34.

Fowler, H. G. and Haines, B. L. (1983). Diversisad de espécies de hormigas cortadoras y termitas de túlmulo en cuanto a la sucesiòon vegetal en praderas paraguayas. In *Social insects in the tropics*, Vol. 2 (ed. P. Jaisson). Université Paris Presses, Paris.

Fowler, H. G. and Stiles, E. W. (1980). Conservative foraging by leaf-cutting ants? The role of foraging trails and territories, and environmental patchiness. *Sociobiology*, **5**, 25–41.

Fowler, H. G., Pereira da Silva, V., Forti, L. C., and Saes, N. B. (1986). Population dynamics of leaf-cutting ants: a brief review. In *Fire ants and leaf-cutting ants: biology and management* (ed. C. S. Lofgren and R. K. Vander Meer). Westview Press, Boulder.

Garling, L. (1979). Origin of ant-fungus mutualism: A new hypothesis. *Biotropica*, **11**, 284–91.

Goldstein, J. L. and Swain, T. (1965). The inhibition of enzymes by tannins. *Phytochemistry*, **4**, 185–92.

Gonçalves de Lima, O., Marino-Bettolo, G. B., Dell Monache, F., Coelho, J. S., Leoncio d'Albuquerque, I., Maciel, G. M., Lacerda, A., and Martins, D. G. (1970). Antimicrobial compounds from higher plants. XXXII. Antimicrobial and antineoplastic activity of 2-methoxy-6-*n*-pentyl-*p*-benzoquinone (primim) isolated from the roots of *Miconia* species (Melastomataceae). *Revisto do Instituto de Antibioticos. Universidade do Recife*, **10**, 29–34.

Grant, W. D. (1976). Microbial degradation of condensed tannins. *Science*, **193**, 1137–9.

Harborne, J. B. (1973). *Phytochemical methods: a guide to modern techniques of plant analysis*, 278 pp. Chapman and Hall, London.

Harborne, J. B. (1977). *Introduction to ecological biochemistry*, 243 pp. Academic Press, London.

Hill, D. L. and Murtha, E. F. (1962). A review of studies on the extractives of the plant *Ryania speciosa*. *US Army Chemical Research and Development Laboratories Special Publication*, 22 pp.

Howard, J. J. (1987). Leafcutting ant diet selection: the role of nutrients, water, and secondary chemistry. *Ecology*, **68**, 503–15.

Howard, J. J. (1988). Leafcutting ant diet selection: relative influence of leaf chemistry and physical features. *Ecology*, **69**, 250–60.

Howard, J. J., Cazin, J., Jr., and Wiemer, D. F. (1988). Toxicity of terpenoid deterrents to the leafcutting and *Atta cephalotes* and its mutualistic fungus. *Journal of Chemical Ecology*, **14**, 59–69.

Hubbell, S. P., Johnson, L. K., Stanislav, E., Wilson, B., and Fowler, H. (1980). Foraging by bucket-brigade in leaf-cutter ants. *Biotropica*, **12**, 210–13.

Hubbell, S. P. and Wiemer, D. F. (1983). Host plant selection by an attine ant. In *Social insects in the tropics*, Vol. 2 (ed. P. Jaisson), pp. 133–54. University of Paris, Paris.

Hubbell, S. P., Wiemer, D. F., and Adejare, A. (1983). An antifungal terpenoid defends a neotropical tree (*Hymenaea*) against attack by fungus-growing ants (*Atta*). *Oecologia (Berlin)*, **60**, 321–7.

Hubbell, S. P., Howard, J. J., and Wiemer, D. F. (1984). Chemical leaf repellency to an attine ant: seasonal distribution among potential host plant species. *Ecology*, **65**, 1067–76.

Johnson, L. K., Hubbell, S. P., and Feener, D. H., Jr. (1987). Defense of food supply by eusocial colonies. *American Zoologist*, **28**, 347–58.

Jonkman, J. C. M. (1977). Biology and ecology of *Atta vollenweideri* Forel 1893 and its impact on Paraguayan pastures. Thesis, Universiteitsbibliothick, Leiden.

Jonkman, J. C. M. (1978). Nests of the leaf-cutting ant, *Atta vollenweideri*, as accelerators of succession in pastures. *Zeitschrift für angewandte Entomologie*, **86**, 25–34.

Kreisel, H. (1972). Pilze aus Pilzgarten von *Atta insularis* in Kuba. *Zeitschrift für allgemeine Mikrobiologie*, **12**, 643–54.

Lampard, J. (1974). Demethylhomopterocarpin: an antifungal compound in *Canavalia ensiformis* and *Vigna unguivculata* following infection. *Phytochemistry*, **13**, 291–2.

Levin, D. A. (1976). The chemical defenses of plants to pathogens and herbivores. *Annual Review of Ecology and Systematics*, **7**, 121–59.

Lewis, T., Pollard, G. V., and Dibley, G. C. (1974). Rhythmic foraging in the leaf-cutting ant *Atta cephalotes* (L.) (Formicidae: Attini). *Journal of Animal Ecology*, **43**, 129–42.

Lighton, J. R. B., Bartholemew, G. A., and Feener, D. H., Jr. (1987). Energetics of load carriage and a model of the energy cost of foraging in the leaf-cutting ant *Atta colombica* Guerin. *Physiological Zoology*, **60**, 524–37.

Littledyke, M. and Cherrett, J. M. (1975). Variability in the selection of substrate by the leaf-cutting ants *Atta cephalotes* (L.) and *Acromyrmex octospinosus* (Reich) (Formicidae, Attini). *Bulletin of Entomological Research*, **65**, 33–47.

Littledyke, M. and Cherrett, J. M. (1976). Direct ingestion of plant sap from cut leaves by the leaf-cutting ants *Atta cephalotes* (L.) and *Acromyrmex octospinosus* (Reich) (Formicidae, Attini). *Bulletin of Entomological Research*, **66**, 205–217.

Lutz, F. E. (1929). Observations on leaf-cutting ants. *American Museum Novitates*, **388**, 1–21.

Martin, M. M. (1974). Biochemical ecology of the attine ants. *Accounts of Chemical Research*, **7**, 1–5.

Martin, M. M. and Boyd, N. D. (1974). Properties, origin and significance of the fecal proteases of the fungus-growing ants. *American Zoologist*, **14**, 1291.

Martin, M. M. and Weber, N. A. (1969). The cellulose utilising capacity of the fungus cultured by the attine ant *Atta colombica tonsipes*. *Annals of the Entomological Society of America*, **62**, 1386–7.

Martin, M. M., Carman, R. M., and MacConnell, J. G. (1969). Nutrients derived from the fungus cultured by the fungus-growing ant *Atta colombica tonsipes*. *Annals of the Entomological Society of America*, **62**, 11–13.

Möller, A. (1893). Die Pilzgarten einiger sudamerikanischer Ameisen. *Botanische Mitteilungen aus den Tropen*, **6**, 1–27.

Mudd, A. and Bateman, G. L. (1979). Rates of growth of the food fungus of the leaf-cutting ant *Atta cephalotes* (L.) (Hymenoptera: Formicidae) on different substrates gathered by the ants. *Bulletin of Entomological Research*, **69**, 141–8.

Mullenax, C. H. (1979). The use of jackbean (*Canavalia ensiformis*) as a biological control for leaf-cutting ants (*Atta* sp.). *Biotropica*, **11**, 313–14.

Powell, R. J. (1984). The influence of substrate quality on fungus cultivation by some attine ants. Ph.D. thesis, University of Exeter.

Powell, R. J. and Stradling, D. J. (1986). Factors influencing the growth of *Attamyces bromatificus*, a symbiont of attine ants. *Transactions of the British Mycological Society*, **87**, 205–13.

Pulliam, H. R. (1975). Diet optimization with nutrient constraints. *American Naturalist*, **109**, 765–8.

Pyke, G. H., Pulliam, H. R., and Charnov, E. L. (1977). Optimal foraging: a selective review of theory and tests. *Quarterly Review of Biology*, **52**, 137–54.

Quinlan, R. J. and Cherrett, J. M. (1977). The role of substrate preparation in the symbiosis between the leaf-cutting ant *Acromyrmex octospinosus* (Reich) and its food fungus. *Ecological Entomology*, **2**, 161–70.

Quinlan, R. J. and Cherrett, J. M. (1978). Aspects of the symbiosis of the leaf-cutting ant *Acromyrmex octospinosus* (Reich) and its food fungus. *Ecological Entomology*, **3**, 221–30.

Quinlan, R. J. and Cherrett, J. M. (1979). The role of fungus in the diet of the leaf-cutting ant *Atta cephalotes* (L.). *Ecological Entomology*, **4**, 151–60.

Rhoades, D. F. and Cates, R. G. (1976). Towards a general theory of plant anti-herbivore chemistry. *Recent Advances in Phytochemistry*, **10**, 168–213.

Ribereau-Gayon, P. (1972). *Plant phenolics*, 252 pp. University Reviews in Botany, Oliver and Boyd, Edinburgh.

Ricklefs, R. E. (1987). Community diversity: relative roles of local and regional processes. *Science*, **235**, 167–71.

Rockwood, L. L. (1976). Plant selection and foraging patterns in two species of leaf-cutting ants (*Atta*). *Ecology*, **57**, 48–61.

Rockwood, L. L. and Glander, K. E. (1979). Howling monkeys and leaf-cutting ants: comparative foraging in a tropical deciduous forest. *Biotropica*, **11**, 1–10.

Rockwood, L. L. and Hubbell, S. P. (1987). Host–plant selection, diet diversity, and optimal foraging in tropical leafcutting ant. *Oecologia (Berlin)*, **74**, 55–61.

Rudolph, S. G. and Loudon, C. (1986). Load size selection by foraging leaf-cutter ants (*Atta cephalotes*). *Ecological Entomology*, **11**, 401–10.

Schildknecht, H. and Koob, K. (1971). Myrmicacin, the first insect herbicide. *Angewandte Chemie, International Edition in English*, **10**, 124–5.

Schlueter, M. and Bordas, E. (1972). Canavanine in *Canavalia paraguayensis, C. gladiata* and *Dioclea paraguayensis. Phytochemistry*, **11**, 3533–4.

Seaman, F. C. (1984). The effects of tannic acid and other phenolics on the growth of the fungus cultivated by the leaf-cutting ant, *Myrmicocrypta buenzlii. Biochemical Systematics and Ecology*, **12**, 155–8.

Southwood, T. R. E. (1971). The insect/plant relationship: an evolutionary perspective. *Symposium of the Royal Entomological Society of London*, **6**, 3–30.

Southwood, T. R. E. (1985). Interactions of plants and animals: patterns and processes. *Oikos*, **44**, 5–11.

Stradling, D. J. (1978). The influence of size on foraging in the ant *Atta cephalotes* and the effect of some plant defense mechanisms. *Journal of Animal Ecology*, **47**, 173–88.

Stradling, D. J. (1987). Nutritional ecology of ants. In *Nutritional ecology of insects, mites, spiders and related invertebrates* (ed. F. Slansky, Jr. and J. G. Rodriguez), pp. 927–69. Wiley, New York.

Stradling, D. J. and Powell, R. J. (1986). The cloning of more highly productive fungal strains: a factor in the speciation of fungus-growing ants. *Experientia*, **42**, 962–4.

Sundman, V. and Nase, L. (1971). A simple plate test for the direct visualization of biological lignon degredation. *Paperi ja Puutavarlehti*, **53**, 67–71.

Swain, T. (1969). Phenols and related compounds. In *Data for biochemical research* (ed. R. M. C. Dawson, D. C. Elliott, W. H. Elliott, and K. M. Jones). Clarendon Press, Oxford.

Swain, T. (1979). Tannins and lignins. In *Herbivores: their interaction with secondary plant metabolites* (ed. G. A. Rosenthal and D. H. Janzen), pp. 657–82. Academic Press, New York.

Tillman, D. (1980). Resources: a graphical-mechanistic approach to competition and predation. *American Naturalist*, **116**, 362–93.

Waller, D. A. (1982*a*). Leaf-cutting ants and avoided plants: defences against *Atta texana* attack. *Oecologia (Berlin)*, **52**, 400–3.

Waller, D. A. (1982*b*). Leaf-cutting ants and live oak: the role of leaf toughness in seasonal and intraspecific host choice. *Entomologia Experimentalis et Applicata*, **32**, 146–50.

Weber, N. A. (1972). Gardening ants: the attines. *Memoires of the American Philosophical Society*, **92**, 1–146.

Wilson, E. O. (1971). *The insect societies*, 548 pp. Harvard University Press (Belknap), Cambridge.

Part 2

Ant–plant interactions involving herbivorous insects

6

Effects of ants on temperate woodland trees

J. B. Whittaker

Introduction

There is a large quantity of literature on the behaviour and activity of ants associated with temperate trees because red wood ants (*Formica rufa* group) in particular, have been considered as potential biological control agents of forest pests. There are excellent reviews of this work by Wellenstein (1954), Cotti (1963), and Adlung (1966), amongst others, which need not be repeated here. Instead, I shall take this opportunity to place more emphasis than they do on deciduous woodlands where more recent work has been carried out, and to consider in more detail the impact of the ants on the herbivore community structure, the dynamics of the insects in the tree canopy, and the effects of ant activity on herbivory.

Opinion is mixed as to the balance of advantages and disadvantages of ant activity to the trees (Adlung 1966). Some studies (e.g. Wellenstein 1952) have shown that *Formica* spp. do not include many 'harmful' insects (i.e. harmful to trees) in their diet and that they are feeding largely on honeydew rather than on solid insect prey (Ploch 1939). In these cases the effect of the ants would be detrimental. Wellenstein (1954) cites 76 per cent of authors who consider *Formica* spp. as having an overall beneficial effect on forestry and 26 per cent who considered them to be detrimental because of their encouragement of sap-feeding insects.

It is known that some ants not only increase the numbers of sap-feeding insects which they tend, but also increase other species from which they are not gaining honeydew. This may be because disturbance by ants reduces oviposition by parasitoids (Bartlett 1961) or because ants remove general predators (De Bach *et al.* 1951). In some cases ants can indirectly protect major herbivores of trees, as was found by Fritz (1983) in his study of *Formica subsericea* (Say) on black locust trees. The ants excluded a major predator of the locust leaf-mining beetle (*Odontota dorsalis* (Thunb.)), and so defoliation could actually be higher on ant-visited branches.

In deciduous woodlands in northern England, at least three *Formica* spp. (all formerly considered to be *Formica rufa* L.) may be active predators. These are *Formica rufa* L. (studied at Cringlebarrow, Lancashire (Skinner 1980*a,b*; Skinner and Whittaker 1981; Warrington and Whittaker 1985*a,b*;

Whittaker and Warrington 1985)), *Formica lugubris* Zett. (studied by
Fowler and MacGarvin (1985) in north Yorkshire), and *Formica aquilonia*
Yarr. This chapter is based on these studies, including comparisons with
other work where appropriate. Much of the discussion is concerned with the
circumstances in which predation can be shown to be effective in reducing
herbivory, to the point where trees can benefit.

Ant activity in tree canopies

The intensity of ant foraging is obviously related to distance from the nest. In
the case of *F. rufa* in northern England the fall-off in the flow rate on ant trails
is exponential (Fig. 6. 1(a)), and usually reaches zero at about 60 m from the
nest. The length of any particular trail, however, depends very much on the
presence of territories of adjacent nests (Skinner 1980*a*) and the availability
of honeydew and prey. Laine and Niemelä (1980) found that *Formica
lugubris* numbers in the tree canopy of mountain birch (*Betula pubescens
tortuosa* (von Ledebour) Nyman) were negatively correlated with distance
(Fig. 6. 1(b)) and reached zero at 15 − 35 m from the nest.

A feature of most, if not all, of the studies of red wood ants on trees in

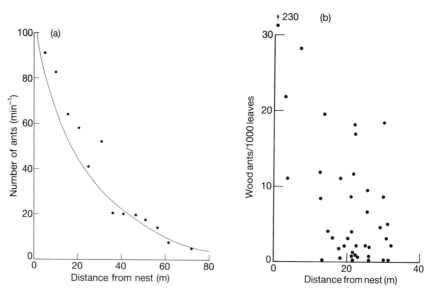

Fig. 6.1. (a) Numbers per minute of *Formica rufa* passing along a trail at different
distances from the nest at Gaitbarrows NNR northern England. The line represents
exponential decay. (Data provided by J. Wright.) (b) Numbers of *Formica aquilonia*
per 1000 leaves of birch foliage at different distances from a nest ($r = 0.485$; $P < 0.01$).
(Redrawn from Laine and Niemelä 1980.)

Europe, is the presence of mutualistic Homoptera as a component of the herbivore community. Laine and Niemelä (1980) pointed out that in a large (530) sample of mountain birch trees in northern Finland the densities of *Formica aquilonia* were positively correlated with those of the aphid *Symydobius oblongus* (von der Heyden), but negatively correlated with a second aphid, *Euceraphis punctipennis* (Zett.). *Symydobius oblongus* was also an important component of the herbivore community in Fowler and MacGarvin's (1985) study of *Betula pubescens.*

In a mixed deciduous woodland in northern England Skinner (1980*b*) showed that honeydew rarely formed less than two-thirds of the energy returned to the nest by *Formica rufa*. However, this was by no means evenly distributed among the different trails radiating from a nest. Figure 6. 2(a) shows the level of activity (as a percentage of total traffic returning to a nest) on two trails of *F. rufa*. One trail was mainly foraging on oak trees and the other mainly on sycamore. The figure illustrates for one nest what Skinner found in the woodland as a whole; namely that intensity of foraging varies from tree species to species as the season progresses (Skinner 1980*b*, Fig. 11), with high activity on oak at the beginning and end of the season and highest activity on sycamore in mid-season.

Figure 6. 2(b) shows the proportion of ants carrying honeydew in those returning on the trails in Fig. 6. 2(a). Peak trail activity is closely associated with peak utilization of honeydew. This can also be seen within the canopy of

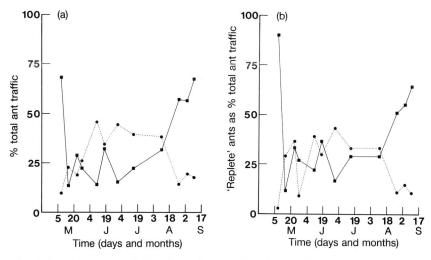

Fig. 6.2. (a) Number of *Formica rufa* returning from oak (■—■) and sycamore (●···●) as a percentage of total numbers returning to a nest, in Thrang Wood. (Redrawn from Skinner 1980*b*.) (b) 'Replete' (i.e. carrying honeydew) *Formica rufa* returning from oak (■—■) and sycamore (●···●) as a percentage of total numbers entering a nest at Thrang Wood. (Redrawn from Skinner 1980*b*.)

individual trees. Activity of *Formica rufa* in May and June in canopy sectors of sycamore is significantly, positively correlated with the density of *Periphyllus testudinaceus* (Fernie) aphids (Fig. 6. 3). On birch, also, ant activity is highly correlated with density of honeydew-tended aphids (Fig. 6. 4(a),(b)), mainly *Symydobius oblongus* (T. Mahdi, personal communication).

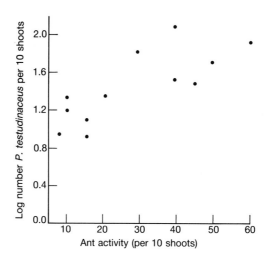

Fig. 6.3. Relationship between ant numbers per ten shoots and the log number of *Periphyllus testudinaceus* aphids per ten shoots, at Cringlebarrow Wood, Lancashire.

The effect of ant tending is often dramatic. *Symydobius oblongus* on *Betula pubescens* was 8200 per cent more abundant in the presence of *Formica lugubris* than on branches from which the ant had been experimentally excluded (Fowler and MacGarvin 1985). In the case of sycamore, honeydew production by *P. testudinaceus* was 40–400 times higher on ant-foraged trees than on unforaged trees (Table 6. 1).

Consequences to other herbivores

The extent to which predation by *Formica rufa* is independent of or, alternatively, conditioned by honeydew collection is questionable. Certainly, the experiments of Bradley and Hinks (1968) in Manitoba, Canada, in which aphids tended by *Formica obscuripes* Forel were removed with insecticide, showed that the ants immediately extended their foraging territory and began to attend new aphid colonies. In addition, when Bradley (1973) prevented access of *Formica obscuripes* to jack pine trees in Manitoba which were infested with the scale insects *Toumeyella numismaticum* Pettit & McDaniel,

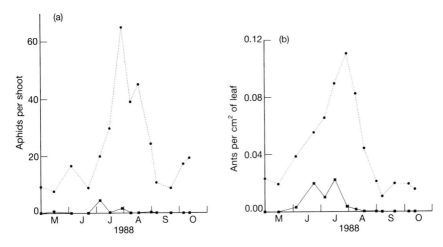

Fig. 6.4. (a) Number of aphids per shoot on *Betula pendula* trees foraged by *Formica rufa* (● · · · ●) and on banded trees (■—■), at Gaitbarrows NNR, north England, 1988. (b) Number of *Formica rufa* per cm² of leaf surface on *Betula pendula* on foraged trees (● · · · ●); and banded trees (■—■); at Gaitbarrows NNR, 1988.

the beetle predator *Hyperaspis congressus* Watson became much more abundant and, consequently, the scale insects were predated and locally eliminated. The ant appeared to be a prerequisite in the maintenance of high scale insect populations, from which the ant gained honeydew.

In our study, there is an overlap between the activities of predation and honeydew collection because the aphid *P. testudinaceus* occurs in the same parts of the sycamore canopy from which prey is being removed. However, by the time *P. testudinaceus* appeared in early spring, ants were already visiting the trees and beginning to remove caterpillars and aphids on other tree species (Skinner 1980*b*). When *P. testudinaceus* became abundant in mid-June it was three to four times as numerous in the lower canopy (1–2 m) as in the upper canopy (6 m). However, the greatest effect of foraging on Lepidoptera populations was in the *upper* canopy (Table 6. 1), where populations were reduced to approximately 6 per cent of those in the control trees. Some of this may be due to disturbance by the ants, causing the caterpillars to move down through the canopy.

Thus, in general terms, patterns of ant activity were conditioned by access to honeydew, and trail strength to particular trees and parts of trees was related to aphid density. Once established, there is a high trail fidelity (Rosengren 1971) and on our study site J. Wright (personal communication) found that, of 800 *F. rufa* marked according to the trail on which they were first found, 100 (98 per cent) of those recaptured were still on the trail of

Table 6.1. Feeding activity of principal herbivores on sycamore in northern England (F = foraged by *Formica*; UF = unforaged trees).

Herbivore		Canopy height (m)	Year	
			1982	1983
(a) *Drepanosiphum platanoidis*, non-tended aphids				
Honeydew excretion	F	1	55	145
(mg dry mass shoot^{-1} yr^{-1})		6	146	176
	UF	1	251	546
		6	208	454
(b) *Periphyllus testudinaceus*, ant-tended aphid				
Honeydew excretion	F	1	39.2	22.5
(mg dry mass shoot^{-1} yr^{-1})		6	10.4	9.2
	UF	1	0.3	1.2
		6	0	0.5
Total Lepidoptera				
(Mean percentage leaf area loss	F	1	1.2	0.3
shoot^{-1} yr^{-1})		6	0.8	0.5
	UF	1	6.9	9.3
		6	6.3	11.8
(d) Typhlocybinae leafhoppers				
(Mean percentage leaf area loss	F	1		0.32
shoot^{-1} yr^{-1})		6		0.32
	UF	1		2.80
		6		1.53

origin up to eight days after marking. Moreover, this fidelity extended to branches within a tree, where 98 per cent (of 44 'recaptures') showed branch fidelity. Nevertheless, active predation of other herbivore groups took place in other parts of the canopy and was, to some extent, independent of aphid distribution.

Effects of predation on community structure

Predation by wood ants is not evenly distributed amongst herbivore species, nor is it necessarily density dependent. It is much more related to the behaviour and other characteristics of the potential prey species. For example, Fowler and MacGarvin (1985) found that the guild structure of herbivores was changed by *Formica lugubris* on birch because the abundance of free-living leaf-chewers was reduced more than those species which had protective refuges in the leaves (leaf-tie-ers), which were in turn reduced by more than the internal feeders, such as leaf-miners. In general, species richness was significantly reduced in the presence of ants. However, the relative safety of leaf miners may itself depend on the structure of the mines. Faeth (1980) found that miners with 'robust' mines survived better in the presence of ants than did those with flimsy mines. Certainly, *Phyllonorycter* (Lepidoptera: Gracillariidae) leaf-miners were found to be substantially reduced by ants in a manipulation experiment by Sato and Higashi (1987) in an oak chaparral in Japan.

Similar differential predation was reported within the Lepidoptera by Warrington and Whittaker (1985*a,b*). The winter moth (*Operophtera brumata* L.) formed between 13 per cent (in 1982) and 16 per cent (in 1981) more of the total Lepidoptera larvae on ant-foraged trees than it did on unforaged trees, whereas mottled umber (*Erannis defoliaria* (Clerk)) declined in foraged compared with unforaged trees (Fig. 6. 5). This occurred despite the fact that winter moth was by far the most abundant caterpillar.

This differential predation may be partly due to the relative sizes of the caterpillars (mottled umber is 180 mg fresh mass at maximum size, whilst winter moth is 60 mg fresh mass) but is also due to the fact that winter moth is cryptically coloured, feeds mainly at night, and retreats by day to a refuge between two overlapping leaves held together with silk, whilst mottled umber is conspicuous and feeds on young expanding leaves during the day and night.

Ants can also affect the age structure of individual species of prey. Warrington (1984) collected all ants carrying captured *Drepanosiphum platanoidis* Schrank aphids down a sycamore trunk over a period of between 10–30

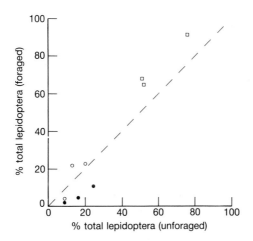

Fig. 6.5. Winter moth (□), mottled umber (●), and scarce umber (○) caterpillars as a percentage of total Lepidoptera on unforaged sycamore trees versus percentage on trees foraged by *Formica rufa*.

minutes on six occasions in April, May, and June. The total sample size was 477. The proportions of aphid instars were not significantly different in these samples from those in samples taken from the canopy above, except that there were significantly fewer adult aphids being captured than were present in the population. Adults were observed to take flight to evade capture and they may have moved off onto trees without ants.

Temporal changes in prey

Availability of prey species in the tree canopy is an important determinant of foraging activity.

Lepidoptera larvae were the most abundant prey (58 per cent of total) returned to the nest by European *Formica lugubris* introduced into Quebec in 1971 (McNeil *et al.* 1978). Most were Tortricidae and of these 88 per cent were spruce budworm (*Choristoneura fumiferana* Clem.), which the introduction was intended to control (Fig. 6. 6). The ants successively returned larvae, pupae, and adults to the nest as the season progressed and availability changed.

There are successive waves of herbivore groups on sycamores in northern England (Whittaker *et al.* 1987). Lepidoptera are abundant during May and

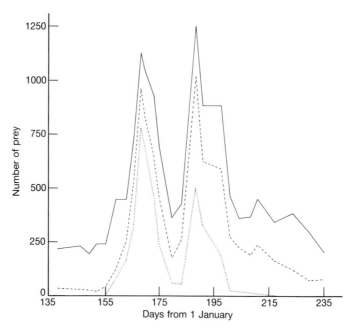

Fig. 6.6. Numbers of total prey (——), total Lepidoptera (---), and Tortricidae (· · ·) collected by *Formica lugubris* at sampling intervals throughout the summer at Valcartier, Quebec in 1976. (Redrawn from McNeil *et al.* (1978).)

early June and during this time they represent 20 per cent of the solid food returned to the nest. The green sycamore aphid *Drepanosiphum platanoidis* then becomes abundant and it too forms a substantial part (80 per cent) of the solid food by mid-June. These two groups correspond in time with the peak of *Periphyllus* activity and, therefore, with the peak of honeydew utilization by the ants. However, leafhoppers (Typhlocybinae), which are most abundant in July and August, are also predated, at a time when there is little honeydew production. There is also some predation of aphids (*Drepanosiphum*) in September at a time when *Periphyllus*, though present, is not attended by the ants. Table 6. 1 shows the effects of this predation on ant-foraged trees compared with the unforaged trees in the near vicinity.

Overall effects on herbivory

It is not possible to generalize about the consequences of ant foraging to herbivore populations. As pointed out by Adlung (1966), the crucial question is the balance between those herbivore populations (some Homoptera species) which are greatly increased by ant tending, and those which are decreased (other Homoptera and other herbivore groups). The question is also complicated by the fact that, whereas damage by chewing insects and some sap-feeding insects such as Typhlocybinae can be readily seen and measured (Warrington and Whittaker 1985*a,b*), the removal of phloem sap by other Homoptera is difficult to measure and its consequences to the tree hard to assess. Few studies have attempted to measure directly the effects of the presence of ants on sap feeders in the tree canopy, or even on chewing insects. Most have simply measured prey returned to the nest.

Campbell and Torgersen (1982) measured predation rates of ants (including *Formica* spp.) on spruce budworm larvae and pupae across a range of budworm densities. They found that at the lower budworm densities, numbers were reduced by 90 per cent or more by ants or birds. At high budworm densities this fell to about 50 per cent. Ants alone caused similar budworm mortality if birds were absent. Markin (in Youngs 1983), however, found no effect on budworms of excluding ants, but this may have been because of compensatory predation by birds. Introductions of European *Formica lugubris* into Quebec were shown by McNeil *et al.* (1978) to result in estimated budworm defoliation in two years of 30 per cent and 43 per cent in the presence of ants compared with 42 and 63 per cent in a control block.

These studies are characterized by high levels of defoliation by a single pest species, spruce budworm. A similar situation was studied by Laine and Niemelä (1980) in northern Finland. Here, mountain birches undergo periodic defoliation by a geometrid moth (*Oporinia autumnata* (Borkhausen)). Among the birch stands are nests of *Formica aquilonia* and surrounding these nests 'green islands' of undamaged trees are frequently found up to a radius of 20 m from the nest. Although these 'green islands' have been attributed to the concentration of soil nutrients around the nest (White 1985),

Niemelä and Laine (1986) do not accept this explanation. In their original study Laine and Niemelä (1980) established a significant negative relationship between distance from the nearest ant-hill and grazing pressure on birch by chewing insects (Fig. 6. 7). Their evidence that this predation is protecting the trees, although not conclusive, is persuasive. The rather special circumstances of geometrid outbreaks may not represent the normal relationship between ants and herbivores of birch. Even in this situation, the tended aphid *Symydobius* was probably the usual attraction for the ants and, unless the geometrid caterpillar is in an outbreak year, trees close to nests may suffer as much if not more than those at a distance, because of sap removal by the aphid.

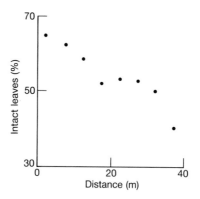

Fig. 6.7. Relationship between distance from nests of *Formica aquilonia* and the percentage of intact leaves of *Betula pubescens* ($r = -0.92$, $P < 0.01$). (Redrawn from Laine and Niemelä (1980).)

Fowler and MacGarvin (1985) also concluded that birch was unlikely to benefit normally from ants because the undoubted reduction in chewing insects which they observed would be likely to be offset by enhanced numbers of *Symydobius*. As Fritz (1983) points out, for ants to provide plant protection, either the ants must be the most important predator of the plant's herbivores or they must not interfere with other predators which may limit these herbivores. In addition, it is essential that any protection afforded to the tree by predation of herbivores by ants should not be offset by any increase in mutualistic Homoptera. These requirements were met in the sycamore woodland studied in northern England. Despite the enhanced numbers of the tended aphid *Periphyllus testudinaceus* on all ant-foraged trees, honeydew production and, therefore, sap removal by all aphids (*P. testudinaceus* and *Drepanosiphum platanoidis*) was two to three times as high on unforaged trees as on trees foraged by *Formica rufa*, with grease-banded trees intermediate. In addition, loss of leaf area due to Lepidoptera feeding and stippling by Typhlocybinae was three to seven times as high on unforaged trees as on foraged ones (Table 6. 1).

We have also measured defoliation of oak trees (*Quercus petraea* (Mattuschka) Liebl.) by Lepidoptera in a woodland by Loch Lomond, Scotland. A small stream bisects the woodland and trees on one side of the stream are extensively foraged by *Formica lugubris* whilst there are no *F. lugubris* on the other side. Random samples of leaves from unforaged trees showed 34 ± 3.0 per cent loss of leaf area by Lepidoptera grazing, whilst leaves from foraged trees showed 11 ± 1 per cent damage. Gosswald and Horstmann (1966) also found that defoliation of oak trees was negatively correlated with ant activity (Fig. 6. 8).

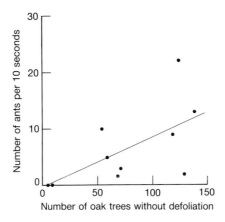

Fig. 6.8. Relationship between number of oak trees free of defoliation and numbers per 10 seconds of *Formica polyctena* climbing the trees. (Redrawn from Gosswald and Horstmann (1966).)

Effects on tree growth

Significant losses of timber growth in the vicinity of ant nests in European coniferous forests have been documented, and this has been attributed to the increase of ant-tended aphid species (Klimetzek and Wellenstein 1978; Wellenstein 1980). Even if there is significant reduction in tree herbivores by ants, it does not necessarily follow that there will be a measurable effect on timber production. However, Varley and Gradwell (1962) and Dixon (1971) have independently calculated that in oak and sycamore, respectively, average levels of invertebrate herbivory may reduce timber production by 40–60 per cent. Our experimental manipulation of herbivory by grease-bands to exclude *Formica rufa*, and our comparisons of naturally ant-free and ant-foraged trees (Whittaker and Warrington 1985) support these figures. Leaf fall occurred earlier from trees unprotected by ants than from foraged trees, and leaf area and leaf dry mass were some 10–15 per cent greater on foraged compared with unforaged trees. Radial growth of timber of mature trees foraged by ants was approximately 35 per cent greater in ant-foraged trees than in unforaged trees, whilst ant-foraged saplings were, on average, 14 per cent heavier than controls protected from ant foraging by grease-bands. There was also a 20 per cent increase in side-shoot mass in the ant-foraged saplings. Other authors have been convinced by similar experiments and observations that there are circumstances in which the net effect of ant activity on trees is beneficial to the trees. Such effects may be quite local, as in the 'green islands' reported by Laine and Niemelä (1980), or of wider economic significance (Bradley and Hinks 1968; Kim and Murakami 1983; Wellenstein 1980). At least for some species of ant, the effect is only likely to be important near to the edges of woodlands, or within 40 m of rides (Sudd *et al.* 1977). For every report of beneficial effects of ants, however, there are several which are neutral (Fritz 1983) or where the ants have such an overwhelming positive influence on honeydew-producing Homoptera that their net effect is detrimental (Müller 1956*a*,*b*; Wellenstein 1961).

Conclusions

Wood ants are unusual predators in that their foraging activity and interaction with potential prey often seems to be more a consequence of their requirement for honeydew than a response to prey distribution and abundance. Moreover, predation is restricted to a limited distance from the nest and nest location is not determined solely by prey availability. The contrast between trees within the territory of a nest and those in the 'enemy free space' (*sensu* Askew 1961) between nest territories is very marked and can lead to the phenomenon of green islands within a wood.

The whole structure of the insect community on foraged trees is changed by the ants and in a mixed woodland this can have completely opposite effects on different tree species depending on the relative proportions of mutualistic aphids and other herbivores (including non-tended aphids). Sycamore may be a somewhat unusual case in that, not only are the predominant sap feeders non-tended aphids (*Drepanosiphum platanoidis*), so that the net effect on the trees of aphids is reduced in the presence of ants, but there is also very extensive predation of other insect herbivores. There is good experimental evidence from grease-banding and comparisons of foraged and unforaged trees that the sycamores in such woodlands benefit from ant activity. Reduction of normal levels of herbivory has a measurable effect on the trees. The same may be true during outbreaks of chewing insects giving relative protection where ants are able to switch to the abundant prey, though in more normal years ant-tended aphids may predominate. For cases like these there are probably many where the overriding effect of ants is neutral or detrimental to tree growth. What is clear is that they offer an excellent opportunity for manipulative field experiments of insect herbivory.

Acknowledgements

I am grateful to the Natural Environment Research Council which funded most of our work on wood ants, and to Gary Skinner, Stuart Warrington, and Thamer Mahdi, who have contributed a great deal to this research.

7

The interaction between red wood ants, *Cinara* aphids, and pines. A ghost of mutualism past?

Rainer Rosengren and Liselotte Sundström

Introduction

A major conceptual problem in ant–plant protection mutualisms is the variability of costs and benefits in space and time (Beattie 1985; Janzen 1985; Addicott 1986). The classical debate about the 'usefulness' of red wood ants (Cotti 1963) provides an illustration of that point. A problem with this old debate, however, is that it tends to confuse benefits in terms of human forestry with benefits in terms of tree fitness.

Red wood ants clearly reduce the numbers of some defoliating tree herbivores (Otto 1967; Wellenstein 1980; Laine and Niemelä 1980; Skinner and Whittaker 1981; Whittaker and Warrington 1985), although this predation should be inversely density-dependent (Horstmann 1977). However, more than one nagging question remains. The cost to the tree of the ant–aphid trophobiosis has to be paid each season. The benefits of ant protection, however, may be both unevenly and unpredictably distributed over time. This is especially true if the main threat to the tree, which is what determines the benefit, is due to maximum rather than to average herbivory (Laine and Niemelä 1980; but see Whittaker and Warrington 1985). Janzen (1985) pointed out that a protection mutualism may be far from an evolutionary equilibrium, due to this lack of synchrony between costs and benefits. Cloudy equilibria will appear even more cloudy in a situation in which air pollutants alter the frequency of other herbivores at the same time as enhancing the growth of ant-tended aphid populations (Flückiger *et al.* 1988). Are we, to paraphrase Connell (1980), already studying 'the ghost of mutualism past'?

The members of the Palaearctic *Formica rufa* group, if *Formica uralensis* Ruzsky and the *F. truncorum–yessensis* complex (Rosengren *et al.* 1985) are excluded, comprises a genetically tight cluster of poorly differentiated species (Pamilo *et al.* 1979). Intraspecific variability however, including ethological differences between local populations, is high (Rosengren and Cherix 1981). It could well be that biogeography and climatology are more relevant than taxonomy when discussing ant–plant interactions in this species group.

Most studies of red wood ant ecology are from the temperate vegetation zone (*sensu* Ahti *et al.* 1968), dominated by deciduous forest. Corresponding studies from the boreal vegetation zone (Ahti *et al.* 1968), dominated by coniferous forests, are notably scarce. The boreal forest covers most of Fennoscandia and continues as the Taiga through Russia and Siberia. The overall impact of red wood ants on the boreal coniferous forest, an area of high nest density (Rosengren *et al.* 1979), can hardly be doubted. However, for this region there is very little factual support for Buckley's (1982) statement that, '*Formica* apparently produces an overall positive effect on European pine plantation production, however, despite tending large aphid populations'.

The case study presented here, perhaps a horror story from the point of view of Panglossian biology, is not intended to replace old tales of ants helping foresters and pines with a new paradigm of red wood ants as forest pests. Red wood ants, an ubiquitous part of the boreal forest, could be potentially useful both for pines and foresters. Their long-lived, robust colonies, which concentrate nutrients from large areas, could and should be used for long-term chemical and biological monitoring of the coniferous ecosystem and the changes imposed on it by man's atmospheric pollutants.

The *Formica–Cinara–Pinus* interaction—a case study

Ants and methods

The ants of the four colonies used for this study in southern Finland resemble *Formica polyctena* Förster with respect to worker phenotype. Sexuals, however, are similar to those of *Formica aquilonia* Yarrow (Rosengren 1977; Collingwood 1979). It is therefore reasonable to consider the material as a species or subspecies related to the '*aquilonia* complex' (C. A. Collingwood, personal communication).

The forest areas studied are all dominated by naturally seeded stands of Scots pine, but differ in their productivity. A description of the main study colony, including the phenology and a map of the foraging area of the ants, is given in Rosengren and Sundström (1987).

Trees visited by honeydew collectors were marked out with numbered labels and the numbers of foragers counted on a 60-cm stretch of each tree trunk, 2–6 times a year during a ten-year survey of the foraging area (1979–1988). These cumulative data were used when selecting pines for drill sampling and year-ring analysis.

All colonies were sampled semi-automatically for solid booty (Chauvin 1966). The total amount of harvested food was estimated for the main study colony only. These estimates were based on regular traffic counts in which both total traffic and the proportion of ants carying solid items were determined. The regression of traffic on temperature is less steep and the proportion of item carriers is lower in the night than in the day (Rosengren and Fortelius 1986). This has been corrected for in the estimates. The amount of liquid food carried in the crops of returning foragers was estimated using weight differences between batches of returning and departing foragers collected in traps close to the nest. The total sugar content in the crop liquids was determined spectrophotometrically (Roberts 1979). The size of the forager population was estimated by marking a known number of foragers and using traffic counts for the Lincoln-index estimate (Kruk-De Bruin *et al.* 1977; Rosengren 1977; Breen 1979; Rosengren *et al.* 1985).

The size of the forager population and the foraging area

The overall impact of an ant species on the vegetation is influenced by its nest density, the number of foragers per nest, and by the size of the area used for foraging. Our main study colony was founded as a small bud-nest in 1967. The size of the forager population was 220 000 in July 1979, 280 000 in July 1980, and 380 000 in July 1981. We did not estimate the population size with the Lincoln-index method between 1982 and 1988 because total traffic rate in that period mostly exceeded our counting capacity. The values of population size are very high compared to data from the westernmost part of the temperate zone (Kruk-De Bruin *et al.* 1977; Breen 1979) but close to some estimates from central Europe (Horstmann 1982).

The area in which trees were visited by the colony for honeydew varies between 0.3 (autumn) and 0.5 hectares (summer), while the total area visited by stray foragers is larger, up to 1 hectare. The ants tend some aphids as far as 50–60 m from the nest but the frequency of visited pines drops off after 30 m (Fig. 7. 1). The 30 most-visited pines (of a total of 91 pines visited for honeydew in some or all years of the ten-year period) had a mean distance of 17 m from the nest (s.d. = 10.6) compared to 28 m (s.d. = 11.4) for the 30 least-visited pines. Red wood ants actively defend tended aphids against ant competitors, which were observed to increase in number toward the periphery of the colony area (R. Rosengren, unpublished work). Defence costs are likely to grow as an exponential function of distance.

The amount of food and the dietary spectrum

The heated debate about whether red wood ants are 'beneficial' or not has centred around the question of whether animal prey or honeydew is more 'important' in the diet of these ants. Wellenstein (1952) may have thought that he had formulated the *modus vivendi* when stating that 33 per cent of the diet consists of insects and 62 per cent of honeydew.

Our data (Table 7. 1), based on estimates for the main study colony during 1981, show that crop liquids constitute 94 per cent, solid arthropod prey 5 per cent, and seeds (mainly *Luzula pilosa*, Juncaceae, in early summer and

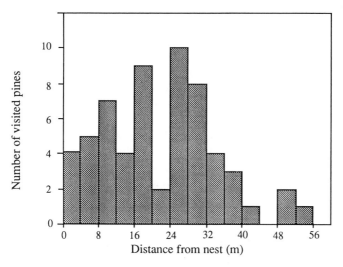

Fig. 7.1. Distribution of the 60 most-visited pines (of a total of 91 pines) along a gradient represented by pine–nest distance in the main study colony of red wood ants in Finland.

Table 7.1. Total amount of food and nest material harvested by foragers of the main study colony of red wood ants during 1981 in Southern Finland.

Length of foraging period	140 days
Size of foraging area (July)	0.6 ha
Size of forager population, July	380 000
September	270 000
October	90 000
Exits/s at 15 °C (daytime, July)	9.6
Foraging journeys/year	84 000 000
Number of arthropod prey items	5 000 000
	(aphids: 2 000 000)
Fresh mass of arthropod prey	12 kg
	(aphids: 0.77 kg)
Dry mass of arthropod prey	7 kg
Number of seeds	300 000
Mass of seeds	1.8 kg
Items of nest material	7 000 000
Mass of nest material	36 kg
Mass of crop liquids (mainly honeydew)	240 kg
Mass of sugars in crop liquids (15%)	36 kg

Melampyrum sp., Scrophulariaceae, in late summer) about 1 per cent of the total fresh food mass. The corresponding dry masses give 82 per cent sugars, 16 per cent arthropods, and 2 per cent seeds. It should be noted, however, that as much as one-third of the prey volume may be brought into the nest in the crop as haemolymph (Horstmann 1974).

We identified a total of 12 tended, aphid species in the colony area (all on trees or large shrubs), six of which belonged to the genus *Cinara*. However, only *Cinara pini* (L.) on pine, *C. pilicornis* (Hart.), and *C. viridescens* (Chol.) on spruce, *Symydobius oblongus* (V. Heyd.) on birch, and *Dysophis sorbi* (Kalt.) on mountain ash were sufficiently common to be important as producers of honeydew for the ants. (*D. sorbi* was common in two out of ten years, but virtually absent otherwise.) The two other tended species on pine, *Cinara pinea* (Mordv.) and *C. nuda* (Mordv.), were both found only occasionally.

The prey spectrum in the different pine-dominated habitats, ranging from a pure pine forest on dry soil to more diverse forest, proved to be very similar; admittedly this was partly due to our crude classification. However, we found seasonal trends in some of the groups (Fig. 7. 2).

Predation of tended aphids

One very intriguing result is the high proportion of aphids amongst the prey material (Fig. 7. 2). The genus *Cinara* comprised 60–90 per cent of the aphid prey. We found no untended species of that genus in our habitats, indicating

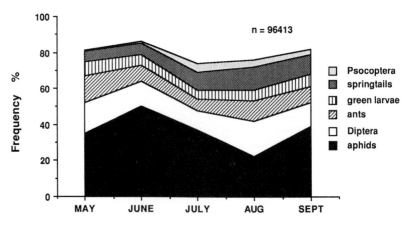

Fig. 7.2. Seasonal distribution of the six most frequent arthropod items carried by foragers of red wood ants. The diagram is based on pooled results for four red wood ant colonies and represents a total of 500 Chauvin-trap catches (> 100 prey insects in each) distributed throughout the season. *n* = Total number of classified arthropods. 'Ants' refers to other species, mainly *Myrmica* sp. 'Green larvae' are phytophagous Lepidoptera and Hymenoptera.

that the bulk of the aphid material, containing very few winged specimens and many immatures, must belong to tended species. The prey samples that have been identified to species so far show that most of the aphids are *Cinara pini*, an obligately tended species. Most of the *Cinara* specimens are apparently undamaged and up to 30 per cent still show signs of life in the trap. Aphids are not carried out from the nest, and are presumably consumed as food, not redistributed in the habitat. Data from the main study colony show that aphids were the preferred food in most years (Fig. 7. 3).

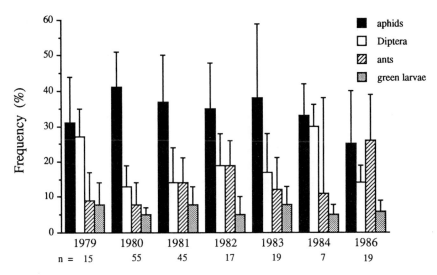

Fig. 7.3. Yearly means, and standard deviations, of the four most frequent prey categories of red wood ants found in the main study colony. *n* = Numbers of Chauvin-trap catches (> 100 booty insects in each) made during each of seven years.

High red wood ant predation on tended aphids has been observed in some subalpine habitats of continental Europe (Cherix 1981, 1987; Torossian 1981), but not in lowland areas of the temperate zone (Sörensen and Schmidt 1987; Skinner and Whittaker 1981). However, this phenomenon may not be of general occurrence in boreal habitats either (Rosengren *et al.* 1985; Larsson 1985).

The impact of red wood ants on pine growth

We divided the pines that were visited by the main study colony for honey-dew into three groups according to our cumulative index for visiting frequency for the ten-year period, and selected the 30 most-visited and the 30 least-visited pines for drill sampling. A third group of pines which were not visited for honeydew (although they were visited to some degree by individual foragers) were selected as a control. The 30 least-visited pines were, on average, younger than those in the other two categories. Age does not influence growth after 50 years, so we corrected for this by excluding all younger trees, ending up with 20 pines in each category.

The year-ring analysis showed a consistently and statistically significant lower mean annual growth rate in the most-visited category as compared to the least-visited and the non-visited categories (Fig. 7.4). The latter two groups were statistically indistinguishable, although they showed the largest difference in mean age (least visited, 68; most visited, 84; non-visited, 101).

Growth rate averages for the total ten-year period were compared for each category using two-way ANOVA: most-visited, least-visited, non-visited ($n = 10 \times 20$ per category). The test gave significant differences between years and significant differences between categories, but no interaction between year and category. This indicates that all three categories responded equally to environmental changes (e.g. the effect of the cold summer of 1982 in Fig. 7.4). The differences between years in all categories were manifested as a decrease in growth rate during the ten-year period.

Assuming that the amount of ant traffic in trees correlates with the amount of honeydew harvested, we estimate that 70 per cent of the honeydew collected in the main study colony emanated from pine. For 1981 (Table 7.1),

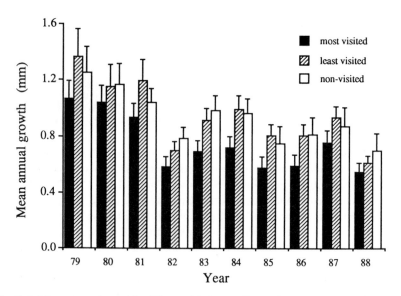

Fig. 7.4. Mean annual growth of Scots pines according to frequency of visiting by red wood ants. Data from year-ring analysis is from 1979 to 1988 of samples of 20 mature pines most visited by red wood ants, 20 least-visited, and 20 not visited. Standard errors of the means are given.

this corresponds to 25 kg sugar (dry mass) as a yearly total for all visited pines. If we assume that 1981 is typical, then this represents an average annual sugar depletion of 708 g per pine (range: 275–1475 g) for the most-visited category and 17 g per pine (range: 8–82 g) for the least-visited category.

Zoebelein (1956b) estimated that red wood ants may collect 70 kg of honeydew each year (corresponding to at least 10 kg of dry sugar) from an individual pine. Our much lower estimate may in itself be too small to account for the observed effect on tree growth rates. It should be noted, however, that the ants may gather only about two-thirds of the honeydew produced by the aphids, the rest being deposited as waste (Wellenstein 1980). It is also clear that the aphids use both sugars and amino acids for their own metabolism.

Wellenstein (1980) determined pine growth by different methods from ours but reached very similar conclusions concerning the negative correlation between red wood ant traffic and pine growth. We are, however, less convinced than Wellenstein that those effects may be attributed to phloem sap depletion. For example, *Cinara pini* (and the tending ants) might prefer to settle on pines which have reduced growth rates for other reasons. This question can be studied by banding experiments preventing ant access to previously visited trees (Whittaker and Warrington 1985; Whittaker Chapter 6, this volume).

General discussion

Are pines and red wood ants mutualists, or should they be?

Our study does not appear to support the case for ant–pine mutualism. But let us play devil's advocate! First there is the point that mutualisms cannot always be happy—or not happy all of the time. Secondly, mutualisms of this labile type may well be an early indicator of environmental disturbance (although our study was made in a pastoral, rural environment). It has been demonstrated that *Cinara pini*, the aphid partner in the triangle, can undergo a dramatic population increase in response to air pollutants (Flückiger *et al.* 1988; Heliövaara and Väisänen 1988). Protection of *Cinara* aphids by ants may be adjusted to the 'normal' constraints on aphid population growth represented by predator pressure and the physiological resistance of the pine. Air pollution which reduces some of those constraints would mean that the activity of ants imposes higher costs than normal on the pine partner. Finally, the ants do indeed mitigate the damage they are doing to the pines by killing large numbers of tended aphids, which they collect as prey. Furthermore, there is the question posed by Buckley (1982) that, 'If these interactions are damaging to the host plants, why have they not evolved an appropriate defence?' An appropriate defence may fail to evolve because of physiological constraints or lack of genetic variability.

However, individual pines do vary in their susceptibility to infestation by *Cinara pini* (see below). Another explanation would be a tendency for mutualistic interactions between Scots pines and the *Formica–Cinara* association. One aspect of this possibility is the fact that both pines and red wood ant colonies are long-lived and likely to be involved in contact that can last for decades and perhaps centuries (in the case of polydomous colonies). Long-lasting contacts between two parties may, according to the Prisoner's Dilemma model of Axelrod and Hamilton (1981), favour a shift in the interaction pattern from parasitism to co-operation.

Red wood ants and pines not only maintain stable contact where they happen to co-occur, but the *frequency* of co-occurrence is likely to be high in the boreal vegetation zone. A national forest inventory made throughout Finland showed that 18 per cent of all forest land consists of red wood ant territories (Rosengren *et al.* 1979), while about 50 per cent of all forests are dominated by Scots pine. This implies that the populations of pines and red wood ants (the bulk of red wood ants in Fennoscandia belong to the *Formica aquilonia* complex) may have a sufficiently broad interface for co-evolution to be feasible.

How red wood ants profit from pines

Red wood ants are territorial 'K-strategists', monopolizing patches of stable 'climax' vegetation. They invest in the local patch on a long-term basis by erecting expensive nest mounds, clearing trails, and removing competitors. In a polydomous species like *Formica aquilonia*, the patch will be inherited by a sequence of daughter nests, reproduced by budding. The local pine stand is consequently an instant food source for a given nest and also a future resource supporting a whole genetic lineage. So, we would expect red wood ants to be fanatical conservationists!

The colonies depend on pine not only for sugar but also as a source of the most usual nest material in the habitat: pine needles, twigs, and lumps of dry resin. Part of the nest material may be used as fuel (through aerobic bacteria) for generating heat within the mound, although this is unlikely to be the main heat source (Rosengren *et al.* 1987). The colonies use the root systems of live or dead pines for sheltering queens and young, replete workers.

How pines profit from red wood ants

Red wood ant protection of pine foliage is particularly well documented in the case of the Lepidopteran defoliators *Panolis flammea* and *Lymantria dispar* (Otto 1967). Red wood ants may, however, be unable to protect conifers against bark-borers and weevils, which are economically the most important of the forest pests in Fennoscandia. Opinions are divided about the effect of red wood ants on sawflies such as *Neodiprion sertifer*, Diprionidae, the most serious pine defoliator in Finland. Observations of *Neodiprion* larvae transferred to small pines near one of our nest mounds show that the ants both located the larvae and hunted them down with great efficiency, although they were obviously harmed by the 'glue' defence of these herbivores. The persistent behaviour of the ants suggested that this aggression was not foraging behaviour (unpublished notes).

Oinonen (1956) observed that some mound builders, including red wood ants, contribute to afforestation of rock in Finland by providing deserted mounds where pine seedlings can grow, a process which has been continuing since the last ice age. Red wood ants do not build their main nests on barren rock, but they often found temporary, 'summer nests' on such sites (Rosengren *et al.* 1985). This process of afforestation may be facilitated by the ants' habit of including pine seeds among their nest material. A pine having its progeny dispersed and surviving among the first colonizers of a virgin patch is likely to experience a substantial increase in fitness.

'Parental care' in pines?

Our finding that pines intensively visited for honeydew show reduced growth rate (Fig.7. 4) is difficult to evaluate in terms of pine fitness. What is the relation between the annual growth of a pine and its reproductive output? The answer depends on the age of the pine; phloem sap depletion and reduced growth rate are likely to reduce fitness more in young than in old pines. Stress may induce or increase reproduction in plants. *Cinara pini* is sometimes found on pines younger than ten years (Wellenstein 1952; Heliövaara and Väisänen 1988), but heavy infestation of young pines by this obligately tended species has not been observed in our sites; infested trees are usually older than 50 years (see Rosengren *et al.* 1985; comparing old, dwarfed pines with young pines of the same size). If we assume a viscous population structure in naturally seeded pine stands, this may suggest a pine strategy based on inclusive fitness maximization (as explained below).

As a rule, pines within the foraging area of a red wood ant colony are also visited by individual foragers, whether visited for honeydew or not. This implies that younger pines, not infested by *Cinara pini*, are likely to profit from the 'insurance premium' paid by older pines in phloem sap currency to the ants. If there is a high degree of genetic relatedness within a stand this would, from the point of view of the old pines, be comparable to costs of 'parental care'.

It is possible that air pollution may have changed the relationship between costs and benefits in a direction that is unfavourable for the pines, but reduced growth rate *per se* need not imply reduced pine fitness, according to the above argument.

Why do ants kill tended aphids and why do aphids apparently allow ants to kill them?

It is easy to find 'adaptationist' explanations of why red wood ants in alpine or northern climates may use tended *Cinara* aphids as food. Aphids are concentrated food packages containing very little chitin and are convenient for handling and carrying, even in cold weather (see Fig. 7. 2; indicating a generally small average size of prey objects). Red wood ant foragers show individual, memory based site allegiance towards tended aphids, and thus do not have to search for their aphids (Horstmann and Geisweid 1978; Rosengren and Sundström 1987). Utilization of tended aphids as prey consequently reduces the searching time and this is important if daily foraging time is short. It is possible that ants of these climatic regions 'over protect' their aphids, and this could combine with other factors favouring growth of aphid populations. A reasonable explanation of why ants do not over exploit tended aphids as prey, is that the amount of predation may be determined by colony-level dietary balance (Pontin 1978).

But why have tended *Cinara* aphids not evolved measures to hinder the ant from combining mutualism with murder? Tended aphids have developed a set of appeasement signals inhibiting aggression in the ant partner (Eckloff 1978). Why should co-evolved aphids fail where some of the ant's worst enemies have succeeded? It was suggested, independently, by Sudd (1987) and by Rosengren and Sundström (1987) that the answer could be altruism evolved through kin selection or selection between clones of parthenogenetic progeny. Sudd (1987) developed the hypothesis that a clonal group of aphids could compete more efficiently for ant attendance by sacrificing some of the clone members as prey for the ants, implying a tradeoff between protein and sugar rewards offered to the ants. However, there is an alternative mechanism.

It was observed by Daniel Cherix that *Cinara* aphids which were captured and killed by ants had all stopped feeding and were walking away from the group (Cherix 1981, 1987), and we have observed this as well. There could be many reasons for such behaviour, including reaction to allelochemicals produced by the pine, seasonal within-crown changes in the availability of phloem sap nutrients (Larsson 1985), too high a density of the aphid group (Eckloff 1978; Cherix 1987), or lack of ant attendance (Bradley and Hinks 1968). High density combined with insufficient ant attendance may lead to a problem of waste honeydew. This problem may reduce the fitness of the clone members, as observed by Way (1954) in homopterans tended by weaver ants. Walking away from the group and taking the risk of predation may be interpreted as altruism, if it improves the fitness of non-dispersers (Aoki 1982). However, from the point of view of the migrant, this behaviour can be considered to be a risk-prone strategy rather than suicide.

Our hypothesis is that the reason why ants use the aphids as prey, is that the aphids are providing honeydew in excess of the requirements of the ant colony. This reaction, originally observed by Way (1954), is compatible with the dietary balance hypothesis of Pontin (1978). The possibility that the pine, through a metabolite which induces unrest in the aphids, could manipulate ants to delouse it, is, perhaps, worth considering. It could well be that we have to deal with a blend of proximate mechanisms reflecting a co-evolved compromise between the three partners: the ant, the aphid, and the pine.

Acknowledgements

We thank Osmo Heikinheimo for checking samples of aphids. The comments by Deborah M. Gordon, Camilla R. Huxley, Pekka Pamilo, and John B. Whittaker are gratefully appreciated. This work was supported by grants from the Foundation for Research on Natural Resources in Finland, Societas Entomologica Helsingforsiensis, and the Finnish Society of Sciences and Letters.

8

Conditional interactions in ant–plant–herbivore mutualisms

J. Hall Cushman and John F. Addicott

Introduction

Ecologists traditionally classify species interactions into categories such as competition, predation, parasitism, and mutualism. While this typological approach has obvious benefits, it ignores the variation that characterizes species associations in nature, and creates significant problems for understanding their ecology and evolution (Thompson 1988). Species interact within a biotic and abiotic matrix that exhibits complex patterns of spatial and temporal heterogeneity, leading to variation in the occurrence, strength, and outcome of species interactions.

In this chapter, we argue that the results of species interactions are strongly dependent upon the ecological settings in which they occur. We refer to such context dependency as 'conditionality', and provide two simple but realistic examples to illustrate our point. First, a given species association may be characterized by mutualism under one set of conditions, commensalism under another set, and predation under a third, such that a single category of interaction does not portray the association accurately. Secondly, under one set of conditions two or more interactions, such as competition and mutualism, may occur simultaneously in the same species association (Rathcke 1983, 1988; Waser 1983; Addicott 1985; Templeton and Gilbert 1985).

Viewing species interactions as conditional is hardly new (Thompson 1982, 1988). For example, many studies of host–parasite and predator–prey interactions emphasize the pattern and causes of spatial and temporal variation in the intensity of attack (see Hassell 1978; Hassell and May 1988). The existence and outcome of competition is also dependent upon the ecological setting (Paine 1966; Brown 1971; Morin 1981; Brown *et al.* 1986). For example, in the same locality competition may be intense one year and absent the next (Weins 1977; Dunham 1980; Pulliam 1986).

Although similar kinds of context-dependent variation exist in mutualistic systems, we are only beginning to accumulate examples (Cushman and Whitham 1989). This situation stems largely from the general lack of attention given to the study of mutualism, which is surprising given the abundance, diversity, and importance of mutualistic associations in nature (Boucher *et al.* 1982; Thompson 1982; Boucher 1985). As a result, most studies focus on demonstrating whether or not a system is mutualistic, rather than considering how and why the occurrence and strength of mutualism varies in time and space.

Our discussion of conditional interactions is divided into three sections. First, we describe five specific ways in which homopteran–ant associations can involve conditional mutualism. We show that the occurrence and strength of mutualism can exhibit temporal, age-specific, and density-dependent variation. We also demonstrate that competition for ants can mediate the strength of mutualism, and present three hypotheses for how host plants mediate these mutualisms. Secondly, we present a general framework for the kinds of conditional mutualisms that occur in nature, with particular emphasis on ant–plant and ant–herbivore associations. Thirdly, we discuss the ecological and evolutionary implications of conditional mutualism, contrasting variation in the strength of interactions with variation in the sign of the interactions.

Conditional mutualism in homopteran–ant associations

Ants commonly tend phloem feeding Homoptera (primarily aphids, membracids and scales) and harvest their energy-rich excretions (honeydew). Through tending, ants can provide a range of beneficial services to homopterans, the most frequently cited being protection from natural enemies (see reviews by Nixon 1951; Way 1963; Beattie 1985; Buckley 1987*b*). Two homopteran–ant associations found in the western USA are the focus of this section. The first system is located in west-central Colorado and consists of the aphid *Aphis varians* Patch tended by the ants *Formica fusca* L. and *Formica cinerea* Wheeler on fireweed (*Epilobium* [= *Chamaenerion*] *angustifolium* L.) (Addicott 1978*a*,*b*,*c*, 1979; Cushman and Addicott 1989). The second system is located in northern Arizona and consists of the membracid *Publilia modesta* Uhler tended by *Formica altipetens* Francour on a herbaceous composite *Helenium hoopesii* Gray (Cushman and Whitham 1989, 1991). We also discuss conditional aspects reported for other homopteran–ant systems and suggest that many of the ideas presented in this section apply equally well to lepidopteran–ant associations.

Temporal variation in mutualisms

Variation between years occurs in the membracid–ant association in northern Arizona. Tending ants had a positive effect on membracid abundance in two years but not in a third. Protection from a predatory salticid spider (*Pellenes* sp.) appears to be the primary mechanism by which ants benefit membracids in this system. In the two years when ants had a positive influence on membracids, ants substantially reduced the abundance of spiders and presumably also reduced their effect on homopterans. However, in the year when ants did not have a positive effect on membracids, spider abundance was much lower and ants did not deter the few that were present. Thus, the mutualism is temporally variable and apparently becomes weak or absent for the membracids when their predators are scarce and/or are not deterred by ants.

Age-specific variation in mutualisms

The membracid–ant association in northern Arizona also provides an example of age-specific variation in mutualisms. Only membracid nymphs directly benefit from ant-tending whereas over-wintering, second-year adults do not. Membracid adults possess heavily sclerotized exoskeletons and are extremely mobile while the nymphs are less sclerotized and fairly sedentary. The nymphs are commonly preyed upon by spiders and thus benefit from being tended by ants, whereas adults can escape such predators without the aid of ants.

Density-dependent variation in mutualisms

At the scale of individual homopteran aggregations, the beneficial effects of ants can increase or decrease as the density of homopterans increases. For instance, Banks (1962) and Banks and Macauley (1967) observed that the ant *Lasius niger* L. provided greater benefit to small colonies of the aphid *Aphis fabae* Scop. than to large colonies. Similarly, Addicott (1979) found that small *Aphis varians* colonies acquire greater benefit from tending than large colonies. Thus these systems are characterized by negative density dependence, where the beneficial effects of ants on homopteran aggregations decrease with homopteran density.

Positive density-dependence at the scale of individual aggregations occurs in the membracid–ant association. Large *Publilia modesta* aggregations acquired 36–46 per cent more benefit from ant tending than small aggregations. McEvoy (1979) and Wood (1982) also provided evidence that large membracid aggregations benefit more from ant tending than small aggregations.

The difference in response of aphids and membracids may be a consequence of different relative densities in the two systems and the effect of these densities on the host plants. For example, high-density *Aphis varians* aggregations on fireweed range from 400 to 10 000 individuals per inflorescence. At such densities, aphids have a large negative effect on fireweed growth and reproduction. Membracid density levels are much lower (usually less than 400), as is their impact on the host plant. On a larger spatial scale, there is also negative density-dependence in the aphid and membracid systems, as the beneficial effects of ants are diluted with increasing density of homopteran-infested plants (see below).

Competition mediating mutualisms

Homopterans can compete intra- and interspecifically for the services of ants, such that the strength of these mutualisms can be altered by the number, kind, and spatial arrangement of neighbouring homopterans. Two conditions must be met for such competition to occur. First, homopteran aggregations on neighbouring host plants must reduce the number of ants that each attracts. Secondly, such reductions in the density of ants must negatively influence homopteran fitness. While numerous studies have used ant-exclusion experiments to demonstrate that homopteran species can benefit from ant tending (Buckley, 1987*a*,*b*), complete exclusion of ants rarely occurs in nature. More realistically, the number of ants tending homopterans varies in space and time in response to numerous factors, and it is the effects of this variation that we focus on in the second condition.

The membracid *Publilia modesta* competes intraspecifically for the services of an ant mutualist, *Formica altipetens*. Experimental increases of membracid densities in the field (using membracid-infested, potted host plants) resulted in three negative effects due to competition for ants. First, the number of ants tending membracids per plant was reduced by 45–59 per cent in the presence of competitors (infested potted plants). Secondly, the abundance of predatory spiders, and presumably the rates of predation, increased by 63 per cent in response to these reduced ant tending levels. Finally, the number of membracid nymphs per plant decreased by 59 per cent in response to reduced tending levels and increased spider abundance. In total, competition for ant mutualists translated into a 92 per cent decrease in the production of newly eclosed membracid adults.

Both intra- and interspecific competition for mutualists occurs in the aphid–ant system in Colorado. First, the longevity of aphid aggregations (*Aphis varians*) on fireweed significantly decreases with increasing proximity to neighbouring shrubs occupied by three ant-tended aphid species, including *A. varians*. Secondly, we used ant-exclusion experiments to document that the presence of heavily aphid-infested fireweed shoots significantly reduced the number of ants tending neighbouring, conspecific populations on fireweed. In addition, the presence of ant-tended aphids (*Cinara* sp.) on Engelmann spruce (*Picea engelmannii* Parry) significantly reduced the number of ants tending neighbouring *A. varians* on fireweed. These reductions in ant-tending levels probably have significant negative effects on fireweed aphids: aphid populations tended by three or more ants exhibited significantly higher probabilities of persisting and growing than aphid populations of the same size tended by one to two ants.

Host plant mediation of mutualisms

Studies of ant–herbivore mutualisms have rarely considered the possibility that host plants mediate these interactions. A noteworthy exception is the work of Pierce (1984, 1985) on lycaenid–ant interactions. She documented that ant-tended lycaenid butterfly larvae usually feed on nitrogen-fixing, protein-rich, host plants whereas non-tended species do not exhibit this pattern. Pierce went on to suggest that natural selection has favoured the use of protein-rich hosts by ant-tended lycaenids because such host-use results in secretions that are more attractive to ant mutualists. However, few studies have directly examined such host-plant mediation in lepidopteran–ant or homopteran–ant systems (see Chapters 9 and 11).

We hypothesize three mechanisms whereby host plants can influence homopteran–ant interactions. First, host plants bearing extrafloral nectaries may affect the strength of homopteran–ant mutualisms by attracting ants independently (see Buckley 1987b). Homopteran aggregations that form on host plants with extrafloral nectaries will be more likely to be located and tended by ants than aggregations on hosts without extrafloral nectaries. This is because ants commonly switch from visiting a plant's extrafloral nectaries to visiting its homopterans (Buckley 1983; Sudd and Sudd 1985; but see also Becerra and Venable 1989).

A second mechanism for host plant mediation of homopteran–ant interactions involves differential recruitment of ants to honeydew of variable quality and quantity (Cushman 1991). Host plants can directly and indirectly affect the fitness of their ant-tended homopteran herbivores. Host plant phloem fluids directly influence the rate of increase of homopterans, as these substances constitute the diet of most ant-tended homopterans (Auclair 1963; Way 1963). Other characteristics, such as plant surfaces and architecture, can influence the ability of homopterans to acquire these food resources (Juniper and Southwood 1986). In addition to the direct effects, host plant phloem fluids can indirectly influence rates of homopteran increase by influencing their interactions with ants. Specifically, ants may be better mutualists towards homopterans feeding on high-quality host plants, since such host use results in the production of honeydew which is more energetically rewarding to ants.

Two conditions must be met in order to conclude that host plant quality mediates the outcome of homopteran–ant interactions. The attractiveness of homopterans to ants must vary predictably with changes in host plant quality and the number of tending ants must have a significant effect on the fitness of homopterans. Variation in the attractiveness of homopterans to ants will occur if, (a) the chemical composition and/or quantity of honeydew produced by homopterans varies with changes in host plant quality, and (b) ants preferentially tend those homopterans producing the most nutritionally rewarding or attractive honeydew. The quality and quantity of honeydew produced by homopterans is strongly influenced by phenotypic and genotypic variation in host plants (Auclair 1963; Holt and Wratten 1986). While no study has explicitly considered the recruitment behaviour of ants to homopterans producing honeydew of variable quality and/or quantity, numerous studies have demonstrated that ants recruit differently to sugar baits varying in the identity and concentration of sugars and amino acids (Taylor 1977; Crawford and Rissing 1983; Lanza and Krauss 1984; Sudd and Sudd 1985). In at least two homopteran–ant systems, tending levels have been shown to have an important influence on measures of homopteran fitness. (Cushman and Addicott (1989), Cushman and Whitham (1991), Inouye and Taylor (1979), and Barton (1986) discuss variable tending levels in ant–plant systems.) Thus, while direct tests of this hypothesis are required, variation in host plant quality should affect the fitness of homopterans through the number of ants recruiting to honeydew of variable quality.

A third way in which host plants could influence homopteran–ant associations emphasizes the needs of tended homopterans in 'marginal' environments (J. F. Addicott, unpublished work). In general, mutualism occurs where one organism encounters limiting conditions and another organism is capable of modifying these adverse conditions (Thompson 1982; Addicott 1984). Accordingly, ant mutualists should be most important to homopterans when the latter are experiencing limiting environments. When homopterans feed on low-quality host plants, they commonly exhibit reduced fecundity, and increased development time and exposure to natural enemies (Auclair 1963). If these homopterans can attract ants under such conditions, then the ants will be especially beneficial, as they may minimize the problems caused by feeding on low-quality host plants. Thus, ant mutualists may allow homopterans to exploit limiting conditions that untended species cannot.

Species-specific variation in mutualisms

Three studies clearly illustrate the importance of interspecific variation in mutualisms. First, in the fireweed–aphid system, the aggressive *Formica cinerea* is more beneficial to aphids (*Aphis varians*) than *F. fusca*, although the latter has a positive effect on aphids. Secondly, Messina (1981) found that ant species vary in their ability to protect a membracid's host plant from a defoliating herbivore. Thirdly, Bristow (1984) showed that two tended homopteran species on the same host plant were affected differently by ant species. The aphid *Aphis veroniae* Thomas benefits most when tended by the ant *Tapinoma sessile* Say, whereas the membracid *Publilia reticulata* Van Duzee benefits most from being tended by *Myrmica lobicornis* Emery and *M. americana* Weber.

General framework for conditionality in mutualisms

In the previous section, we considered five ways in which homopteran–ant interactions can involve conditional mutualism. Now we recast this variation into a more comprehensive scheme, applicable to mutualisms in general, but particularly those involving ant–plant and ant–herbivore interactions. We consider three distinct kinds of conditionality, each of which can vary in

space and time: variation in the ecological 'problems' that organisms in mutualisms experience; variation in the possible 'solutions' that partners can
provide to these problems; and variation in the availability of mutualists.

Variation in ecological problems

Organisms face a wide range of ecological problems for which other species
can provide solutions. For example, plants require movement of gametes
(pollination), protection from herbivores, and dispersal of seeds to favourable germination sites, while homopterans and lepidopterans require protection from natural enemies. However, the occurrence or magnitude of these
problems rarely remains constant in space or time. For example, as suggested
earlier the membracid *Publilia modesta* benefits from being tended by protective ants only in those places and at those times when their natural enemies
are abundant. When natural enemies are absent or in low abundance, the
problem disappears and associating ants have no beneficial effect on membracids. In addition, only membracid nymphs have a problem with predatory
spiders, whereas membracid adults do not. Similarly, plants bearing extrafloral nectaries may benefit from protective ants only when herbivory is high
(O'Dowd and Catchpole 1983; Boecklen 1984; Beattie 1985; Rashbrook *et
al.*, Chapter 16, this volume; Mackay and Whalen, Chapter 17, this volume).

Variation in ecological problems and solutions also comes into play when
evaluating those conditions under which plants will benefit indirectly from
being infested with ant-tended homopterans (Beattie 1985). For example,
Messina (1981) has shown that a herbaceous composite can benefit from
being infested by ant-tended membracids because ants greatly reduce the
effects of a defoliating chrysomelid beetle during outbreak years. However,
such indirect benefits will not occur in times or places where beetle densities
are low (see also Buckley 1983 and Fritz 1983). In addition, Compton and
Robertson (1988, Chapter 10, this volume) demonstrate that an ant-tended
homopteran can indirectly benefit its host plant, because the homopterans
attract ants which in turn provide solutions to two of the plant's problems:
ants reduce seed predation and parasitism of its pollinator. Thus, in some
cases, the mutualism occurring between ants and a host plant's herbivore can
provide the plant with benefits.

Variation in quality of mutualists

Species vary in their ability to provide solutions to ecological problems. We
have already discussed species-specific variation in the ability of ants to
benefit homopterans. Beattie (1985) also reviews a variety of ways in which
ant species can vary as seed-dispersal agents and such species-specific effects
undoubtedly occur in EFN (extrafloral nectary) systems (see review by
Keeler 1989 and Rashbrook *et al.* Chapter 16, this volume).

Species-specific variation in the ability of ants to act as mutualists will commonly result in the existence of 'hot spots' where mutualism is intense, sur-

rounded by areas where mutualism is weak or non-existent. These hot spots are likely to persist because ant colonies are often long-lived (see Wilson 1971). For example, in west central Colorado, for at least 12 years, sites dominated by *Formica cinerea* have many aphid-infested fireweed shoots, with each shoot bearing large aphid aggregations. Conversely, sites dominated by *F. fusca* have fewer aphid-infested fireweed shoots, each of which supports smaller *Aphis* populations. This pattern results from species-specific variation in the tending effect of ants.

Variation in availability of mutualists

There are several ways in which conditional interactions result from a shortage of mutualists. For example, tended homopterans and plants often compete with neighbouring individuals for the beneficial services of ants (Addicott 1978*b*; Davidson and Morton 1981; Buckley 1983; Cushman and Addicott 1989; Cushman and Whitham (1991)). Host plants can mediate herbivore–ant interactions through their influence on the ability of herbivores to attract ants. Further, the density dependence reported in our homopteran–ant systems may result from variation in ant tending levels due to homopteran aggregation size.

The density of tending ants is also important for various ant–plant mutualisms. For example, Inouye and Taylor (1979) have shown that the number of ants tending the EFNs of a herbaceous composite, strongly influences the level of pre-dispersal seed predation (see also Chapters 16 and 17).

The strongest and most persistent source of spatial variation in the availability of mutualists is likely to be generated by the dispersion patterns of the ants. Central-place foraging of ant colonies results in mutualistic 'hot spots' where the activity of ants is most intense. For example, in homopteran–ant (McEvoy 1979; Sudd 1983) and ant–plant systems (Inouye and Taylor 1979; Laine and Niemelä 1980), tending levels, and presumably mutualism strength, increases with increasing proximity to the nests of servicing ant colonies.

Consequences of conditional mutualism

Various types of conditional responses, and the patterns that result from them, can undoubtedly have profound effects on the ecology and evolution of mutualistic systems. What follows is a preliminary attempt to identify some of the potential consequences of conditional mutualisms.

Ecological consequences

Conditional interactions may play a significant role in the population dynamics of mutualistic systems, both at the local and regional levels. For example, negative density-dependent variation is stabilizing but positive density-dependent variation is not (May 1973, 1981; Addicott 1981; Wolin and Lawlor 1984; Wolin 1985). As the intensity of mutualism declines with density, either at a local level (e.g. the density of homopterans on individual plants) or a regional level (e.g. the density of homopteran-infested host plants in a given area), the tendency of mutualistic populations to increase in size declines (Wolin 1985) and systems return to equilibrium (Addicott 1981).

The conditionality of mutualism may also be important at low densities if there are low-density threshold points below which populations have increased probabilities of becoming extinct (May 1981). If mutualism affects population growth rates primarily at low population sizes, this could lead to strong buffering of population dynamics, because mutualistic populations temporarily depressed to low densities would increase more rapidly towards equilibrium densities than would non-mutualistic populations. For example, while fireweed aphid populations are extremely susceptible to extinction at low densities, tending ants increase the growth rates of these low-density populations, thereby reducing the probability of extinction.

Evolutionary consequences

Conditional mutualisms undoubtedly involve highly variable selection pressures that arise from ecological processes being intense in some years, localities, densities, or life stages, and absent or weak in others. Species interactions, and the selection pressures that accompany them, can vary in two fundamentally different ways. First, the strength of interactions can fluctuate from strong to weak ($+++$ to $+$) and from present to absent ($+$ to 0). Such is the case with competition for mutualists in homopteran–ant and ant–plant associations, where competition for ants reduces the beneficial effects of ants on homopterans and plants. An extreme case of variation in strength occurs when associations change from being mutualistic to commensalistic (or non-mutualistic), as illustrated by the temporal and age-specific variation reported for the membracid–ant system. The second form of variation involves actual fluctuations in the sign of interactions (i.e. from mutualism to predation). While sign reversals in homopteran–ant interactions have been suggested by Way (1963), Pontin (1978), and Edinger (1985), few studies have examined this extreme form of conditionality (see Chapter 7).

These different forms of variation in interactions and selection pressures have very different evolutionary implications. For example, selection pressures will be weak but consistent (i.e. in the same direction) when only the magnitude of interactions change. Conversely, when the sign of an interaction is reversed, the actual direction of selection changes and selection becomes distruptive, favouring traits (behavioural and morphological) that facilitate the association in certain situations and favouring different and perhaps opposing traits in other times and/or places.

Changes in the strength and sign of ant–plant and ant–herbivore interactions often result from variation in behaviour of ants. Such variation will be determined in part by the nutritional rewards that ants receive and their need for such resources. We hypothesize that differences in the value of these rewards (value being some joint function of nutritional reward and need) can be used to predict whether ants will act as mutualists or predators towards homopterans or plants, and the strength of such interactions. In Fig. 8.1, we present a simple graphical model that illustrates these predicted relationships (Cushman 1991). When tended homopterans, lepidopterans, or plants offer rewards of low value (as perceived by the tending colony), ants will acquire greater fitness by acting as predators than as mutualists. When the value of rewards is high, ant colonies will gain more by acting as mutualists (e.g. tending herbivores and harvesting their honeydew). The value of herbivore and plant rewards will be a function of various factors, including the composition and quantity of rewards, distance of rewards from the ant nest, nutritional status of ant colonies, and availability of alternative resources.

Edinger (1985) has conducted one of the few studies that addresses these ideas on 'switch points' in conditional mutualisms. He found that the ant *Lasius neoniger* Emery preyed upon aphid species that produced less honeydew three times more often than it preyed upon the aphid *Chaitophorous populicola* Panzar, which produced large amounts of honeydew. In addition, when *C. populicola* was experimentally deprived of phloem fluids (by girdling the vascular tissue of its host), the ant preyed upon it twice as often as on controls.

Similar kinds of switch points may occur in dispersal systems involving

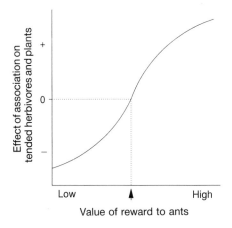

Fig. 8.1. Proposed relationship between the value of rewards provided to ants and the behaviour of ants towards tended herbivores and/or plants. When the value of rewards is low, ants are predicted to act as predators whereas when the value of rewards is high, ants are predicted to act as mutualists. The arrow indicates a hypothetical value of reward after which ant associates switch from acting as predators to acting as mutualists. (Modified from Cushman 1991.)

elaiosomes. Ants can either act as dispersal agents by consuming the elaiosome and discarding the seed, or they can act as predators and consume both. Given such a dichotomy, we hypothesize that protein- and lipid-stressed ant colonies will act primarily as predators towards such seeds, and as dispersal agents when colonies are less nutrient-stressed. Unfortunately, we do not know of any data concerning this kind of intraspecific variation in the behaviour of ants.

We make two brief comments on the evolutionary implications of conditionality here. First, given the variation in mutualistic systems, it seems probable that selection will favour the evolution of flexible, non-obligate associations between species, such that the costs of having an unpredictable mutualist are minimized. Secondly, the preponderance of conditionality in nature suggests that species-specific co-evolution in mutualistic systems will be the exception rather than the rule, due to the disruptive effects of spatial and temporal variation in the strength and direction of selection pressures.

Conclusions

'What ecologists look for' (Risch and Boucher 1976) is still an important topic in population ecology. Only by focusing on the variation characterizing most species associations in nature, will we gain a realistic picture of the processes operating within populations and communities. In addition to questions like 'does a particular interaction occur?', it is crucial to ask 'what factors influence the strength, symmetry, and outcome of this interaction?'

When studies fail to detect an expected mutualism in a particular system it is essential to examine the conditions which led to this result. For example, in the case of studies that fail to detect a positive effect of ants on plants bearing EFNs, the following questions should be asked:

1. Was there a lack of herbivores such that there was no problem for the plant?
2. Are the herbivores in the system ones that can readily escape ant guards?
3. Are there insufficient numbers of ants to provide adequate defence of the plant, and if so, is this because of (a) inadequate rewards provided by the plant, (b) availability of more rewarding alternative resources nearby which attract ants away from the plant, or (c) a general absence of adequate resources within the region such that ant populations are absent or in low abundance?
4. Are the plants located within the territory of a relatively poor ant mutualist?

We are not suggesting that the hypothesis of mutualism should be clung to doggedly until finally one set of conditions is found in which the expected mutualism occurs. Rather, we suggest that there are many reasons to expect that interactions will vary in space and time. Therefore, as ecologists we must carefully specify the conditions under which observations are made, as these may provide clues as to why an interaction is variable, and conduct studies at several localities and times. With this kind of information, we can address the degree to which interactions vary for a wide range of systems and begin to develop our ability to predict the occurrence and degree of conditional interactions.

Acknowledgements

We thank Alan Harvey, Mike Hochberg, John Lawton, Peter Price, Vanessa Rashbrook, Phil Warren, and Tom Whitham for stimulating discussion of conditional mutualisms and comments on this manuscript. J.H.C. gratefully acknowledges the support and insight of Tom Whitham, especially in regard to the development of ideas on conditional interactions and switch points. Our research has been generously supported by grants from the National Science Foundation (to T. G. Whitham and J.H.C.), NATO/NSF Fellowship Program (to J.H.C.), US Department of Agriculture (to T. G. Whitham), and the Natural Sciences and Engineering Research Council of Canada (to J.F.A.).

9

Why are so few aphids ant-tended?

Catherine M. Bristow

Introduction

The relationships between sap-feeding Homoptera and ants have been widely studied (Buckley 1987; Way 1963). However, our understanding of them is still limited, and the factors that predispose some aphid species to enter into mutualistic associations with ants, while other species fail to do so, remain largely a mystery. Theoretical studies (Roughgarden 1975; Wilson 1983; Keeler 1985) suggest that mutualisms will be restricted to situations where the cost of maintaining the association is low to each participant, while the benefits are relatively great. Ant–aphid interactions would seem to fit these restrictions. The main attractant that aphids offer to ants is honeydew, a carbohydrate-rich excretion that derives from their mode of feeding almost exclusively on phloem sap (Miles 1987). Unlike plants that must entice ants with specialized structures such as extrafloral nectaries (O'Dowd 1979), aphids will excrete honeydew as a waste product whether or not they form myrmecophilous associations. Thus, the presumed cost to the aphid of honeydew production is negligible. However, excretion of honeydew may generate indirect costs since it serves both to attract natural enemies (Zoebelein 1956*a,b*; 1957) and as a substrate for the growth of contaminants such as sooty moulds (Fokkema *et al.* 1983). Similarly, we might infer that the cost to ants of honeydew collection would also be relatively low. A honeydew source, once located, can be exploited for days or even weeks (Bristow 1983, 1984), thus reducing search costs, and further, unlike large prey items, it entails low pursuit or capture costs (MacArthur and Pianka 1966). However, there are costs of defence. For example, Dreisig (1988) found that foraging ants may incur a significant cost in the defence behaviour required to monopolize homopteran or plant resources. None the less we can make the following generalizations:

1. aphids are common in many habitats, particularly in the temperate zones (Eastop 1973);
2. most aphids produce honeydew (Miles 1987);
3. ants co-occur in virtually all of the same habitats (Wheeler 1910); and
4. many ants collect honeydew (Wheeler 1910; Gregg 1963).

From this we might conclude that ants would often be found in association with aphid colonies. To test this conclusion, I have compiled data on ant–aphid associations from the Rocky Mountain region of the United States. Only 117 of 479 aphids, fewer than a quarter of the species, have ever been recorded as myrmecophiles. Why do three-quarters of aphid species forego the potential benefits of associating with ants? There are several hypotheses regarding factors which could limit the formation of mutualisms. I have identified five broad classes of constraints which are summarized in Table 9. 1. These classes are neither exhaustive nor (entirely) mutually exclusive. However, they do provide a framework for asking: why are so few aphids tended by ants?

Table 9.1. Potential factors contributing to or hindering the formation of ant–aphid associations.

Ant associations are limited by phylogenetic constraints
   Ant associations may be limited to one or several aphid families. Absence of
   association may simply reflect evolutionary circumstance.
Ant attendance is incompatible with alternative predator defences
   Morphological adaptation to ant attendance
      Reduced cornicle development
      Reduced leg length to body length ratio
   Behavioural adaptation to ant attendance
      Tolerance of approach by other insect species
   Alternative defences unavailable
      Hiding (roots, leaf axils, curled leaves, galls), armour (wax, armour)
Ant attendance offers greater benefit for some aphid species
   Large aphids—more vulnerable to predation while withdrawing stylets
   Ants as predators vs. ants as mutualists
   Variability in complexes of, or pressures from, predators and parasitoids
Host plants mediate aphid attractiveness to ants
   Host plant species alters aphid attractiveness (e.g. honeydew)
   Host plant development alters aphid attractiveness
   Host plant augments or detracts from aphid attractiveness (e.g. extrafloral nectar)
Ant attendance is limited by the availability or predictability of suitable ants
   Distribution and availability of suitable ants
   Competition for mutualists
   Predictability of associations in time and space
      Habitat predictability, host plant growth form

Hypotheses concerning factors which limit mutualisms

Ant associations are limited by phylogenetic constraints

Myrmecophily could be limited in distribution if it were restricted to one or only a few major aphid taxa. If the suite of characteristics required to attract ants appeared late in the evolutionary history of aphids, then we might expect to find ant–aphid associations limited to lineages derived from that event. A problem presents itself at this point: there is no consensus regarding the phylogeny of aphids (Eastop 1973; Heie 1980, 1987; Shaposhnikov 1987a,b).

Regardless of which suggested phylogeny one prefers, however, association with ants does not seem to be strongly constrained to limited taxa. Table 9. 2 shows the number of genera and species forming associations with ants across ten subfamilies recognized by Quist (1978). Little can be inferred from the absence of recorded myrmecophily in the Mindarinae, Thelaxinae, and Greenideinae, since all three groups are very rare in the region, and little is known of their biology. Six of the remaining seven groups support at least some tending, suggesting either a tended common ancestor for all groups or, more probably, independent evolution of myrmecophily.

The frequency of ant associations does vary between groups, but not in a manner consistent with any major phylogenetic scheme. For example, the Pemphiginae are considered to be among the most primitive of groups in virtually all phylogenies (Ilharco and van Harten 1987), while the Lachninae are placed either near them or elevated to the same status as the Aphidinae. However, the tending rate of the Pemphiginae (33 per cent) is intermediate between that found in the Lachninae (38 per cent), and the Aphidinae (22 per cent). Even within a subfamily there is little consistency. In the Lachninae tending rates range from 0 (Anoecina) to 100 per cent (Tramina). In the Aphidinae, 32 per cent of the species in the subtribe Aphidina form associations with ants while only 12 per cent of Macrosiphina are tended. Myrmecophily appears to be a highly labile trait evolutionarily.

Is ant attendance incompatible with alternative predator defences?

Although aphids possess many adaptations that are suitable to their mode of living, they are generally not well protected against predators and parasitoids. Their soft bodies, often flightless condition, and frequent formation of large aggregates, make aphid colonies large, slow moving targets. Indeed, the greatest positive role played by ants in the life of the aphid is in this area: as a defence agent against predators and parasitoids (Way 1963). However, many aphid species employ a variety of other defences. These may be morphological adaptations, such as long cornicles which release both waxy secretions and alarm pheromones, or long or saltatorial legs associated with escape behaviour. Additionally, some aphids retain behaviour patterns that

Table 9.2. Phylogenetic distribution of ant associations among aphids of the Rocky Mountain region, USA.

Subfamily
  Tribe
    Subtribe
	Total
	genera
		Genera
		with
		tending
			Total
			species
				Spp. with
				tending
					Per cent
					spp.
					tended

Lachninae					
Lachnini	**10**	**6**	**66**	**25**	**38**
Anoecina	1	0	3	0	0
Eulachnina	3	2	9	2	22
Cinarina	1	1	49	20	41
Lachnina	4	2	4	2	50
Tramina	1	1	1	1	100
Callaphidinae	**16**	**2**	**25**	**2**	**8**
Callipterini	2	0	4	0	0
Callaphidini	14	2	21	2	10
Saltusaphidinae					
Saltusaphidini	**4**	**0**	**8**	**0**	**0**
Drepanosiphinae					
Drepanosiphini					
Drepanosiphina	**2**	**2**	**8**	**2**	**25**
Chaitophorinae	**6**	**3**	**23**	**9**	**39**
Chaitophorini					
Chaitophorina	3	2	15	4	27
Pterocommatini					
Pterocommina	1	1	5	5	100
Fullwayina	2	0	3	0	0
Aphidinae	**62**	**28**	**310**	**67**	**22**
Aphidini					
Aphidina	25	12	122	39	32
Rhopalosiphina	12	5	26	8	31
Macrosiphini					
Macrosiphina	25	11	162	20	12
Pemphiginae	**15**	**8**	**36**	**12**	**33**
Pemphigini	8	4	24	5	21
Eriosomatini	6	3	10	5	50
Fordini	1	1	2	2	100
Mindarinae					
Mindarini	**1**	**0**	**1**	**0**	**0**
Thelaxinae	**2**	**0**	**2**	**0**	**0**
Thelaxini	1	0	1	0	0
Hormaphidini	1	0	1	0	0
Greenideinae					
Setaphidini	**1**	**0**	**1**	**0**	**0**
Total	**119**	**49**	**479**	**117**	**24**

Data from Jones 1929; Palmer 1952; Quist 1978; Addicott 1978a, 1979a, b; Russell 1989. Figures in bold indicate totals for that subfamily.

provide protection. They may be solitary and cryptic, or excitable and prone to jumping, kicking, or dropping. Other species gain some defence by living in a protected environment. This can be a simple covering of thick flocculent wax, a hidden or subterranean lifestyle, or it can be as complex as the induction of an enclosed gall. In each of these cases, it is possible that one line of defence is achieved at the expense of others; namely, aphids which rely on such alternative defences may be unable to attract ants.

It has been widely cited that aphids which are regularly associated with ants will have some or all of the following characteristics (Way 1963; Skinner 1980b; Sudd 1987; Sudd and Franks 1987):

cornicle length reduced
leg length reduced, not saltatorial
caudal length reduced
a perianal ring of hairs or trophobiotic organ often present

Using the Rocky Mountain aphids for comparison, we can examine several of these generalizations. Figure 9. 1(a) shows the ratio of cornicle length to body length by subfamily. Cornicles produce both waxy, defensive secretions and alarm pheromones. It has been postulated that these may actually be disadvantageous to aphids in ant associations, since their use would hinder or upset the tending relationship. The pattern of cornicle length to tending is equivocal. The only group which fits the predicted pattern are the Aphidinae, which are the largest group. In all of the other subfamilies, however, tended species have longer cornicles than untended species. Similarly, the pattern of leg length is also equivocal. Figure 9. 1(b) shows the ratio of mean tibial length to the square of body length, presuming that longer legs are more efficient for moving the greater volume of a large individual. In most subfamilies, untended species have the longer legs as predicted. However, the opposite is true for the Lachninae, which have one of the most well-developed interactions with ants. These results suggested that generalizations should be approached with some caution.

Generalizations have also been made regarding behavioural adaptations to ant attendance. Dixon (1958) suggested that tended species did not avoid approaching predators and parasitoids. Untended species like *Microlophium viciae* Buckton and *M. carnosum* (= *evansi*) Theobald walk away, drop from the plant, or kick at any intruder, including ants (Sudd 1967). Aphids which exhibit these wandering tendencies, or that indulge in any violent movement, are perceived as prey by ants (Way 1963; Cherix 1981), a consequence that would strongly select for placid behaviour among tended species.

A related behaviour that enhances ant association is gregariousness (Dixon 1958; Pierce *et al.* 1987). Aphids, such as *Aphis fabae* Scopoli, that actively form dense colonies are often found to be associated with ants (Ibbotson and Kennedy 1951; Hayamizu 1982). Conversely, certain species

(a)

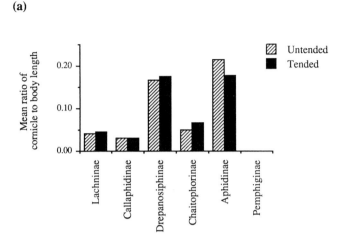

(b)

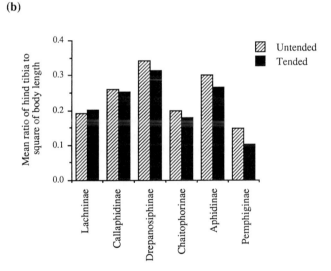

Fig. 9.1. Morphological correlates of ant-attendance of aphids. (a) Mean ratio of cornicle length to body length for the tended and untended species of the major aphid subfamilies. (b) Mean ratio of hind tibia length to the square of body length for the tended and untended species of the major aphid subfamilies. Based on ant–aphid associations of the Rocky Mountain region, USA.

of *Cinara* that are restricted to small colonies on young growing tips are only sporadically tended (Sudd 1987). *Drepanosiphum platanoidis* Schrank, the sycamore aphid, is an untended species which forms loose associations rather than dense clusters (Kennedy and Crawley 1967). This could reduce the colony profitability and defensibility, thus making it less attractive to ants. It is also likely, however, that the aphids' restless behaviour when crowded contributes at least as much to its failure to form associations with ants. However, no data are available on these behavioural modifications, and counter examples can be found for most of the above behaviour types. For example, among *Cinara* species in the Rockies, species with a range of colony sizes can be found associated with ants. *Cinara apini* Gillette & Palmer (large sluggish colonies), *C. atra* Gillette & Palmer (small colonies), and *C. flexilis* Gillette & Palmer and *C. glabra* Gillette & Palmer (both solitary) are all ant-tended. Conversely, *C. pinena* Mordvilko and *C. pruinosa* Hartig, both form large colonies but do not attract ants. Patterns of behavioural adaptation to ant attendance, like patterns of morphological adaptation, would benefit from further study.

Some aphids adopt a cryptic lifestyle which might render them simultaneously less available to both predators and ants. Adaptations associated with cryptobiosis include heavy wax, root-feeding, or living in leaf axils, curled leaves, or galls. Gall-forming aphids, in particular, face tradeoffs that could affect ant attendance. Bodenheimer and Swirski (1957) report that aphids living in closed galls produce little honeydew, are not usually ant tended, and are often preyed upon by ants. Other galls which are not closed do not limit ant associations in this way. Several studies (Setzer 1980; Shannon and Brewer 1980; Whitham 1980) suggest that galls provide little if any protection from predators, and that their primary function is to extend and enhance the quality of the host plant and to provide favourable micro-climatic conditions. Indeed, rather than being mutually exclusive, several defence tactics are employed simultaneously by many Rocky Mountain aphids (Table 9. 3). This suggests two possible hypotheses: (1) that 'defences' like galls are not primarily defensive, but rather function in providing favourable micro-climate, or (2) predation pressure on some aphid species is so intense that it favours multiple defences. We might then predict that predation pressures on free-living forms are quite different (see below).

Does ant attendance offer greater benefit for some aphid species than for others?

In the framework of cost–benefit analyses, the incompatibility of ant attendance with other predator defences could be a factor which decreases the benefit of the mutualism to the aphids. Other factors could also decrease the benefit of association with ants, including the size of the aphid, mutualistic vs. predaceous tendencies of the ant, and variation and variability in the pressure from natural enemies.

Table 9.3. Occurrence of predator defences of untended and tended aphids of the Rocky Mountain region, USA.

Defences	Untended spp.	Tended spp.
None	242	87
Agility	4	1
Crypsis	12	1
Root feeding	18	10
Wool	7	2
Galls	22	13
Roots and wool	1	2
Roots and galls	2	2
Wool and galls	6	1
Roots, wool, and galls	0	1

Dixon (1985) has suggested that large aphids are more likely to form associations with ants than small ones because they benefit more from their protective services. Large aphids require much longer to remove their stylets when disturbed (*Stomaphis quercus* L. takes up to 50 minutes to withdraw from feeding), so they are not able to drop or leave the plant when attacked. Consequently, the ant defence is critically important. There is little evidence, however, that this is true across a wide spectrum of aphid species. Although tended aphids include the larger species in the Drepanosiphinae, Chaitophorinae, and Pemphiginae, there is no such trend among other taxa, including those with the most species and those with the highest tending ratios (Aphidinae and Lachninae). Such a trend might become apparent if length of rostrum rather than length of body was calculated.

Ant predation on aphids is an area of ant–aphid mutualisms which deserves more attention, since ants will readily prey upon untended aphids. Skinner and Whittaker (1981) showed that the untended aphid *Drepanosiphum platanoidis* Schrank on sycamore trees suffers heavy predation by the ant *Formica rufa* L. A second species found on the same host plant, *Periphyllus testudinaceus* Fernie, is usually tended by *F. rufa* (see Chapter 6) and is rarely carried back to the nest as prey. This is consistent with the observation by Wheeler (1910) that ant-tended aphids were somehow immune to ant predation. However, ants can vary in their predacious activities, even upon the species that they tend (see Chapter 7). Predation on root aphids is believed to contribute significantly, if not totally, to the protein needs of associated ants (Neilsen *et al.* 1976; Pontin 1978), and aerial aphids may make up as much as 30 per cent of the prey collected by their attendant wood ants in some cases (Cherix 1981). Way (1954) found that he could shift the nature of the interaction between the ant *Oecophylla longinoda* Latreille and its tended scale insect, *Saissetia zanzibarensis* Williams, from mutualism to predation simply by providing the ants with a richer source of sugars. It is plausible that aphids differ in the quality and/or quantity of the honeydew that they produce (see below); and thus in the quality of sugar vs. protein that they provide to ants. Any factor that could cause higher predation on aphids by the tending ants would serve to reduce the overall benefits from the association, and could well prevent the formation of a mutualistic interaction.

Variation in predation depends on both the absolute number of predators and on their species composition. Syrphid larvae (Diptera), particularly young ones, appear to be relatively immune to ant disruption (Way 1963; C. M. Bristow, personal observation), as do some parasitoids. This immunity may vary with the ant species in attendance. Thus, *Iridomyrmex humilis* Mayr attacks and displaces larval Neuroptera and Diptera, but seldom molests coccinellids. The effectiveness of a mutualism, therefore, will be determined both by the complex of tending ants and by the complex of predators. The degree of predation pressure can vary from site to site and from year to year. When predator pressures are low, the mutualism could be reduced to a commensalism (see Chapter 8). Since predation pressure is probably one of the key factors that generates and maintains the association of aphids with ants, this area merits much more work.

Does the host plant mediate aphid attractiveness to ants?

In an era of abiding fascination for plant–herbivore interactions (Scriber and Ayers 1988) and three trophic-level interactions (Barbosa and Letourneau 1988), it seems curious that the importance of the host plant in the formation and maintenance of ant–aphid associations has escaped scrutiny. Both secondary plant compounds and differences in available nutrients have major impacts not only on herbivorous species but on those insects which feed upon the herbivores (Price *et al.* 1980; Pierce *et al.* 1987). Plants may influence aphid–ant associations either directly, by altering the quality of honeydew produced, or indirectly, through the production of extrafloral nectar or other ant attractants.

Our knowledge of the relationship between phloem sap and honeydew composition, although limited to a few species, suggests that strong correlations exist. In some pioneering work, Mittler (1958) determined that the phloem sap of willow contained high concentrations of sucrose (5–10 per cent w/v) and a mixture of amino acids that changed during the season from a high of 0.2 per cent during bud growth to a low of 0.03 per cent at leaf maturity. Nitrogen was re-mobilized during leaf senescence and regained levels of about 0.13 per cent. The composition of honeydew produced by a willow-feeding aphid, *Tuberolachnus salignus* Gmelin, closely reflected the host phloem sap. Sugar concentration did not differ from that found in phloem sap by more than 5 per cent, although the specific sugars excreted had been modified to include not only sucrose but also glucose, fructose, and melezitose. The latter sugar is characteristic of homopteran honeydew and may be an ant attractant (Kiss 1981). The amino acid concentrations were reduced from phloem sap to honeydew by about 35 per cent. However, the relative proportions of each amino acid were virtually identical. Additional work on willow by Leckstein and Llewellyn (1975) confirmed that amino acid composition changes considerably over the season. By comparison, the phloem sap available to the pea aphid, *Acyrthosiphon pisum* Harris, feeding on pea (*Pisum sativum* L.) had a much lower sugar concentration and much higher available nitrogen (up to 4.5 per cent) than the levels found in willow (Barlow and Randolph 1978). Unfortunately, the honeydew values for pea aphid were not measured. Both aphids occur in the Rocky Mountain region. *Tuberolachnus salignus* feeds on *Salix alba* L. and *S. elegantissima* K. Koch, and is ant-tended on both. *Acyrthosiphon pisum* occurs on *Astragalus* sp., *Lathyrus odoratus* L., *Medicago sativa L., Melilotus* sp., *Pisum sativum* L., and *Trifolium* sp., but is not associated with ants on any host.

In addition to nutrients, phloem sap may also contain secondary plant compounds, although it is generally thought that most secondary compounds are not phloem-transported (Klingauf 1987). Some aphids, such as *Aphis nerii* Fonscolombe, are known to sequester plant toxins for their own defence (Rothschild *et al.* 1970). Further evidence supports the hypothesis that substances such as cardenolides are acquired by feeding on phloem (Janzen 1978), and may be passed out in the honeydew (Bailey 1974). Quantitative factors may also play a role in ant–aphid interactions. The rate of honeydew production can differ for the same aphid species feeding on different species of plants (Sudd 1983), and this has been implicated in affecting whether aphids will attract ants. Although we lack specific information on both feeding rates and phloem and honeydew composition for most aphid–plant interactions, the data suggest that host plants vary in quality, and that at least some of this variability may affect the quality or quantity of honeydew produced.

Whether host plants directly affect the probability of aphids attracting ants can be examined at several levels. There is no overall correlation between percentage of species tended and total number of species infesting a host ($r^2 = 0.003$). However, we can compare the distribution of tended vs. untended species across hosts. If the host plant is instrumental in determining an aphid's attractiveness to mutualists, then some plants may be viewed as permissive hosts, supporting high levels of tending. Other plants may be prohibitive hosts, with aphids that form few or no ant associations. This may be true at the level of the plant family, where there is a significant tendency for some families to support high levels of tending compared to others (ANOVA; $F = 2.467$; $P < 0.001$; on arcsine-transformed percentages of tended species). Families with low overall percentages of ant–associated aphids include the Geraniaceae and Juglandaceae (0 per cent), Betulaceae and Fagaceae (2 per cent), Cyperaceae (4 per cent), Ranunculaceae (6 per cent), and Solanaceae (15 per cent). Families rich in ant–aphid associations include Poaceae (49 per cent), Asclepiadceae and Salicaceae (52 per cent), Lamiaceae (57 per cent), Onagraceae (61 per cent), Polygonaceae (62 per cent), and Anacardaceae (100 per cent).

In addition to patterns of ant association at the host-family level, many plant genera also seem to involve differential ant association patterns. Table 9.4 compares *Pinus* (pines), a genus which is rich in ant-associated aphid species with *Acer* (maples), a poor genus. Each genus contains eight plant species that support aphids. Overall, 41 per cent of the aphids on pine are found in association with ants, while only 20 per cent of those on maple are tended. Even within the genus, however, there can be considerable difference in attractiveness to ants. Seventy-one per cent of the species on

Table 9.4. Comparison of ant associations across species of host plants in *Pinus*, an ant-rich genus, and *Acer*, an ant-poor genus.

Genus Species	Untended species	Tended species	Per cent tended spp.
Pinus			
P. contorta latifolia	2	5	71
P. edulis	9	2	18
P. flexilis	2	2	50
P. monophylla	1	0	0
P. nigra	1	0	0
P. palustris	1	0	0
P. ponderosa	7	8	53
P. sylvestris	1	0	0
Total	24	17	41
Acer			
A. campestre	1	0	0
A. glabrum	3	0	0
A. grandidentatum	5	0	0
A. negundo	0	2	100
A. platanoides	2	0	0
A. pseudoplatanus	2	1	33
A. saccharinum	1	1	50
A. saccharum	2	0	0
Total	16	4	20

Pinus contorta latifolia are tended compared to 18 per cent on *P. edulis*. Similarly, both aphids occurring on *Acer negundo* are tended, while five other maple species have no tended species.

Several potential drawbacks to such broad comparisons should be noted. First, they overlook taxonomic affiliations between the aphids. For example, much of the tending on pines involves *Cinara* aphids which possess a well-developed filter chamber which may concentrate the honeydew (Kunkel and Kloft 1977). Secondly, the host plants grow in dissimilar habitats which may contain key differences, such as different ant faunas (see below). Pines, for example, tend to be found in more xeric conditions than maples. Geographical differences can play a role as well. In the Rocky Mountain region, the birch genus, *Betula*, supports only one sporadically ant-tended species, *Wahlgreniella nervata* Gillette on *B. occidentalis*. However, *Betula* in Great Britain supports significant ant–aphid associations (see Chapter 6). Such differences, although poorly understood, are by no means unheard of. The well-documented differences in control of the winter moth, *Operophtera brumata* L., by its natural enemies in Wytham Wood, England compared to Canada is a case in point (Hassell 1978).

As another test, we can use the aphids themselves to assess permissive and non-permissive hosts. A number of aphid species will be ant-tended on one host but not on another. If the host plant affects the aphids' ability to attract mutualists, then the former plant species is serving as a permissive host and the latter as a non-permissive one. We would then predict that a comparison of other aphid species utilizing the same set of plants would reveal a higher rate of ant-attendance on the permissive hosts. Twenty-eight sets of permissive and non-permissive hosts were identified using these criteria. A comparison of the number of tended vs. untended aphid species (Fig. 9. 2) shows that aphids sharing a permissive host were almost three times as likely to form their own associations with ants. Again, there are caveats associated with this approach. For example, one or several aphid species may be extremely attractive, drawing so many ants onto the plant that co-occurring aphids are fortuitously associated with ants which they did not attract themselves. This difficulty should not be insurmountable, however, since it is easy to differentiate tended from untended species co-occurring in the field (Ewart and Metcalf 1956; Addicott 1978*a*; Skinner and Whittaker 1981; personal observation).

We can test the host permissiveness hypothesis more rigorously by examining plants which support tended aphids under some circumstances but not under others. Host factors such as seasonal changes in sap constituents, nutrient or water stress, or even changes in feeding site may alter the plant's quality to both aphids and ants. In a study of *Aphis nerii* on oleander (*Nerium oleander* L.) (C. M. Bristow, unpublished work), I examined the effect of the feeding site (floral or leaf tip) on ant–aphid associations. The feeding site appeared to be instrumental in determining

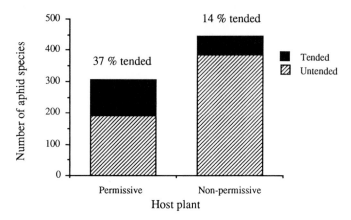

Fig. 9.2. The number of tended and untended aphid species on permissive and non-permissive hosts. (See text for explanation of host permissiveness.) Based on ant-aphid associations of the Rocky Mountain region, USA.

the ability of an aphid colony to attract Argentine ants, *Iridomyrmex humilis*. This was evident for colonies within the same plant, and was independent of colony size. So, ants often bypassed large, dense colonies on leaf tips to visit smaller, sparser colonies on flower buds. No ants were attracted to the flowers in the absence of aphid colonies. The differential attractiveness probably reflects chemical differences in the honeydew itself, since honeydew was produced by aphids feeding at both sites.

Another way in which the host plant could infuence ant–aphid associations is by offering its own rewards to ants, in the form of extrafloral nectar or some other food. This could enhance the ant–homopteran association in a manner analogous to that observed for ants and butterflies (Atsatt 1981; Pierce and Elgar 1985; DeVries Chapter 12, this volume). Alternatively, rich extrafloral resources might lure ants away from the aphids, possibly even shifting the association from mutualism to predation (Way 1954). It would be interesting to learn whether there is any connection between ant-tended plants and ant-tended Homoptera. Namely, do aphids serve as temperate-zone extrafloral nectaries (Messina 1981)? Oliveira and Brandão (see Chapter 14), for example, indicate that 25 per cent of plant species in the Brazilian cerrados have extrafloral nectaries and attract ants. In the Rockies, 25 per cent of aphids attract ants. Are the same plants likely to enter into both types of interaction? Or do they share some other ecological characteristics such as a habitat with an abundance of ants, or a particularly harsh community of predators? In many cases, host permissiveness may be an important factor contributing to the establishment of ant–aphid associations. However, the host plant is unlikely to be the sole factor involved, since many plants support both tended and untended species simultaneously.

Is ant attendance limited by the availability or predictability of suitable ants?

The habit of attending aphids is widespread among ants, particularly those in the subfamilies Dolichoderinae and Formicinae (Sudd 1987). However, suitable ants may be a limited resource for aphids. Some anecdotal evidence supporting this contention comes from an introduction of the wood ant *Formica lugubris* Zett. into Canada (McNeil *et al.* 1977). After two years, the ant was established and found to be tending 21 aphid species, seven of which had not been found previously in Quebec. It seems probable that the addition of a suitable attendant species altered the carrying capacity for these aphids, and allowed populations to increase to detectable levels.

Considerable work has been carried out recently on competition between aphids for the services of ants (Addicott 1978*b*; Cushman and Addicott 1989). This new direction in the study of the ecology of mutualisms may itself generate new predictions regarding the overall distribution of ant–aphid associations (see Chapter 8).

Finally, habitat or host predictability may also influence the formation of ant–aphid associations, perhaps in much the same way that plant architecture and habitat structure have been instrumental in shaping plant–herbivore interactions (Feeny 1976; Rhodes and Cates 1976; Coley *et al.*

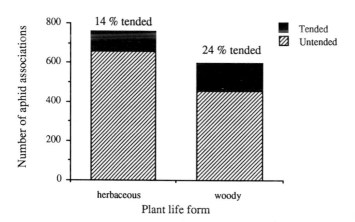

Fig. 9.3. Effect of plant life form on the proportion of ant-tended aphids. Based on 1354 specific aphid–plant associations involving 479 aphid species associated with 448 plant species in the Rocky Mountain region, USA. Woody plants include trees and shrubs. Herbaceous plants include both annual and perennial herbs and grasses. Associations on aquatic plants and vines are omitted.

1985). Certainly, ant–aphid associations are more common on woody perennials (trees and shrubs) than on herbaceous hosts (perennial and annuals) (Fig. 9. 3). However, we do not know the mechanisms involved. Late successional stages may simply support richer faunas of both participants (Southwood 1977; Lawton 1986). Alternatively, herbaceous plants may possess hairs or other features which make them less accessible to ants (Southwood 1986; Harley Chapter 28, this volume) or may be suitable for only a short period in time and space, thus decreasing the likelihood that an aphid colony will be discovered during its most vulnerable early days. Similarly, the aphid resources would be more transient over time, selecting for ants that are highly opportunistic in their associations. Trees, on the other hand, present literal 'trunk' trails to guide foraging ants to a limited search area. Thus, once attracted to a tree, possibly to other honeydew producers, ants are more likely to find new aphid colonies. Certainly, evidence suggests that wood ants maintain fidelity to particular routes, trees, and even homopteran colonies (Rosengren 1971; Ebbers and Barrow 1980). This sort of behaviour would favour the rapid establishment and continued tending of aphid colonies on these plants. Finally, forest ants may include larger, more aggressive species which can increase the protective benefit derived by aphids (Addicott 1979; Bristow 1984).

Conclusions

Ant–aphid associations involve only about a quarter of all aphids. Several commonly accepted explanations for this have been called into question. For example, morphological adaptation, size, and alternative predator defences do not seem to be strongly correlated with ant attendance. Host plants may play a critical role in allowing aphids to attract ant associates. Both variation in the quality of different hosts and of different parts of the same host can be involved. Finally, ant associations are more common for species feeding on woody plants.

This survey of factors that influence the formation and maintenance of ant–aphid associations has generated a shopping list for future research:

the temporal and spatial changes in the distribution and nutritional requirements of ants;

competition among aphids for effective ant mutualists;

changes in host-plant quality and its effects on both honeydew production and aphid attractiveness to ants; and

predator profiles for tended and untended aphids, perhaps compared to herbivore profiles for tended and untended plants.

The recent rise in interest in ant–plant and ant–homopteran interactions is heartening. We are moving from an era of descriptive natural history to an examination of how mutualistic interactions evolve and are maintained. With such understanding, mutualisms may advance from the level of 'evolutionary curiosities' to their proper status as one of the major forces in population and community structure.

10

Effects of ant–homopteran systems on fig–figwasp interactions

S. G. Compton and H. G. Robertson

Introduction

The pest status of many homopterans is often attributable to the ants that tend the bugs for honeydew and protect them from natural enemies (Buckley, 1987*a*). If the ants tending the homopterans incidentally protect other herbivores, then the negative impact of the ants will be compounded (Fritz 1983). In South Africa, for example, red scale (*Aonidiella aurantii* Mask.) can devastate citrus orchards if ants are allowed to interrupt the biological control by parasitoids. The ants are attracted by honeydew produced by soft scales, not by *A. aurantii* (Samways *et al.* 1982).

Protection by ants also favours the proliferation of homopterans in natural ecosystems, with a corresponding increase in the damage caused to their host plants (Fritz 1982; Tilles and Wood 1982; Way 1963). Sometimes, however, the ants may attack or interfere with other herbivores on the plants, and damage may be reduced. This reduction in herbivory may outweigh the damage caused by the honeydew producers and the net effect is that the ants, and indirectly the homopterans, benefit the plant (Carroll and Janzen 1973; Beattie 1985; Buckley 1987*b*).

The Cape fig (*Ficus sur* Forssk., Moraceae) is one example of a plant which benefits from having homopteran-tending ants on its fig-bearing branches (Compton and Robertson, 1988). The presence of the ant *Pheidole megacephala* Fabr. reduces seed predation and increases the numbers of pollen-carrying wasps emerging from the figs. In this chapter we describe the figwasp and ant assemblages associated with *F. sur* and examine whether the ants generally benefit the trees. The results are discussed in the context of the overall complex of interactions which take place between ants and the other animals occupying the trees.

The Cape fig and its figwasps

Ficus sur (= *F. capensis* Thunb.) is distributed throughout most of the non-arid areas of Africa (Berg *et al.* 1985). The tree bears figs (syconia) on modified, leafless branches which hang from the trunk or major branches. Figs are produced throughout the year and often occur in varying stages of development on the same tree. When mature the figs reach a maximum diameter of about 35 mm, while the crop size on different trees varies from less than ten to many thousands of figs.

The sole pollinator of *F. sur* in southern Africa is the figwasp *Ceratosolen capensis* Grandi (Chalcidoidea, Agaonidae, Agaoninae *sensu* Bouček 1988). Adult female wasps enter unpollinated figs via the bract-covered ostioles, pollinate the flowers which line the inside of the fig, then lay their eggs in some of the flowers and die. Larval development is synchronized with the development of the seeds, and emergence of the next generation of wasps coincides with changes in the figs which make them attractive to vertebrate dispersal agents.

While *Ceratosolen* is the only pollinator, in southern Africa about nine other wasps also utilize the figs (S. G. Compton, unpublished work). In the eastern Cape the figwasp assemblage consists of *Sycophaga cyclostigma* Waterston, three species of *Apocryptophagus* (= *Idarnes*) (all Agaonidae, Sycophaginae), and *Apocrypta guineensis* Grandi (Agaonidae, Sycoryctinae) (Fig. 10. 1). *Sycophaga* females enter the figs of *F. sur* in the same way as the pollinator, but gall the flowers without pollinating them. The *Apocrypto-phagus* species also gall the fig flowers, but they reach the ovules from the outside, by inserting their long ovipositors through the outer wall of the fig. *Apocrypta* is a parasitoid/inquiline which, like the *Apocryptophagus* spp., oviposits through the walls of the fig. Its larvae destroy those of *Ceratosolen* and the other figwasps. Normally, only one wasp larva develops inside each gall.

As the wasps enter the figs in order to oviposit, the females of *Sycophaga* and *Ceratosolen* are less likely to come into contact with ants than the

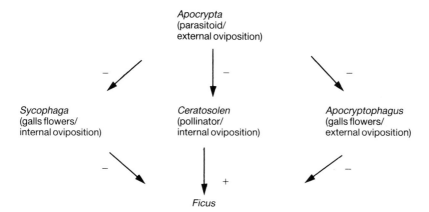

Fig. 10.1. Interactions among the figwasps associated with the Cape fig, *Ficus sur*.

Apocryptophagus spp. and *Apocrypta*, which oviposit from the surface of the figs. However, all the figwasps are equally vulnerable when leaving the figs after completing their development.

Homopterans and ants on the Cape fig tree

The gregarious tettigometrid *Hilda patruelis* Stal produces honeydew and is tended by ants (Weaving 1980). It is a polyphagous species recorded from numerous hosts and is an economic pest of groundnuts in southern Africa (Weaving 1980; P. Hewitt, personal communication).

Observations were made on *F. sur* growing in semi-natural or disturbed habitats around Grahamstown, South Africa. The presence of figs, *Hilda*, and honeydew-producing scale insects, as well as the behaviour of the ants, was noted between February and September 1988. Twenty trees were monitored, of which 17 produced figs during the study and eight had figs present continuously. *Hilda* was present on 11 trees; it showed a distinct preference for the fig-bearing branches and was usually found only when figs were present (Fig. 10. 2). Scale insects provided ants with honeydew on three trees, but were only abundant on tree 6, where *Hilda* was absent. The occurrence of ants was linked with the presence of figs and homopterans. Ant densities increased when figs were on the trees, and between crops ants were often absent. Ants and *Hilda* were almost always present on the eight trees which had figs continuously.

A total of 12 ant species were recorded on the fig-bearing branches of *F. sur*, with *Pheidole megacephala* the most widely distributed species and also the most consistently active ant on the trees where it occurred (Table 10. 1). Species such as *Polyrachis schistacea* Grst. and *Anoplolepis custodiens* Smith were found on fewer trees, but were also consistently recorded where they occurred, while the *Tetraponera* spp., *Crematogaster* sp., and others were quite widely distributed, but their presence was erratic. In terms of potential interactions with other insects on the figs *Pheidole* (a native species) was therefore the most important ant in the area, although other species were also important on certain trees.

The behaviour of the ants on the fig-bearing branches is summarized in Table 10. 2. Not all the species were observed tending *Hilda* and exudates from the figs were at least as attractive. These exudates were found in close association with *Hilda* and may have resulted from the feeding activity of the bug. The ants also congregated around the ostioles of ripe figs and preyed on emerging wasps.

Ant exclusion from developing figs

Ant exclusion experiments were carried out on six crops from five of the trees, as described by Compton and Robertson (1988). Pairs of branches

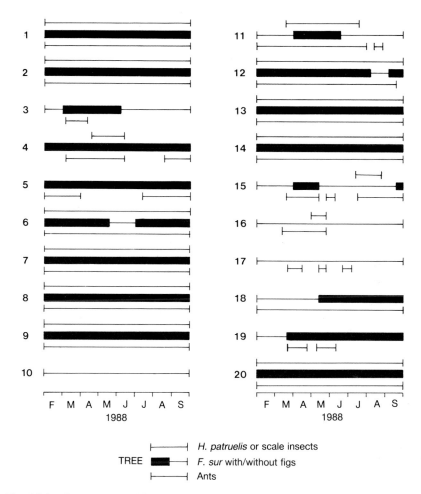

Fig. 10.2. The presence of figs, homopterans, and ants on the reproductive branches of the Cape fig, *Ficus sur*. The homopterans on trees 6, 15, and 16 were scale insects, otherwise *Hilda patruelis*.

bearing young, unpollinated figs were selected, one as a control (with ants) and the other with ants excluded by a band of Formex®. We noted the number of ants on the control branches at intervals and as the figs matured they were harvested and their contents recorded.

The ants were tending scale insects on crop 2, and *Hilda* elsewhere, with mean ant densities varying from 1.2 to 8.1 ants per fig (Table 10. 3). On crop 1, where the highest ant densities were recorded, ants significantly increased the numbers of *Ceratosolen* and *Sycophaga* emerging from the figs at the

Table 10.1. The ant species recorded on the reproductive branches of *Ficus sur* near Grahamstown (33° 22' S, 26° 29' E) between February and September 1988. Species indicated by ++ were present on more than 25 per cent of the observation dates.

Tree no.	1	2	3	4	5	6	7	8	9	10	11	12	13	14	15	16	17	18	19	20	Total
No. of observations	16	19	15	18	14	16	14	14	14	14	14	16	15	16	15	13	14	14	14	15	
Pheidole megacephala		++		++	++	++	++	++	++		++	++	++	++	++			++			13
Tetraponera sp. 1		+		++	++	+		+	+					+					+		8
Crematogaster liengmei		+		+	+			++	++		++					+					7
Camponotus sp. 3		+			++			+	+				+							+	6
Tetraponera sp. 2								++	++			+	++							+	5
Acantholepis capensis								++					++	++						++	4
Polyrachis schistacea	++										+										2
Camponotus sp. 1									+			+									2
Camponotus sp. 4											+						+				2
Camponotus sp. 2			+																		1
Cataulacus intrudens									++												1
Anoplolepis custodiens												++									1
Total no. of species	1	4	1	3	4	2	1	6	7	0	4	4	4	3	1	1	1	1	1	3	

Table 10.2. The behaviour of ants towards homopterans and figwasps on the reproductive branches of *Ficus sur*.

Ant species	Behaviour					
	Tending Hilda	Tending scales	Feeding at fig exudates	Capturing wasps entering figs	Capturing wasps exiting figs	Capturing wasps ovipositing
Pheidole megacephala	+	+	+	+	+	+
Tetraponera sp. 1	+		+			
Crematogaster liengmei	+		+	+		
Camponotus sp. 3			+			+
Tetraponera sp. 2			+	+		+
Acantholepis capensis	+					
Polyrachis schistacea	+					
Camponotus sp. 1						
Camponotus sp. 4			+			
Camponotus sp. 2	+					
Cataulacus intrudens			+			
Anoplolepis custodiens	+					
Total no. of species	6	1	7	3	1	3

Table 10.3. The figwasps emerging from figs on branches where ants were excluded (E) or control branches with ants present (A). Z = normal deviate of U (Mann–Whitney test). The dominant ant species were *Anoplolepis custodiens* on crop 1, *Pheidole megacephala* on crops 2–4, *Acantholepis capensis* on crop 5, and *Polyrachis schistacea* on crop 6. The males of *Apocryptophagus* and *Sycophaga* could not be distinguished (Mean no. of wasps emerging from the figs.).

Fig wasps which oviposit from inside the figs

Crop	Tree	Mean ants/fig	No. of figs		Ceratosolen			Sycophaga		
			E	A	E	A	Z	E	A	Z
1	12	8	5	6	2	45	2.57*	168	561	2.65*
2	8	7	6	2	5	9	1.83	13	88	1.51
3	6	4	6	6	383	472	1.20	0	0	–
4	12	1	4	5	0	16	1.60	248	261	0.61
5	20	1	5	5	65	52	0.52	187	305	1.67
6	1	1	7	7	6	38	0	23	90	2.37*
Mean/fig					82	129		91	226	

Species which oviposit from the outside of the figs

Crop	*Apocryptophagus* spp.			Apocrypta		
	E	A	Z	E	A	Z
1	0	0	—	403	1	2.71*
2	15	25	0.50	58	18	1.83
3	0	0	—	13	1	2.08†
4	6	14	0.64	91	141	0.61
5	56	7	1.67	55	27	1.25
6	6	4	0	124	85	1.02
Mean/fig	13	6		120	48	

*$P < 0.01$. †$P < 0.05$.

expense of the parasitoid *Apocrypta*. Effects of the same magnitude were detected on crop 2, but losses of many of the control figs meant that the results just failed to be statistically significant. On trees with lower ant densities, either no significant differences were present, or they only involved changes in one of the wasp species. None the less, the general pattern was consistent, with figs from ant-excluded branches producing fewer externally ovipositing wasps and more internally ovipositing ones.

The results of the exclusion experiments confirm our earlier findings that ants tending *Hilda* can reduce the levels of parasitism of fig tree pollinators. This results in more pollinators carrying pollen from the trees and so increases their male reproductive potential. However, the degree of protection afforded to the pollinators varied considerably between crops. As these were randomly chosen at various times of the year they provide an estimate of the general impact of ants on the figwasps in the Grahamstown area. Overall, the presence of ants increased the numbers of pollinators emerging from the figs by over 50 per cent and the numbers of the more abundant, internally ovipositing seed predator *Sycophaga* by about 150 per cent. The externally ovipositing species showed a contrasting reduction in numbers, by about 50 per cent for both *Apocrypta* and *Apocryptophagus* spp.

Honeydew-producing scale insects were only numerous on one tree. The abortion of experimental figs on this tree prevented the detection of statistically significant effects of ants on the figwasps, but the results strongly suggest that scale insects can provide *F. sur* with benefits analagous to those of *Hilda*. The abortion of the figs was not typical of the tree as a whole and cannot be ascribed to the scales.

A complexity of interactions: is there a multi-species mutualism?

Whether or not a particular crop will benefit from ants will be determined by factors such as the composition of the figwasp fauna and the densities of the ants. The trees had little to gain from ants when *Apocryptophagus* spp. and *Apocrypta* were rare (for example crop 3). Similarly, when *Sycophaga* and not the pollinator had entered most of the figs (for example crop 4) the ants were favouring a seed predator. When ant densities were low the workers tended to be aggregated around certain figs, rather than spread over the whole branch. Consequently, individual figs could still gain protection, while others on the same branch were ant free and ovipositing figwasps went unmolested.

Although the ants on the fig trees can protect the seeds and pollinators, this does not necessarily mean that the overall effects of the ants are beneficial to the plant and would lead to a tree–bug–ant mutualism. Relationships between the ants and other animals on the trees are complex and only some of the interactions are beneficial to the trees. Ants can influence the reproductive success of *F. sur* via their effects on pollination, seed predation,

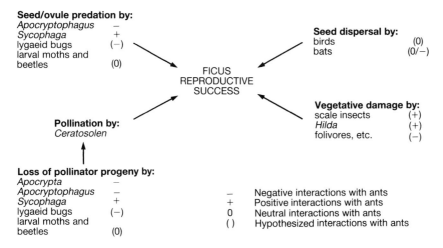

Fig. 10.3. A summary of the ways in which ants can directly or indirectly influence the reproductive success of the Cape fig, *Ficus sur*.

parasitism/predation of pollinator progeny, seed dispersal, and vegetative damage (Fig. 10. 3).

In addition to destroying seeds by galling them, the *Apocryptophagus* spp. also indirectly reduce the numbers of seeds and pollinators produced by the figs. This is because their galls are large, develop rapidly, and expand into the fig lumen, where they interfere with the pollinating females' access to the flowers, and even trap pollinators trying to get through the ostiole (C. Zachariades, personal communication). Ants can therefore improve pollination by reducing oviposition by *Apocryptophagus*, but while doing so may reduce parasitism of *Sycophaga* which, in addition to being a seed predator, competes with the pollinator for oviposition sites. Furthermore, ants also have direct effects because they attack some of the pollinators before they enter the figs.

Other insects which feed on the figs include lygaeid bugs (*Dinomachus* sp.) which pierce the walls of the figs with their extremely long mouthparts, and the larvae of *Botyodes hirsutalis* Walker (Lepidoptera) and *Omophorus* sp. (Coleoptera), which develop inside the figs and destroy most of the contents. The adults of these species are large and, at least in the case of *Omophorus*, appear to ignore the ants. These seed predators not only destroy the healthy seeds, but also kill any figwasp larvae developing in the galls, acting as 'accidental' predators of the wasps (Bronstein 1988). Invertebrate predators of adult figwasps include spiders and flies (Phoridae and Asilidae). Ants, none the less, appear to be the most important predators because they tend to congregate around the exit holes as the wasps are emerging from the figs.

The seeds of *F. sur* are dispersed mainly by birds (Dowsett-Lemaire 1988; I. Waters, personal communication) and bats (Thomas 1988). Ants (including *Pheidole* spp.) are also known to remove fallen fig seeds and may be important additional dispersal agents (Roberts and Heithaus 1986). Conversely, ants on *F. sur* in west Africa have been shown to interfere with bats feeding on figs and thereby reduce dispersal (Thomas 1988). The ant species concerned (*Oecophylla longinoda* Latr.) is absent from the eastern Cape and most of the ants we found on *F. sur* were smaller and appear less aggressive towards vertebrates.

Hilda, like other honeydew-producing homopterans, is presumably protected from parasitoids and predators by its attendant ants. The ants tending the bugs congregated on the fig-bearing branches, were less dense elsewhere on the trees, and did not appear to affect leaf-feeding insects. Scale insects do not show the same preference for fruiting branches and, while this means that attendant ants may be more widely distributed, the scales may be less beneficial to the reproductive biology of the plant.

We have observed *Hilda* on *F. sur* throughout southern Africa, and one of us (S.G.C.) found the bug (or its cogener *H. undata* Walk.) feeding on the same species in Cameroon, west Africa. The effects of *H. patruelis* on *F. sur* in Grahamstown may therefore apply to the whole geographical range of the tree. In addition, we have recorded the bug in large numbers on *F. ingens* (Miq.) Miq., *F. sycomorus* L., *F. trichopoda* Baker, and *F. abutilifolia* Miq. On the latter two species the homopteran favoured the terminal leaf buds as well as the figs, raising the possibility that attendant ants may also lower the levels of leaf damage in these species.

A wider perspective

The insects associated with the figs of *Ficus sur* interact in diverse ways. The complexity of these interactions is increased because the relationships are not constant in time or space, and vary with the local combination of species. How exceptional is this degree of complexity in community interactions? The obligate mutualism between agaonids and fig trees is unusual, but analogues of the other interactions are widespread (see Chapter 8). Spiders feeding in flower heads, for example, can either increase seed survival by feeding on seed predators, or reduce seed setting by feeding on pollinators (Louda 1982). As Addicott (1984) has commented, indirect mutualisms are difficult to detect and may be more prevalent than has been realized.

Acknowledgements

We would like to thank R. Kansky, P. E. Hulley, V. K. Rashbrook, and C. Zachariades for their help and advice.

11

Variation in the attractiveness of lycaenid butterfly larvae to ants

N. E. Pierce, D. R. Nash, M. Baylis, and E. R. Carper

Introduction

Although this volume is primarily concerned with ant–plant interactions, the association between lycaenid butterflies and ants brings an extra dimension to bear on such relationships by introducing an additional trophic level, or, to be slightly more accurate, an additional organism that links two levels. As has been argued by May (1973), the length of a food web can be related to its stability. Communities with many species are likely to be less stable following perturbation than those with fewer species, and variations are amplified as they propagate through the system when the food web is more complex (Pimm 1982). Discussions of stability and complexity have largely focused on food webs whose third-level trophic interactions are essentially antagonistic: for example, the role of predation and parasitism in structuring herbivorous insect communities has been considered at length (Price *et al.* 1980; Strong *et al.* 1984). Somewhat less attention has been paid to how mutualisms may influence community structure and diversity (Addicott 1986; Cushman and Addicott, Chapter 8, this volume).

Mutualism can be defined as an interaction in which the genetic fitness of each participant is increased by the action of its partner. One of the intriguing aspects of mutualisms is that they vary widely, from facultative, indirect and/or diffuse relationships, to highly co-evolved, complex, and often obligate associations. This same variability makes it difficult to derive abstract models of mutualism that are generally applicable. The range in the strength of interactions suggests that the relationship between mutualistic partners is often fragile, subject to environmental fluctuations in which alternative lifestyles may well be more attractive.

The association between lycaenid butterflies and ants represents an ideal system for research on mutualism. The family Lycaenidae is characterized by extreme variety in life histories, ranging from facultative and obligate mutualisms to parasitism. Lycaenids seem to specialize in complex interactions involving other groups of insects, usually ants. Thus, the larvae of some species are carried by attendant ants into the nest where they feed on the ant brood; the males of others use ants as cues in finding pupae that are about to eclose and then engage in frenzied battles for access to mates; the larvae of others are entirely carnivorous, specializing on species of soldier-forming aphids whose soldiers are relatively ineffective against the marauding caterpillars (Cottrell 1984; Pierce 1987).

In a great many lycaenids, larval–ant relationships appear to be mutualistic in ways that are analogous to homopteran–ant associations. While comparisons are useful, it is worth keeping in mind that these interactions differ in at least one essential feature: the secretions produced by lycaenid caterpillars for their attendant ants are derived from specialized glands, and are not a modified excretion, as is the honeydew of homopterans. Lycaenid secretions are expensive to produce, and any caterpillar that achieves the maximum benefit for the minimum cost will be at a distinct advantage relative to its conspecifics. Thus, in some species, selection has favoured the evolution of secretions that mimic ant recognition signals (Henning 1983; Thomas *et al.* 1989).

In this study we investigate how variation in three features of the environment shapes the strength of the relationship between a lycaenid butterfly and ants. We have chosen a complex mutualism that is obligate for one of the partners because it allows us to measure some of the costs and benefits of the interaction in a straightforward manner, and because it enables us to tease apart contributing variables that might be less obvious or more difficult to quantify in a more diffuse interaction. Manipulating various aspects of an obligate mutualism and observing the consequences also provides us with insights as to the mechanisms by which obligate mutualisms are able to persist through such environmental fluctuations. In particular, we describe how variability in the host plants, the attendant ants, and the social organization of the butterflies can affect the mutualism. Having identified some of the ecological hurdles that the butterflies face, we then discuss ways in which they are adapted to survive these challenges.

The system we have been studying is that of the Australian lycaenid, *Jalmenus evagoras* Donovan, the larvae of which feed on species of *Acacia*, and both the larvae and pupae associate with several species of ants in the genus *Iridomyrmex*. While predators and parasitoids of *J. evagoras* are patchily distributed, and attendant ants are more effective against some enemies than against others, the net effect of ant removal is essentially the same: larvae and pupae of this butterfly cannot survive without attendant ants. The colonies that tend larvae and pupae of *J. evagoras* harvest secretions that contain both sugars and concentrated free amino acids; these nutritious rewards result in higher rates of colony growth (Pierce 1983; Nash 1989). Colonies of attendant ant species have alternative food sources and may be found without butterfly associates (Smiley *et al.* 1988). Thus, the relationship appears to be obligate from the point of view of the lycaenids, whereas the attendant ant species receive a net benefit from the association, but can survive without the butterflies.

The natural history of *Jalmenus evagoras*

Jalmenus evagoras has a widespread distribution within Australia, ranging from Melbourne, Victoria in the south to Gladstone, Queensland in the north, and occurring both along the coast and inland on the ranges (Common and Waterhouse 1981). Larvae feed on the foliage of about twenty different species of *Acacia*. Females lay eggs in clusters beneath loose bark, and both the larvae and pupae form large and visible aggregations that are attended by myriad, small, black ants. To date, we have observed at least five different species of *Iridomyrmex* ants tending the juveniles in the field, although two of these are particularly common, a small black one in the *vicinus* Clarke species group (here called *I. vicinus* for convenience), and a larger black one in the *anceps* Roger species group (referred to as *I. anceps*). All the ants that tend *J. evagoras* are closely related, and the majority of other *Iridomyrmex* species found in the same regions are predators of the lycaenid caterpillars.

Despite the broad geographical range of this species, populations of *J. evagoras* are discrete and highly localized, probably as a result of their dependence upon attendant ant colonies and suitable host plants (Smiley *et al.* 1988). Thus, at one field site we were able to observe 74 out of 80 marked, individual butterflies almost daily for their entire estimated lifespans (about three days for females and seven days for males) (Elgar and Pierce 1988). Males of *J. evagoras* tend to eclose several days earlier than females, and the effective sex ratio in the field is usually strongly male biased. Adult males search for mates by regularly investigating trees containing juveniles of the species, and they use ants as cues during this searching process. They hover around clusters of pupae, occasionally appearing to tap them with their antennae. When a pupa is about to eclose, it is not uncommon to find a dozen males clustered around it: after a fierce struggle among the males, the female is mated before she has even expanded her wings.

Both the larvae and pupae of *J. evagoras* secrete food for ants from special-ized exocrine glands, and in return the ants protect them against parasites and predators. Five field experiments conducted over three years have demonstrated that predation and parasitism, largely by invertebrates, is so intense that the larvae and pupae of *J. evagoras* cannot survive in the field without attendant ants, although they can be successfully reared without them in the laboratory (Pierce *et al.* 1987; D. R. Nash and M. Baylis, unpublished work). Not surprisingly, female butterflies use ants as cues when they lay eggs (Pierce and Elgar 1985). Moreover, when given a choice, ovi-positing females can distinguish between different kinds of ants, and respond preferentially to the appropriate species (N. E. Pierce, unpublished work). In addition to guarding juveniles, attendant ants can shorten larval duration by as much as 14 per cent, thereby reducing the time that larvae are exposed to predators and parasitoids (Pierce *et al.* 1987).

The cost of associating with ants is expressed as a reduction in adult weight and size, presumably due to the material given up to ants (Baylis 1989). For example, in one experiment, females that were tended by ants pupated at a weight which was approximately 20 per cent lighter than their untended counterparts (Pierce *et al.* 1987). This difference in size definitely represents a cost: relative size and weight are important components of both lifetime mating success in males, and fecundity in females (Elgar and Pierce 1988).

Attendant ants benefit nutritionally from their association with *J. evagoras*. In an experiment designed to examine the effect of larval secretions on the growth rate of colonies of *Iridomyrmex vicinus*, we found that the size of egg masses laid by queens of colonies which foraged on an artificial diet supplemented with secretions from five caterpillars was larger than that laid by queens of colonies which were supplemented with secretions from only one caterpillar. Both were larger than egg masses laid by queens whose colonies foraged on an artificial diet alone. Colony growth rates, as measured by the relative increase in numbers of larvae, pupae, and workers over a three-month period were higher for colonies fed artificial diet supplemented with secretions from either one or five caterpillars, than for colonies fed the artificial diet alone (Nash 1989).

The effect of host plant quality on larval attractiveness to ants

In 1878, Edwards commented upon the larval secretions of the North American lycaenid, *Celastrina* (*Lycaena*) *pseudargiolus*, 'it is probable that the quality of this . . . secreted fluid . . . and perhaps its attractiveness depends on the nature of the food plant' (p. 8). Since the food rewards that lycaenid larvae provide for their attendants are derived from their host plants, it is reasonable to suspect that variation in host plant quality might alter the quality of the rewards and hence the number of attendant ants (see Chapter 9). As the secretions of some lycaenids are known to be rich in proteins as well as carbohydrates, the protein concentration of the host plant may be of particular importance to the relationship. For example, the Lycaenidae as a whole are well known for their predilection for feeding on flowers, seed pods, and terminal foliage (all the protein rich parts of host plants) (Mattson 1980; Robbins and Aiello 1982). The larvae of *Glaucopsyche lygdamus* Doubleday attract more attendant ants when feeding on the seed pods of their host plant, *Lupinus floribundus* Greene, than they do when feeding on other parts of the plant, perhaps because the seed pods represent higher quality food (Pierce and Easteal 1986).

We investigated this question experimentally by manipulating the quality of the host plants consumed by larvae of *J. evagoras*, and observing the subsequent effect on ant attendance and larval survivorship. Young, potted food plants of *Acacia decurrens* Willd. were either given water containing a nitrogenous fertilizer, or were given water alone. Fertilized plants had a higher nitrogen content than their unfertilized counterparts. Under natural, field conditions, fifth instar larvae of *J. evagoras* feeding on fertilized plants attracted a significantly larger number of attendant ants (mean ± s.e.; 7.9 ± 0.8) than those feeding on unfertilized plants (6.2 ± 0.7). In the absence of caterpillars, ants were not differentially attracted to either plant type. Not only did caterpillars on fertilized plants attract a larger ant guard, but they also survived better in the field over a ten-day period than did larvae on unfertilized plants (Fig. 11. 1). However, larvae reared in a screened bush house in the absence of ants and predators survived equally on fertilized and unfertilized plants. We concluded from this that larvae on fertilized plants survived better than their counterparts on unfertilized plants by attracting a larger ant guard (Baylis and Pierce 1991). Not surprisingly, we also found that females of *J. evagoras* preferred to lay egg batches on fertilized, rather than unfertilized plants (Baylis and Pierce 1991).

The effect of host plant quality on ant attendance and subsequent larval survival was both strong and immediate. Considering the importance of attendant ants to the survival of the larvae, we might have expected selection to favour larvae whose secretions were more constant and predictable despite fluctuations in host plant quality. However, the extreme sensitivity of the female butterflies to the condition of prospective host plants indicates that the butterflies do indeed possess a powerful mechanism for responding to this level of variation in their environment.

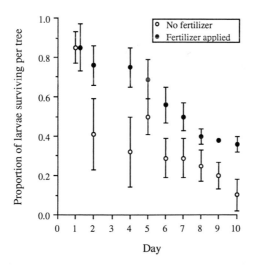

Fig. 11.1. Survival of larvae of *Jalmenus evagoras* as a function of host-plant quality. Points represent means, and error bars the standard errors of the proportion of larvae surviving on each of three plants per treatment (data for only one site shown; two-way ANOVA of treatment and site on proportion surviving per plant: day six, variance ratio = 4.40, d.f. 1,8; $P = 0.07$; day seven, variance ratio = 4.00, d.f. 1,8; $P = 0.08$). (Redrawn from Baylis and Pierce 1991.)

The effect that host plant quality can have on lycaenid–ant interactions is reflected in a more evolutionary sense by a comparison of the host plant usage of species that do and do not associate with ants. A survey of 297 species of Lycaenidae revealed that ant association is strongly correlated with the consumption of protein-rich, nitrogen-fixing plants. Thus the larvae of lycaenid species that feed on legumes are much more likely to associate with ants than the larvae of lycaenids that feed on other kinds of plants. Similarly, of the larvae of lycaenids that feed on non-leguminous plants, those on nitrogen-fixing plants are more likely to be ant-tended than those species that feed on non-nitrogen fixing plants (Pierce 1985).

The effect of ant nutrition and foraging behaviour on levels of tending

Variation in ant attendance of individual lycaenid larvae depends in part upon the spatial arrangement of each partner. Since ants that tend larvae are rewarded by food secretions, foraging theory would predict that the time spent by ants tending lycaenids, and hence the rates of ant traffic to and from those lycaenids, will vary as a function of foraging distance. We found this to be the case. Field experiments using larvae on potted host plants, placed at different distances from a central ant nest, showed that larvae placed at greater distances elicited lower flow rates of ants, and were more likely to be preyed upon than larvae placed closer to the ant nest (Carper 1989; Fig. 11. 2).

In addition to spatial effects, colony size may be an important determinant in the number of ants available to tend larvae and pupae. Laboratory experiments demonstrated that colony size had a strong effect upon levels of attendance of larvae (Nash 1989). Presumably for young colonies, colony size is often correlated to colony age.

Finally, the nutritional state of the attendant ant colony can influence the attractiveness of individual larvae and pupae in complex ways. Some of our field work has focused on the behaviour of individual ant colonies as a function of their overall diet. In order to do this, we observed ant colonies in the field that were engaged in tending larvae and pupae of *J. evagoras*, and we recorded whether these colonies were also tending homopterans. As a rough measure of how much food each colony was harvesting from each food source, we scored the number of workers foraging on homopterans and the number foraging on larvae and pupae of *J. evagoras*, and in cases where a resource had more than thirty attendant ants, we measured the rate of flow of ants travelling to and from the resource. We then laid out a grid of one meter squares covering the area occupied by the ant colony, baited the vertices of the grid with sugar and canned tuna baits, and scored the number of ants foraging from these baits. We found that colonies that were predominantly tending larvae and pupae of *J. evagoras* sent out more workers to sugar than to tuna baits, whereas colonies that were foraging pre-

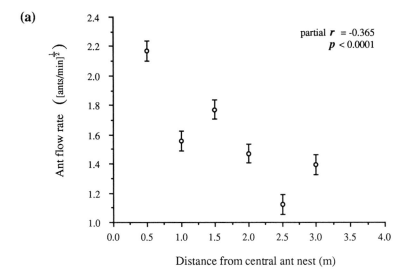

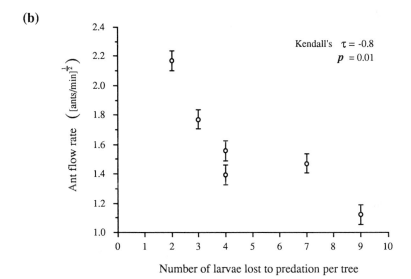

Fig. 11.2. The correlations between distance, flow rate of ants, and survival of larvae of *Jalmenus evagoras*. (Redrawn from Carper 1989.) (a) Variation in flow rates of ants to and from trees bearing *J. evagoras* as a function of ant foraging distance. Ant flow rate (transformed as the square root of the numbers of ants running to and from a tree per minute) was measured for potted host plants of *Acacia decurrens* containing six fifth-instar larvae of *J. evagoras* and placed at different distances from a nest of the ant, *Iridomyrmex anceps*. (b) Predation as a function of flow rates of *Iridomyrmex anceps* ants to and from *Acacia decurrens* trees bearing *Jalmenus evagoras*. The total number of larvae of *J. evagoras* lost by the trees described above over a two-week period are plotted as a function of ant flow rate to each tree. Whenever a larva disappeared, it was replaced with another. The total number of larvae lost per tree was negatively correlated with the mean ant flow rate to that tree.

dominantly from homopterans were more attracted to the tuna baits (Nash 1989; Fig. 11. 3).

From these observations we tentatively conclude that the nutritional needs of the colony as a whole can best be met by feeding on both carbohydrates (homopteran honeydew and sugar) and proteins (lycaenid larvae secretions and tuna fish) (Brian 1973), and that the colonies we examined appeared to be compensating for a preponderance of one or the other of these nutrients in order to achieve a balanced diet. However, it is possible that the secretions of lycaenid larvae contain nutrients other than proteins that are important for colony growth, and that these nutrients are analogous in some way to nutrients provided by the canned tuna baits.

While this experiment did not address individual attractiveness of larvae and pupae, the results have clear implications for an important source of variation: the nutritional state of an ant colony (in which the brood in the nest can be regarded as a collective stomach) may influence the relative attractiveness of individual lycaenid larvae to potential attendant ants if there is varia-

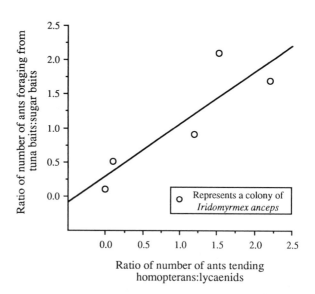

Fig. 11.3. Ant-foraging preferences as a function of the availability of alternative food resources. The five points represent five colonies of the ant *Iridomyrmex anceps* mapped at a field site at Cloud's Creek, New South Wales, Australia. The ratio of workers foraging from homopteran honeydew and from lycaenid larvae secretions is plotted against the ratio of the number of ants from the same colony that were sent to sugar and tuna fish baits. When colonies were foraging primarily on homopteran honeydew, they preferred canned tuna baits to sugar ones; however when they were foraging primarily on the secretions of the lycaenid, *J. evagoras*, they preferred sugar baits to canned tuna ones ($y = 0.81x + 0.246$; $r^2 = 0.824$; $P < 0.05$). (Redrawn from Nash 1989.)

tion in the nutritional composition of their secretions. Lycaenids such as *J. evagoras* whose juveniles rely upon ants for survival are in competition with the other food sources available to their attendant ants, and alternative food sources may thus exert profound indirect effects on the survival of the butterflies. Presumably, there is strong selection for lycaenids to secrete nutritional rewards that are somehow limiting in their attendant ant's diet, and therefore valuable. There may also be selection on the lycaenids to produce a complete diet for the ants (see Chapter 18) so that they do not need to spend time elsewhere. It might also be interesting to investigate whether colonies that are in the process of producing alates have different dietary requirements than those producing workers, and whether these needs are reflected in the numbers of workers attracted to lycaenid larvae and pupae (Brian 1973).

Cumulatively, these results suggest that the characteristics that make an ant a valuable mutualistic partner are not only interspecific differences such as the aggressiveness, size, and/or abundance of available workers. In addition, variables such as colony size, age, alternative food sources, growth rate, and/or ratio of foragers to brood may all contribute to the level of attendance by ants. It is reasonable to conclude that populations of obligate mutualists such as *J. evagoras* are constrained not only spatially by the distribution of appropriate host plants and attendant ants, but also temporally by the phenology and growth characteristics of both host plants and attendant ant colonies, and in a more indirect way by the spatial and temporal distribution of alternative food sources such as homopterans. Although we know that females of *J. evagoras* can detect the presence or absence of appropriate ant associates and use them as cues in oviposition, we would predict from these results that they must also be sensitive to differences in ant densities and perhaps to the distribution of other trophobionts.

The effect of conspecifics on individual variation in larval attractiveness

As described earlier, the larvae and pupae of *J. evagoras* aggregate, and this poses some interesting questions about individual variation in attractiveness to ants. If a threshold number of ants is necessary to protect the larvae and pupae, then aggregating may be one mechanism by which *J. evagoras* could simultaneously increase its collective defence and decrease the amount of food that each individual would need to produce to attract that defence. For example, our field observations showed that first-instar larvae could conceivably gain more ants by joining a group of any size than by remaining alone, and second or third instars could have a higher number of attendant ants by joining a mean sized group of about four larvae (Pierce *et al.* 1987).

In order to examine whether larvae selectively form groups in response to ant densities, we investigated the decisions made by final-instar larvae which were about to pupate. We first determined that aggregations of pupae are indeed tended by more ants per pupa than equivalent numbers of pupae placed a similar mean distance from a central ant-nest, but in dispersed rather than clumped configurations (Carper 1989). Thus, a final-instar larva can increase its level of ant attendance as a pupa by pupating beside another individual rather than pupating alone. We then compared the pupation choices made by larvae on potted host plants where larval and/or pupal number was kept constant, but the flow rates of ants foraging to and from the plants varied as a function of distance. We found that larvae were extremely sensitive to the flow rates of ants on plants and adjusted their behaviour accordingly. Given a choice between pupating beside another individual or pupating alone, a larva was more likely to join a cluster when the mean ant flow rate to the tree was relatively low (about three ants per minute), whereas it pupated by itself when the mean flow rate was high (about six ants per minute) (Table 11. 1). Similarly, a larva faced with the choice of pupating with a smaller or larger cluster of pupae joined the larger cluster when ant flow rates were low, but joined the smaller of the two clusters when ant flow rates were high (Carper 1989).

Table 11.1. Pupation decisions of larvae on trees with differing densities of attendant ants.

	Low ant flow rate	High ant flow rate
Pupate alone	0	4
Pupate in a group	15	7

When the ant flow rate to a tree was low (< 3.18 ants/min, the mean ant flow rate), a larva was more likely to pupate beside other pupae. When the ant flow rate was high (> 3.18 ants/min), a larva was more likely to pupate alone (Chi-square = 3.96 with continuity correction; $P = 0.047$; $N = 26$) (from Carper 1989).

In the field, it is not uncommon to observe that small larvae have pupated near the base of a host plant, directly in the trail of ants leading to large aggregations of pupae. Presumably these individuals receive a benefit from pupating in a location with a higher ant density than they might have been able to attract on their own. However, it would also appear that these larvae are simply making the best of a bad job. There are likely to be benefits gained by pupating at a higher position in a tree, perhaps related to a tradeoff in the costs and benefits of ant attendance, or perhaps to other factors such as the ease of discovery upon eclosion by potential mates. We found that large larvae which attracted many ants were more likely to pupate in a relatively high position on a host plant than their smaller, less attractive counterparts (Carper 1989).

These complex behavioural responses suggest equally complex interactions between ant attendance and social context for the juveniles of *J. evagoras*. Consider the following possibility. The cost to larvae of ant attendance is pupating at a smaller size. If individual larvae and pupae vary in their attractiveness to ants, then there may be a kind of auto-mimicry taking place within aggregations of larvae and/or pupae whereby certain 'defecting' individuals in the population take advantage of 'co-operating' individuals. Defectors are those larvae and/or pupae that aggregate with their cooperative counterparts and enjoy the benefits of ant attendance, but do not secrete food in return and hence do not pay the costs. Such defection might be less likely if co-operative individuals aggregated preferentially with their kin. Nevertheless, selection could still favour an individual that adopted a strategy that involved not secreting, but aggregating with non-kin. This could result in a mixed ESS (evolutionary stable strategy) in which the ratio of defecting to co-operating individuals would be governed, in a frequency-dependent fashion, by the effectiveness of the ant guard and the predator and parasitoid pressure of the environment. Of course, this scenario would sound less far-fetched and more interesting if we could establish whether the ants can respond directly to the rates of secretion of caterpillars, and whether attractiveness to ants is a genetically variable trait in *J. evagoras*. This is work that is currently in progress.

In summary, the aggregation behaviour of larvae and pupae of *J. evagoras* allows the larvae to exert some influence over the level of ant attendance on the particular tree upon which they find themselves. Thus, if ant densities are low, they can join together to increase their per capita tending levels, whereas if ant densities are high, they can spread out, perhaps thereby reducing some of the costs.

Conclusions

The experiments described here illustrate how variation in three aspects of the interaction between lycaenid butterflies, their host plants, their attendant ants, and their predators and parasitoids can have ramifications at other levels in the system. Our analysis has taken the perspective of the butterfly. Since larval survival is correlated with the level of ant attendance, any perturbation that subsequently alters the level of attendance provides us with a powerful bioassay.

In each case, the butterflies described in this study are buffered against variability in their environment by their behavioural flexibility. This complexity is partly the result of the holometabolous life history of butterflies, in which the constraints of the relatively sessile larval stage are compensated for by the free-flying, and hence dispersive, adult stage. Thus, we determined that if one alters the quality of the host plant, the level of ant attendance and hence larval survival also changes. The primary mechanism by which the butterflies appear to cope with this variation is through their selection of host plant. Females are extremely sensitive to the quality of their prospective host plants.

Similarly, if one alters the colony size, foraging distance, or nutritional state of the attendant ant colony, one may also affect the level of ant attendance and larval survival. Again, the mechanism that enables the butterfly to have some control over this variability is in its choice of oviposition sites: females can distinguish between different ant species and use ants as cues in laying eggs. We would further predict that females may respond to ant density and the distribution of alternative food sources (such as homopterans) when ovipositing.

Finally, the social behaviour of the larvae and the pupae also appears to affect levels of ant attendance in complex ways. Variation in the host plant quality and attendant ant colonies may affect the butterfly's overall distribution and clumping patterns, since females are more likely to lay eggs on high-quality host plants that also contain ants. However, even when a larva hatches out upon a particular host plant, its propensity to aggregate with conspecifics can affect its numbers of attendant ants. Joining or not joining a group is an additional fine-tuning mechanism by which a larva can adjust the level of ant attendance.

Acknowledgements

We thank A. J. Berry, R. L. Kitching, J. R. Krebs, R. M. May, D. I. Rubenstein, T. R. E. Southwood, and M. F. J. Taylor for their assistance during various stages of the work. M. J. Magrath assisted with the preparation of the manuscript, and R. Bondi kindly made the illustrations. N.E.P. was supported by NSF grant BSR-8705483, M.B. and D.R.N. held studentships from the Natural Environmental Research Council, and E.R.C. was sponsored by Princeton University's Field Studies Program.

12

Evolutionary and ecological patterns in myrmecophilous riodinid butterflies

P. J. DeVries

Introduction

The ability to form mutualisms with ants has been well documented in many groups of plants, and amongst insect herbivores in the Lepidoptera and Homoptera. In general, in both plants and insects the ability to provide secretions to ants is a fundamental part of maintaining these mutualistic associations. In these systems, insects provide ants with secretions directly, through specialized organs (Way 1963; Cottrell 1984), and plants generally provide secretions to ants directly through extrafloral nectaries or pearl bodies (Beattie 1985; Schupp and Feener Chapter 13, this volume; Koptur Chapter 15, this volume).

In contrast to all other butterfly lineages, only the monophyletic lycaenoids have evolved the ability to form associations with ants. The lycaenoid butterflies comprise two monophyletic groups–the lycaenids and the riodinids (Ehrlich 1958; Eliot 1973; Kristensen 1976; Harvey 1987), both of which include myrmecophilous and non-myrmecophilous taxa. Differences of opinion exist only with respect to whether lycaenids and riodinids should be considered as separate families (*sensu* Eliot 1973) or as subfamilies (*sensu* Ehrlich 1958); but see also Robbins (1988).

The lycaenoids account for a major fraction of all butterfly species (Vane-Wright 1978), and their diversity is reflected in their life histories. Lycaenoid caterpillars may be phytophagous, carnivorous, or both. If associated with ants, their relationship may be parasitic, mutualistic, commensal, or combinations of these (Pierce 1987; Thomas *et al.* 1989). Alternatively, they may have nothing to do with ants at all (Cottrell 1984). In the Lycaenidae, myrmecophily and caterpillar biology has been documented extensively for many species (see reviews and citations in Lamborn 1915; Farquarson 1922; Jackson 1937; Hinton 1951; Malicky 1970; Henning 1983; Cottrell 1984; Pierce and Elgar 1985; Pierce 1987; Pierce *et al.* 1987; Fiedler and Maschwitz 1988*a,b*, 1989; Thomas *et al.* 1989). In the Riodinidae, however, caterpillar myrmecophily and biology has been documented in relatively few species (Bruch 1936; Borquin 1953; Ross 1966; Callaghan 1977, 1986; Schremmer 1978; DeVries 1987, 1988*a*; see also Harvey 1987).

The framework for our understanding of the evolution of myrmecophily in butterflies comes from general studies that, typically, lump the riodinids and lycaenids together (Hinton 1951; Malicky 1970; Atsatt 1981; Pierce 1984, 1985, 1987). As a result, our perception of myrmecophily in riodinids has, with few exceptions, been inferred from what we know about lycaenids. My principal aim here is to use the natural history of riodinid butterflies to explore three inter-related ideas concerned with the evolution of ant association in butterflies and other organisms. The first is to examine morphological and biological features of riodinids and lycaenids which suggest that myrmecophily in butterflies has evolved independently at least twice–once in the riodinids and once in the lycaenids. The second is to focus on ecological patterns found in riodinid butterflies and describe aspects of the host plant use by myrmecophilous species. Finally, features typical of myrmecophilous riodinid and lycaenid biology are used to suggest patterns that may be common to ant associations found in Homoptera and plants.

Larval ant organs

Myrmecophilous lycaenid and riodinid caterpillars typically possess adaptations for associating with ants. The most familiar of these adaptations are various derived organs (hereafter called ant organs) that provide ants with food or stimuli that modify ant behaviour (summarized in Malicky 1970; Henning 1983; Cottrell 1984; DeVries 1988a; Fiedler and Maschwitz 1988a,b, 1989). The complement of larval ant organs is variable among myrmecophilous lycaenid taxa (Cottrell 1984), but much less so among riodinid taxa. In other words, a lycaenid caterpillar of one species may bear only one organ, the caterpillar of another lycaenid species may bear two or more organs, whereas the myrmecophilous caterpillars of all riodinid species (excepting the genus *Eurybia*) typically bear a full complement of ant organs. However, it is clear that the ant organs of both lycaenids and riodinids play important roles in mediating the mutualisms between caterpillars and ants. Following the terminology of Cottrell (1984), the major ant organs in lycaenid and riodinid caterpillars and their functions are briefly described here.

1. The *dorsal nectary organ* in lycaenids (Malicky 1970; Clark and Dickson 1971; Maschwitz *et al.* 1975; Claassens and Dickson 1977; Pierce 1983; Henning 1983; Fiedler and Maschwitz 1988a, 1989) and *tentacle nectary organs* in riodinids (Borquin 1953; Ross 1966, Callaghan 1977, 1982; Schremmer 1978; Horvitz *et al.* 1987; DeVries 1988a; De Vries and Baker 1989) provide secretions of sugars and amino acids upon solicitation by ants. This type of ant organ is possessed by all myrmecophilous taxa of riodinids, but not all lycaenid taxa (Cottrell 1984; Harvey 1987).

2. The *tentacle organs* in lycaenids (Malicky 1970; Claassens and Dickson 1977; Henning 1983; DeVries 1984; Fiedler and Maschwitz 1988b) and the

anterior tentacle organs in riodinids (Ross 1966; Callaghan 1977; De Vries 1988*a*) appear to produce chemical stimuli that modify the behaviour of the attending ants. This type of organ has been found on caterpillars of all myrmecophilous riodinids, except the genus *Eurybia* (Harvey 1987; Horvitz *et al.* 1987), but not all lycaenids (Cottrell 1984).

3. The *vibratory papillae* are two mobile, chitinized rods found on caterpillars of all known myrmecophilous riodinid taxa, except the genus *Eurybia* (Harvey 1987; Horvitz *et al.* 1987). These organs produce sounds that appear to attract ants (Ross 1966; DeVries 1988*a*, 1990). Vibratory papillae are not found on any species of lycaenid caterpillar (Cottrell 1984).

4. The modified setae termed *perforated cupola organs* (PCOs) are minute pits found scattered across the epidermis of both lycaenid and riodinid caterpillars, which are thought to secrete appeasement substances or food to attending ants (Malicky 1970; Pierce 1983, 1984). These organs have been found on all lycaenid and riodinid taxa (both myrmecophilous and non-myrmecophilous) that have been examined (Malicky 1970; Kitching 1983; Cottrell 1984; Kitching and Luke 1985; DeVries *et al.* 1986; Harvey 1987).

5. A few species of myrmecophilous lycaenid and riodinid caterpillars possess derived organs whose function is not well understood; *dish organs* and *bladder setae*. These are mentioned here for completeness, but will not be discussed. The dish organs are a series of depressions found on the dorsum of some aphnaeine lycaenids (e.g. *Spindasis*, *Crudaria*) that produce a liquid imbibed by ants (Cottrell 1984). The bladder setae (see Fig. 12. 1(b)) are found on segment T-1 (tergite-1) of the nymphidiine riodinid genera *Theope* and *Nymphidium*. These setae are sometimes antennated or grasped by the mandibles of ants (Harvey *et al.* 1990; P. J. DeVries, personal observation).

The fact that riodinid and lycaenid ant organs with the same functions are found on different segments has, apparently, not been fully appreciated by investigators concerned with lycaenoid evolution. For example, the single, non-eversible dorsal nectary organ found on segment A-7 of some myrmecophilous lycaenids may be analogous, but not homologous, to the pair of eversible tentacle nectary organs on segment A-8 of all myrmecophilous riodinids (Fig. 12. 1(a(–(d)). The pair of eversible tentacle organs found on segment A-8 of lycaenids may be analogous, but not homologous, to the pair of eversible anterior tentacle organs on segment T-3 of riodinids (Fig. 12. 1(a),(b),(d)). Although the vibratory papillae are found only in riodinid caterpillars (Fig. 12. 1(a),(b)), both myrmecophilous lycaenids and riodinids have the ability to produce low-amplitude sound (DeVries 1988*a*, 1990, unpublished work). Thus, the ability to produce sound in lycaenids and riodinids is analogous, but not homologous. The only organs that are clearly homologous in lycaenids and riodinids are the PCOs, but these are also found among non-lycaenoid Lepidoptera (see below).

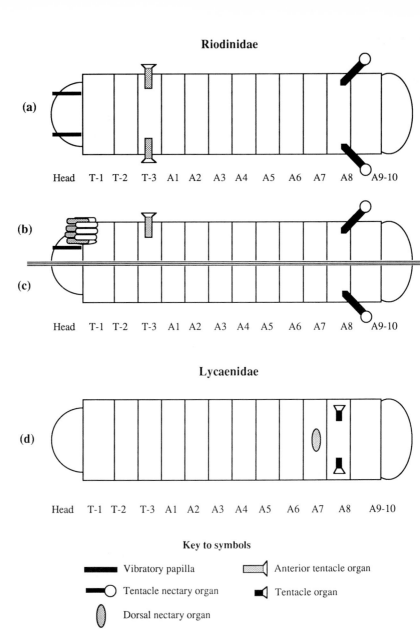

Fig. 12.1. Schematic diagrams of myrmecophilous riodinid and lycaenid caterpillars showing positions of the ant organs. All diagrams present a dorsal view. (a) Typical riodinid caterpillar of the tribe Lemoniini showing a full complement of ant organs: vibratory papillae (attached to segment T-1), a pair of everted anterior tentacle organs (segment T-3), and a pair of everted tentacle nectary organs (segment A-8). (b) One-half of a typical riodinid caterpillar of the tribe Nymphiniini showing the full complement of ant organs, plus the corona of bladder setae (segment T-1). (c) One-half of a caterpillar of the genus *Eurybia* (Riodinidae) showing no vibratory papillae or anterior tentacle organs. (d) Typical lycaenid caterpillar showing the full complement of ant organs: a single dorsal nectary organ (segment A-7) and a pair of everted tentacle organs (segment A-8).

The evolution of butterfly myrmecophily

Considering their function and correlation with myrmecophily, a defensible conjecture is that derived ant organs have evolved under selection for caterpillar–ant symbioses. Comparing the homologies of ant organs on lycaenoid caterpillars allows at least four hypotheses about the evolution of myrmecophily in butterflies (Fig. 12. 2) to be evaluated. There are three assumptions:

1. based on all non-lycaenoid Lepidoptera as an outgroup, most traits associated with myrmecophily are derived;
2. the lycaenids and riodinids are sister taxa (Kristensen 1976; Harvey 1987; but also see Robbins 1988); and
3. both lycaenids and riodinids contain myrmecophilous and non-myrmecophilous taxa.

The four hypotheses are as follows.

Hypothesis 1: ancestral myrmecophily. Myrmecophily is ancestral in the lycaenoids and a minimum of two independent losses have occurred—one in the lycaenids and one in the riodinids (Fig. 12. 2(a)).

The possession of perforate cupola organs (PCOs) has been considered explicitly (Hinton 1951; Malicky 1970) and implicitly (Vane-Wright 1978; Pierce 1987) as support for ancestral myrmecophily: the most widely held hypothesis about lycaenoid evolution. However, two reasons suggest that PCOs may be inappropriate characters. First, the possession of homologous, secretory PCOs by the large, entirely non-myrmecophilous family Hesperiidae shown by Franzl *et al.* (1984) weakens PCOs as a lycaenoid synapomorphy, and as a strong character indicating myrmecophily in butterflies. (Incidentally, Scudder (1889) and Frowhawk (1934) recognized that lycaenid and hesperiid caterpillars bear PCOs.) Secondly, despite the importance placed on them, the role played by PCOs in myrmecophily remains to be demonstrated for the vast majority of lycaenid and riodinid species (e.g. DeVries 1988*a*; Fiedler and Maschwitz 1989). Since most lycaenid and riodinid ant organs are analogous, but not homologous, this hypothesis is rejected.

Hypothesis 2: once arisen. Myrmecophily has evolved once from non-myrmecophilous lycaenoid ancestors and gave rise to both lycaenids and riodinids (Fig. 12. 2(b)); implying polyphyly of both riodinids and lycaenids. This hypothesis is rejected because lycaenid and riodinid ant organs are not homologous.

Hypothesis 3: twice arisen. Myrmecophily evolved twice in non-myrmecophilous lycaenoid ancestors; independently in the lycaenids and in the riodinids (Fig. 12. 2(c)). This simple hypothesis is consistent with the

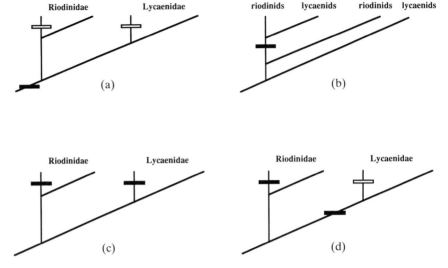

Fig. 12.2. Four alternative hypotheses for the evolution of myrmecophily in lycaenoid butterflies. Solid bars indicate the acquisition of myrmecophily, hollow bars indicate a loss of myrmecophily. (a) The most widely held hypothesis indicating that myrmecophily was primitive to the lycaenoids, and then lost in some lineages. This hypothesis is at variance with the homologies of larval ant organs. (b) A variation of hypothesis (a) indicating that lycaenoids evolved once from non-myrmecophilous ancestors. This hypothesis is also at variance with the homologies of larval ant organs. (c) An hypothesis indicating that myrmecophily evolved twice from non-myrmecophilous ancestors. This is the simplest hypothesis that is consistent with evidence from the homologies of larval ant organs. (d) A more complicated hypothesis than (c) suggesting that myrmecophily evolved twice from non-myrmecophilous ancestors, but was lost in some lycaenid lineages. This is also consistent with the homologies of larval ant organs.

homologies of ant organs, and agrees with the current classifications of riodinids and lycaenids (Eliot 1973; Kristensen 1976; Harvey 1987).

Hypothesis 4: twice arisen, once lost. An alternative hypothesis that is consistent with the homologies of ant organs suggests that myrmecophily evolved twice, independently, but that the trait was lost again in some lycaenids (Fig. 12. 2(d)). Note that more intricate hypotheses can be constructed that are all equally parsimonious and consistent with the homologies of ant organs. Such hypotheses might argue that myrmecophily could have arisen at least twice, independently, and then been lost two, three, or more times.

The hypotheses presented in Fig. 12. 2 ignore complications that might arise with a larger, and more complete character set. The hypotheses are not meant to represent resolved phylogenies, but are intended to question our current views about the evolution of myrmecophily in lycaenoid butterflies. Tracing the evolutionary history of myrmecophily will ultimately depend on support provided by a well wrought phylogeny of the lycaenoids and butterflies as a whole. Unfortunately, none of the current classifications (Ehrlich 1958; Kristensen 1976; Eliot 1973; Harvey 1987; Robbins 1988) provide such a base. None the less, comparing shared–derived larval ant organs suggests that butterfly myrmecophily has evolved independently at least twice from non-myrmecophilous lycaenoid ancestors (Fig. 12. 2). In the light of this it is reassuring to note that myrmecophily has evolved several times within several families of Homoptera, Coleoptera, and plants (e.g. Beattie 1985; Woodward *et al.* 1970; Wilson 1971; Schupp and Feener Chapter 13, this volume), all of which bear analogous, but not homologous organs for ant association.

Distribution of riodinid and lycaenid species

Zoogeographical patterns suggest further differences between lycaenids and riodinids. Lycaenid and riodinid species are found on all major temperate and tropical land masses and account for approximately 40 per cent of the total butterfly species richness (Vane-Wright 1978; Robbins 1982). Available estimates show that lycaenid species richness is greatest in the tropics, and that different tropical regions have similar total numbers of species (Robbins 1982, personal communication). A world-wide estimate for riodinids assembled from Stichel (1930–31), D'Abrera (1977, 1980, 1986), Scott (1986), and Harvey (1987) also shows a high tropical species richness, but a completely different distributional pattern from the lycaenids (Table 12. 1). Of over 1200 known riodinid species, only 8 per cent occur outside the neotropics.

Proportion of myrmecophilous lycaenoid species

It has been assumed that myrmecophily is as widespread within the riodinids as it is in the lycaenids (Hinton 1951; Callaghan 1977; Robbins and Aiello 1982; Pierce 1987). If myrmecophily has been a major force in amplifying diversity and species richness in lycaenids (Pierce 1984, 1987), it might be expected that the same is true of the Riodinidae and, therefore, that myrmecophilous riodinid species would outnumber those species that do not associate with ants. Two recent studies suggest the opposite.

As in caterpillars of the lycaenid subfamily Lipteniinae (Cottrell 1984), the density and length of body setae on riodinid caterpillars, typically, separate species that associated with ants from those species that do not (e.g. DeVries 1988*b*); 'naked' riodinid caterpillars are myrmecophilous and bear ant organs, 'fuzzy' caterpillars have no ant organs, except pore cupola

Table 12.1. The zoogeography of lycaenid and riodinid butterfly species richness.

	Australian	Asian	Palaearctic	African	North American	Neo-tropical
Lycaenidae	420	1200	>95	1300	>100	1100
Riodinidae	21	32	10	13	20	1200

organs, and are non-myrmecophilous (Fig. 12. 3). Using morphological characteristics of neotropical caterpillars one comparison between eight of the largest myrmecophilous genera and eight of the largest non-myrmecophilous genera suggested that there were almost twice as many species in non-myrmecophilous genera of riodinids than myrmecophilous ones (DeVries 1987). A second, and broader method of estimating myrmecophily in riodinids used taxonomic affinity. Of the five currently recognized subfamilies of the Riodinidae, only three of the 11 tribes in the entirely American subfamily Riodininae are known to be myrmecophilous, and all other riodinid subfamilies (which account for 75 per cent of the species richness) were predicted to be non-myrmecophilous (Harvey 1987). In other words, only 25 per cent of all riodinid species are myrmecophilous, and myrmecophilous species occur *only* in the neotropics. Together, these two methods of comparison strongly imply that ants have played a very different role in the evolution of the riodinids and lycaenids. Thus, previous assumptions about riodinid myrmecophily based on data from lycaenids may be invalid.

Patterns of host use in riodinids

Whereas the life histories of phytophagous lycaenids are richly varied with respect to the plant parts that their caterpillars eat (Ehrlich and Raven 1965; Robbins and Aiello 1982; Cottrell 1984), a summary of widely scattered host records by Harvey (1987) showed that riodinids are comparatively boring (they typically feed only on leaf tissues). However, riodinid host records are less boring if divided into host plants of myrmecophilous and non-myrmecophilous riodinid species, and then compared with respect to whether these plants have extrafloral nectaries (EFNs) or not. This was done by supplementing the records in Harvey (1987) with additional records from Costa Rica, Panama, and Ecuador (P. J. DeVries, unpublished work), pooling all multiple host records of a given riodinid species into a single category (either with or without EFNs), and discarding records where the presence or

Fig. 12.3. Neotropical riodinid caterpillars. (A) Non-myrmecophilous caterpillar of *Anteros formosus* from Panama. Note the long hairs that keep ants away from the body of the caterpillar. (B) Two fourth-instar *Thisbe irenea* caterpillars in Panama being tended by an *Ectatomma ruidum* ant. Both caterpillars are drinking from the extrafloral nectary (EFN) that is visible just to the left of where the leaf petiole meets the leaf blade. (C) A third-instar Panamanian *Synargis mycone* caterpillar drinking from the EFN while being guarded by an *E. ruidum* ant. (D) A fifth-instar Ecuadorian *Nymphidium* sp. caterpillar drinking from an EFN while being tended by *Solenopsis* sp. ants.

absence of EFNs could not be ascertained. The result suggests that myrmecophilous riodinid species use hosts with EFNs, whereas non-myrmecophilous species use hosts without EFNs (Fig. 12. 4). The feeding habits of caterpillars may explain part of this pattern.

The myrmecophilous caterpillars of *Thisbe irenea* spend significant amounts of time with their heads over EFNs (Fig. 12. 3(b)). In addition to feeding on leaf tissue, *T. irenea* caterpillars gain substantial growth benefits by drinking the extrafloral nectar of their host plant. Unlike the extrafloral nectar of its host plant, larval secretions of *T. irenea* contain almost no sugars, suggesting that the sugars in extrafloral nectar are used for larval growth (DeVries and Baker 1989). Drinking extrafloral nectar is probably wide-spread. For example, Panamanian *Synargis mycone* has at least 13 species of larval hosts in 8 distinct plant families on Barro Colorado Island and surrounding areas, and their caterpillars are found with their heads over the EFNs of all 13 species. Since the caterpillars of many other riodinid species typically rest with their heads over EFNs (Fig. 12. 3(b)–(d)), it is likely that drinking extrafloral nectar is a common habit among myrmecophilous riodinids (DeVries 1987, unpublished work).

It is worth noting that 'nectaring' by caterpillars also occurs in the lycaenids. Although the best known examples of fluid feeding by lycaenid caterpillars are the African genera *Euliphyra* and *Lachnocnema*, which

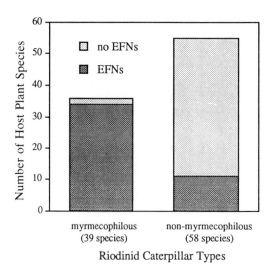

Riodinid Caterpillar Types

Fig. 12.4. Summary of host plant records for 39 myrmecophilous and 58 non-myrmecophilous New World riodinid species. Myrmecophilous species tend to feed on plants with extrafloral nectaries (EFNs), whereas non-myrmecophilous species tend to feed on plants without EFNs. The number of host plants is shown on the vertical axis.

apparently feed on regurgitations solicited from ants (Cottrell 1984), Australian *Jalmenus evagoras* caterpillars have been observed to drink extrafloral nectar (Pierce 1983); Malaysian *Allotinus unicolor* caterpillars drink homopteran secretion (Maschwitz *et al.* 1985), and the illustration in Gilbert (1976) of the adult lycaenid *Megalopalpus zymna* feeding on caterpillar dorsal nectary organ secretion in Africa also clearly shows the lycaenid caterpillar with its head on an extrafloral nectary.

The cost to lycaenid caterpillars of providing ants with a nitrogen-rich secretion has resulted in a strong tendency for them to feed on protein-rich host plants, especially legumes (Pierce 1985). However, even though some riodinid species use legumes as host plants, one reason why EFN plants (including legumes) are prominent in myrmecophilous riodinid diets (Fig. 12. 4) is that their caterpillars use extrafloral nectar for growth. When more life histories are known it will be of interest to see if myrmecophily and protein-rich host plants are correlated in riodinids, and if the presence of EFNs on lycaenoid host plants is correlated with an ability to fix nitrogen.

The interaction between plants bearing EFNs and the myrmecophilous herbivores that feed on them may be a conflicting one; both use the same ants as protection against enemies. Since many myrmecophilous insect species feed on plants with EFNs (Homoptera, lycaenoids), one consequence of the evolution of insect myrmecophily seems to be that it allowed these herbivores to exploit mutualisms between ants and plants (DeVries and Baker 1989). For example, the herbivorous caterpillars of *Thisbe irenea* provide attending ants with tentacle nectary organ secretions significantly higher in amino acids than the secretions provided by the plant EFNs (Fig. 12. 5). The provision of amino acid-rich tentacle nectary organ secretion and the concerted use of other ant organs (anterior tentacle organs, vibratory papillae) ensures that *T. irenea* caterpillars maintain the constant attention of ants, resulting in the ants tending the caterpillars more assiduously than the plant EFNs (DeVries 1988*a*). Because ants confer protection on *T. irenea* caterpillars against predators (DeVries 1987, in press), this allows caterpillars to continue eating leaf tissues and drinking the extrafloral nectar that has ostensibly been produced by the plant to attract ants that deter herbivores. Thus, *T. irenea* host plants may have a significant fraction of their leaf area removed by herbivores that use the mutualism between its EFN host plant and ants to their own benefit. The work of Buckley (1983) and Horvitz and Schemske (1984) shows, convincingly, that in such three-way interactions, ants may indeed confer more benefits on myrmecophilous herbivores than to the EFN plants which they feed on (see also Fritz 1982, 1983; Maschwitz *et al.* 1984).

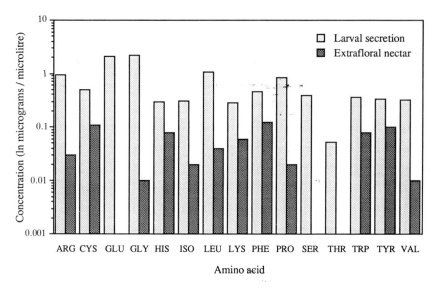

Fig. 12.5. The concentration of 15 amino acids found in the tentacle nectary organ secretions of the riodinid caterpillar *Thisbe irenea* and its host plant, *Croton billbergianus*. (Redrawn from DeVries and Baker 1989.)

The ant taxa involved with myrmecophiles

Caterpillar–ant interactions range from associations between ant and butterfly genera to associations that are species specific. For example, Australian lycaenid species of *Jalmenus* associate with the ant genus *Iridomyrmex* (Pierce 1987), European lycaenid species of *Maculinea* associate only with certain species of *Myrmica* ants (Thomas *et al.* 1989), and some neotropical riodinid species of *Theope* appear to associate only with *Azteca* ants (P. J. DeVries, personal observation). On the other hand, the Panamanian riodinids *Thisbe irenea* and *Synargis mycone* are tended primarily by the ponerine ants *Ectatomma ruidum* or *E. tuberculatum* but, depending on their host plant location, their caterpillars are also tended by ants from three other subfamilies: *Camponotus* (Formicinae), *Pheidole*, *Wasmannia* (Myrmicinae), and *Tapinoma* (Dolichoderinae). Although detailed, long-term studies are necessary for a better understanding of the dynamics of particular ant associations (e.g. Thomas *et al.* 1989), focusing on ant feeding biology has been useful in highlighting the overall patterns of mutualisms involving ants.

The evolution of myrmecophily does not appear to have involved just ants, but rather particular kinds of ant. Carroll and Janzen (1973) categorize the ecology of ant species into predators, herbivores and seed eaters, scavengers, and secretion harvesters. Two major traits appear to characterize all caterpillar–ant associations: the ability for caterpillars to produce secretions and the ability for the recipient ants to utilize them. When world-wide literature records indicating which ant genera tend riodinid or lycaenid caterpillars, Homoptera, and EFNs are assembled (summarized in DeVries, in press) three patterns become evident:

1. there is almost complete congruence between those ant genera reported to tend caterpillars and those that tend EFNs and Homoptera, suggesting that these ant genera tend any organisms that produce secretions;
2. ant genera that are reported to harvest secretions represent a small fraction of the total generic diversity of ants; and
3. ant taxa with specialized diets, such as *Atta*, *Eciton*, *Dorylus*, *Strumgenys*, *Pogonomyrmex*, are never found associating with insects or plants that produce secretions.

Hence, it is likely that the secretory ant organs of caterpillars evolved in response to selection by the small proportion of ant genera that forage on secretions (DeVries, in press). At a general level, it seems likely that benefits resulting from secretion-foraging ants have not only selected for the ant organs of butterfly larvae, but also the ant organs of Homoptera and plant EFNs.

Secretions and specificity with ants

The ability to produce secretions appears to be crucial for maintaining mutualisms with ants for many organisms (e.g. Beattie 1985; Way 1963; DeVries 1988a; Fiedler and Maschwitz 1989; Schupp and Feener Chapter 13, this volume; Koptur Chapter 15, this volume; Oliviera and Brandao Chapter 14, this volume). The secretions, however, are not all the same. The concentration and presence of specific amino acids and sugars in secretions vary among Lepidoptera (Maschwitz *et al.* 1975; Pierce 1983; Fiedler and Maschwitz 1988a; DeVries and Baker 1989), among Homoptera (Gray 1952; Maltais and Auclair 1952; Auclair 1963; Salama and Rizk 1969; Adamoko 1972), and among EFN plants (Baker *et al.* 1978).

Lanza (1988) and Lanza and Krauss (1984) made significant contributions by showing variation in the responses of different ant species to different sugar and amino acid mixtures, thus providing the impetus for exploring the taxonomic and chemical patterns of myrmecophily in insects and plants. Characterization of secretions and preference testing among ant taxa may demonstrate 'syndromes' for ant-tended organisms similar to those described by Baker and Baker (1975) for pollinators and floral nectar constituents. Knowing that ant species vary in response to nutrients, we may now ask how the constituents of secretions affect mutualisms with ants in a species-specific and generalist context. Broad level comparisons of the secretion contents of generalist and specialist insect myrmecophiles, and the extrafloral nectars of plant species occurring in different habitats may prove very illuminating indeed.

Taxonomic and temporal variation of mutualisms

It is both exciting and daunting to see that variation and multiple levels of environmental complexity have begun to be recognized as important aspects in the study of mutualisms with ants. In North America, Bristow (1984) has shown that the benefits accrued by myrmecophilous homopterans may differ depending upon the ant species that tend them. In another temperate zone study it was shown that age of plants, density of insects, and inter-colony competition have a significant effect on homopteran–ant systems (Cushman and Addicott Chapter 8, this volume). I wish to conclude by voicing something that every field biologist working on EFN plant–ant or insect–ant systems knows: the ant species tending an organism may change through time. In the neotropics the ant species occupying extrafloral nectaries, homopterans, or butterfly caterpillars may change dramatically within a 24-hour period. Such rapidly changing ant faunas will almost certainly affect the benefits provided to plants and myrmecophilous insects. Thus, accounting for the variation in ant species through time, at least in tropical systems, should be adding to our growing list of factors influencing mutualisms and other interactions involving ants.

Acknowledgements

I thank B. Crothers, D. H. Feener, D. Grimaldi, N. Greig, D. Hillis, I. J. Kitching, J. T. Longino, J. Miller, D. R. Nash, R. I. Vane-Wright, and P. S. Ward for discussions of various parts of this manuscript. Thanks to P. Harvey and R. M. May for logistical support during my visit to Oxford. Parts of this study were supported by the Smithsonian Tropical Research Institute and the MacArthur Foundation. This paper is dedicated to the mutualisms of D. Ellington, B. Strayhorn, B. Webster, and L. Young.

References to Part 2

Adamoko, D. (1972). Studies on mealybug (*Planococcoides njalensis*) (Laing)) nutrition: a comparative analysis for the free carbohydrate and nitrogenous compounds in cocoa bark and mealybug honeydew. *Bulletin of Entomological Research*, **61**, 523–31.

Addicott, J. F. (1978*a*). Niche relationships among species of aphids feeding on fireweed. *Canadian Journal of Zoology*, **56**, 1827–41.

Addicott, J. F. (1978*b*). Competition for mutualists: aphids and ants. *Canadian Journal of Zoology*, **56**, 2093–6.

Addicott, J. F. (1978*c*). The population dynamics of aphids on fireweed: a comparison of local populations and metapopulations. *Canadian Journal of Zoology*, **56**, 2554–64.

Addicott, J. F. (1979). A multispecies aphid–ant association: density dependence and species-specific effects. *Canadian Journal of Zoology*, **57**, 558–69.

Addicott, J. F. (1981). Stability properties of 2-species models of mutualism: simulation studies. *Oecologia*, **49**, 42–9.

Addicott, J. F. (1984). Mutualistic interactions in populations and communities. In *A new ecology* (ed. P. W. Price, C. N. Slobodchikoff, and W. S. Gaud). John Wiley and Sons, New York.

Addicott, J. F. (1985). Competition in mutualistic systems. In *The biology of mutualism* (ed. D. H. Boucher), 388 pp. Oxford University Press, New York.

Addicott, J. F. (1986). On the population consequences of mutualism. In *Community ecology* (ed. J. Diamond and T. J. Case), pp. 437–55. Harper and Row, New York.

Adlung, K. G. (1966). A critical evaluation of the European research on use of red wood ants (*Formica rufa* group) for the protection of forests against harmful insects. *Zeitschrift für angewandte Entomologie*, **57**, 167–89.

Ahti, T., Hämet-Ahti, L., and Jalas, J. (1968). Vegetation zones and their sections in northwestern Europe. *Annales Botanici Fennici*, **5**, 169–211.

Aoki, S. (1982). Soldiers and altruistic dispersal in aphids. In *Biology of social insects* (ed. M. D. Breed, C. D. Michener, and H. E. Evans). Westview Press, Boulder, Colorado.

Askew, R. R. (1961). On the biology of the inhabitants of oak galls of Cynipidae (Hymenoptera) in Britain. *Transactions of the Society for British Entomology*, **14**, 237–68.

Atsatt, P. R. (1981). Lycaenid butterflies and ants: selection for enemy-free space. *American Naturalist*, **118**, 638–54.

Auclair, J. L. (1963). Aphid feeding and nutrition. *Annual Review of Entomology*, **8**, 439–90.

Axelrod, R. and Hamilton, W. D. (1981). The evolution of cooperation. *Science*, **211**, 1390–6.

Bailey, M. P. (1974). Cardenolides (Cardiac glycosides) in the food and the honeydew of the oleander aphid, *Aphis nerii*. B.A. Honours thesis, University of Oxford.

Baker, H. G. and Baker, I. (1975). Studies of nectar constitution and plant–pollinator coevolution. In *Coevolution of animals and plants* (ed. L. E. Gilbert and P. H. Raven), pp. 100–40. University of Texas Press, Austin.

Baker, H. G., Opler, P. A., and Baker, I. (1978). A comparison of the amino acid compliments of floral and extrafloral nectars. *Botanical Gazette*, **139**, 322–32.

Banks, C. J. (1962). Effects of the ant *Lasius niger* (L.) on insects preying on small populations of *Aphis fabae* Scop. on bean plants. *Annals of Applied Biology*, **50**, 669–79.

Banks, C. J. and Macauley, E. D. (1967). Effects of *Aphis fabae* Scop. and its attendant ants and insect predators on yield of field beans (*Vicia faba* L.). *Annals of Applied Biology*, **60**, 445–53.

Barbosa, P. and Letourneau, D. K. (1988). *Novel aspects of insect–plant interactions*, 362 pp. John Wiley and Sons, New York.

Barlow, C. A. and Randolph, P. A. (1978). Quality and quantity of plant sap available to the pea aphid. *Annals of the Entomological Society of America*, **71**, 46–8.

Bartlett, B. R. (1961). The influence of ants upon parasites, predators, and scale insects. *Annals of the Entomological Society of America*, **54**, 543–51.

Barton, A. M. (1986). Spatial variation in the effect of ants on an extrafloral nectary plant. *Ecology*, **67**, 495–504.

Baylis, M. (1989). *The role of nitrogen in an ant–lycaenid–host plant interaction.* D.Phil. thesis, University of Oxford.

Baylis, M. and Pierce, N. E. (1991). The effect of host-plant quality on the survival of larvae and oviposition by adults of an ant-tended lycaenid butterfly, *Jalmenus evagoras. Ecological Entomology*, **16**, 1–9.

Beattie, A. J. (1985). *The evolutionary ecology of ant–plant mutualisms*, 182 pp. Cambridge University Press, Cambridge.

Becerra, J. X. I. and Venable, D. L. (1989). Extrafloral nectaries: a defense against ant–homopteran mutualisms? *Oikos*, **55**, 276–80.

Berg, C. C., Hijman, M. E. E., and Weerdenburg, J. C. A. (1985). *Flore du Cameroun 28 Moracees*. Ministry of Higher Education and Scientific Research MESRES, Yaounde.

Bodenheimer, F. S. and Swirski, E. (1957). *The Aphidoidea of the Middle East*, 378 pp. Weizmann Science Press of Israel, Jerusalem.

Boecklen, W. J. (1984). The role of extrafloral nectaries in the herbivore defense of *Cassia fasiculata. Ecological Entomology*, **9**, 243–9.

Borquin, F. (1953). Notas sobre la metamorfosis de *Hamearis susanae* Orfila, 1953 con orgua mirmecofila (Lep.: Riodin.). *Revista del la Sociedad Entomologica Argentina*, **16**, 83–7.

Bouček, Z. (1988). *Australasian Chalcidoidea*. C.A.B. International, Wallingford.

Boucher, D. H. (ed.) (1985). *The biology of mutualism*, 388 pp. Oxford University Press, New York.

Boucher, D. H., James, S., and Keeler, K. H. (1982). The ecology of mutualism. *Annual Review of Ecology and Systematics*, **13**, 315–47.

Bradley, G. A. (1973). Effect of *Formica obscuripes* (Hymenoptera: Formicidae) on the predator–prey relationship between *Hyperaspis congressis* (Coleoptera: Coccinellidae) and *Toumeyella numismaticum* (Homoptera: Coccidae). *Canadian Entomologist*, **105**, 1113–18.

Bradley, G. A. and Hinks, J. D. (1968). Ants, aphids and jack pine in Manitoba. *Canadian Entomologist*, **100**, 40–50.

Breen, J. (1979). Worker populations of *Formica lugubris* Zett. nests in Irish plantation woods. *Ecological Entomology*, **4**, 1–7.

Brian, M. V. (1973). Feeding and growth in the ant *Myrmica. Journal of Animal Ecology*, **42**, 37–53.

Bristow, C. M. (1983). Treehoppers transfer parental care to ants: a new benefit of mutualism. *Science*, **220**, 532–3.

Bristow, C. M. (1984). Differential benefits from ant attendance to two species of Homoptera on New York ironweed. *Journal of Animal Ecology*, **53**, 715–26.

Bronstein, J. L. (1988). Predators of fig wasps. *Biotropica*, **20**, 215–19.

Brown, J. H. (1971). Mechanisms of competitive exclusion between two species of chipmunks (*Eutamias*). *Ecology*, **52**, 306–11.

Brown, J. H., Davidson, D. W., Munger, J. C., and Inouye, R. S. (1986). Experimental community ecology. In *Community ecology* (ed. J. Diamond and T. J. Case). Harper and Row, New York.

Bruch, C. T. (1936). Orugas mirmecofilas de *Haemaeris epuulus* signatus Stich. *Revista del la Sociedad Entomological Argentina*, **1**, 2–9.

Buckley, R. C. (1982). Ant–plant interactions: a world review. In *Ant–plant interactions in Australia* (ed. R. C. Buckley), pp. 111–41. Dr. W. Junk, The Hague.

Buckley, R. C. (1983). Interaction between ants and membracid bugs decrease growth and seed set of a host plant bearing extrafloral nectaries. *Oecologia*, **58**, 132–6.

Buckley, R. C. (1987*a*). Interactions involving plants, Homoptera, and ants. *Annual Review of Ecology and Systematics*, **18**, 111–35.

Buckley, R. C. (1987*b*). Ant–plant–homopteran interactions. *Advances in Ecological Research*, **16**, 53–85.

Callaghan, C. J. (1977). Studies on Restinga butterflies. 1. Life cycle and immature biology of *Menander felsina* (Riodinidae), a myrmecophilous metalmark. *Journal of the Lepidopterists Society*, **20**, 36–42.

Callaghan, C. J. (1982). Notes on the immature biology of two myrmecophilous Lycaenidae: *Juditha molpe* (Riodininae) and *Panthiades bitias* (Lycaeninae). *Journal of Research on the Lepidoptera*, **20**, 36–42.

Callaghan, C. J. (1986). Notes of the biology of *Stalachtis susanna* (Lycaenidae: Riodininae) with a discussion of riodinine larval strategies. *Journal of the Lepidopterists Society*, **24**, 258–63.

Campbell, R. W. and Torgersen, T. R. (1982). Some effects of predaceous ants on western spruce budworm pupae in north central Washington. *Environmental Entomology*, **11**, 111–14.

Carper, E. R. (1989). The effects of varying levels of ant attendance on the aggregation behaviour and survivorship of larvae of *Jalmenus evagoras* (Lepidoptera: Lycaenidae). Senior thesis, Princeton University, Princeton.

Carroll, C. R. and Janzen, D. H. (1973). Ecology of foraging by ants. *Annual Review of Ecology and Systematics*, **4**, 231–57.

Chauvin, R. (1966). Un procede pour recolter automatiquement les proies que les *Formica polyctena* rapportent au nid. *Insectes Sociaux*, **13**, 59–68.

Cherix, D. (1981). Contribution a la biologie et a l'ecologie de *Formica lugubris* Zett. Le probleme des supercolonies. Thesis, Universite de Lausanne, Switzerland.

Cherix, D. (1987). Relation between diet and polyethism in *Formica* colonies. *Experientia Supplementum*, **54**, 93–115.

Claassens, A. J. M. and Dickson, C. G. C. (1977). A study of the myrmecophilous behaviour of the immature stages of *Aloeides thyra* (L.) (Lep.: Lycaenidae) with special reference to the function of the retractile tubercles and with additional notes on the general biology of the species. *Entomologists' Record and Journal of Variation*, **19**, 195–215.

Clark, G. C. and Dickson, C. G. C. (1971). *Life histories of southern African lycaenid butterflies*. Cape Town, Purnell.

Coley, P. D., Bryant, J. P., and Chapin, F. S. (1985). Resource availability and plant antiherbivore defense. *Science*, **230**, 895–9.

Collingwood, C. A. (1979). The Formicidae (Hymenoptera) of Fennoscandia and Denmark. *Fauna Entomologica Scandinavica*, **8**, 1–174.

Common, I. F. B. and Waterhouse, D. F. (1981). *Butterflies of Australia*, 2nd edn, 682 pp. Angus and Robertson, Sydney.

Compton, S. G. and Robertson, H. G. (1988). Complex interactions between mutualisms: ants tending homopterans protect fig seeds and pollinators. *Ecology*, **69**, 1302–5.

Connell, J. H. (1980). Diversity and the coevolution of competitors, or the ghost of competition past. *Oikos*, **35**, 131–8.

Cotti, G. (1963). *Bibliografia ragionata 1930–61 del gruppo* Formica rufa *in Italiano, Deutsch, English*. 413 pp. Ministero Dell' Agricoltura Delle Foreste, Collana Verde 8, Italy.

Cottrell, C. B. (1984). Aphytophagy in butterflies: its relationship to myrmecophily. *Zoological Journal of the Linnean Society*, **79**, 1–57.

Crawford, D. L. and Rissing, S. W. (1983). Regulation of recruitment by individual scouts of *Formica oreas* Wheeler (Hymenoptera: Formicidae). *Insectes Sociaux*, **30**, 117–83.

Cushman, J. H. (in press). Host-plant mediation of insect mutualisms: variable outcomes in herbivore–ant interactions. *Oikos*. (In press.)

Cushman, J. H. and Addicott, J. F. (1989). Intra- and interspecific competition for mutualists: ants as a limited and limiting resource for aphids. *Oecologia*, **79**, 315–21.

Cushman, J. H. and Whitham, T. G. (1989). Conditional mutualism in a membracid–ant association: temporal, age-specific, and density-dependent effects. *Ecology*, **70**, 1040–7.

Cushman, J. H. and Whitham, T. G. (in press). Competition mediating the dynamics of a mutualism: protective services of ants as a limiting resource for membracids. *American Naturalist*. (In press.)

D'Abrera, B. (1977). *Butterflies of the Australian region*. Melbourne.

D'Abrera, B. (1980). *Butterflies of the Afrotropical region*. Melbourne.

D'Abrera, B. (1986). *Butterflies of the Oriental region*. Melbourne.

Davidson, D. W. and Morton, S. R. (1981). Competition and dispersal in ant-dispersed plants. *Science*, **2132**, 1259–61.

De Bach, P., Fleschner, C. A., and Dietrick, E. J. (1951). A biological check method for evaluating the effectiveness of entomophagous insects. *Journal of Economic Entomology*, **44**, 763–6.

DeVries, P. J. (1984). Of crazy ants and the Curetinae: are *Curetis* butterflies tended by ants? *Zoological Journal of the Linnean Society*, **80**, 59–66.

DeVries, P. J. (1987). Ecological aspects of ant association and hostplant use in a riodinid butterfly. Ph.D. thesis, University of Texas, Austin.

DeVries, P. J. (1988*a*). The ant associated larval organs of *Thisbe irenea* (Riodinidae) and their effects on attending ants. *Zoological Journal of the Linnean Society*, **94**, 379–93.

DeVries, P. J. (1988*b*). The use of epiphylls as larval hostplants by the neotropical riodinid butterfly, *Sarota gyas*. *Journal of Natural History*, **22**, 1447–50.

DeVries, P. J. (1990). Enhancement of symbioses between butterfly caterpillars and ants by vibrational communication. *Science*, **248**, 1104–6.

DeVries, P. J. (in press). Mutualism between *Thisbe irenea* larvae and ants, and the role of ant ecology in the evolution of myrmecophilous butterflies. *Biological Journal of the Linnean Society*. (In press.)

DeVries, P. J. and Baker, I. (1989). Butterfly exploitation of a plant–ant mutualism: adding insult to herbivory. *Journal of the New York Entomological Society*, **97**, 332–40.

DeVries, P. J., Harvey, D. J., and Kitching, I. J. (1986). The ant associated epidermal organs on the larva of the lycaenid butterfly *Curetis regula* Evans. *Journal of Natural History*, **20**, 621–33.

Dixon, A. F. G. (1958). Escape responses shown by certain aphids to the presence of the coccinellid, *Adalia decempunctata* (L.). *Transactions of the Royal Entomological Society of London*, **10**, 319–34.

Dixon, A. F. G. (1971). The role of aphids in wood formation. 1. The effect of the sycamore aphid *Drepanosiphum platanoidis* (Schr.) (Aphididae) on the growth of sycamore, *Acer pseudoplatanus* L. *Journal of Applied Ecology*, **8**, 165–79.

Dixon, A. F. G. (1985). *Aphid ecology*, 157 pp. Blackie, New York.

Dowsett-Lemaire, F. (1988). Fruit choice and seed dissemination by birds and mammals in the evergreen forests of upland Malawi. *Revue d'Ecologie la Terre et la Vie*, **43**, 251–85.

Dreisig, H. (1988). Foraging rates of ants collecting honeydew or extrafloral nectar, and some possible constraints. *Ecological Entomology*, **13**, 143–54.

Dunham, A. E. (1980). An experimental study of interspecific competition between the iguanid lizards *Sceloporus merriami* and *Urosaurus ornatus*. *Ecological Monographs*, **50**, 309–30.

Eastop, V. F. (1973). Deductions from the present day host plants of aphid and related insects. In *Insect/plant relationships* (*Symposium of the Royal Entomological Society of London. No. 6*) (ed. H. F. Van Emden), pp. 157–81.

Ebbers, B. C. and Barrow, E. M. (1980). Individual ants specialize on particular aphid herds. *Proceedings of the Entomological Society of Washington*, **82**, 405.

Eckloff, W. (1978). Wechselbeziehungen zwischen Pflanzenläusen und Ameisen. *Biologie in unserer Zeit*, **8**, 48–53.

Edinger, B. B. (1985). Conditional mutualism in three aphid-tending ants. *Bulletin of the Ecological Society of America*, **66**, 168.

Edwards, W. H. (1878). Notes on *Lycaena pseudargiolus* and its larval history. *Canadian Entomologist*, **10**, 1–14.

Ehrlich, P. R. (1958). The comparative morphology, phylogeny, and higher classification of the butterflies (Lepidoptera: Papilionoidea). *University of Kansas Science Bulletin*, **39**, 305–70.

Ehrlich, P. R. and Raven, P. (1965). Butterflies and plants: a study in coevolution. *Evolution*, **18**, 596–604.

Elgar, M. A. and Pierce, N. E. (1988). Mating success and fecundity in an ant-tended lycaenid butterfly. In *Reproductive success: studies of selection and adaptation in contrasting breeding systems* (ed. T. H. Clutton-Brock), pp. 59–77. Chicago University Press, Chicago.

Eliot, J. N. (1973). The higher classification of the Lycaenidae (Lepidoptera): a tentative arrangement. *Bulletin of the British Museum (Natural History)*, **28**, 371–505.

Ewart, W. H. and Metcalf, R. L. (1956). Preliminary studies of sugars and amino acids in the honeydews of five species of coccids feeding on citrus in California. *Annals of the Entomological Society of America*, **49**, 441–7.

Faeth, S. H. (1980). Invertebrate predation of leaf miners at low densities. *Ecological Entomology*, **5**, 111–14.

Farquarson, C. O. (1922). Five years observations (1914–1918) on the bionomics of southern Nigerian insects, chiefly directed to the investigtion of lycaenid life-histories and the relation of Lycaenidae, Diptera, and other insects to ants. *Transactions of the Entomological Society of London*, **1921**, 319–448.

Feeny, P. P. (1976). Plant apparency and chemical defense. *Recent Advances in Phytochemistry*, **10**, 1–40.

Fiedler, K. and Maschwitz, U. (1988*a*). Functional analysis of the myrmecophilous relationships between ants (Hymenoptera: Formicidae) and lycaenids (Lepidoptera: Lycaenidae). II. Lycaenid larvae as trophobiotic partners of ants—a quantitative approach. *Oecologia*, **75**, 204–6.

Fiedler, K. and Maschwitz, U. (1988*b*). Functional analysis of the myrmecophilous relationships between ants (Hymenoptera, Formicidae) and lycaenids (Lep.: Lycaenidae). III. New aspects of the function of the retractile tentacular organs of lycaenid larvae. *Zoologische Beiträge*, **31**, 409–16.

Fiedler, K. and Maschwitz, U. (1989). Functional analysis of the myrmecophilous relationships between ants (Hymenoptera, Formicidae) and lycaenids (Lepidoptera: Lycaenidae). 1. Release of food recruitment in ants by lycaenid larvae and pupae. *Ethology*, **80**, 71–80.

Flückiger, W., Braun, S., and Bolsinger, M. (1988). Air pollution: effect on hostplant–insect relationships. In *Air pollution and plant metabolism* (ed. S. Schulte-Hostede, N. M. Darral, L. W. Blank, and A. R. Wellburn). Elsevier Applied Science, London, New York.

Fokkema, N. J., Riphagen, I., Poot, R. J., and de Jong, C. (1983). Aphid honeydew, a potential stimulant of *Cochliobolus satirus* and *Septoria nodorum* and the competitive role of saprophytic mycoflora. *Transactions of the British Mycological Society*, **81**, 355–63.

Fowler, S. V. and MacGarvin, M. (1985). The impact of hairy wood ants, *Formica lugubris*, on the guild structure of herbivorous insects on birch, *Betula pubescens*. *Journal of Animal Ecology*, **54**, 847–55.

Franzl, S., Locke, M., and Huie, P. (1984). Lenticles: innervated secretory structures that are expressed at every other larval moult. *Tissue and Cell*, **16**, 251–68.

Fritz, R. S. (1982). An ant–treehopper mutualism: effects of *Formica subsericea* on the survival of *Vanduzea arquata*. *Ecological Entomology*, **7**, 267–76.

Fritz, R. S. (1983). Ant protection of a host plant's defoliator: consequence of an ant–membracid mutualism. *Ecology*, **64**, 789–97.

Frowhawk, F. W. (1934). *The complete book of British butterflies*. Ward, Lock & Co., London and Melbourne.

Gilbert, L. E. (1976). Adult resources in butterflies: African lycaenid *Megalopalpus* feeds on larval nectary. *Biotropica*, **8**, 282–3.

Gösswald, K. and Horstmann, K. (1966). Untersuchungen über den Einfluss der Kleinen Roten Waldameise (*Formica polyctena* Foerster) auf den Massenwechsel des Grünen Eichenwicklers (*Tortrix viridana* L.). *Waldhygiene*, **6**, 230–55.

Gray, R. A. (1952). Composition of honeydew excreted by pineapple mealybugs. *Science*, **115**, 129–30.

Gregg, R. E. (1963). *The ants of Colorado*, 792 pp. University of Colorado Press, Boulder.

Harvey, D. J. (1987). The higher classification of the Riodinidae (*Lepidoptera*). Ph.D. thesis, University of Texas, Austin.

Harvey, D. J., Mallet, J. A., and Longino, J. (1989). Biology and morphology of immature *Nymphidium cachrus* (Riodinidae), a myrmecophilous metalmark butterfly. *Journal of the Lepidopterists' Society*, **43**, 332–3.

Hassell, M. P. (1978). *The dynamics of arthropod predator–prey systems*, 237 pp. Princeton University Press, Princeton.

Hassell, M. P. and May, R. M. (1988). Spatial variation in the dynamics of parasitoid–host systems. *Annales Zoologici Fennici*, **25**, 55–61.

Hayamizu, E. (1982). Comparative studies on aggregations among aphids in relation to population dynamics. I. Colony formation and aggregation behavior of *Brevicoryne brassicae* L. and *Myzus persicae* (Sulzer) (Homoptera: Aphididae). *Applied Entomology and Zoology*, **17**, 519–29.

Heads, P. A. (1986). Bracken, ants and extrafloral nectaries. IV. Do wood ants (*Formica lugubris*) protect the plant against insect herbivores? *Journal of Animal Ecology*, **55**, 795–809.

Heie, O. E. (1980). The Aphidoidea (Hemiptera) of Fennoscandia and Denmark. I. General part. The Families Mindaridae, Hormaphididae, Thelaxidae, Anoeciidae, and Pemphigidae. *Fauna Entomologica Scandinavica*, **9**, 1–216.

Heie, O. E. (1987). Palaeontology and phylogeny. In *Aphids: their biology, natural enemies, and control*, Vol. A. (ed. A. K. Minks and P. Harrewijn). Elsevier, Amsterdam.

Heliövaara, K. and Väisänen, R. (1988). Interactions among herbivores in three polluted pine stands. *Silva Fennica*, **22**, 283–92.

Henning, S. F. (1983). Chemical communication between lycaenid larvae (Lepidoptera: Lycaenidae) and ants (Hymenoptera: Formicidae). *Journal of the Entomological Society of Southern Africa*, **46**, 341–66.

Hinton, H. E. (1951). Myrmecophilous Lycaenidae and other Lepidoptera—a summary. *Proceedings and Transactions of the South London Entomological and Natural History Society*, **1949–1950**, 111–75.

Holt, J. and Wratten, S. D. (1986). Components of resistance to *Aphis fabae* in faba bean cultivars. *Entomologica Experimentalis Applica*, **40**, 35–40.

Horstmann, K. (1974). Untersuchungen über den Nahrungserwerb der Waldameisen (*Formica polyctena* Foerster) im Eichenwald III. Jahresbilanz. *Oecologia*, **15**, 187–204.

Horstmann, K. (1977). Waldameisen (*Formica polyctena* Foerster) als Abundanzfaktorem für den Massenwechsel des Eichenwicklers *Tortrix viridiana* L. *Zeitschrift für angewandte Entomologie*, **82**, 421–35.

Horstmann, K. (1982). Die Energiebilanz der Waldameisen (*Formica polyctena* Foerster) in einem Eichenwald. *Insectes Sociaux*, **29**, 402–21.

Horstmann, K. and Geisweid, H. J. (1978). Untersuchungen zu Verhalten und Arbeitsteilung von Waldameisen an Rindenlauskolonien. *Waldhygiene*, **12**, 157–68.

Horvitz, C. C. and Schemske, D. W. (1984). Effects of ant-mutualists and an ant-sequestering herbivore on seed production of a tropical herb, *Calathea ovandensis* (Marantceae). *Ecology*, **65**, 1369–78.

Horvitz, C. C., Turnbull, C., and Harvey, D. J. (1987). Biology of immature *Eurybia elvina* (Lepidoptera Riodinidae), a myrmecophilous metalmark butterfly. *Annals of the Entomological Society of America*, **80**, 513–19.

Ibbotson, A. and Kennedy, J. S. (1951). Aggregation in *Aphis fabae* Scop. I. Aggregation on plants. *Annals of Applied Biology*, **38**, 65–78.

Ilharco, F. A. and van Harten, A. (1987). Systematics. In *Aphids: their biology, natural enemies, and control*, Vol. A (ed. A. K. Minks and P. Harrewijn). Elsevier, Amsterdam.

Inouye, D. W. and Taylor, O. R. (1979). A temperate region ant–plant–seed predator system: consequences of extrafloral nectar secretions. *Ecology*, **60**, 1–7.

Jackson, T. H. E. (1937). The early stages of some African Lycaenidae (Lepidoptera), with an account of the larval habits. *Transactions of the Royal Entomological Society*, **86**, 201–38.

Janzen, D. H. (1978). Cicada (*Diceroprocta apache* (Davis)) mortality by feeding on *Nerium oleander*. *The Pan-Pacific Entomologist*, **54**, 69–70.

Janzen, D. H. (1985). The natural history of mutualisms. In *The biology of mutualism* (ed. D. H. Boucher). Croom Helm, London, Sydney.

Jones, C. R. (1929). Ants and their relation to aphids. *Colorado Experiment Station Bulletin*, **341** (February), 96 pp.

Juniper, B. E. and Southwood, T. R. E. (ed.) (1986). *Insects and the plant surface*, 360 pp. Edward Arnold, London.

Keeler, K. H. (1985). Cost:benefit models of mutualism. In *The biology of mutualism, ecology and evolution* (ed. D. H. Boucher). Oxford University Press, New York.

Keeler, K. H. (1989). Ant–plant interactions. In *Plant–Animal Interactions*, ed. W. G. Abrahamson. McGraw Hill, New York.

Kennedy, J. S. and Crawley, L. (1967). Spaced-out gregariousness in Sycamore aphids *Drepanosiphum platanoidis* (Schrank) (Hemiptera, Callaphididae). *Journal of Animal Ecology*, **36**, 147–70.

Kim, C. H. and Murakami, Y. (1983). Ecological studies on *Formica yessensis* with special reference to its effectiveness as a biological control agent of the pine caterpillar moth *Dendrolimus spectabilis* in south Korea. 5. Usefulness of *Formica yessensis*. *Journal of the Faculty of Agriculture Kyushu University*, **28**, 71–82.

Kiss, A. (1981). Melezitose, aphids, and ants. *Oikos*, **37**, 382.

Kitching, R. L. (1983). Myrmecophilous organs of the larva and pupa of the lycaenid butterfly *Jalmenus evagoras* (Donovan). *Journal of Natural History*, **17**, 417–81.

Kitching, R. L. and Luke, B. (1985). Myrmecophilous organs of the larvae of some British Lycaenidae (Lepidoptera): a comparative study. *Journal of Natural History*, 259–76.

Klimetzek, D. and Wellenstein, G. (1978). Assimilateentzug und Zuwachsminderung auf Forstpflanzen durch Baumlause (Lachnidae) unter dem Einfluss von Waldameisen (Formicidae). *Forstwissenschaftliches Centralblatt*, **97**, 1–12.

Klingauf, F. A. (1987). Feeding, adaptation, and excretion. In *Aphids: their biology, natural enemies, and control*, Vol. A (ed. A. K. Minks and P. Harrewijn). Elsevier, Amsterdam.

Kristensen, N. P. (1976). Remarks on the family-level phylogeny of butterflies. *Zeitschrift für Zoologische Systematik und Evolutionsforschung*, **14**, 25–33.

Kruk-De Bruin, M., Röst, L., and Draisma, F. (1977). Estimates of the number of foraging ants with the Lincoln-index method in relation to the colony size of *Formica polyctena*. *Journal of Animal Ecology*, **46**, 457–70.

Kunkel, H. and Kloft, W. (1977). Fortschritte auf dem Gebiet der Honigtau Forschung. *Apidologie*, **8**, 369–91. [Cited in Sudd, 1987.]

Laine, K. J. and Niemelä, P. (1980). The influence of ants on the survival of mountain birch during an *Oporinia autumnata* (Lep., Geometridae) outbreak. *Oecologia*, **47**, 39–42.

Lamborn, W. A. (1915). On the relationship between certain West African insects, especially ants, Lycaenidae and Homoptera. *Transactions of the Entomological Society of London*, **1913**, 436–98.

Lanza, J. (1988). Ant preferences for *Passiflora* nectar mimics that contain amino acids. *Biotropica*, **20**, 341–4.

Lanza, J. and Krauss, B. R. (1984). Detection of amino acids in artificial nectars by two tropical ants, *Leptothorax* and *Monomorium*. *Oecologia*, **63**, 423–5.

Larsson, S. (1985). Seasonal changes in the within-crown distribution of the aphid *Cinara pini* on Scotch pine. *Oikos*, **45**, 217–22.

Lawton, J. H. (1986). Surface availability and insect community structure: the effects of architecture and fractal dimension of plants. In *Insects and the plant surface* (ed. B. E. Juniper and T. R. E. Southwood), pp. 317–31. Edward Arnold, London.

Leckstein, P. M. and Llewellyn, M. (1975). Quantitative analysis of seasonal variation in the amino acids in phloem sap of *Salix alba* L. *Planta*, **124**, 89–91.

Louda, S. A. (1982). Inflorescence spiders: a cost/benefit analysis for the host plant, *Haplopappus venetus* Blake (Asteraceae). *Oecologia*, **55**, 185–91.

MacArthur, R. H. and Pianka, E. R. (1966). On the optimal use of a patchy environment. *American Naturalist*, **100**, 603–9.

McEvoy, P. B. (1979). Advantages and disadvantages to group living in treehoppers (Homoptera: Membracidae). *Miscellaneous Publications of the Entomological Society of America*, **11**, 1–13.

McNeil, J. N., Delisle, J., and Finnegan, R. J. (1977). Inventory of aphids on seven conifer species in association with the introduced red wood ant *Formica lugubris*. *Canadian Entomologist*, **109**, 1199–203.

McNeil, J. N., Delisle, J., and Finnegan, R. J. (1978). Seasonal predatory activity of the introduced red wood ant, *Formica lugubris* (Hymenoptera: Formicidae) at Valcartier, Quebec, in 1976. *Canadian Entomologist*, **110**, 85–90.

Malicky, H. (1970). New aspects of the association between lycaenid larvae (Lycaenidae) and ants (Formicidae, Hymenoptera). *Journal of the Lepidopterists' Society*, **24**, 190–202.

Maltais, J. B. and Auclair, J. L. (1952). Occurrence of amino acids in the honeydew of the crescent-marked lily aphid, *Myzus circumflexus* (Buck). *Canadian Journal of Zoology*, **30**, 190–3.

Maschwitz, U., Schroth, M., Hanel, H., and Pong, T. Y. (1984). Lycaenids parasitizing symbiotic plant–ant partnerships. *Oecologia*, **64**, 78–80.

Maschwitz, U., Schroth, M., Hänel, H., and Tho. Y. P. (1985). Aspects of the larval biology of myrmecophilous lycaenids from West Malaysia. *Nachrichten der entomologische Verein, Frankfurt*, **6**, 181–200.

Maschwitz, U., Würst, M., and Schurian, K. (1975). Blaulingraupen als Zuckerlieferanten für Ameisen. *Oecologia*, **18**, 17–21.

Mattson, W. J. (1980). Herbivory in relation to plant nitrogen content. *Annual Review of Ecology and Systematics*, **11**, 119–61.

May, R. M. (1973). *Stability and complexity in model ecosystems*, 235 pp. Princeton University Press, Princeton.

May, R. M. (1981). Models of two interacting populations. In *Theoretical ecology* (ed. R. M. May). Blackwell Press, Oxford.

Messina, F. J. (1981). Plant protection as a consequence of an ant–membracid mutualism: interactions on goldenrod (*Solidago* sp.). *Ecology*, **62**, 1433–40.

Miles, P. W. (1987). Feeding process in Aphidoidea in relation to effects on their food plants. In *Aphids: their biology, natural enemies, and control*, Vol. A (ed. A. K. Minks and P. Harrewijn), pp. 321–39. Elsevier, Amsterdam.

Mittler, T. E. (1958). Studies on the feeding and nutrition of *Tuberolachnus salignus* (Gmelin) (Homoptera, Aphididae). II. The nitrogen and sugar composition of ingested phloem sap and excreted honeydew. *Journal of Experimental Biology*, **35**, 74–84.

Morin, P. J. (1981). Predatory salamanders reverse the outcome of competition among three species of anuran tadpoles. *Science*, **212**, 1284–86.

Müller, H. (1956*a*). Konnen Honigtau liefernde Baumläuse (Lachnidae) ihre Wirtspflanzen Schadigen? *Zeitschrift für angewandte Entomologie*, **39**, 168–77.

Müller, H. (1956*b*). Massenwechsel einiger Honigtau liefernden Baumlause im Jahre 1954. *Insectes Sociaux*, **3**, 75–90.

Nash, D. R. (1989). *Cost-benefit analysis of a mutualism between lycaenid butterflies and ants*. D.Phil. thesis, University of Oxford.

Neilsen, M. G., Skyberg, N., and Winther, L. (1976). Studies on *Lasius flavus*. I: Population density, biomass and distribution of nests. *Entomologiske Meddedelser, Entomologisk Forening Kobenhavn*, **44**, 65–75.

Niemelä, P. and Laine, K. J. (1986). Green islands—predation not nutrition. *Oecologia*, **68**, 476–8.

Nixon, G. E. J. (1951). *The association of ants with aphids and coccids*. Commonwealth Institute of Entomology, London.

O'Dowd, D. J. (1979). Foliar nectar production and ant activity on a neotropical tree *Ochrama pyramidalis*. *Oecologia*, **43**, 233–48.

O'Dowd, D. J. and Catchpole, E. A. (1983). Ants and extrafloral nectaries: no evidence for plant protection in *Helichrysum* spp.–ant interactions. *Oecologia*, **59**, 191–200.

Oinonen, E. A. (1956). On the ants of the rocks and their contribution to the afforestation of rocks in southern Finland. *Acta Entomologica Fennica*, **12**, 1–212.

Otto, D. (1967). Die Bedeutung der *Formica*–Völker für die Dezimierung der wichtigsten Schadeinsekten—Ein Literaturbericht. *Waldhygiene*, **7**, 65–90.

Paine, R. T. (1966). Food web complexity and species diversity. *American Naturalist*, **100**, 65–76.

Palmer, M. A. (1952). *Aphids of the Rocky Mountain region*, Vol. V, pp. 1–452. The Thomas Say Foundation.

Pamilo, P., Vepsäläinen, K., Rosengren, R., Varvio-Aho, S.-L., and Pisarski, B. (1979). Population genetics of *Formica* ants II. Genic differentiation between species. *Annales Entomologici Fennici*, **45**, 65–76.

Pierce, N. E. (1983). *The ecology and evolution of symbiosis between lycaenid butterflies and ants*. Ph.D. thesis, Harvard University, Cambridge.

Pierce, N. E. (1984). Amplified species diversity: a case study of an Australian lycaenid butterfly and its attendant ants. *The biology of butterflies* (ed. R. I. Vane-Wright and P. R. Ackery), pp. 196–200. Symposium of the Royal Entomological Society of London, no. 11.

Pierce, N. E. (1985). Lycaenid butterflies and ants: selection for nitrogen-fixing and other protein-rich food plants. *American Naturalist*, **125**, 888–95.

Pierce, N. E. (1987). The evolution and biogeography of associations between lycaenid butterflies and ants. In *Oxford surveys in evolutionary biology*, Vol. 4 (ed. P. H. Harvey and L. Partridge), pp. 89–116. Oxford University Press, Oxford.

Pierce, N. E. and Elgar, M. A. (1985). The influence of ants on host plant selection by *Jalmenus evagoras*, a myrmecophilous lycaenid butterfly. *Behavioral Ecology and Sociobiology*, **16**, 209–22.

Pierce, N. E. and Easteal, S. (1986). The selective advantage of attendant ants for the lycaenid butterfly, *Glaucopsyche lygdamus*. Journal of Animal Ecology, **55**, 451–62.

Pierce, N. E., Kitching, R. L., Buckley, R. C., Taylor, M. F. J., and Benbow, K. (1987). Costs and benefits of cooperation between the Australian lycaenid butterfly, *Jalmenus evagoras* and its attendant ants. *Behavioral Ecology and Sociobiology*, **21**, 237–48.

Pimm, S. L. (1982). *Food webs*. Chapman and Hall, New York.

Ploch, L. (1939). Über die Nahrung und den Nahrungserwerb der Roten Waldameise, eine wissenschaftliche Klarstellung. *Entomologische Zeitschrift*, **53**, 239–44.

Pontin, A. J. (1978). The numbers and distribution of subterranean aphids and their exploitation by the ant *Lasius flavus* (Fabr.). *Ecological Entomology*, **3**, 203–7.

Price, P. W., Bouton, C. E., Gross, P., McPherson, B. A., Thompson, J. N., and Weis, A. E. (1980). Interactions among three trophic levels: influence of plants on interactions between insect herbivores and natural enemies. *Annual Review of Ecology and Systematics*, **11**, 41–65.

Pulliam, H. R. (1986). Niche expansion and contraction in variable environments. *American Zoologist*, **26**, 71–9.

Quist, J. A. (1978). The revised list of aphids of the Rocky Mountain region. *Colorado state agricultural experiment station, bulletin 567–S*, February, 77 pp.

Rathcke, B. (1983). Competition and facilitation among plants for pollination. In *Pollination biology* (ed. L. A. Real). Academic Press, New York.

Rathcke, B. (1988). Interaction for pollination among coflowering shrubs. *Ecology*, **69**, 446–57.

Rhodes, D. F. and Cates, R. H. (1976). Toward a general theory of plant antiherbivore chemistry. *Recent Advances in Phytochemistry*, **10**, 168–213.

Risch, S. and Boucher, D. H. (1976). What ecologists look for. *Bulletin of the Ecological Society of America*, **57**, 8–9.

Robbins, R. K. (1982). How many butterfly species? *News of the Lepidopterists' Society*, **1982**, 40–1.

Robbins, R. K. (1988). Comparative morphology of the butterfly foreleg coxa and trochanter (Lepidoptera) and its systematic implications. *Proceedings of the Entomological Society of Washington*, **90**, 133–54.

Robbins, R. K. and Aiello, A. (1982). Foodplant and oviposition records for Panamanian Lycaenidae and Riodinidae. *Journal of the Lepidopterists Society*, **36**, 65–75.

Roberts, R. B. (1979). Spectrophotometric analysis of sugars produced by plants and harvested by insects. *Journal for Apicultural Research*, **18**, 191–5.

Roberts, J. T. and Heithaus, E. R. (1986). Ants rearrange the vertebrate-generated seed shadow of a neotropical fig tree. *Ecology*, **67**, 1046–51.

Rosengren, R. (1971). Route fidelity, visual memory and recruitment behavior in foraging wood ants of the genus *Formica* (Hymenoptera: Formicidae). *Acta Zoologica Fennica*, **133**, 1–102.

Rosengren, R. (1977). Foraging strategy of wood ants (*Formica rufa* group). I. Age polyethism and topographic traditions. *Acta Zoologica Fennica*, **149**, 1–30.

Rosengren, R. and Cherix, D. (1981). The pupa-carrying test as a taxonomic tool in the *Formica rufa* group. In *Biosystematics of social insects* (ed. P. E. Howse and J.-L. Clement). Academic Press, London, New York.

Rosengren, R. and Fortelius, W. (1986). Light–dark induced activity rhythms in *Formica* ants (Hymenoptera: Formicidae). *Entomologica Generalis*, **11**, 221–8.

Rosengren, R. and Sundström, L. (1987). The foraging system of a red wood ant colony

(*Formica* s. str.)—collecting and defending food through an extended phenotype. *Experientia supplementum*, **54**, 117–37.

Rosengren, R., Vepsäläinen, K., and Wuorenrinne, H. (1979). Distribution, nest densities, and ecological significance of wood ants (the *Formica rufa* group) in Finland. *Bulletin Srop* (Organisation Internationale de Lutte Biologique contre les Animaux et les Plantes Nuisibles), **2**, 181–213.

Rosengren, R., Cherix, D., and Pamilo, P. (1985). Insular ecology of the red wood ant *Formica truncorum* Fabr. I. Polydomous nesting, population size and foraging. *Mitteilungen der Scweizerischen Entomologischen Gesellschaft*, **58**, 147–75.

Rosengren, R., Fortelius, W., Lindström, K., and Luther, A. (1987). Phenology and causation of nest heating and thermo-regulation in red-wood ants of the *Formica rufa* group studied in coniferous forest habitats in southern Finland. *Annales Zoologici Fennici*, **24**, 147–55.

Ross, G. N. (1966). Life history studies of a Mexican butterfly. IV. The ecology and ethology of *Anatole rossi*, a myrmecophilous metalmark. *Annals of the Entomological Society of America*, **59**, 985–1004.

Rothschild, M., von Euw, J., and Reichstein, T. (1970). Cardiac glycosides in the oleander aphid, *Aphis nerii*. *Journal of Insect Physiology*, **16**, 1141–5.

Roughgarden, J. (1975). Evolution of marine symbiosis: a simple cost-benefit model. *Ecology*, **56**, 1201–8.

Russell, L. M. (1989). Addition and corrections to data on types of aphids in Palmer' *Aphids of the Rocky Mountain Region* (*Homoptera: Aphididae*). *Annals of the Entomological Society of America*, **82**, 407–13.

Salama, H. S. and Rizk, A. M. (1969). Composition of the honeydew in the mealybug, *Sacchariococcus sacchari*. *Journal of Insect Physiology*, **15**, 1873–5.

Samways, M., Nel, M., and Prins, A. J. (1982). Ants (Hymenoptera: Formicidae) foraging in citrus trees and attending honeydew producing Homoptera. *Phytophylactica*, **14**, 155–7.

Sato, H. and Higashi, S. (1987). Bionomics of *Phyllonorycter* (Lepidoptera: Gracillariidae) on *Quercus*. II. Effects of ants. *Ecological Research*, **2**, 53–60.

Schoener, T. W. (1983). Field experiments on interspecific competition. *American Naturalist*, **122**, 240–85.

Schremmer, F. (1978). On the bionomy and morphology of the myrmecophilous larva and pupa of the neotropical butterfly species *Hamaeris erostratus* (Lep.: Riodinidae). *Entomologica Generalis*, **4**, 113–21.

Scott, J. A. (1986). *The butterflies of North America. A natural history field guide.* Stanford University Press, Stanford.

Scriber, J. M. and Ayers, M. P. (1988). Leaf chemistry as a defense against insects. *ISI Atlas of Science: Plants and Animals*, **1**, 117–23.

Scudder, S. H. (1889). *Butterflies of the eastern United States and Canada with special reference to New England.* 3 Vols. Published by the author, Cambridge.

Setzer, R. W. (1980). Intergall migration in the aphid genus *Pemphigus*. *Annals of the Entomological Society of America*, **73**, 327–31.

Shannon, R. E. and Brewer, J. W. (1980). Starch and sugar levels in three coniferous insect galls. *Zeitschrift für angewandte Entomologie*, **89**, 526–33.

Shaposhnikov, G. C. (1987a). Evolutionary estimation of taxa. In *Aphids: their biology, natural enemies, and control*, Vol. A (ed. A. K. Minks and P. Harrewijn). Elsevier, Amsterdam.

Shaposhnikov, G. G. (1987b). Evolution of aphids in relation to evolution of plants. In

Aphids: their biology, natural enemies, and control, Vol. A (ed. A. K. Minks and P. Harrewijn). Elsevier, Amsterdam.

Skinner, G. J. (1980a). Territory, trail structure and activity patterns in the wood-ant *Formica rufa* (Hymenoptera: Formicidae) in limestone woodland in north-west England. *Journal of Animal Ecology*, **49**, 381–94.

Skinner, G. J. (1980b). The feeding habits of the wood-ant *Formica rufa* in limestone woodland in Northwest England. *Journal of Animal Ecology*, **49**, 381–94.

Skinner, G. J. and Whittaker, J. B. (1981). An experimental investigation of the interrelationships between the wood ant (*Formica rufa*) and some tree canopy herbivores. *Journal of Animal Ecology*, **50**, 313–26.

Smiley, J. T., Atsatt, P. R., and Pierce, N. E. (1988). Local distribution of the butterfly, *Jalmenus evagoras*, in response to host ants and plants. *Oecologia*, **76**, 416–22.

Sörensen, U. and Schmidt, G. H. (1987). Das Beutespektrum der Waldameisen (Genus: *Formica*, Hymenoptera) in der Bredstedter Geest (Schleswig-Holstein) um Jahre 1980. *Waldhygiene*, **17**, 59–84.

Southwood, T. R. E. (1977). Habitat, the templet for ecological strategies. *Journal of Animal Ecology*, **46**, 337–65.

Southwood, T. R. E. (1986). Plant surface and insects—an overview. In *Insects and the plant surface* (ed. B. E. Juniper and T. R. E. Southwood), pp. 1–22. Edward Arnold, London.

Stichel, H. (1930–1931). *Lepidopterorum Catalogus*, Parts 40, 41, 44. W. Junk, Berlin.

Strong, D. R., Lawton, J. H., and Southwood, T. R. E. (1984). *Insects on plants: community patterns and mechanisms*, 313 pp. Blackwell, Oxford.

Sudd, J. H. (1967). *An introduction to the behaviour of ants*, 200 pp. Edward Arnold, London.

Sudd, J. H. (1983). The distribution of foraging wood-ants (*Formica lugubris* Zett.) in relation to the distribution of aphids. *Insectes Sociaux*, **30**, 298–307.

Sudd, J. H. (1987). Ant–aphid mutualism. In *Aphids: their biology, natural enemies, and control*, Vol. A (ed. A. K. Minks and P. Harrewijn). Elsevier, Amsterdam.

Sudd, J. H. and Franks, N. (1987). *The behavioural ecology of ants*, 206 pp. Blackie, New York.

Sudd, J. H. and Sudd, M. E. (1985). Seasonal changes in the response of wood-ants to sucrose baits. *Ecological Entomology*, **10**, 89–97.

Sudd, J. H., Douglas, J. M., Gaynard, T., Murray, D. M., and Stockdale, J. M. (1977). The distribution of wood-ants (*Formica lugubris* Zeff.) in a northern English forest. *Ecological Entomology*, **2**, 301–13.

Taylor, F. (1977). Foraging behaviour of ants: experiments with two species of myrmecine ants. *Behavioral Ecology and Sociobiology*, **2**, 147–67.

Templeton, A. R. and Gilbert, L. E. (1985). Population genetics and mutualism. In *The biology of mutualism* (ed. D. H. Boucher). Oxford University Press, New York.

Thomas, D. W. (1988). The influence of aggressive ants on fruit removal in the tropical tree, *Ficus capensis* (Moraceae). *Biotropica*, **20**, 49–53.

Thomas, J. A., Elmes, G. W., Wardlaw, J. C., and Woyciechowski, M. (1989). Host specificity among *Maculinea* butterflies in *Myrmica* ant nests. *Oecologia*, **79**, 452–7.

Thompson, J. N. (1982). *Interaction and coevolution*. John Wiley Press, New York.

Thompson, J. N. (1988). Variation in interspecific interactions. *Annual Review of Ecology and Systematics*, **19**, 65–87.

Tilles, D. A. and Wood, D. L. (1982). The influence of carpenter ants (*Camponotus modoc*) (Hymenoptera: Formicidae) attendance on the development and survival of aphids (*Cinara* spp.) (Homoptera: Aphididae) in a Giant Sequoia forest. *The Canadian Entomologist*, **114**, 1133–42.

Torossian, C. (1981). L'alimentation proteidique des colonies de *Formica lugubris* Zett. de la cerdagne Orientale. *Bulletin de la Société de l'Histoire Naturelle de Toulouse*, **116**, 207–11.

Vane-Wright, R. I. (1978). Ecological and behavioural origins of diversity in butterflies. *Symposium of the Royal Entomological Society of London*, **9**, 56–70.

Varley, G. C. and Gradwell, G. R. (1962). The effect of partial defoliation by caterpillars on the timber production by oak trees in England. *Proceedings of the XIth International Congress of Entomology* (Vienna 1960), **2**, 211–14.

Warrington, S. (1984). An experimental field study of different levels of insect herbivory induced by wood-ant (*Formica rufa*) predation on sycamore. Ph.D. thesis, University of Lancaster.

Warrington, S. and Whittaker, J. B. (1985*a*). An experimental field study of different levels of insect herbivory induced by *Formica rufa* predation on sycamore (*Acer pseudoplatanus*). I. Lepidoptera larvae. *Journal of Applied Ecology*, **22**, 775–85.

Warrington, S. and Whittaker, J. B. (1985*b*). An experimental field study of different levels of insect herbivory induced by *Formica rufa* predation on sycamore (*Acer pseudoplatanus*). II. Aphidoidea. *Journal of Applied Ecology*, **22**, 787–96.

Waser, N. M. (1983). Competition for pollination and floral character differences among sympatric plant species: a review of evidence. In *Handbook of experimental pollination biology* (ed. C. E. Jones and R. J. Little). Van Nostrand Reinhold, New York.

Way, M. J. (1954). Studies of the association of the ant *Oecophylla longinoda* and the scale insect *Saissetia zanzibarensis*. *Bulletin of Entomological Research*, **45**, 113–34.

Way, M. J. (1963). Mutualism between ants and honeydew-producing Homoptera. *Annual Review of Entomology*, **8**, 307–44.

Weaving, A. J. S. (1980). Observations on *Hilda patruelis* Stal. (Homoptera: Tettigometridae) and its infestation of the groundnut crop in Rhodesia. *Journal of the Entomological Society of Southern Africa*, **43**, 151–67.

Weins, J. A. (1977). On competition in variable environments. *American Scientist*, **65**, 590–7.

Wellenstein, G. (1952). Zur Ernährungsbiologie der Roten Waldameise (*Formica rufa* L.). *Zeitschrift für Pflanzenkrankheiten*, **59**, 430–51.

Wellenstein, G. (1954). Die Insektenjagd der Roten Waldameise (*F. rufa* L.). *Zeitschrift für angewandte Entomologie*, **36**, 185–217.

Wellenstein, G. (1961). Weitere Ergebnisse uber die bienenwirtschaftliche Bedeutung der Waldameisen der *F. rufa*—Gruppe. Atti IV Congresso U.I.E.I.S. Pavia, September 9–14, vol XII of the Symposium Gen. and Biol. Italiana, pp. 60–73.

Wellenstein, G. (1980). Auswirkung hügelbauender Waldameisen der *Formica rufa*—Gruppe auf forstschädliche Raupen und das Wachstum der Waldbaurne. *Zeitschrift für angewandte Entomologie*, **89**, 144–57.

Wheeler, W. M. (1910). *Ants, their structure, development, and behaviour*, 663 pp. Columbia University Press. (Reprinted 1960.)

White, T. C. R. (1985). Green islands—nutrition not predation—an alternative hypothesis. *Oecologia*, **67**, 455–6.

Whitham, T. G. (1980). The theory of habitat selection: examined and extended using *Pemphigus* aphids. *American Naturalist*, **115**, 449–66.

Whittaker, J. B. and Warrington, S. (1985). An experimental field study of different levels of insect herbivory induced by *Formica rufa* predation on sycamore (*Acer pseudoplatanus*). III. Effects on tree growth. *Journal of Applied Ecology*, **22**, 797–811.

Whittaker, J. B., Warrington, S., and Moore, R. (1987). Patterns of typhlocybine feeding on sycamore and oak in relation to that of other herbivores, with comments on induced defences. *Proceedings of the 6th Auchenorrhyncha Meeting, Turin, Italy, 7–11 September, 1987*, pp. 539–44.

Wilson, D. S. (1983). The effect of population structure on the evolution of mutualism: a field test involving burying beetles and their phoretic mites. *American Naturalist*, **121**, 851–70.

Wilson, E. O. (1971). *The insect societies*, 548 pp. Harvard University Press, Cambridge.

Wolin, C. L. (1985). The population dynamics of mutualistic systems. In *The biology of mutualism* (ed. D. H. Boucher). Oxford University Press, New York.

Wolin, C. L. and Lawlor, L. R. (1984). Models of facultative mutualism: density effects. *American Naturalist*, **124**, 843–62.

Wood, T. K. (1982). Ant-tended nymphal aggregations in the *Enchenopa binotata* complex (Homoptera: Membracidae). *Annals of the Entomological Society of America*, **75**, 649–53.

Woodward, T. E., Evans, J. W., and Eastop, V. F. (1970). Hemiptera. In *Insects of Australia*. C.S.I.R.O., Melbourne University Press, Melbourne.

Youngs, L. C. (1983). Predaceous ants in biological control of insects pests in North American forests. *Bulletin of the Entomological Society of America*, **29**, 47–50.

Zoebelein, G. (1956*a*). Der Honigtau als Nahrung der Insekten, Teil I. *Zeitschrift für angewandte Entomologie*, **38**, 369–416.

Zoebelein, G. (1956*b*). Der Honigtau als Nahrung der Insekten, Teil II. *Zeitschrift für angewandte Entomologie*, **39**, 129–67.

Zoebelein, G. (1957). Die Rolle des Waldhonigtaus im Nahrungshaushalt forstlich nützlicher Insekten. *Forstwissenschaftliches Zentralblatt*, **76**, 24–34.

Part 3

Extrafloral nectary-mediated interactions

13

Phylogeny, lifeform, and habitat dependence of ant-defended plants in a Panamanian forest

Eugene W. Schupp and D. H. Feener, Jr.

Introduction

Facultative defence mutualisms, in which many species of ants visit and defend many species of plants but nest elsewhere, are abundant (Bentley 1977*b*; Inouye and Taylor 1979; Koptur 1979; Keeler 1980*a*; Oliveira and Brandão Chapter 14, this volume). These ant-defended plants attract and maintain their 'bodyguard' of ants (and parasitoids; see Chapter 15) with extrafloral (non-pollination) nectar and solid pearl bodies, proffered primarily on young, vulnerable, actively growing tissues (e.g. expanding leaves and shoots, flower buds, and young fruits) (Beattie 1985). In most cases, ants collecting these rewards or defending the sites of their production either actively or passively reduce herbivore damage.

Most experimental studies of ant-defended species have demonstrated that some correlate of plant fitness is increased by ant visitation (see Buckley 1982; Beattie 1985). Such studies provide convincing evidence for the adaptive significance of ant defence, but are less useful in accounting for the ecological and phylogenetic distributions of these mutualisms. Accordingly, other studies have used comparative surveys and models to investigate the ecological and evolutionary conditions under which ant defence is favoured (Bently 1976, 1981; Keeler 1981; McKey 1984, 1988, 1989; Davidson Chapter 20, this volume).

Comparative studies have identified two possible ecological correlates of ant–plant mutualisms in tropical forests:

1. ant-defended species should be more common in disturbed, open habitats (e.g. tree fall gaps) than in undisturbed, closed habitats (e.g. forest under-storey) (Bentley 1976, 1977*a*; Buckley 1982; Keeler 1989); and
2. climbing lianas and vines are more likely to be ant-defended than free-standing trees and shrubs (Bentley 1981; Keeler 1981*c*).

However, these surveys were based on percentage cover and conclusions may be biased by a few extremely abundant species. Secondly, they failed to control for interactions such as that between lifeform and habitat (Putz 1984; McKey 1989), which may give spurious correlations. Thirdly, because these studies did not explicitly account for the phylogenetic relationships of surveyed species, associations between variables may be more related to phylogeny than ecology (Ridley 1983; Pagel and Harvey 1988).

With these problems in mind, we surveyed 243 native plant species on Barro Colorado Island, Panama, for attributes attractive to ants. By analysing three-way contingency tables we were able to test both the habitat and lifeform dependence of ant-defended species, and identify potentially confounding interactions. Moreover, by examining these associations at several taxonomic levels we could determine which, if any, are the result of phylogenetic interdependence.

The study area

Barro Colorado Island (BCI) is a biological reserve in the Panama Canal administrated by the Smithsonian Tropical Research Institute. The island receives about 2616 mm of rain per year, mostly in May to December (50-year mean; Rand and Rand 1982). Except for small human clearings, the 15-km² island is covered with tropical moist forest (Holdridge Life-Zone System) supporting at least 1212 native vascular plants (Croat 1978; Leigh *et al.* 1982). The forest is about half secondary (> 100 yr) and half mature (> 400 yr) (Croat 1978; Foster and Brokaw 1982; Hubbell and Foster 1986).

Species selection, lifeforms, and habitat requirements

Over a period of 3.5 years (June 1985 to January 1989) we examined 243 species and categorized them by lifeform, habitat distribution, and the presence of ant attractants and ant defence. We restricted our study to terrestrially rooted dicotyledons in five of Croat's (1978) lifeform (habit) classes: tree, small tree or shrub, strictly shrub, woody liana, and herbaceous vine. Our survey encompassed approximately 35 per cent of the BCI flora in these lifeform classes, and included most of the common species. Species were included in the survey as encountered in the field, although we enhanced the taxonomic diversity of our sample by specifically searching for species in unrepresented families and by not including more than four species per genus.

We assigned species to three lifeform categories. Large trees ($n = 109$ species, 52 per cent of the BCI flora in this lifeform class) are greater than 10 m tall when mature and generally reproduce as part of the canopy. Small trees and shrubs are less than 10 m tall ($n = 64$ species, 26 per cent) and often reproduce in the understorey. Preliminary analyses revealed no differences between woody lianas and herbaceous vines in the frequency of ant defence or in habitat preference. We, therefore, combined these two classes into one liana category ($n = 70$ species, 28 per cent).

Although tropical trees vary continuously in the conditions they require for establishment, two broad habitat categories are generally recognized

(Coley 1983*a*). Gap species, also referred to as light-demanding, shade-intolerant, or pioneer species, generally need the high light environment of a treefall gap to establish and grow. In contrast, shade-tolerant species, also called persistent or mature forest species, survive and grow in deep shade, as well as in the high light of gaps and clearings. We were able to assign 228 of our species to these two habitat categories based on the opinions of eight plant systematists and ecologists with extensive experience on BCI.

Ant attractants

Using a dissecting microscope we carefully inspected actively flushing vegetative growth for the presence of nectaries, nectar accumulation without obvious nectaries, and pearl bodies. Whenever possible we inspected more than one individual per species through several cycles of leaf flush. We restricted our survey to vegetative tissue because ant attractants on repro-ductive structures are more difficult to sample reliably and selective pressures maintaining them need not be the same as those maintaining ant defence on vegetative tissues. Similarly, we excluded myrmecophytes, species in which ants nest within domatia. We classified 99 species using field collected material only, and 144 species using actively growing potted plants isolated from ants in a screened growing house. For 62 of the latter species we also made observations on field material. With few exceptions, both field and growing house samples were from individuals between 0.25 and 2.0 m tall.

We distinguished extrafloral nectaries from other glands by taste (sweet-ness = nectary), comparison with homologous structures in related taxa, and observations on ant behaviour in the field. We divided glandular trichomes that were possible pearl bodies into two distinct categories:

1. large bodies tended to be oily, cream-coloured, and visible to the naked eye; while
2. small bodies were generally clear, and visible only with magnification.

While our observations and those of O'Dowd (1982) suggest that large bodies are harvested by ants, it is unclear which, if any, of the small bodies are ant-harvested, and which serve some other function of glandular trichomes (Rodriguez *et al.* 1984; Duffey 1986; Jeffree 1986). In most analyses we conservatively defined ant-defended species as those with extrafloral nectaries and/or large bodies.

We classified 81 species (33 per cent) as ant-defended (Table 13. 1) with 74 species (30 per cent) possessing extrafloral nectaries and 4 species (2 per cent) with both nectaries and large pearl bodies. In addition to the seven species with large pearl bodies, we found 37 species (15 per cent) with small bodies. The presence of extrafloral nectaries (EFNs) was independent of both large pearl bodies alone (generalized exact test, $P > 0.65$) and of large and small bodies combined ($G = 0.53$, d.f. $= 1$, $P > 0.45$).

Table 13.1. The percentage of species with extrafloral nectaries and of species that are ant-defended (with extrafloral nectaries and/or large pearl bodies), in woody dicotyledons and herbaceous vines on Barro Colorado Island. n = Number of species surveyed in that category.

Lifeform	n	% Species	
		Extrafloral nectaries	Ant-defended
Small trees and shrubs	64	14	16
Large trees (> 10 m)	109	34	37
Lianas and vines	70	44	44
Total	243	32	33

For several reasons our list (Appendix 13.1) underestimates the full extent of ant attractants. Firstly, the field surveys sometimes missed ant attractants, especially pearl bodies, detected in isolated plants. Correct classification of field samples was 92 per cent (57/62) for small bodies and 98 per cent (61/62) for both nectaries and large bodies. However, field surveys missed 4 per cent (1/22) of the species with nectaries, 33 per cent (1/3) of the species with large bodies, and 21 per cent (4/19) of the species with small bodies. Secondly, because the breadth of our survey precluded extensive observations on individual species, those producing nectar or pearl bodies only occasionally, or in very small quantities, may have been missed. Thirdly, some species may produce ant attractants only under ideal conditions (Bentley, 1977a; E. W. Schupp and D. H. Feener, personal observation).

Accounting for phylogenetic effects

Although the comparative method is a powerful tool for studying evolutionary hypotheses, it is faced with the problem of distinguishing between similarities due to phylogenetic relatedness and similarities due to convergent evolution in response to similar selection pressures (Pagel and Harvey 1988). When character states are discrete, as in the present study, cladistic outgroup comparison is a preferred method for counting the number of independent evolutionary origins of particular character combinations (Ridley 1983, 1986; Pagel and Harvey 1988; McKey Chapter 21, this volume; Ward Chapter 22, this volume). This method requires a presumed phylogeny, preferably of monophyletic groups of small to moderate size.

Since our survey comprised many taxa with unresolved phylogenetic relationships, we could not strictly apply out-group comparison. Instead, we first assumed that all species evolved independently, and then, successively, that all genera, families, and orders evolved independently. For each taxonomic level we counted the number of unique combinations of character states within a taxon and assumed each combination to be independently evolved associations. For example, all four *Inga* species in our survey were classified as ant-defended, shade-tolerant large trees. For the generic level analysis, therefore, *Inga* represented one independent evolutionary event. If, however, one *Inga* species had been a liana, we would have counted *Inga* as representing two independently derived character combinations. While this method may still overestimate the number of independent evolutionary events, it is preferable to assuming all species are independent, or that a higher taxonomic category can contain only one combination of character states.

Phylogenetic distribution of ant defence

The 243 species in our survey included members of 186 genera, 68 families, 34 orders, and all six subclasses of the Magnoliopsida (dicotyledons) (see Appendix 13. 1). The distribution of ant defence was not independent of phylogeny. Ant-defended species tended to have ant-defended congeners. For the 21 genera with more than a single surveyed species and at least one ant-defended species, 76 per cent included only ant-defended species (Fig. 13. 1). The frequency of taxa composed entirely of ant-defended species

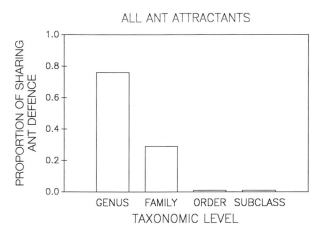

Fig. 13.1. The proportion of taxa sharing ant defence as a function of taxonomic level; based on taxa with a minimum of two species surveyed. The proportion sharing ant defence is the proportion of taxa with at least one ant-defended species in which all species surveyed were ant-defended.

declined at higher taxonomic categories (Fig. 13. 1). This decline was not simply a function of more species per taxa at higher levels. The same decline in phylogenetic influence held if calculations were based only on those families and orders with two to four species surveyed; the same range of numbers as in the generic comparison.

One striking pattern that emerged from our survey is the prevalence of ant defence in some families and its near absence from others (see Appendix 13. 1). The frequency of ant-defended species among the subclasses was not equal (Table 13. 2). There was a highly significant association between ant defence and subclass whether we considered all ant-defended species or only those with extrafloral nectaries (generalized exact tests, $P < 0.001$). This association was due to a virtual lack of ant defence in the three primitive subclasses; Magnoliidae, Hamamelidae, and Caryophyllidae (Table 13. 2).

Table 13.2. Phylogenetic distribution of woody plants and vines with extrafloral nectaries and species that are ant-defended (with extrafloral nectaries and/or large pearl bodies) among the subclasses of the Magnoliopsida on Barro Colorado Island. n = Number of species surveyed in that subclass.

Subclass	n	No. of species	Extrafloral nectaries % Species	No. of species	Ant-defended % Species
Magnoliidae	22	0	0	1	5
Hamamelidae	11	0	0	0	0
Caryophyllidae	4	0	0	0	0
Dilleniidae	51	19	37	20	39
Rosidae	108	46	43	48	44
Asteridae	47	12	26	12	26

When these three subclasses were eliminated from the analyses no significant association existed between the remaining subclasses and either all ant-defended species or only those species with extrafloral nectaries.

Association among ant defence, lifeform, and habitat

When all 228 species classified with respect to ant defence, lifeform, and habitat were assumed to be independent, log–linear analysis indicated that all two-way interaction terms were significant (Table 13. 3) and should be retained in the best-fit model ($G = 0.13$, d.f. = 2, $P > 0.90$). Although only 40 per cent of the surveyed species were gap species, 61 per cent of the ant-defended species were in this habitat category, a highly significant over-representation (Fig. 13.2 (a)). Relative to large trees, ant-defended species were over-represented in the lianas and under-represented among small trees and shrubs (Fig. 13. 2(b)). Similarly there was an excess of gap-requiring lianas and a deficit of gap-requiring small trees and shrubs (Fig. 13. 2(c)).

Only the interaction between ant defence and habitat remained significant at all taxonomic levels (Table 13. 3). When genera, families, or orders were assumed to be independent, the [ant defence × lifeform] interaction was no

Table 13.3. Significance of the interaction terms of ant defence, habitat, and lifeform at different taxonomic levels in log–linear models of the three-way contingency tables. The level of analysis is the taxonomic level at which we assumed independent evolution. These levels represent increasing data reduction that accounts for phylogenetic relatedness at increasingly higher taxonomic levels.

Level of analysis	n	Interactions					
		Ant defence × habitat		Ant defence × lifeform		Lifeform × habitat	
		d.f	G	d.f.	G	d.f.	G
Species	228	1	15.62*	2	6.61†	2	8.62†
Genus	198	1	14.29*	2	2.97	2	6.09†
Family	132	1	7.72‡	2	0.54	2	6.75†
Order	109	1	6.61†	2	0.42	2	4.53

*$P < 0.001$. ‡$P < 0.01$. †$P < 0.05$.
n is the number of observations in the table.
The significance of each interaction term was determined with G, or log-likelihood ratio, tests (Bishop *et al.* 1975; Sokal and Rohlf 1981). For each table we first fit the log–linear model that included all two-way interactions, then, one at a time, we removed each two-way interaction term. An interaction was significant if the difference in G values between the model with all three interactions and the model with the one term removed was greater than the critical Chi-square at $P = 0.05$ with the different degrees of freedom.

longer significant (Table 13. 3). The [lifeform × habitat] interaction continued to be significant at the generic and familial levels, but not at ordinal levels.

These results indicate that gap species are more likely to be ant-defended than are shade-tolerant species, and that this association is largely independent of phylogeny. Similarly, the persistent association between lifeform and habitat appears to be mostly independent of phylogeny. In contrast, the

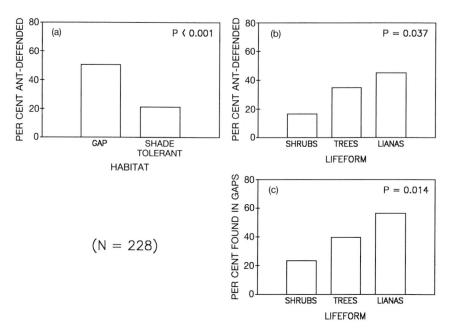

(N = 228)

Fig. 13.2. Plots of two-way interaction terms from the three-way [ant defence × habitat × lifeform] contingency table (species level analysis, $n = 228$ species). (a) The percentage of ant-defended species as a function of habitat requirements. (b) The percentage of ant-defended species as a function of lifeform. (c) The percentage of gap species as a function of lifeform.

absence of a significant association between ant defence and lifeform at all taxonomic levels, except that of species, suggests that the excess of ant-defended lianas and the deficit of ant-defended small trees and shrubs is an artefact due to phylogenetic relatedness. The greatest effect of controlling for phylogenetic relatedness is a reduction in the differences among lifeforms in the prevalence of ant defence (Figs. 13.2 and 13.3). The remaining differences are easily explained by the associations between ant defence and habitat, and habitat and lifeform.

Additional support for the conclusion that ant defence is more common in gap than in shade-tolerant species, comes from separate [ant defence × habitat] contingency tables for individual families. These analyses revealed a significant excess of tables with residuals indicating a positive association between ant defence and gap habitats (49/64 families, binomial test with $p = q = 0.5$, $P < 0.001$; four families were excluded through lack of habitat information or because their residuals were zero).

The pattern of prevalence of ant defence in some families, and its near or

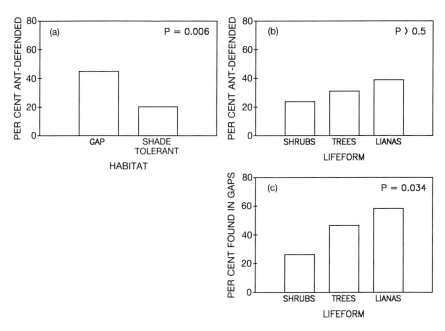

Fig. 13.3. Plots of two-way interaction terms from the three-way [ant defence × lifeform × habitat] contingency table (family level analysis, $n = 132$ observations). (a) The percentage of ant-defended species as a function of habitat requirements. (b) The percentage of ant-defended species as a function of lifeform. (c) The percentage of gap species as a function of lifeform.

total absence from others, may be due directly to a phylogenetic effect on the distribution of ant defence, or indirectly to a phylogenetic effect on habitat requirements and a correlation between habitat requirements and ant defence. The fact that families predominated by ant-defended species are also more likely to be predominated by gap species (Table 13. 4) is compatible with the notion that indirect effects are partially responsible for the observed phylogenetic distribution. This association between ant defence and habitat, however, does not account for subclass differences in ant defence. When primitive (Magnoliidae, Hamamelidae, and Caryophyllidae) and advanced (Dilleniidae, Rosidae, and Asteridae) subclasses were compared, the association between ant defence and subclass category was highly significant ($G = 12.05$, d.f. $= 1$, $P < 0.001$; for family-level analysis). In contrast, habitat was not significantly associated with subclass category ($G = 0.96$, d.f. $= 1$, $P > 0.30$; for family-level analysis).

Table 13.4. Association between ant defence and habitat requirements for those plant families with a minimum of four sampled species and predominated by one or the other category for each variable. Generalized exact test, $P < 0.01$.

Family predominantly	Family predominantly	
	Gap	Shade-tolerant
Not ant-defended	2	12
Ant-defended	7	3

Discussion

Our estimate that one-third of the woody dicotyledons, lianas, and herbaceous vines are ant-defended is the highest figure yet reported for any flora (Table 13. 5). This high frequency of ant-defended species is especially significant since we did not inspect the reproductive tissues of the plants in our survey. For this reason, and those given above, we suspect that our strikingly high frequency actually underestimates the true frequency of ant-defended species on BCI by several per cent, or more. The prevalence of ant defence in monocotyledons and non-vining herbs on Barro Colorado Island is unknown, but may be less than in the dicotyledons sampled.

Our results support the contention that extrafloral nectaries are more common in moist tropical than in temperate floras (Bentley 1977a; Buckley 1982; Oliveira and Leitão-Filho 1987; Coley and Aide 1989). However, we note that careful species counts in temperate communities may also uncover a high frequency of species with extrafloral nectaries (Table 13. 5). Clearly, more surveys based on numbers of species, rather than on the number of individuals or percentage cover, are necessary to resolve biogeographical patterns in ant defence against herbivores.

In our survey, species with extrafloral nectaries were not independently distributed among taxa, a pattern which has been found in other surveys as well (Elias 1983; Lersten and Brubaker 1987; McKey 1989). The vast majority of genera (87 per cent) represented by more than one species in our survey contained either all ant-defended species or no species that were ant-defended. Phylogenetic dependency occurred at higher taxonomic levels as well. Elias (1983) proposed that species with extrafloral nectaries were most abundant in the three advanced subclasses of the Magnoliopsida, the Dilleniidae, the Rosidae, and the Asteridae. Published lists of species with extrafloral nectaries from local floras (Keeler 1979b; Oliveira and Leitão-Filho 1987; Pemberton 1988) qualitatively support this contention, but

Table 13.5. The frequency of ant-defended plants in temperate and tropical communities of the New World.

Site (vegetation)	% Ant-defended species		Reference
Temperate			
Nebraska, USA (mixed)	2.3	a,e	Keeler 1979*b*
Iowa, USA (deciduous forest)	25.9	a,f,h	E. W. Schupp, unpublished work
Virginia, USA (deciduous forest)	25.0	a,e,h	E. W. Schupp, unpublished work
Warm Desert			
California, USA (Mojave)	1.0	b,e	Pemberton 1988
California, USA (Mojave)	2.6	b,e	Pemberton 1988
California, USA (Colorado)	3.2	b,e	Pemberton 1988
Tropical			
Hawaii, USA (mixed)	1.2	c,e	Keeler 1985
Sao Paulo, Brazil (cerrado)	18.3	a,e,g	Oliveira and Leitão-Filho 1987
Barro Colorado Is., Panama (moist forest)	33.3	d,f,h	Present study

a = Woody dicotyledons
b = All species
c = Hawaiian native vascular species
d = Dicotyledonous trees, shrubs, lianas, and vines
e = Extrafloral nectaries only
f = Includes species with large pearl bodies
g = Mean of 5 sites, s.d. = 2.1 per cent
h = Ant rewards on reproductive structures not considered

provide no estimates of the relative frequencies of ant defence in the various subclasses. Since we considered absence as well as presence, this survey is the first to quantify the relative rarity of ant-defended species in primitive subclasses.

Several lines of evidence suggest that the taxonomic distribution of ant defence is the product of both phylogeny and convergent selection. Extrafloral nectaries clearly show structural and positional homologies frequently at the generic, often at the familial, and rarely at the ordinal levels (Elias 1983). However, the apparent ease with which extrafloral nectaries can evolve or be lost within a lineage (Bentley 1977*a*; McKey 1989) (though this seems not to be true of bracken, see Chapter 16) suggest a role for selection. Our finding that families which tend to be ant-defended also tend to be gap-requiring, further supports the notion that the taxonomic distribution of ant defence is not simply an outcome of phylogeny.

The rarity of ant defence in the primitive dicotyledonous subclasses is not easily explained by either habitat requirements or phylogenetic 'primitiveness'. We found no evidence that advanced and primitive subclasses differ in habitat requirements, and species from the primitive subclasses, as well as even more primitive taxa such as ferns (Koptur *et al.* 1982), have extrafloral nectaries. Perhaps the difference is ultimately due to phylogenetic constraints associated with such attributes as physiology, resource allocation patterns, or leaf-flushing phenology.

A mixture of dietary nutrients such as the sugars and amino acids of nectar with the lipids and proteins of pearl bodies should be the most effective means of attracting and maintaining an efficient bodyguard of ants (Beattie 1985). Given the complementary nature of the nutrients in extrafloral nectar and pearl bodies, O'Dowd (1982) suggested that pearl bodies would be most common among species with extrafloral nectaries. Our survey of both pearl bodies and extrafloral nectaries, however, did not suggest that these ant attractants are positively associated.

Two distinct, although not mutually exclusive, hypotheses have been proposed to account for the ecological distribution of ant-defended plants: 'ant limitation' and 'resource limitation'. According to the 'ant limitation hypothesis' the origin and maintenance of ant defence is limited by aspects of the ant community, such as the abundance of foragers, species richness, or the frequency of aggressive species (Bentley 1976, 1977*a*, 1981; Keeler 1980*b*, 1981*c*, 1989). Ant-defended plants should be most abundant where large numbers of effective ant defenders are reliably present.

In formulating the 'ant limitation hypothesis', Bentley (1976, 1977*a*, 1981) presented two lines of evidence in its favour. First, she found higher percent covers of ant-defended plants and higher abundances of ants at baits in human clearings and forest edges than in undisturbed forest. However, it is unlikely that such human-disturbed habitats closely resemble naturally occurring tree fall gaps, which were probably the major disturbance regime in which lowland neotropical plants evolved (Brokaw 1985, 1987). Moreover, the low-diversity ant assemblages of tropical human-disturbed habitats, composed of widespread species found less frequently within the forest, may differ extensively from assemblages found in natural treefall gaps (J. T. Longino, personal communication). Intensive surveys in natural tree fall gaps and forest understorey on Barro Colorado Island suggest that ant assemblages in these habitats do not differ substantially in species diversity, abundance or composition (D. H. Feener and E. W. Schupp, unpublished work). While we question the role of ants in determining the local distribution of ant-defended plants, they do appear to affect the large-scale biogeographical distributions of these plants (Keeler 1985).

Bentley (1981) argued that the prevalence of ant-defended lianas also supported the 'ant limitation hypothesis'. Presumably, the climbing habit of these plants greatly increases branch to branch contact, providing greater access for foraging ants. Bentley found that ant activity on natural as well as experimental 'soda-straw' lianas and free-standing plants, supported this interpretation. However, our analyses suggest that the climbing habit itself has not been a major factor promoting the evolution of ant defence.

Recent evidence and newly emerging theory on the quantitative and qualitative allocation of resources to anti-herbivore defence suggest that the distributions of particular defence strategies are functions of resource availability (McKey 1979, 1984; Keeler 1981c; Coley et al. 1985; Huxley 1986; Coley, 1987; Davidson and Fisher Chapter 20, this volume). In accord with this general 'resource limitation hypothesis', the evolution of ant defence may be favoured in gap species even in the absence of differences in the characteristics of ant assemblages inhabiting tree fall gaps and closed canopy forest. Gap species, growing in high-light, carbon-rich environments have rapid vertical growth with relatively continuous production of short-lived leaves (Bazzaz and Pickett 1980; Brokaw 1985; Coley et al. 1985). These characteristics may promote ant defence in two ways. The sugar- and lipid-based rewards of facultatively ant-defended plants appear to be energetically and metabolically cheap in general (O'Dowd 1979, 1980; Beattie 1985), but this will be especially true in carbon-rich habitats (Keeler 1981c). Further, since extrafloral nectar and pearl body production on vegetative structures is mainly on young, expanding shoots and leaves (Beattie 1985), the more continuous growth and leaf production of gap species should lead to more continual ant guarding than the periodic, synchronized leaf flush typical of shade-tolerant species (Aide 1988). Ant defence is more likely to be a viable form of defence for gap species than for shade-tolerant species simply because of the contemporary and evolutionary responses to the availability of light. This argument is compatible with McKey's (1989) suggestion, supported by our results, that the prevalence of ant-defended lianas may simply reflect the tendency of lianas to be gap species (Putz 1984) with more continuous leaf production than trees and shrubs (Putz and Windsor 1987).

Acknowledgements

For identifying species, classifying habitat requirements, and discussion we are extremely grateful to T. M. Aide, C. Augspurger, N. V. L. Brokaw, L. Coley, R. B. Foster, N. Garwood, A. Gentry, and J. Putz. C. Herrera, P. Jordano, J. T. Longino, E. Marity, K. McLeod, K. Morehead, R. Sharitz and R. Soriguer were extremely helpful during the analysing and writing. We thank A. Smith and the Whitehall Foundation for access to growing-house space, N. Garwood for plants, J. T. Longino for information on ants, J. Beam for inspiration, and the Government of Panama and the Smithsonian Tropical Research Institution for financial and logistical support. This study was initiated while the authors held Smithsonian Fellowships. Manuscript preparation was assisted by contract DE-AC09-765R00-819 between the

U.S. Department of Energy and the University of Georgia's Savannah River Ecology Laboratory.

Appendix 13.1.

Characteristics of 243 woody dicotyledons and herbaceous vines of Barro Colorado Island surveyed for the presence of ant attractants. Genera and species follow Croat (1978), higher classification follows Cronquist (1981). Sample: I = evidence of ant attractants based on observations of potted saplings growing isolated in a screened growing house; F = evidence based on field-collected material. Ant attractants: N = extrafloral nectaries, LB = large bodies, SB = small bodies, and A = ants observed to harvest attractants. Lifeform categories: LT = large trees, ST = small trees and shrubs, and L = lianas and vines. Habitat requirements: G = gap, S = shade-tolerant, U = unclassified; * = those species for which a minimum of three experts had an opinion on habitat requirements and at least 75 per cent of these respondents were in agreement; analyses of only these species yield the same conclusions as those based on the entire data set.

Classification	Sample	Potential ant attractants	Lifeform	Habitat
Subclass Magnoliidae				
Order Magnoliales				
Annona acuminata	F	—	ST	S
A. hayesii	F	—	ST	S
A. spraguei	I	—	LT	G*
Desmopsis panamensis	I	—	ST	S*
Guatteria amplifolia	I	—	ST	S
G. dumetorum	F	—	LT	S*
Xylopia macrantha	I	—	ST	S*
Myristacaceae				
Virola sebifera	I	—	LT	S*
V. surinamensis	F	—	LT	S
Order Laurales				
Monimiaceae				
Siparuna pauciflora	F	—	ST	S
Lauraceae				
Beilschmiedia pendula	I	—	LT	S*
Nectandra purpurascens	I	—	LT	S
Ocotea skutchii	F	—	LT	S

Classification	Sample	Potential ant attractants	Lifeform	Habitat
Order Piperales				
Piperaceae				
Piper aequale	I	LB	ST	S
P. cordulatum	I	SB	ST	S
P. marginatum	I	SB	ST	G*
Order Aristolochiales				
Aristolochiaceae				
Aristolochia chapmaniana	F	—	L	G*
A. pilosa	F	—	L	G
Order Ranunculales				
Menispermaceae				
Abuta racemosa	F	—	L	S*
Chondrodendron tomentosum	F	—	L	S
Cissampelos pareira	F	—	L	G
Odontocarya truncata	F	—	L	S
Subclass Hamamelidae				
Order Urticales				
Ulmaceae				
Celtis iguanaeus	F	—	L	G
Trema micrantha	F	SB	LT	G
Moraceae				
Brosimum alicastrum	I	SB	LT	S*
Castilla elastica	I	SB	LT	G*
Ficus insipida	I	—	LT	G*
Maquira costaricana	I	—	LT	S
Olmedia aspera	I	—	LT	S
Poulsenia armata	I	—	LT	S*
Sorocea affinis	I	SB	ST	S*
Cecropiaceae				
Pourouma guianensis	I	—	LT	G
Urticaceae				
Urera baccifera	I	SB	ST	S
Subclass Caryophyllidae				
Order Caryophyllales				
Nyctaginaceae				
Neea amplifolia	I	—	ST	S
Order Polygonales				
Polygonaceae				
Coccoloba acuminata	I	—	ST	S

Appendix 13.1. (*cont.*)

Classification	Sample	Potential ant attractants	Lifeform	Habitat
C. manzanillensis	I	—	LT	S*
C. parimensis	F	—	L	S*
Subclass Dilleniidae				
Order Dilleniales				
Dilleniaceae				
Doliocarpus dentatus	F	—	L	S*
D. major	F	—	L	
Tetracera portobellensis	F	SB	L	S*
Order Theales				
Ochnaceae				
Ouratea lucens	I	N, A	ST	S*
Clusiaceae				
Calophyllum longifolium	I	—	LT	S*
Rheedia acuminata	F	—	LT	S
R. edulis	I	—	LT	S*
Symphonia globulifera	F	—	LT	S
Vismia baccifera	I	—	ST	G*
Order Malvales				
Elaeocarpaceae				
Aloanea terniflora	F	—	LT	S*
Tiliaceae				
Apeiba membranacea	I	LB	LT	G*
A. tribourbou	I	SB	LT	G*
Luehea seemannii	I	—	LT	G*
Trichospermum mexicanum	I	N, SB	LT	G*
Triumfetta lappula	I	N	ST	G
Sterculiaceae				
Byttneria aculeata	F	N, A	L	G
Guazuma ulmifolia	I	SB	LT	G*
Herrania purpurea	I	SB	LT	U*
Melochia lupulina	I	SB	LT	
Sterculia apetala	I	—	LT	G*
Bombacaceae				
Bombacopsis quinata	I	N, SB, A	LT	G*
Cavanillesia platanifolia	I	—	LT	G*
Ceiba pentandra	I	N, SB, A	LT	G*
Ochroma pyramidale	I	N, LB, A	LT	G*

Classification	Sample	Potential ant attractants	Lifeform	Habitat
Pseudobombax septenatum	I	N, A	LT	G*
Quararibea asterolepis	I	—	LT	S*
Malvaceae				
Hampea appendiculata	F	N, A	LT	G*
Order Lecythidales				
Lecythidaceae				
Gustavia superba	I	N, LB, A	LT	S
Order Violalels				
Flacourtiaceae				
Casearia arborea	I	—	LT	G*
Hasseltia floribunda	I	N, SB, A	LT	S
Laetia thamnia	I	—	LT	S
Lindackeria laurina	F	—	LT	S*
Zuelania guidonia	I	—	LT	G*
Bixaceae				
Cochlospermum vitifolium	I	N, SB, A	LT	G*
Lacistemataceae				
Lacistema aggregatum	I	—	LT	S*
Violaceae				
Hybanthus prunifolius	I	—	ST	S*
Rinorea sylvatica	I	—	ST	S*
Turneraceae				
Turnera panamensis	F	N, A	ST	G
Passifloraceae				
Passiflora auriculata	F	N, A	L	G
P. biflora	F	N	L	G*
P. coriacea	I	N	L	G
P. nitida	I	N	L	U
Cucurbitaceae				
Gurania makoyana	F	SB	L	G
Melothria pendula	F	—	L	G
Order Capparales				
Capparaceae				
Capparis frondosa	I	—	ST	S*
Order Ebenales				
Sapotaceae				
Chrysophyllum cainito	I	—	LT	S*
Cynodendron panamense	I	—	LT	S*
Pouteria unilocularis	I	—	LT	S*
Ebenaceae				
Diospyros artanthifolia	I	N	LT	U

Appendix 13.1. (*cont.*)

Classification	Sample	Potential ant attractants	Lifeform	Habitat
Order Primulales				
Myrsinaceae				
Ardisia bartlettii	F	—	ST	S
Stylogyne standleyi	I	N, SB, A	LT	S
Subclass Rosidae				
Order Rosales				
Connaraceae				
Connarus panamensis	F	—	L	S*
C. turczaninowii	F	—	L	S*
Chrysobalanaceae				
Hirtella americana	I	N, A	LT	S*
H. racemosa	F	N, A	ST	S*
H. triandra	I	N, A	LT	S*
Licania hypoleuca	F	SB	LT	S*
L. platypus	I	N, A	LT	S*
Order Fabales				
Mimosaceae				
Acacia hayesii	F	N, A	L	G*
Albizia guachapele	I	N	LT	U
Inga marginata	I	N, A	LT	S
I. sapindoides	I	N, A	LT	S
I. spectabilis	I	N, A	LT	S
Pithecellobium rufescens	I	N, SB, A	LT	S
Caesalpiniaceae				
Bauhinia guianensis	F	—	L	S
Cassia fruticosa	I	N, SB, A	ST	G
Prioria copaifera	I	—	LT	S*
Swartzia simplex	I	—	LT	S*
Tachigali versicolor	I	SB	LT	S
Fabaceae				
Adira inermis	I	—	LT	S*
Clitoria javitensis	F	—	LT	S*
Dioclea wilsonii	F	—	L	S
Erythrina costaricensis	I	—	ST	S
Machaerium arboreum	F	—	L	G
M. milleflorum	F	—	L	G
M. seemannii	F	—	LT	S
Platypodium elegans	I	—	LT	S*

Classification	Sample	Potential ant attractants	Lifeform	Habitat
Pterocarpus rohrii	I	—	LT	S*
Teramnus uncinatus	F	—	L	G
Vatairea erythrocarpa	I	—	LT	S
Order Myrtales				
Myrtaceae				
Calycolpus warscewiczianus	I	—	ST	S
Eugenia coloradensis	F	—	LT	S
E. oerstedeana	I	—	LT	S
Myrcia fosteri	I	—	ST	S
M. gatunensis	I	—	LT	S
Melastomataceae				
Conostegia speciosa	F	—	ST	G
Miconia argentea	I	—	LT	G*
M. impetiolaris	I	—	ST	U
M. lacera	I	—	ST	U
M. prasina	I	—	ST	U
Mouriri myrtilloides	F	—	ST	S*
Combretaceae				
Terminalia amazonica	I	N, A	LT	G*
Order Rhizophorales				
Rhizophoraceae				
Cassipourea elliptica	I	—	LT	S*
Order Santalales				
Olacaceae				
Heisteria concinna	I	—	LT	S*
Order Celastrales				
Celastraceae				
Maytenus schippii	I	—	ST	S
Hippocrateaceae				
Anthodon panamense	F	—	L	S
Hippocratea volubilis	F	—	L	U
Hylenaea praecelsa	F	—	L	S
Prionostemma aspera	F	—	L	U
Tontelea richardii	I	—	L	U
Order Euphorbiales				
Euphorbiaceae				
Acalypha diversifolia	I	SB	ST	S
Alchornea costaricensis	I	N, A	LT	G*
A. latifolia	I	N	LT	U
Croton billbergianus	I	N, A	ST	G*

Appendix 13.1. (*cont.*)

Classification	Sample	Potential ant attractants	Lifeform	Habitat
C. panamensis	I	N	LT	G
Dalechampia cissifolia	F	N	L	G
Drypetes standleyi	I	—	LT	S*
Hura crepitans	I	N, A	LT	G*
Hyeronima laxiflora	I	—	LT	G*
Mabea occidentalis	I	N, A	ST	S*
Margaritaria nobilis	I	—	LT	S
Omphalea diandra	F	N, A	L	G
Sapium caudatum	I	N, A	LT	G*
Order Rhamnales				
Rhamnaceae				
Gouania lupuloides	F	N	L	G*
Vitaceae				
Cissus microcarpa	F	N	L	G*
C. pseudosicyoides	F	N	L	G
Vitis tiliifolia	I	—	L	G*
Order Linales				
Erythroxylaceae				
Erythroxylum panamense	I	—	ST	S
Order Polygalales				
Malpighiaceae				
Byrsonima crassifolia	F	—	LT	G*
Hiraea grandifolia	F	N	L	S*
H. reclinata	F	N	L	S*
Mascagnia hippocrateoides	F	N	L	U
M. nervosa	F	N	L	U
Spachea membranacea	I	—	ST	S
Stigmaphyllon ellipticum	F	N	L	G
S. hypargyreum	F	N	L	G
S. lindenianum	F	N	L	G
Vochysiaceae				
Vochysia ferruginea	I	—	LT	G
Order Sapindales				
Sapindaceae				
Allophyllus psilospermus	I	SB	ST	S
Cupania rufescens	I	SB	LT	S*
C. sylvatica	I	SB	ST	S*
Paullinia fibrigera	F	SB	L	S

Classification	Sample	Potential ant attractants	Lifeform	Habitat
P. glomerulosa	F	—	L	S*
P. pterocarpa	F	N	L	S*
P. turbacensis	F	N	L	S*
Serjania decapleuria	F	N	L	G
S. mexicana	F	N	L	G*
Talisia nervosa	F	—	ST	S*
T. princeps	I	—	ST	S*
Burseraceae				
Protium panamense	I	—	LT	S*
P. tenuifolium	I	—	LT	S*
Tetragastris panamensis	I	SB	LT	S*
Trattinnickia aspera	I	SB	LT	G*
Anacardiaceae				
Anacardium excelsum	I	—	LT	G*
Spondias mombin	I	LB, A	LT	G*
S. radlkoferi	I	LB, A	LT	G*
Simaroubaceae				
Picramnia latifolia	I	N	ST	S
Quassia amara	I	N, SB	ST	S*
Simarouba amara	I	N, LB, A	LT	S*
Meliaceae				
Cedrela odorata	I	N	LT	G
Guarea glabra	I	N, SB, A	LT	S*
G. 'hairy'	F	N, SB	LT	S
Trichilia cipo	I	N, SB	LT	S*
T. montana	F	N, A	LT	S
Rutaceae				
Zanthoxylum belizense	I	N	LT	G*
Z. panamense	I	N, SB	LT	G*
Z. setulosum	I	N	LT	G
Order Apiales				
Araliaceae				
Dendropanax arboreum	I	—	LT	S
Didymopanax morototoni	F	—	LT	G*
Subclass Asteridae				
Order Gentianales				
Loganiaceae				
Strychnos brachistantha	F	—	L	S
Apocynaceae				
Aspidosperma cruenta	I	—	LT	S*
Lacmellea panamensis	I	—	LT	S*

Appendix 13.1. (*cont.*)

Classification	Sample	Potential ant attractants	Lifeform	Habitat
Odontadenia macrantha	I	—	L	U
Prestonia acutifolia	F	—	L	G
Tabernaemontana arborea	I	—	LT	S
Thevetia ahouai	I	—	ST	S
Order Solanales				
Solanaceae				
Cyphomandra hartwegii	I	—	ST	G
Solanum hayesii	F	—	LT	G*
S. lanciifolium	F	—	L	G*
S. subinerme	F	—	ST	G
S. umbellatum	F	—	ST	G
Witheringia solanacea	F	—	ST	G
Convolvulaceae				
Ipomoea phillomega	F	N	L	G*
I. quamoclit	F	—	L	G
Maripa panamensis	F	—	L	S*
Order Lamiales				
Boraginaceae				
Cordia bicolor	F	—	LT	S
C. lasiocalyx	F	SB	ST	S*
Tournefortia cuspidata	F	—	ST	S
Verbenaceae				
Aegiphila cephalophora	F	N	L	G
A. panamensis	I	N	LT	G
Petrea aspera	F	—	L	S*
Order Scrophulariales				
Acanthaceae				
Aphelandra sinclairiana	I	SB	ST	S
Justicia graciliflora	F	—	L	G
Mendoncia gracilis	I	—	L	S
Bignoniaceae				
Arrabidaea chica	F	N, A	L	G
A. verrucosa	F	N	L	S
Ceratophytum tetragonolobum	F	N	L	G
Jacaranda copaia	F	N, A	LT	G*
Macfadyena unguis-cati	F	N	L	G
Martinella obovata	F	N	L	S
Pachyptera kerere	F	N	L	S

Classification	Sample	Potential ant attractants	Lifeform	Habitat
Pleonotoma variabilis	F	N	L	S
Tabebuia rosea	I	N	LT	G
Order Rubiales				
Rubiaceae				
Alibertia edulis	I	—	ST	S
Alseis blackiana	I	—	LT	S*
Coussarea curvigemmia	F	—	ST	S*
Faramea occidentalis	I	—	ST	S*
Macrocnemum glabrescens	I	—	LT	S
Palicourea guianensis	I	—	ST	G*
Pentagonia macrophylla	I	—	ST	S*
Posoquiera latifolia	I	—	ST	S
Psychotria deflexa	I	—	ST	S*
P. furcata	F	—	ST	S
P. horizontalis	I	—	ST	S
P. pubescens	I	—	ST	G
Order Asterales				
Asteraceae				
Vernonia canescens	F	—	ST	G*

14

The ant community associated with extrafloral nectaries in the Brazilian cerrados

Paulo S. Oliveira and Carlos R. F. Brandão

Introduction

Schemske (1983) stressed that plants providing 'limited' resources for ants, such as extrafloral nectar, generally have a higher diversity of ant associates than those providing 'major' resources, such as nest sites and diverse foods. Since extrafloral nectar comprises only a part of the nutritional requirements of an ant colony, ant-guard systems mediated solely by the supply of nectar are usually very generalized (Schemske 1983; Beattie 1985).

Since ants visiting extrafloral nectaries (EFNs) can differ strikingly with respect to their effectiveness as anti-herbivore agents (Bentley 1977*a,b*; Schemske 1980; Oliveira *et al.* 1987), most studies have focused on 'key' ant visitors rather than on whole ant assemblages. The study conducted by Schemske (1982) on *Costus* spp. (Zingiberaceae) in Panama is one of the few providing a more complete analysis of the species composition of ants at EFNs.

There is no information on the species composition of the ant faunas visiting EFN plants in South America, and the present paper discusses data on ant assemblages at EFNs in cerrado vegetation of Brazil (Oliveira *et al.* 1987; Oliveira and Oliveira-Filho 1990). Surveys were made of ants taking extrafloral nectar and attacking live termite baits on two typical cerrado shrubs, *Caryocar brasiliense* Camb. (Caryocaraceae) and *Qualea grandiflora* Mart. (Vochysiaceae). The range of ant taxa visiting these two cerrado plants is discussed, and a preliminary comparison of ant assemblages at EFNs from other tropical and temperate habitats is made on the basis of a literature review.

Extrafloral nectaries in cerrado vegetation

The cerrados are a savannah-like formation that covers two million km², approximately 25 per cent of the Brazilian territory (Fig. 14. 1). The cerrados vary physiognomically and Goodland (1971) recognized four main structural types:

1. forest with more or less merging canopy ('cerradão');
2. dense scrub of shrubs and trees ('cerrado' *sensu stricto*);
3. open scrub ('campo cerrado'); and
4. open grassland with scattered shrubs ('campo sujo').

Surveys conducted in nine areas of different types of cerrado showed that EFNs are widely distributed amongst the woody flora (Oliveira and Leitão-Filho 1987). A total of 44 cerrado species from 17 families were found to have EFNs; the Mimosaceae, Vochysiaceae, and Bignoniaceae were the families most frequently bearing these glands. In south-east Brazil EFNs were present on 15–22 per cent of the species sampled in five cerrado areas, with local abundance of the individuals varying from 8–20 per cent. In the state of Mato Grosso, West Brazil, 21–26 per cent of the woody species sampled in four cerrado sites had EFNs, representing 22–31 per cent of the individuals examined (Oliveira and Oliveira-Filho 1990). These values are in general much higher than those obtained in temperate habitats (see Chapter 13).

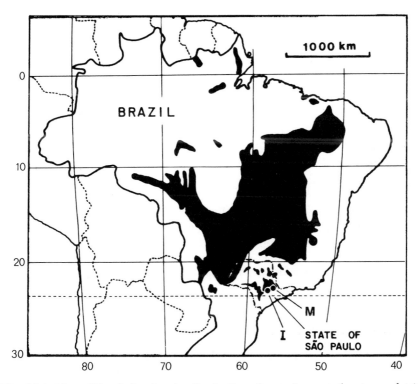

Fig. 14.1. Map of Brazil showing the distribution of cerrado vegetation (*sensu lato*) (dark area). The localities of Itirapina (I) and Mogi-Guaçu (M) are indicated.

Ants at extrafloral nectaries in cerrado vegetation

Extrafloral nectaries are intensively visited by several ant species in the cerrado, and there is evidence that mutualism with ants may constitute an important anti-herbivore tactic in this vegetation type. Possession of nectaries on the stem was shown to increase ant density on shrubs of *Qualea grandiflora* over that of non-nectary plants, resulting in many more live termite baits being attacked by foraging ants (Oliveira *et al.* 1987). Similar results were obtained with *Caryocar brasiliense*, which has nectaries on the outer surface of the sepals that are visited day and night by many ant species. Ant-excluded shrubs of *Caryocar* were infested by significantly more phytophagous insects than plants to which the ants had free access (Oliveira 1988).

Ant censuses on *Caryocar brasiliense*

The ant fauna foraging on the sepals of *Caryocar brasiliense* was censused in an area of cerrado (*sensu stricto*) in the county of Itirapina (22°15'S, 47°49'W), state of São Paulo, SE Brazil. Daytime censuses were conducted at 08.00 and 14.00 h, and night censuses at 20.00 and 02.00 h. Three replicate censuses were performed from September to November 1986 (the growing season of *Caryocar*), totalling 12 samples of the ant fauna. Sampling was conducted on sunny days and clear nights on the same 40 shrubs (1.0–1.5 m tall). At each plant 20 seconds were spent recording the number of ant workers from different species. Ant specimens were then collected for taxonomic identification.

Figure 14. 2 summarizes the species composition of ants at the EFNs of *Caryocar*. Twenty-seven ant species in 12 genera and five subfamilies were recorded. Ants collected extrafloral nectar on the plants by day and night, but the species composition of the principal ant visitors changed from one period to the other. Both in number of species and in individuals the genus *Camponotus* was far better represented than the remaining 11 genera. Nine *Camponotus* species together accounted for 54 per cent of the ant individuals recorded on *Caryocar* (Fig. 14. 3). *Camponotus* was also the most frequent genus on the plants, comprising nearly 60 per cent of all generic occurrences. However, frequency of visits may be a poor indicator of mutualistic activity; for example *Camponotus rufipes* accounted for 22 per cent of the records of ant visitors to *Caryocar*, but only 4 per cent of attacks on live termite baits.

Other common ant genera on *Caryocar* were *Azteca* (4 species), *Pheidole* (3 species) and *Zacryptocerus* (1 species), each accounting for 11–12 per cent of the ants recorded and 8–15 per cent of the generic observations on the plants monitored (Fig. 14. 3). Eight other genera were rarely seen on

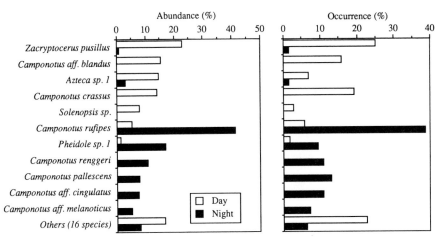

Fig. 14.2. Daily variation in species composition of the main ant visitors to the extra-floral nectaries of *Caryocar brasiliense* in cerrado vegetation of SE Brazil. The abundance of a given species is defined as the percentage of all individuals observed, and occurrence as the percentage of all species observed in each period. A total of 291 ants were observed during the day and 268 during the night; total species observations in each period were 176 and 137, respectively.

Caryocar and together comprised only 10 per cent of all ants observed, with individual frequencies around 2 per cent or less (Fig. 14. 3).

The ant fauna on *Caryocar* was also evaluated by placing live workers of the termite *Armitermes euamignathus* on the plants as bait for ants. Six termites were placed on three different parts of a plant: buds, young leaves, and adult leaves (two baits per location). Once baiting was completed on a plant, attacks on termites by foraging ants were monitored during a 15-minute period. Sixty different plants (1.0–1.5 m tall) were baited; 30 during the day (08.00 to 12.00) and 30 at night (19.00 to 24.00).

Twenty-nine per cent (103 of 360) of the termites placed on *Caryocar* plants were attacked by foraging ants. A total of 20 ant species from 12 genera were recorded attacking the termites (Fig. 14. 3), all of which were also seen collecting extrafloral nectar on the plants. Seven of these species, however, had not been recorded in the ant censuses conducted previously. The surveys with baits confirmed the importance of *Camponotus* over the other ant genera: seven *Camponotus* species together attacked 52 per cent of the termites (Fig. 14. 3). Two common species, *Camponotus crassus* and *C. aff. blandus*, together conducted 38 per cent of the attacks. Ants in the genera *Pheidole* (2 species), *Zacryptocerus*, *Wasmannia*, and *Brachymyrmex* (1 species each) together accounted for 36 per cent of the termites attacked. As

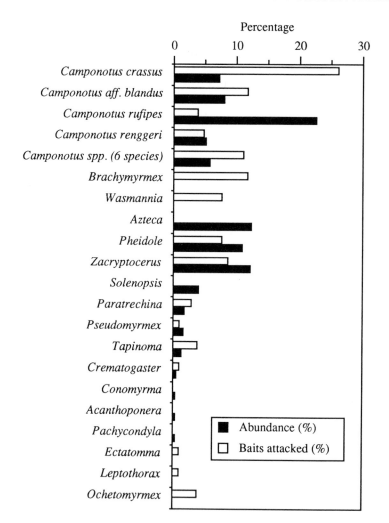

Fig. 14.3. Comparison of the abundance at nectaries and percentage of attacks carried out by the most common ant species and genera found on *Caryocar brasiliense* in cerrado vegetation. Abundance is defined as the percentage of all individuals observed day and night (559 ants on 40 shrubs). A total of 360 termite baits were placed on 60 plants, 103 of which were attacked by foraging ants.

in the previous censuses, some species (e.g. *Camponotus crassus* and *Zacryptocerus pusillus*) presented a typically diurnal activity pattern on the plants, while others (e.g. *Brachymyrmex* sp. and *Pheidole* sp. 3) visited mainly at night.

The combined data on the ant fauna associated with *Caryocar* provides an

Table 14.1. Ant species observed attacking live termite-baits on leaves of *Qualea grandiflora* in cerrado vegetation of São Paulo, SE Brazil. Baiting was done during the day; three baits were each placed on three different leaves of 100 shrubs. In parentheses is the percentage of baits attacked by different ant species. The scores of *Camponotus crassus*, *C. rufipes*, and *C. aff. blandus* are grouped. Original data from Oliveira *et al.* (1987).

Ant species	No. of baits attacked (%)
Formicinae	
Camponotus crassus, *C. rufipes* and *C.* aff. *blandus*	167 (87.4)
Camponotus sp.	1 (0.5)
Unidentified genus	1 (0.5)
Myrmicinae	
Zacryptocerus depressus	7 (3.7)
Z. pusillus	6 (3.1)
Wasmannia sp.	1 (0.5)
Dolichoderinae	
Unidentified genus	3 (1.6)
Pseudomyrmecinae	
Pseudomyrmex flavidulus	2 (1.0)
P. gracilis	2 (1.0)
Ponerinae	
Ectatomma quadridens	1 (0.5)
Total baits attacked	191
Total baits	300

overall picture of 34 species and 17 genera of ants visiting the plant for its nectary secretions. The subfamilies Formicinae and Myrmicinae dominated the ant fauna, comprising, together, 23 species of visiting ants. The genera *Camponotus*, *Brachymyrmex*, *Pheidole*, *Zacryptocerus*, and *Azteca* were the most frequent at EFNs of *Caryocar*. However *Brachymyrmex*, which conducted 12 per cent of attacks on baits, was not recorded by the survey at EFNs.

Ant censuses on *Qualea grandiflora*

As part of a study on ant foraging and herbivore deterrence (Oliveira *et al.* 1987), the ant fauna associated with the extrafloral nectary plant *Qualea grandiflora* was evaluated on plants 0.6–1.2 m tall using the same method of live termite baiting employed for *Caryocar*.

A total of 12 ant species from seven genera and five subfamilies were recorded attacking termites on *Qualea*, and all of them were also seen at EFNs on the plants (Table 14. 1). As in the censuses on *Caryocar brasiliense*, ants in the genus *Camponotus* (4 species) were the most predaceous visitors to *Qualea*, conducting 88 per cent of the termite attacks. The ant fauna visiting *Qualea* is probably underestimated, since we did not census nocturnal visitation. The censuses with termite baits on *Caryocar* also recorded 12 ant species visiting the plants during the day, five of which were recorded on *Qualea*-as well (*Camponotus crassus*, *C.* aff. *blandus*, *C. rufipes*, *Zacryptocerus pusillus*, and *Pseudomyrmex flavidulus*). We believe that common nocturnal species observed on *Caryocar*, such as *Camponotus renggeri*, *C. pallescens*, *Brachymyrmex* sp., and *Pheidole* spp., also collected extrafloral nectar on *Qualea* during the night.

As a whole, the surveys of the ant fauna on *Caryocar brasiliense* and *Qualea grandiflora* showed the presence of 40 species from 19 genera of ants which collect extrafloral nectar in cerrado vegetation. Many of the species recorded on *Caryocar* and *Qualea* are arboreal, and occupy stems previously hollowed out by boring beetles. The ant genera *Camponotus*, *Zacryptocerus*, *Pseudomyrmex*, *Azteca*, *Solenopsis*, *Crematogaster*, and *Leptothorax* are among the most common inhabitants of stem galleries in cerrado vegetation (Morais and Benson 1988; Oliveira 1988). During a study in the cerrado of Mogi-Guaçu, São Paulo, Morais (1980) recorded 27 stem-nesting ant species and 13 ground-nesting ones, and two which nested in either substrate. Although all species were observed foraging both on foliage and on the ground, stem-nesting ants were more frequent at baits placed on leaves than on the ground. The fauna in the cerrado of Mogi-Guaçu was also dominated by *Camponotus*, which comprised ten of the 42 species recorded by Morais, with *Camponotus rufipes* and *C. crassus* being the most abundant.

Ant assemblages at extrafloral nectaries in tropical and temperate habitats

Tables 14. 2 and 14. 3 summarize the ant genera, and number of species per genus, observed at extrafloral nectaries of different plants in tropical, subtropical, and temperate habitats. Although different authors have employed varying methods to estimate the ant fauna at EFNs, it is evident from the data that a wide variety of ant taxa are attracted to many different nectary-bearing plants, with no apparent specialization being required. As stressed by Carroll and Janzen (1973), extrafloral nectar is a generalized food source 'eaten by practically any ant that encounters it'. The combined data from Tables 14. 2 and 14. 3 provide a total of 38 ant genera recorded at

EFNs; 33 of which occur in tropical and subtropical areas and 17 in temperate localities. Twelve genera occur in both major regions (these scores assume that the three unidentified genera are different from all others). The subfamilies Myrmicinae and Formicinae comprised most of the genera in both tropical and temperate studies, whereas the Ponerinae and Pseudo-myrmecinae were not recorded on temperate EFN-bearing plants (Fig. 14.4).

In tropical and subtropical habitats 9.4 ± 3.2 ant genera ($\bar{x} \pm$ s.d.) and 15.3 ± 7.2 species were found associated with EFN-bearing plants, while in temperate areas these values were 5.5 ± 3.2 genera and 7.3 ± 4.8 species. These differences are not surprising as many more species and genera of ants are found in the tropics than in temperate regions (Kusnezov 1957). The results in Tables 14.2 and 14.3 are certainly conservative with respect to the taxonomic diversity of ant assemblages at EFNs, since only our study has included nocturnal censuses of the ant fauna visiting these glands. Moreover, myrmecophytes bearing EFNs and their ant inhabitants, as well as scattered reports of particular ant species visiting EFNs were not included in this review. Nevertheless, some general trends in the prevalence of different ant genera at EFNs can be detected with certainty. Table 14.4 presents a ranking list of the ant genera most frequently recorded visiting EFNs in 23 studies

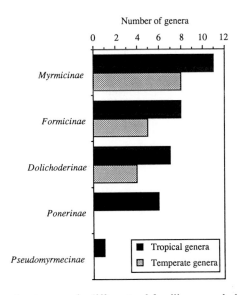

Fig. 14.4. Number of ant genera in different subfamilies recorded visiting extrafloral nectaries worldwide. Data are from 15 studies conducted in tropical and subtropical areas (black bars), and eight studies conducted in temperate habitats (hatched bars). Three unidentified genera in the Formicinae, Dolichoderinae, and Ponerinae were considered as different from all the others. See Tables 14.2 and 14.3.

Table 14.2. Ant genera, by subfamily, observed at extrafloral nectaries from different plant species in New World tropical and subtropical localities. Myrmecophytes (plants with ant domatia) bearing EFNs are not included. Values in parentheses are the number of species. Part of the data from *Caryocar* (7 species in 7 genera) and all data from *Qualea* were obtained using live termites as baits (see text). Only the study on *Caryocar* includes night-time observations.

Plant species	Formicinae	Dolichoderinae	Myrmicinae	Ponerinae	Pseudomyrmecinae	Total no. genera (species)	Source
Bixa orellana (Costa Rica)	Camponotus (3)	Monacis (1) Azteca (1)	Crematogaster (1) Zacryptocerus (1)[a]	Ectatomma (1) Pachycondyla (1)[b]	Pseudomyrmex (3)	8 (12)	Bentley (1977b)
Bytneria aculeata (Costa Rica)	Camponotus (2) Paratrechina (2) Brachymyrmex (2)	Tapinoma (1)	Crematogaster (2) Solenopsis (1) Pheidole (1) Procryptocerus (1) Zacryptocerus (1) Cyphomyrmex (1)	Ectatomma (2) Pachycondyla (3) Paraponera (1)	Pseudomyrmex (4)	14 (24)	Hespenheide (1985)
Caryocar brasiliense (SE Brazil)	Camponotus (10) Paratrechina (2) Brachymyrmex (1)	Azteca (4) Tapinoma (1) Conomyrma (1)	Zacryptocerus (1) Pheidole (3) Crematogaster (2) Solenopsis (1) Leptothorax (1) Ochetomyrmex (1) Wasmannia (1)	Acanthoponera (1) Pachycondyla (1) Ectatomma (1)	Pseudomyrmex (2)	17 (34)	Oliveira (1988)
Cassia fasciculata (Florida, USA)	Camponotus (1) Formica (1) Paratrechina (1)	Conomyrma (1) Forelius (1)[c]	Crematogaster (2) Monomorium (1) Pheidole (2) Solenopsis (1)		Pseudomyrmex (1)	10 (12)	Barton (1986)
Costus allenii (Panama)	Camponotus (4)	Azteca (1) Tapinoma (1)	Pheidole (3) Solenopsis (1) Wasmannia (1) Crematogaster (1)	Ectatomma (2) Pachycondyla (1) Paraponera (1) Odontomachus (1)		11 (17)	Schemske (1982)
Costus laevis (Panama)	Camponotus (4) Dendromyrmex (1)	Monacis (2) Azteca (1) Tapinoma (1)	Pheidole (2) Wasmannia (1) Crematogaster (5)	Ectatomma (2) Pachycondyla (2) Paraponera (1) Odontomachus (2)		12 (24)	Schemske (1982)

Plant species							Reference
Costus pulverulentus (Panama)	Paratrechina (1)	Azteca (1)	Pheidole (1) Crematogaster (3)	Ectatomma (2) Pachycondyla (1) Paraponera (1) Odontomachus (1)	Pseudomyrmex (1)	9 (12)	Schemske (1982)
Costus scaber (Panama)	Camponotus (3)	Monacis (1) Azteca (2)	Pheidole (2) Wasmannia (1)	Ectatomma (1) Pachycondyla (1) Paraponera (1) Odontomachus (1)		9 (13)	Schemske (1982)
Costus woodsonii (Panama)	Camponotus (4)		Wasmannia (1) Crematogaster (1)	Ectatomma (1)		4 (7)	Schemske (1980)
Ferocactus gracilis (Baja California, Mexico)	Brachymyrmex (1) Camponotus (1) Myrmecocystus (1)	Conomyrma (1) Forelius (1)[c]	Pheidole (2) Crematogaster (1) Monomorium (1) Solenopsis (1) Tetramorium (1)[d]			10 (11)	Blom and Clark (1980)
Inga densiflora (Costa Rica)	Camponotus (1) Myrmelachista (1) Paratrechina (1)		Pheidole (1) Solenopsis (1)	Ectatomma (1)		6 (6)	Koptur (1984)
Inga punctata (Costa Rica)	Camponotus (3) Myrmelachista (1)	Monacis (1)	Pheidole (2) Crematogaster (3) Procryptocerus (1)			6 (11)	Koptur (1984)
Passiflora quadrangularis (Costa Rica)	Camponotus (2)	Dolichoderus (1) Conomyrma (1)	Pheidole (1) Crematogaster (3) Solenopsis (2)		Pseudomyrmex (5)	7 (15)	Smiley (1986)
Passiflora vitifolia (Costa Rica)	Camponotus (3) Paratrechina (1)	Azteca (1)	Pheidole (1) Crematogaster (7) Leptothorax (1) Wasmannia (1)	Ectatomma (1) Pachycondyla (1) Unidentified genus (1)	Pseudomyrmex (2)	11 (20)	Smiley (1986)
Qualea grandiflora (SE Brazil)	Camponotus (4) Unidentified genus (1)	Unidentified genus (1)	Zacryptocerus (2) Wasmannia (1)	Ectatomma (1)	Pseudomyrmex (2)	7 (12)	Oliveira et al. (1987)

[a] Zacryptocerus (= Paracryptocerus; Kempf 1973).
[b] Pachycondyla (= Neoponera; Brown 1973).
[c] Forelius pruinosus (= Iridomyrmex pruinosus; S. P. Cover, personal communication).
[d] Tetramorium (= Xiphomyrmex; Bolton 1976).

Table 14.3. Ant genera by subfamily observed at extrafloral nectaries from different plant species in temperate localities. Values in parentheses are number of species. The subfamilies Ponerinae and Pseudomyrmecinae were not recorded in any survey and are not included.

Plant species	Formicinae	Dolichoderinae	Myrmicinae	Total no. genera (species)	Source
Cassia fasciculata (Iowa, USA)	*Lasius* (2) *Formica* (1) *Paratrechina* (1)		*Myrmica* (1) *Leptothorax* (1) *Crematogaster* (1) *Monomorium* (1) *Aphaenogaster* (1)	8 (9)	Kelly (1986)
Catalpa speciosa (Michigan, USA)	*Camponotus* (2) *Formica* (1) *Prenolepis* (1)		*Crematogaster* (1)	4 (5)	Stephenson (1982)
Helianthella quinquinervis (Colorado, USA)	*Formica* (3)	*Tapinoma* (1)	*Myrmica* (1)	3 (5)	Inouye and Taylor (1979)

Ipomoea pandurata (North and South Carolina, USA)	Camponotus (2) Formica (2)	Forelius (1)[a]	Crematogaster (2) Monomorium (1) Pheidole (1) Solenopsis (1)	7 (10)	Beckmann and Stucky (1981)
Mentzelia nuda (Nebraska, USA)	Lasius (1) Formica (2)	Conomyrma (1)[b]	Pheidole (1)	4 (5)	Keeler (1981b)
Passiflora incarnata (Georgia, USA)	Camponotus (1) Formica (1)	Forelius (1)[a]	Solenopsis (1) Crematogaster (1)	5 (5)	McLain (1983)
Pteridium aquilinum (New Jersey, USA)	Camponotus (2) Formica (2) Lasius (2)	Hypoclinea (2) Tapinoma (1)	Aphaenogaster (1) Crematogaster (2) Leptothorax (1) Monomorium (1) Myrmica (2) Solenopsis (1) Tetramorium (1)	12 (18)	Tempel (1983)
Vicia sativa (Yorkshire, England)			Myrmica (1)	1 (1)	Koptur and Lawton (1988)

[a] Forelius pruinosus (= Iridomyrmex pruinosus; S. P. Cover, personal communication).
[b] Conomyrma insana (= Dorymyrmex pyramicus; Snelling 1973).

Table 14.4. Records of different ant genera at extrafloral nectaries in tropical, subtropical, and temperate habitats. See also Tables 14.2 and 14.3. All studies were from the New World except a single European study.

Genera	Tropical and subtropical (15 studies)	Temperate (8 studies)	Total (23 studies)
Camponotus	14	4	18
Crematogaster	12	5	17
Pheidole	12	2	14
Ectatomma	11	0	11
Solenopsis	7	3	10
Pachycondyla	8	0	8
Pseudomyrmex	8	0	8
Formica	1	7	8
Wasmannia	7	0	7
Azteca	7	0	7
Paratrechina	6	1	7
Tapinoma	4	2	6
Paraponera	5	0	5
Conomyrma	4	1	5
Odontomachus	4	0	4
Forelius	2	2	4
Myrmica	0	4	4
Monomorium	1	3	4
Zacryptocerus	4	0	4
Monacis	4	0	4
Leptothorax	2	2	4
Brachymyrmex	3	0	3
Lasius	0	3	3

The following genera were recorded only once or twice: *Acanthoponera*, *Aphaenogaster*, *Ochetomyrmex*, *Cyphomyrmex*, *Dendromyrmex*, *Dolichoderus*, *Prenolepis*, *Hypoclinea*, *Myrmecocystus*, *Myrmelachista*, *Procryptocerus*, *Tetramorium*, and three unidentified genera.

conducted at varying localities (indicated in Tables 14. 2 and 14. 3). On the whole *Camponotus*, *Crematogaster*, and *Pheidole* were the genera most frequently occurring at EFNs. *Ectatomma*, *Pachycondyla*, and *Pseudomyrmex* also ranked highly but their records are confined to tropical nectary plants, since they are either rare or absent from temperate habitats (Brown 1973). Similarly, *Formica* and *Myrmica* are limited to temperate habitats (Brown 1973).

Wilson (1976) defined 'prevalence' with respect to ant genera based on four different components: species diversity as measured by number of species, extent of geographic range, diversity of adaptations, and local abundance. Using these criteria Wilson (1976) recognized *Camponotus*, *Pheidole*, and *Crematogaster* as the most prevalent ant genera in the world, each occurring with high local abundances, and large numbers of species in most zoogeographical regions. Based upon independent analysis, Brown (1973) arrived at the same conclusion and selected these three ant genera as 'world dominating'. Censuses on diverse nectary plant species have revealed that *Camponotus*, *Pheidole*, and *Crematogaster* usually represent a higher number of species than other nectarivorous ant genera, especially in tropical and subtropical habitats. For example, 10 species of *Camponotus* were recorded on *Caryocar* in Brazilian cerrados (Oliveira 1988), seven species of *Crematogaster* were recorded on *Passiflora* in Costa Rica (Smiley 1986), and three species of *Pheidole* were observed on *Costus allenii* in Panama (see Table 14. 2) (Schemske 1982). Moreover, *Camponotus* ants usually outnumber other ant genera in the number of individuals on both tropical (Koptur 1984, Oliveira 1988) and temperate nectary plants (Stephenson 1982; McLain 1983). However, Schemske (1982) reported *Ectatomma* (Ponerinae) and *Monacis* (Dolichoderinae) as the principal visitors of *Costus* spp. in Panama, and indicated that the composition of ant assemblages at *Costus* varied according to the plant species and the height of the inflorescence above the ground.

Conclusion

Besides the need for more extensive round the clock censuses of the ant fauna, the plant architecture should also be taken into account in future research on the species composition of ant assemblages at EFN-bearing plants (see Chapter 8). Studies on the interaction between ants and EFNs have usually emphasized the role of 'key' ant visitors as anti-herbivore agents on the plants, and additional effort should be directed towards more complete censuses of the associated ant fauna. Comparison of visits to EFNs with ant attacks on live baits suggests that the levels of visitation by different ant species may give a misleading impression of their effectiveness as anti-herbivore agents. The generalized nature of this type of mutualistic association (cf. Schemske 1983), and its widespread occurrence in many different environments, may provide promising insights for faunistic studies of ants, both on the local and global scale. Moreover, correlative data on ant abundance, abundance of plants with EFNs, herbivore pressure, and efficacy of ant species in deterring different herbivores would help to elucidate the selective pressures involved in the evolution and maintenance of this type of mutualism in varying geographical contexts.

Acknowledgements

We thank D. L. Perlman for helpful suggestions on the manuscript, and S. P. Cover and E. O. Wilson for advice on the taxonomy of temperate ants. The preparation of the manuscript was greatly facilitated by the logistic support provided by B. Hölldobler at the Museum of Comparative Zoology, Harvard University. Financial support to P. S.Oliveira was provided by the Conselho Nacional de Desenvolvimento Cientifico e Tecnológico (Brazil), CNPq, and Fundo de Apoio à Pesquisa da Universidade Estadual de Campinas, FAP/ UNICAMP, during field work, and as a post-doctoral fellowship from the CNPq (proc. 200512/88.9). C. R. F. Brandão was supported by the CNPq and Fundação de Amparo à Pesquisa do Estado de São Paulo, FAPESP.

15

Extrafloral nectaries of herbs and trees: modelling the interaction with ants and parasitoids

Suzanne Koptur

Plant tissue lost to consumers is a physiological cost to a plant. Over evolutionary time, a variety of plant defences have evolved which enable plants to avoid or better survive herbivory. Plants are rooted, and cannot flee, so their defences operate at the site of attack. Much research has been done on the mechanical and chemical defences of plants, but biotic defences have received less attention, perhaps because they require greater field study. Biotic defences of plants are dynamic, and there are compelling reasons to study them. Most plant species have multiple defences, since no single line of defence works against all herbivores. By gaining a better understanding of how plants benefit from mutualisms with biotic protective agents we gain insight into complex interactions that shape community structure.

Extrafloral nectaries and plant defence

Evidence for ants visiting extrafloral nectaries and providing protection against herbivores is abundant (Bentley 1977a; Buckley 1982; Beattie 1985; Jolivet 1986). Ants visiting nectaries on vegetative parts of the plant body may protect foliage (Janzen 1966, 1967; Bentley 1976; Tilman 1978; Koptur 1979, 1984; Kelly 1986) which can translate into greater seed set and increased fitness for the plant (Koptur 1979; Stephenson 1982; Barton 1986). Nectaries on or near reproductive structures can provide ant protection of ovules and seeds (Elias and Gelband 1975; Bentley 1977b; Deuth 1977; Inouye and Taylor 1979; Schemske 1980, 1982; Keeler 1981b; Horvitz and Schemske 1984). However, ant protection is by no means universal (e.g. O'Dowd and Catchpole 1983; Tempel 1983; Boecklen 1984; Heads and Lawton 1984; Lawton and Heads 1984; Whalen and Mackay 1988; Rashbrook *et al.* Chapter 16, this volume; Mackay and Whalen Chapter 17, this volume), and the benefits to plants from ants vary both geographically (Koptur 1985; Barton 1986; Kelly 1986; Koptur and Lawton 1988) and temporally (Tilman 1978; Schemske and Horvitz 1988; Cushman and Addicott Chapter 8, this volume).

Extrafloral nectaries also attract predators and parasitoids of the plant's herbivores (Leius 1967, Price *et al.* 1980; Washburn 1984; Hespenheide 1985). Parasitoids can be important biological control agents in agricultural systems (de Bach 1964; Crepps 1975; Altieri *et al.* 1977; Hassell 1980, 1982) and in natural situations (Washburn and Cornell 1981; Hassell and Waage 1984; Hassell 1985; Weis and Abrahamson 1985; Price and Clancy 1986), especially in areas where ants are not abundant (Keeler 1985; Koptur 1985; Koptur and Lawton 1988).

The nature of the mutualism

A mutualism is 'any interaction in which two (or more) species reciprocally benefit from the presence of the other species' (Addicott 1984). Benefits of a mutualism may be either direct (e.g. plants feeding ants) or indirect (involving additional species, e.g. removal of herbivores from the plant surface). In some cases it is probable that the indirect benefits of a mutualistic interaction extend beyond the individual mutualist paying for the interaction (see later). Mutualistic interactions in which a number of species serve on one or both sides of the interaction are called diffuse (Addicott 1984), or non-specific (Law 1985), and are usually facultative (Koptur 1979). A number of authors have noted that interactions between mutualistic species tend to non-specificity (Schemske 1983; Harley and Smith 1983; Howe 1984; Law 1985). The tendency for non-specificity or specificity to evolve depends on the interplay between antagonistic and mutualistic interactions in the community; interactions evolve towards non-specificity when there is strong differentiation between phenotypes in their mutualistic interactions, causing the equilibrium to be unstable (Law and Koptur 1986).

Mutualisms mediated by extrafloral nectaries are non-specific in that nectar can be taken by a variety of animals, whose behaviour varies widely, and protective abilities range from positive to negative (as certain herbivores themselves may be attracted or nourished by extrafloral nectar: DeVries and Baker 1989; DeVries Chapter 12, this volume). These facultative interactions will be distinguished here from the often obligate, and more specific mutualisms of myrmecophytes, in which there are a greater number of mediating factors (domatia, food bodies, and sometimes also extrafloral nectaries). In extrafloral nectar-mediated interactions, not only can different species of the same guild, e.g. different species of nectar-drinking ants ('guild' in the strictest sense used by Hawkins and MacMahon (1989)), serve on one side of the interaction, but also species from different guilds, e.g. ants, parasitoids, and predatory insects. The consequences of these different interactions on plant fitness are measurable, and are important to quantify in order to understand the dynamics of mutualisms in variable environments (Thompson 1988).

Extrafloral nectaries: a quantitative defence

Chemical defences of plants have been categorized as either qualitative (toxins, effective against specialist herbivores) or quantitative (e.g. digestibility reducers, effective against generalists) (Rhoades and Cates 1976); plants that are apparent (long-lived and/or abundant) tend to have quantitative defences, while non-apparent plants may utilize qualitative defences more effectively (Feeny 1976). Typically, there are specialist herbivores that can circumvent qualitative chemical defences, even utilizing protective chemicals for their own nutrition or defence (Bernays and Woodhead 1982).

McKey (1988) compares and contrasts chemical and biotic plant defences and notes that, in situations where extrafloral nectaries function on young leaves, biotic protection replaces phenological defence. In many plant species, however, extrafloral nectaries function on mature leaves and/or other parts of the plant body. In general, biotic protection supported by extrafloral nectaries can be considered quantitative for three reasons:

1. More nectar supports greater visitor activity and increases protection.
2. Ants can control a wide variety of phytophagous insects.
3. Parasitoids attracted to nectaries are often taxonomic generalists.

Many plants with extrafloral nectaries are abundant, widespread, and/or long-lived, and could certainly qualify as 'apparent'. The generalized defence supported by extrafloral nectaries may facilitate colonization of new areas, where the plant can strike up a facultative mutualism with non-co-evolved ants, e.g. Old World vetches that grow in north America and are protected by an introduced ant (Koptur 1979).

In habitats where ants are scarce, plants with extrafloral nectaries are less abundant (Keeler 1979a, 1980b, 1981a). Certain plants lose their ant-related traits when ants are absent (Rickson 1977). Removing nectaries from plants substantially reduces the number of ants on a plant (Koptur 1979) unless there are few ants to begin with, or ant-tended homopterans are involved (see below). The more nectar that is offered, the more carbohydrate is available, and many ant species recruit to good resources (Carroll and Janzen 1973; Bentley 1976; Taylor 1977).

As the exception that proves the rule, plants with extrafloral nectary-mediated defence have a small number of specialist herbivore species (in their native habitats) that circumvent the biotic protection. The presence of ants on a plant may create 'enemy-free space' (Lawton 1978; Atsatt 1981; Jeffries and Lawton 1984) for herbivores specialized for survival among ants (Janzen 1966, 1967; Koptur 1984; Heads and Lawton 1985; Jolivet Chapter 26, this volume). Certain insect herbivores secrete honeydew and are tended by ants (Way 1963; Hill and Blackmore 1980; Messina 1981; Briese 1982; Fritz 1982; Sudd 1983; Takeda et al. 1983; Whalen and Mackay 1988;

Chapters 8–10, this volume) and may thus be protected from their parasitoids (Pierce and Mead 1981; Horvitz and Schemske 1984; Maschwitz *et al.* 1984; Smiley *et al.* 1988; Pierce *et al.* Chapter 11, this volume). Other herbivores may obtain protection by feeding inside structures which are tended outside by nectar-drinking ants (Washburn 1984; Compton and Robertson 1988, Chapter 10, this volume; Koptur and Lawton 1988).

Specialized herbivores that are immune to ant defence may be especially devastating to plant fitness in situations where ant protection is so complete that the natural enemies of these herbivores are virtually excluded (Compton and Robertson 1988; Koptur and Lawton 1988). The effects of the different guilds (herbivores, ants, and parasitoids) on plants depends on their interactions with each other, which can vary spatially (Gilbert 1975; Gilbert and Smiley 1978; Koptur 1985; Smiley 1985, 1986; Barton 1986; Kelly 1986; Koptur and Lawton 1988) and temporally (Clancy and Price 1986; Schemske and Horvitz 1988; DeVries Chapter 12, this volume).

Habitat considerations

The diversity and abundance of herbivores on a plant are related to many factors, including habitat heterogeneity and structural diversity; host plant abundance, distribution, density, and architecture or structure; and seasonal phenology (Strong *et al.* 1984). In addition, biotic interactions between plants and non-phytophagous insects can influence the herbivores utilizing a given host plant (Price *et al.* 1980, 1986; Hassell and Waage 1984). The complex interactions between herbivores and host plants vary in different situations (Denno and McClure 1983; Thompson 1988). The relative benefit from different guilds of protective agents may therefore vary, resulting in different patterns of protection, herbivore damage, and diversity in different situations.

The structural diversity of plants influences herbivore diversity in plants of different lifeforms (Lawton 1983) and successional stages (Southwood *et al.* 1983). Pioneer plant species experience greater herbivore pressure, especially on mature foliage (Coley 1983*b*; Newbury and Foresta 1985). Bentley (1976) found ant abundance and plants with nectaries to be more common in clearings and forest edges than in closed forest; Schupp and Feener (Chapter 13) find nectar-mediated ant defence to be more common in light-dependent, tropical woody plant species than climax species. Balsa (*Ochroma pyrimidale* Bombacaceae) (O'Dowd 1979) and certain *Inga* (Mimoscacae) are examples of such successional plants.

Life history considerations

Annual plants have herbaceous tissues, most of which actively photosynthesize and are therefore appealing to herbivores. Herbs that utilize biotic protection usually have the extrafloral nectaries in positions that enhance protection of flowers and fruits, to maximize seed protection during the sole reproductive opportunity. Temperate annuals with extrafloral nectaries near the inflorescences include the common vetch, *Vicia sativa*, which has axillary flowers on short pedicels and stipular nectaries in very close proximity (Koptur 1979), and the partridge pea, *Cassia fasciculata*, which has inflorescences in the axil of each leaf that bears a petiolar nectary (Barton 1986, Kelly 1986).

Tropical herbs are usually perennials, but many of these also have inflorescence-associated extrafloral nectaries: the neotropical ginger *Costus pulverulentus* (Zingiberaceae) has nectaries on the inflorescence bracts, which are visited by ants which repel dipteran ovule predators (Schemske 1980). Other tropical herbs with similar strategies include *Aphelandra deppeana* (Acanthaceae) (Deuth 1977) and *Calathea ovandensis* (Marantaceae) (Horvitz and Schemske 1984). There are also examples of temperate perennial herbs which use this strategy: *Helianthella quinquenervis* (Asteraceae) has nectaries on the phyllary bracts of the composite inflorescence (Inouye and Taylor 1979) and *Vicia sepium* has stipular nectaries quite similar to those of the annual *V. sativa* (Koptur, personal observation).

It is in perennials, especially woody ones, that extrafloral nectaries are more commonly associated with vegetative parts of the plant body, presumably because, even when ovules and seeds are not directly protected, a reduction in the amount of damage to leaf tissue can have a significant impact on eventual seed reproduction in either the same year or a following year. *Catalpa speciosa* (Bignoniaceae) has foliar nectaries, and Stephenson (1982) found that ants visiting these nectaries decreased loss of leaf tissue and increased seed setting on protected branches. In some species of *Inga*, nectaries function at a time other than flowering (Koptur 1984), yet the resulting damage reduction may increase fruit production.

Biotic defences are 'mobile' defences (Coley *et al.* 1985), for although secretory structures involve a non-returnable investment, their secretory activity is regulated by the plant and is often temporary (e.g. on young leaves only while they are expanding). The same long-lived leaves may economize by investing in generalized defences that involve one-time investments (e.g. lignins and tannins) rather than prolonging the biotic protection (nectaries) that involves high maintenance costs (McKey 1988). Examples of this strategy include rainforest legume trees *Leonardoxa africana* in the Old World (McKey 1984) and *Inga* spp. in the New World (Koptur 1984). It is more common for herbaceous plants to have extrafloral nectaries which function for the major part of their lives (e.g. *Vicia sativa* (Koptur and Lawton 1988) than for woody plants.

Extrafloral nectaries, establishment, and survival

Extrafloral nectaries are probably more important in the successful survival of perennials to reproductive age than that of annuals. Some annuals do not have functional nectaries until the time of flower formation (e.g. *Vicia sativa*). Nectaries can play a significant role in the survival of juvenile *Inga* trees to reproductive maturity: if an unlucky sapling is totally defoliated (entirely possible if one or two pierid caterpillars spend their fifth instar feeding on the leaves) there is a high probability of death (Table 15. 1). Ants visiting nectaries on seedlings and saplings can repel ovipositing Lepidoptera, remove eggs, harass caterpillars, and thereby minimize damage; parasitoids fed by nectaries may also shorten the feeding life of certain herbivores. Certain herbaceous species, however, have extrafloral nectaries functioning on the cotyledons (e.g. the castor bean *Ricinus communis*, Euphorbiaceae), suggesting an adaptive advantage of nectar production in seedling establishment.

Costs and inducibility of extrafloral nectar production

Herbs have a larger relative proportion of their biomass actively photosynthesizing than do woody plants. Therefore, they have greater vulnerability to herbivores. Herbs have lower maintenance costs than woody plants, and there is generally greater net productivity in herbaceous vegetation. An acceptable short-term approach to measuring fitness in plants is to calculate cost as the amount of carbon required to produce and maintain a structure (Mooney 1972; Chapin 1989); benefit is the carbon gain resulting from the structure, and can be estimated experimentally or by simulation (Mooney and Gulmon 1982). Using biomass as a rough indication of carbon investment, woody plants invest more in their structures than do perennial herbs, which in turn invest more than do annuals. A given volume of nectar (of a certain sugar concentration or amount of carbon) will therefore be less expensive for a herb than a woody plant, and for an annual than a perennial.

Perhaps this is why inducible extrafloral nectar production has been found (so far) to be more common in herbs than in trees (see below). Many plants increase levels of chemical defence in response to herbivore attack (Fowler and Lawton 1985). There is preliminary evidence that some plants may respond to damage from herbivores by secreting more extrafloral nectar (Koptur 1989) and nectar of different constitution (Smith *et al.* 1990), which are potentially adaptive responses if more biotic protection results. Smith *et al.* (1990) have found that herbaceous *Impatiens* secrete nectar higher in amino acid concentration subsequent to artificial herbivory. My own experiments with herbaceous *Vicia* have shown increases in nectar volumes and sugar concentrations with moderate levels of damage; but experiments with woody *Ipomoea* and two *Inga* species have had negative results for induction

Table 15.1. Effects of artificial and natural defoliation on survival of *Inga* saplings.

% Leaf area destroyed	Number of individuals	Survival 1 year later	
		No.	%
(a) Artificial defoliation of *I. densiflora* in Monteverde, Costa Rica			
25	8	8	100
50	10	9	90
75	9	6	67
100	10	2	20

% Leaf area lost	Number of individuals	Survival 1 year later	
		No.	%
(b) Natural defoliation of *Inga* by caterpillars of *Dismorphia* spp. (Pieridae)			
<50	15	14	93
80–95	15	3	20

(Koptur 1989). However, Stephenson (1982) observed that nectar production increased in woody *Catalpa* with greater natural defoliation.

Ants versus parasitoids: trees versus herbs

Ants remove or deter herbivores on the surface of plants; protection takes place before the herbivores cause much damage, and the benefit from producing the nectar accrues largely to the nectar-producing individual. Parasitoids visiting nectaries are more likely to encounter and parasitize herbivores on plants secreting nectar. The parasitoid larvae eventually emerge, killing the herbivore egg, larva, or pupa. If herbivore lifespan is shortened and plant consumption is reduced, there will be direct benefit to the individual nectar-producing plant. However, an indirect benefit of parasitized herbivores not becoming reproductive adults is experienced by all plants of that species in the vicinity, and indeed by other plants susceptible to that herbivore.

My work has led me to wonder how these traits are selected for by the various guilds of biotic protective agents, especially in the situations where parasitoids are more important than ants as nectary visitors. Two examples contrast the life history and ecological situations.

1. *Inga* are legume trees, with foliar nectaries that support fairly effective ant-guards in neotropical, lowland wet forests. The same species of *Inga* occur in upland wet forests where ants are rare, and do not regularly visit nectaries: here there can be no ant protection, but most trees of the species secrete extrafloral nectar. Caterpillars on upland trees are parasitized more frequently than their lowland counterparts, suggesting that parasitoids visiting nectaries are reducing herbivore numbers in upland populations. It is of interest that not every upland individual secretes nectar, while virtually every lowland tree does.

2. *Vicia sativa* (vetch) is a temperate annual legume herb, with stipular nectaries in close proximity to the flowers. Ants visit these nectaries, and are effective in reducing the numbers of surface-feeding Lepidoptera, but are ineffectual against the many phytophagous adult beetles. Some of the lepidopteran and beetle herbivores oviposit in the flowers and young fruit, and their larvae eat the seeds as they develop in the pods. These species may be controlled by parasitoids, the adults of which visit the nectaries, and oviposit in/on the larvae or in flowers and developing fruit. The nectaries support the two guilds of protective agents simultaneously.

Modelling extrafloral nectary mutualisms

It is easy to see how ants that deter herbivores could select for traits which promote associations between ants and plants. Consider a plant species in which a novel type that secretes extrafloral nectar arises: Type-A plants secrete nectar, Type-B plants do not. The two types have a combined density of N, in proportions p and q ($p + q = 1$).

Let d = an individual of Type-A's effect on itself
z = its effect on every other member of the group
b = benefit due to nectary visitors
c = cost of nectaries and nectar

So $d = b - c$ (benefit of action minus the cost).

When $b > c$, the effect on self will be positive. There is no need to assume any benefit to Type-B plants, as ants will tend to stay on nectaried plants; however, indirect benefits or disadvantages may also exist. For instance, in the short term, phytophagous insects disturbed on Type-A plants may be more likely to oviposit and/or feed directly on Type-B plants.

The fitness of a Type-A plant is the net benefit to itself (d) plus the effect on every other member of the group ($Np - 1$; 1 is that individual itself). Therefore, changes in fitness from the nectar secreted by Type-A plants can be calculated as follows (f_X = fitness of Type-X plants):

$$f_A = d + (Np - 1) z \tag{15.1}$$

The effect on Type-B plants from Type-A plants is:

$$f_B = Npz \qquad (15.2)$$

The type with the higher fitness increases its proportionality in the next generation. So, for Type-A to be selected,

$$f_A > f_B, \text{ and}$$

$$d + (Np - 1)z > Npz \qquad (15.3)$$

$$d > z$$

Namely, net benefit to self must be greater than benefit of action to other members of the group for the trait to be selected. Ants will remove more herbivores from Type-A plants, thereby increasing f_A, and the nectar-producing type will be selected.

If, however, there are no ants present, some parasitoids may select for nectaries (via the same model) if they reduce herbivore pressure on individual plants that secrete nectar (e.g. by shortening the life of parasitized herbivores, especially eliminating the late instars that eat more, or by parasitizing eggs that will then not hatch, or produce short-lived larvae). Not all parasitoids will do this, and if they do not, the cost of nectar production will not be offset by an immediate benefit to the individual nectar producer:

$$c > b, d \text{ will be negative}$$

Therefore, $d < z$.

If the cost of nectar production is non-zero ($c > 0$), d will be less than z, and Type-B plants (non-producers) will be selected. Even if nectar secretion and parasitoids reduce overall herbivore numbers (by preventing some larvae from becoming reproductive adults), this will benefit all members of the population equally (both Types A and B) so that the fitness of Type-B will be greater (as they do not have any cost to bear) and the nectar secreting trait will be selected against. Again, $d < z$, because c will detract from b for Type-A, whereas $c = 0$ for Type-B).

How, then, are nectaries selected and/or maintained in areas where ants are not present in numbers sufficient to afford protection? In a population of upland *Inga*, with no ants present the nectaries are visited by parasitoids and herbivores are parasitized. It may be possible to understand why most plants secrete nectar if we consider the basic selection model with the homogeneity assumption relaxed. Consider that the population may be sub-structured and that selection may operate between groups within the deme (the large breeding population).

The plant that secretes extrafloral nectar will increase the likelihood of its herbivores being parasitized. The parasitized herbivores may continue feeding on the plants for some time, so the benefit is not necessarily immediate, unless the herbivore eats less than its non-parasitized counterparts. This problem (having to sustain parasitized herbivores) may not be that important as *all* individual plants in the area will have herbivores, i.e. no extra cost is involved from foliage feeding on nectar-secreting individuals over and above that cost to non-secretors. The greatest protective benefit may arise from the parasitized herbivore not living to reproduce, therefore decreasing the numbers of herbivores in the next generation. As many herbivorous insects have adults that move some distance (e.g. Lepidoptera), the eggs that might have been laid by a given parasitized larva would have produced larvae feeding not only on the plant that produced the nectar, but on other individuals of that species in the vicinity. There is, therefore, the potential for all plants susceptible to that herbivore and growing in the vicinity of the nectar-producing individual to benefit from the nectar.

For *Inga*, the deme is determined by strong-flying, far-ranging pollinators (hawkmoth and hummingbird) and seed dispersers (mammal and bird). Trait groups are smaller groups within which ecological interactions take place: for *Inga* we will consider the groups defined by herbivore ovipositing range. (This is certainly a continuous function, the oviposition frequency falling off with distance from the original larval food plant; but here we will consider it, for simplicity, as discrete.) These trait groups are small compared to the deme: herbivorous Lepidoptera on these trees are mostly smaller and less vagile than the pollinators and seed dispersers.

Let each trait group be characterized by a certain density of Types A and B. The more Type-A (nectar secretors) there are in a trait group, the greater that trait group's fitness will be, and that trait group will be represented proportionately more in the next generation.

Let f_A and f_B represent fitnesses within a trait group (as in Wilson (1980)).

Within a deme, there will be a large number of trait groups (T) each characterized by a single density (of Type-A) and frequency (of Type-A).

Let p_{mn} = the proportion of trait groups containing m Type-A and n Type-B; $m + n = N$.

The fitnesses of the two types over the entire deme (F) consist of the weighted averages of fitness for each type over all trait groups (mixing occurs throughout the whole deme, so this is justified). As defined previously, d = a Type-A individual's effect on itself, and z = its effect on every other member of the group:

$$F_A = \frac{T \sum_{0}^{\infty} p_{mn} [d + (m-1)z]}{T \sum_{0}^{\infty} p_{mn} m} \qquad (15.4)$$

$$F_B = \frac{T \sum_{0}^{\infty} p_{mn} nmz}{T \sum_{0}^{\infty} p_{mn} n} \qquad (15.5)$$

Even if individual fitnesses indicate otherwise, trait groups with more Type-A individuals in them can have enhanced fitness under certain circumstances, which will increase the fitness of Type-A over the entire deme (Wilson 1980).

Consider a four-species system, with the two types of plant (A and B), one parasitoid, and one herbivore. The interactions between all these entities are shown in Table 15.2. Self effects are negative (intraspecific competition). The direct effect of one entity on another is expressed as positive, negative, or none (+, −, or 0).

Table 15.2. Interactions between nectar-secreting plants, non-secreting plants, parasitoids, and herbivores.

Y	The effect of X on Y			
	X:Type-A plant (with nectar)	Type-B plant (without nectar)	Parasitoids	Herbivores
Type-A plant	−	−	0	−
Type-B plant	−	−	0	−
Parasitoids	+	0	−	+
Herbivores	+	+	−	−

We can specify basic difference equations for each species, using

P = density of parasitoids (1,2—from herbivores on Type-A,B plants)
H = density of herbivores (1,2—on Type-A,B plants)
N = density of plants (1,2—Type-A and Type-B plants)
A = attack rate of parasitoids on herbivores (1,2—for herbivores on Type-A,B plants)
r = rate of increase (r_1—for plants; r_2—for herbivores)
K = carrying capacity (K_1—plants; K_2—herbivores per individual plant)
L = effect of herbivores on plants

The plant population growth equation has density dependence, so that growth slows as the carrying capacity is approached. The effect of herbivores on plants (L) also limits population growth, relative to the number of herbivores at time t:

$$N_{t+1} = N_t \exp\left[r_1 \{1 - (N_t/K_1)\} - LH_t\right]$$
(15.6)

For herbivores, population growth is positively affected by plant density (N) at time t. Herbivore carrying capacity is defined on a per plant basis, and limits the increase of herbivores. Herbivore population growth is negatively affected by parasitoid density and the efficiency of parasitoids in finding the herbivores (A):

$$H_{t+1} = H_t \exp\left[r_2 \{1 - (H_t/N_t K_2)\} - AP_t\right]$$
(15.7)

Using the Nicholson–Bailey equation (Nicholson and Bailey 1935), parasitoid population growth is directly related to herbivore density at time t, and is self-limiting by both the attack rate on herbivores (A) and the parasitoid density at time t. As in equation (15.7), these two entities are inversely related: with a low rate of attack, more parasitoids can be accommodated in available herbivore hosts.

$$P_{t+1} = H_t (1 - \exp[-AP_t])$$
(15.8)

The Nicholson–Bailey model assumes that parasitoids encounter prey in direct proportion to prey density, and that these encounters are distributed randomly among the available prey (Hassell 1978). The Nicholson–Bailey model shows increasing oscillations, but the density dependence in the basic herbivore equation renders the parasitoid equation (above) stable.

For the four-species system, with two types of plants, the equations are elaborated as follows. First, the densities at time t are defined (Part I, equations (15.9)–(15.12)). Then, the densities at time $t + 1$ are defined (Parts II–IV, equations (15.13)–(15.16).

Part I

(a) Parasitoid density on Type-A plants at time t is a function of the proportion of Type-A plants in the population. The more Type-A plants in a group, the greater the attraction to parasitoids. The combined attraction is greater than that expected by a simple linear relationship up to the maximum of a pure population of Type-A. If $x =$ the proportion of Type-A plants, the function is defined as

$$fnQ(x) = 2x - x^2$$

This gives the following equation for parasitoids on Type-A plants:

$$P_{1,t} = P_t fnQ (N_{1t}/(N^{1t} + N_{2t}))$$
(15.9)

Parasitoid density on Type-B plants is simply the remainder of the total number of parasitoids at time t less the density on Type-A:

$$P_{2,t} = P_t - P_{1,t} \tag{15.10}$$

(b) Herbivore densities at time t are similarly divided between the two types of plants, with the number of herbivores on Type-A plants directly related to the proportion of Type-A plants in the population. Herbivores on Type-B plants are the remainder of the total less the density on Type-A plants:

$$H_{1,t} = H_t (N_{1t}/(N_{1t} + N_{2t})) \tag{15.11}$$

$$H_{2,t} = H_t - H_{1,t} \tag{15.12}$$

Part II

Plant densities at time $t + 1$ (the next generation) are expressed for Type-A and Type-B plants separately. The plant rate of increase and carrying capacity are, of course, the same for each type of plant, as is the effect of herbivore damage on both types of plants:

$$N_{1,t+1} = N_{1,t} \exp [r_1 \{1-((N_1 + N_2)/K_1)\} - LH_{1,t}] \tag{15.13}$$

$$N_{2,t+1} = N_{2,t} \exp [r_1 \{1-((N_1 + N_2)/K_1)\} - LH_{2,t}] \tag{15.14}$$

Part III

Herbivore density at time $t + 1$ is a combination of herbivores supported on Type-A plants and Type-B plants. Both groups of herbivores have the same rate of increase and the same carrying capacity per plant, but the attack rate of parasitoids is greater on Type-A than on Type-B plants:

$$\begin{aligned} H_{t+1} = {} & H_{1,t} \exp [\{r_2(1-H_{1,t}/(N_1 K_2)) - A_1 P_{1,t}\}] \\ & + H_{2,t} \exp [\{r_2(1-H_{2,t}/(N_2 K_2)) - A_2 P_{2,t}\}] \end{aligned} \tag{15.15}$$

Part IV

The density of parasitoids at time $t + 1$ is determined by using the Nicholson–Bailey equation on the subdivided population. The first part of the following equation represents parasitoids from herbivores on Type-A plants, the second part represents those from herbivores on Type-B plants:

$$P_{t+1} = H_{1,t} (1-\exp[-A_1 P_{1,t}]) + H_{2,t} (1-\exp[-A_2 P_{2,t}]) \tag{15.16}$$

The equations were used to construct a simple simulation model. The BASIC program (shown in Appendix 15. 1) then does the following:

1. Uses the above difference equations, and specifying initial population densities (N) for each type of plant, and the values for the various parameters.

2. Creates a number (T) of trait groups (using a random number generator), with mean density N_i and variance VN_i for each species (plant type). The method of setting up the proportion of each plant type used here is not ideal (see Appendix 15. 1).

3. For each trait group, the equations are iterated a number (I) of times to obtain densities at $t + I$. I = The amount of time spent in the trait group, and the simplest case is used here, i.e. one generation ($I = 1$).

4. The densities of each trait group are added, and divided by T, to obtain average densities for the global population.

5. Repeating from step 2 above a number of times (generations), the simulation is achieved.

Behaviour of the model

The community undergoes stable oscillations, i.e. all entities persist, within certain limits of the various parameters of the model. Starting with equal numbers of Type-A and Type-B plants at a designated carrying capacity for plants and herbivores, stability is achieved when the magnitude of the effect of herbivores on plants is less than the magnitude of the attack rate of parasitoids on Type-A plants ($L < A_1$). Stability is also achieved when the relative magnitude of parasitoid attack rate on Type-A vs. Type-B plants is more than 10:1 ($A_1 > 10(A_2)$).

Starting with a large number of Type-B plants (no nectaries), the community can be invaded by Type-A. The speed of invasion of Type-A is not affected by the magnitude of A_1: for both $A_1 = 0.1$ and 0.3, the numbers of Type-A plants match Type-B numbers after 10 generations. This is probably due to the self-regulating effects of A in the parasitoid population growth equation (15. 8).

The reverse situation (many Type-A plants, few Type-B) is less easily invaded by Type-B. This is because the small size of A_2, the attack rate of parasitoids on herbivores on Type-B plants, does not enhance the representation of trait groups with the same magnitude as Type-A plants can upon invasion (above).

When the effect of herbivores on the plants is too great ($L > 0.05$) relative to the reproductive capacity of plants and the parasitoid attack rates, plants do not persist and the simulation is unstable, with all species going to zero.

If the initial number of Type-A plants is zero, after a number of generations the parasitoids do not persist, with $A_2 = 0.01$, but if A_2 is increased to 0.1, parasitoids do persist.

Other considerations

The cost of producing extrafloral nectar could also be included in this model, by using two different values for plant reproductive capacity in equations (15. 13) and (15. 14): r_{11} for Type-A plants, and r_{12} for Type-B plants. The cost of nectar production could be experienced as a direct fitness cost that lowers the reproductive rate very slightly, so that $r_{11} < r_{12}$. Existing estimates of cost of extrafloral nectar production are 1–2 per cent or less (e.g. O'Dowd 1979); the effect of this cost on the reproductive capacity would presumably be even less than that (though there is no experimental basis for this assumption), suggesting that the appropriate difference in r values might be as little as 0.0001. This elaboration of the model has not yet yielded satisfactory results.

No benefit to parasitoids from nectar is included in the parasitoid growth equation. The beneficial effects on adult parasitoid longevity and reproductive fitness (egg production) could be expressed only if there were a parasitoid 'r'. However, most biologists do not consider this parameter to be of importance in parasitoid population growth.

Differences in generation time are always a consideration in modelling plant–herbivore interactions, and it is interesting here to consider the implications for annuals (where, simplistically, there could be a one-for-one match with the herbivore species) versus perennials. For perennial plants, there is always a component of future herbivory on that individual plant: the cost incurred in secretion of extrafloral nectar and support of parasitized caterpillars one year may not yield any benefit that same year, but would in the future (a delayed-reward type of individual selection). For annuals, extrafloral nectar production to support parasitoids may appear to be an altruistic act. Altruistic, that is, unless one considers the potential kin selection for increase in the number of Type-A (nectaried) plants, another dimension to the structured deme.

Discussion

The model presented here is in an early state of development and the results are only preliminary. Moreover, the full biological situation is far more complex than has been shown here. Many kinds of selection operate at once. There is direct selection by some parasitoids (e.g. those that kill eggs). There are different generation times involved with trees and herbivores; would it be more realistic to hold herbivore numbers as a constant? The model points out certain parameters that it is essential to measure accurately:

1. What is the cost to plants of producing nectar? How does this cost affect their rate of increase?
2. What is the effect of herbivores on plant fitness?

3. What are attack rates of parasitoids on herbivores? Do they differ on plants with nectaries and without, and what is the magnitude of that difference?

Modelling is a useful tool for field ecologists for two reasons. Firstly, one can test with a model whether or not a particular scenario (selective process, population change, etc.) can theoretically occur, and within what limits. Secondly, the results one obtains indicate what parameters of natural populations are the most important to measure in order to evaluate the theoretical predictions. This exercise has indicated the necessity of measuring several parameters, enumerated above, encountered in the study of plant-ant-parasitoid-herbivore interactions. It has compared and contrasted the interactions between plants bearing extrafloral nectaries with their two main guilds of protective agents, ants and parasitoids. And most importantly, this exercise has revealed the different ways in which selection might operate between ants and plants versus parasitoids and plants. The first is direct, straightforward, and easily demonstrated; while the second is indirect, more complex, and needs to be tested.

Acknowledgements

The ideas presented here were clarified through discussions with Diane Davidson, Carol Horvitz, Deborah Letourneau, Jonathan Majer, and Doyle McKey. Discussions with Richard Law, Peter Shaw, and John H. Lawton were helpful in applying David Sloan Wilson's structured deme model to plants with extrafloral nectaries.

Appendix 15.1. BASIC program for simulation described in this chapter

Note: The variable designations here differ in some cases from the equations in the paper.

Designated variables in this model

A1	attack rate of parasitoids for herbivores on Type-A plants
A2	attack rate of parasitoids for herbivores on Type-B plants
D	generation
H	density of herbivores
H1	density of herbivores on Type-A plants
H2	density of herbivores on Type-B plants
K1	plant carrying capacity
K2	herbivore carrying capacity per plant
L	effect of herbivores on plants
M	average density (initial)
N1	density of Type-A plants
N2	density of Type-B plants

P   density of parasitoids
R1   plant *r* (reproductive capacity)
R2   herbivore *r* (reproductive capacity)
T   number of trait groups
T1   total density of plants

```
00010 DIM N1(20),N2(20),P1(20),P2(20),P(20),H1(20),H2(20),H(20),S1(4),
  S2(5),M(4)
00020 OPEN "DATA.1" FOR OUTPUT AS #1
00030 INPUT "DO YOU WANT ALL TRAIT GROUPS LISTING (Y/N)";A$
00040 if not(a$="y" or a$="Y" or a$="n" or a$="N" ) then 30
00050 M(1)=60:M(2)=60:M(3)=100:M(4)=160
00054 PRINT#1, "M(1)="M(1);"M(2)="M(2);"M(3)="M(3);"M(4)="M(4)
00055 PRINT "M(1)="M(1);"M(2)="M(2);"M(3)="M(3);"M(4)="M(4)
00060 T = 20
00070 DEF FNB(I)=EXP(R1*(S-L*I))
00080 DEF FNC(I)=EXP((R2*(1-I/K2)-Q1*Q2))
00090 DEF FNA(X) = INT(0.5+(X+1)*RND(1))-1
00100 DEF FNZ(I)=INT(0.5+I*10)/10
00105 DEF FNQ(X)=2*X-X*X
00110 A1=.2:A2=.02:A3=0:R1=1:K1=600:L=.05:R2=1:K2=10
00115 PRINT#1,"A1="A1;"A2="A2;"R1="R1;"K1="K1;"L="L;"R2="R2;"K2="K2
00116 print "A1="A1; "A2="A2;"R1="R1;"K1="K1;"L="L;"R2="R2;"K2="K2
00120 PRINT #1, "PLANT A, PLANT B, PTOID, HERBV"
00122 PRINT "PLANT A, PLANT B, PTOID, HERBV"
00124 PRINT #1, "GEN, POP SIZES, MEANS, VAR."
00125 PRINT "GEN, POP SIZES, MEANS, VAR."
00126 PRINT #1, " 1 ";
00127 PRINT " 1 ";
00130 GOSUB 150
00140 GOTO 300
00145 H(T)=P(T)=N1(T)=N2(T)=0: GOTO 150
00150 FOR G = 1 TO 19
00160 N1(G)=FNA(M(1)): IF N1(G) < 0 OR N1(G) > M(1) THEN 160
00170 N2(G)=FNA(M(2)): IF N2(G) <0 OR N2(G) >M(2) THEN 170
00180 P(G)=FNA(M(3)): IF P(G) <0 OR P(G) >M(3) THEN 180
00190 H(G)=FNA(M(4)): IF H(G) <0 OR H(G) >M(4) THEN 190
00210 NEXT G
00220 GOSUB 700
00230 N1(T)=INT(0.5+(T*M(1)/2)-S1(1))
00240 N2(T)=INT(0.5+(T*M(2)/2)-S1(2))
00250 P(T)=INT(0.5+(T*M(3)/2)-S1(3))
00260 H(T)=INT(0.5+(T*M(4)/2)-S1(4))
00280 IF H(T)<=0 OR P(T)<=0 OR N2(T)<=0 OR N1(T)<=0 GO TO 145 ELSE GOTO
  290
00290 RETURN
00300 REM NOW ALL 20 TRAIT GROUPS ARE SET UP
00310 GOSUB 510
00320 D=2
00330 FOR G=1 TO T
00340 if a$="y" or a$="Y" then PRINT#1,  N1(G); N2(G); P(G); H(G)
00345 if a$="y" or a$="Y" THEN PRINT N1(G); N2(G); P(G); H(G)
00346 T1=N1(G)+N2(G)
00347 IF T1>0 THEN H1(G)=H(G)*N1(G)/T1: H2(G)=H(G)-H1(G) ELSE
  H1(G)=0:H2(G)=0:H(G)=0:
00350 IF T1>0 THEN P1(G)=P(G)*FNQ(N1(G)/T1):P2(G)=P(G)-P1(G) ELSE
  P1(G)=0:P2(G)=0:P(G)=0
00355 P(0)=INT (0.5+H1(G)*(1-EXP(-A1*P1(G)))+H2(G)*(1-EXP(-A2*P2(G))))
00360 S=1-(N1(G)+N2(G))/K1
00370 N1(0)=N1(G)*FNB(H1(G))
00380 N2(0)=N2(G)*FNB(H2(G))
00390 IF N1(G)>0 THEN Q1=A1:Q2=P1(G):H(0)=H1(G)*FNC(H1(G)/N1(G)) ELSE
  H1(G)=0:H(0)=0
00400 IF N2(G)>0 THEN Q1=A2: Q2=P2(G): H(0)=H(0)+H2(G)*FNC(H2(G)/N2(G))
  ELSE H2(G)=0
00410 N1(G)=N1(0):N2(G)=N2(0):P(G)=P(0):H(G)=H(0)
00440 NEXT G
00450 PRINT#1, D;
00455 PRINT D;
00460 GOSUB 620
00470 FOR J9=1 TO 4
00472 M(J9)=INT(2*(S1(J9)/t))
```

```
00474 NEXT J9
00480 GOSUB 150
00485 GOSUB 510
00490 D=D+1
00500 IF D=21 THEN GOTO 750 ELSE GOTO 330
00510 GOSUB 620
00512 FOR G=1 TO 4
00513 PRINT #1,S1(G);
00514 PRINT S1(G);
00515 NEXT G
00520 FOR G=1 TO 4
00530 PRINT#1, FNZ(S1(G)/T);
00535 PRINT FNZ(S1(G)/T);
00540 NEXT G
00550 FOR G=1 TO 4
00560 PRINT#1, FNZ(ABS((S2(G)-S1(G)^2)/T)/(T-1));
00565 PRINT FNZ (ABS((S2(G)-S1(G)^2)/T)/(T-1));
00570 NEXT G
00571 PRINT#1,"."
00572 PRINT"."
00580 RETURN
00590 '
00600 '
00610 '
00620 GOSUB 700
00623 FOR G=1 TO 4
00624 S2(G)=0
00625 NEXT G
00630 FOR G=1 TO T
00632 S2(1)=S2(1) +N1(G)*N1(G)
00640 S2(2)=S2(2) +N2(G)*N2(G)
00650 S2(3)=S2(3) +P(G)*P(G)
00660 S2(4)=S2(4) + H(G)*H(G)
00665 NEXT G
00670 RETURN
00680 '
00690 '
00700 FOR G=1 TO 4:
00702 S1(G)=0
00704 NEXT G ' SOLVES SUM OF X FOR EACH POP
00710 FOR G=1 TO T
00712 S1(1)=S1(1) + N1(G)
00714 S1(2)=S1(2) +N2(G)
00716 S1(3)=S1(3) +P(G)
00718 S1(4)=S1(4) +H(G)
00720 NEXT G
00740 RETURN
00750 END
```

16

Bracken and ants: why is there no mutualism?

V. K. Rashbrook, S. G. Compton, and J. H. Lawton

Introduction

One of the advantages of studying a plant as cosmopolitan and ubiquitous as bracken fern (*Pteridium aquilinum* (L.) Kuhn), is that ecological comparisons can be made between different areas of its world-wide range (Lawton 1982, 1984; Compton *et al.* 1989). Bracken was initially assumed to be relatively immune to herbivory and thought to possess a scant arthropod fauna (Darwin 1877). However, although the plant does contain an array of secondary plant compounds (Lawton 1976; Cooper-Driver *et al.* 1977), it is by no means under-exploited by insect herbivores in comparison with other common plant species, and supports a diverse arthropod fauna in different parts of its range. Areas surveyed so far include continental Europe (Simmonds 1967), the UK (Lawton 1976, 1982), Papua New Guinea (Kirk 1977, 1982), south-west USA (Lawton 1982), Hawaii (Lawton 1984), Australia (E. Shuter and M. Westoby, unpublished work), New Zealand (Winterbourne 1987), and South Africa (Compton *et al.* 1989).

The extrafloral nectaries of bracken were first noted by Darwin (1877) and Lloyd (1901) and a comprehensive account of their variation in structure, distribution, and activity is given by Page (1982). Ants are by far the most frequently observed visitors to bracken nectaries. Although Tempel (1983) and Lawton and Heads (1984) also recorded parasitic hymenopterans at the nectaries in the USA and UK respectively, it was not established whether they utilized bracken herbivores as hosts. No parasitic wasps or predators other than ants have been observed at the nectaries in South Africa (Rashbrook 1989). Therefore, ants are the only group likely to offer the prospect of defending bracken from herbivores and hence to enter into a mutualism with the plant.

As Bentley (1977*a*) pointed out, no defence mutualism can exist between ants and plants unless the plants are actually vulnerable to their herbivores and ants constitute a meaningful defensive force. Testing the importance of herbivory is most readily achieved in plants for which an accurate measurement of sexual reproductive output can be made (Waloff and Richards 1977; Whitham and Mopper 1985). Due to its vegetative mode of reproduction, an entire bracken patch can consist of a single clone (Wolf *et al.* 1988), making the effects of damage to individual fronds on the overall fitness of the clone hard to quantify. Changes in herbivore abundance, or levels of herbivory, must necessarily continue to be used as an indication of possible ant protection. Clearly, extrapolation of these measurements to the costs or benefits to the plant must be made with caution.

Studies of bracken–ant interactions

Investigations of bracken–ant interactions now span three continents, enabling comparisons to be made of the benefits of attracting ants in North America, England, and South Africa (Douglas 1983; Tempel 1983; Lawton and Heads 1984; Heads and Lawton 1984, 1985; Heads 1986; Rashbrook 1989).

Douglas (1983) studied bracken in Michigan (USA) and noted that several species of ants systematically patrolled the plant and drove off or killed 'intruding' arthropods. He stated that 'the croziers are protected [by a] mobile, predaceous arthropod defence community'. Experimental data to confirm such claims was lacking, but these observations were suggestive of a mutualistic interaction between bracken and ants. Exclusion experiments carried out in New Jersey (USA) revealed a very different scenario (Tempel 1983). Although ant activity was correlated with seasonal availability of nectar, levels of herbivory were found to be unaffected by the presence of ants. In England, although many aspects of the biology of the ant–plant association appeared to be potentially beneficial to the plant (Lawton and Heads 1984), exclusion experiments showed that ants had no effect on the densities of herbivores or on the number of species found per frond (Heads and Lawton 1984; Heads 1986). Exclusion experiments in the eastern Cape (South Africa), using natural and artificially augmented ant densities, again failed to detect evidence of a bracken–ant defence mutualism (Rashbrook 1989). Two species of moths were found to be the most widespread and destructive bracken herbivores on the sub-continent (Compton *et al.* 1989). The abundance and feeding damage of these species did not differ between ant-accessible and ant-free fronds (Rashbrook 1989). Other widespread herbivores were less abundant and appeared to be immune from the ants patrolling the plants (Rashbrook 1989). Thus, with the exception of the qualitative observations of Douglas (1983), studies in three temperate regions all indicate that bracken gains no protective advantage from its relationship with ants.

Factors influencing bracken–ant interactions

Effective plant defence by ants depends on a number of factors, including the species of ants involved, herbivore community structure, and the densities of insects on the plants. The following sections discuss the importance of these variables in the bracken–ant systems studied to date and draw comparisons with other associations between ants and plants with extrafloral nectaries.

Ant faunal composition

The defence hypothesis necessitates active foraging by ants: those species that fail to exhibit predatory behaviour are, in effect, acting as nectar thieves. A wide diversity of ant species are attracted to bracken nectaries. For example, at least 17 species were recorded on bracken in South Africa (Rashbrook 1989). Of the four species active during exclusion experiments, the most common, *Crematogaster peringueyi* Emery, was seen attacking and removing lepidopteran larvae. It also proved capable of reducing numbers of larvae in laboratory experiments conducted under simulated field conditions (Rashbrook 1989). Thus, the potential for a mutualism between bracken and at least one ant species existed at these sites. The biology of the system is similar in New Mexico and at two sites in northern England, where a variety of ants visit nectaries, patrol the fronds, and rapidly attack and remove experimentally introduced prey (Lawton and Heads 1984; Heads 1986). Conversely, in New Jersey Tempel (1983) observed no foraging by ants on the fronds (other than at the nectaries), or any suggestion of aggressive behaviour between ants and herbivores, even though two of the ant species involved in this study had been noted by Douglas (1983) to systematically patrol the plant in Michigan.

The size of ant workers, relative to that of the herbivores, also affects the likelihood of predation (Tilman 1978; Barton 1986). The final-instar larvae of the South African moth *Appana cinisigna* (de Joannis) (Noctuidae) are some four to six times longer than *Crematogaster* and are either ignored or require minimal effort to avoid attack, yet early instars are more vulnerable (Rashbrook 1989). Heads (1986) found that large, aggressive wood ants (*Formica lugubris* Zett.) can have an impact, albeit weak, on individual bracken herbivores at sites in Yorkshire. The ants were effective at finding and removing most external herbivores in short-term introduction experiments, though a species of highly distasteful sawfly remained immune. Thus, the same herbivores may have less chance of escape where very large ants are present.

Herbivore community composition

The value of a plant's nectaries will vary according to the degree of susceptibility of its herbivores to ant predation. Certain species, for example endophages, may be largely immune to ants during their active feeding stages, while other insects may even benefit from the ants. For example, lepidopterans feeding on the pods of common vetch in Yorkshire caused greatest damage when ants were abundant, because their larval parasitoids were deterred (Koptur and Lawton 1988). Honeydew-producing homopterans also gain protection from ants. However, by attracting more ants to a plant, they can increase predation of other herbivores (V. K. Rashbrook, unpublished work).

There are major differences between the herbivore communities on bracken in different regions of the world (Lawton 1984; Compton *et al.* 1989). Of the 14 herbivore species recorded in Papua New Guinea, the majority are coleopterans which feed by boring into the rachis or rhizome (Kirk 1977, 1982). The fauna in New Zealand is dominated by fluid-sucking hemipterans (Winterbourne 1987), while the relatively depauperate (3–5 species) community in New Mexico (USA) consists mainly of pinnae chewers (Lawton 1982). Lepidopteran chewers predominate in South Africa (Compton *et al.* 1989), whereas the UK herbivore community is rich in free-living sawflies (Symphyta) (Lawton 1976), but also contains a relatively high proportion of species in other orders which are gall formers, miners, or concealed in other ways.

Outcomes for the plant may vary with these differences in community structure. Internal and other concealed feeders will often be immune from attack by ants, and may in part explain why ant exclusion experiments had no effect on British bracken-feeding insects (Heads and Lawton 1985). The failure of ants to reduce herbivory by free-living species suggests that many of these apparently more vulnerable herbivores possess defences which make them immune to ant predation. Ant-avoidance strategies (for example, distasteful haemolymph and escape behaviours) have been demonstrated for bracken-feeding lepidopteran and sawfly larvae in the UK (Heads and Lawton 1985). The two South African lepidopterans also use physical and chemical means to reduce their susceptibility to ants (Rashbrook 1989). The responses of *Appana* larvae to ant attack are instar related. Early instars leave the frond and hang beneath, suspended on silk, while larger larvae writhe and produce a fluid repellent to attacking ants.

Insect densities

Several studies have referred to the importance of temporal and spatial variations in ant activity and herbivore densities, such that the same plant species is unlikely to receive equal benefits from ants between years, during different seasons, or in different areas of its range (Tilman 1978; Boecklen 1984; Koptur 1985; Barton 1986; Koptur and Lawton 1988). High ant densities have been linked with increased plant growth (Messina 1981) and reproductive success (Schemske 1980) and with more effective herbivore deterrence (Compton and Robertson 1988). Ant abundance at extrafloral nectaries is affected by, amongst other things, the distance from ant nests (Tilman 1978; Inouye and Taylor 1979), alternative, richer sources of sugar (Sudd and Sudd 1985), and altitude (Koptur 1984; Lawton and Heads 1984). Koptur (1984) showed that reduced ant densities at high altitudes translated into less effective defence of *Inga* species, and Barton (1986) found that inter-site variation in insect densities (both of ants and herbivores) was responsible for correspondingly variable protection of *Cassia fasciculata* Michx. The graphical model proposed by Boecklen (1984) envisages protection by ants as a continuum, whereby the net gain or reduction in benefits to the plant is a function of its density, herbivore density, and ant activity.

Insect activity on bracken varies with plant phenology in temperate regions (Tempel 1983; Lawton and Heads, 1984; Rashbrook 1989). The free-living lepidopteran larvae associated with bracken in the UK do not feed in the early spring, when the plant is most nutritious (Lawton 1976). Heads and Lawton (1985) attributed this to the influence of ants, as the greatest ant densities occurred on the plant at this time (Lawton and Heads 1984). In contrast, the two major lepidopteran larvae in South Africa are abundant on young fronds from early spring to late summer, during the peak period of ant activity (Rashbrook 1989). Daily patterns in ant activity also influence the likelihood of effective plant defence. Most ant species on South African bracken are diurnal and will come into contact with grazing moth larvae, but not the nocturnally ovipositing adults. In the USA, Tempel (1983) found no nocturnal ant species on bracken. Lawton and Heads (1984) did not sample at night, although ant activities at their study sites varied with temperature and were lowest in the early morning.

Daytime ant densities obtained on bracken fronds in the USA, England, and South Africa are compared in Table 16. 1. Although average values do not necessarily give the complete picture, because of considerable inter-plant variation (Lawton and Heads 1984), average levels of ant activity varied by more than an order of magnitude between sites. However, even the relatively high average ant densities in England (2–3 ants per frond) did not affect numbers of bracken herbivores. In South Africa it was only under conditions of very high ant densities in laboratory field-simulation experiments (up to 23.5 ants per frond) that significant larval removal was obtained. These results suggest that the potential for ant defence of bracken is present, but only when there are exceptionally high densities of ants.

Comparisons between ant densities on bracken and other extrafloral nectary-bearing plants are difficult because of differences in plant morphology. It is unclear whether the general absence of protection for bracken is due to it having fewer ants than those species where significant protection has been demonstrated. Barton (1986) recorded significant increases in seed set on *Cassia* with mean densities of 5.22 ants per plant, but not on those with mean densities of 2.05 ants per plant. Whalen and Mackay (1988) also noted relatively low densities of ants on five euphorbiaceous saplings in Papua New Guinea (ranging between 0.04 and 1.43 ants per leaf). Of the three species where ant exclusion experiments were carried out, significant effects were only obtained on the one with the highest *number* of ants per leaf. Surprisingly, the plant species (with small leaves) that had the highest *densities* of herbivores and ants per unit leaf area gained no benefits. Clearly, other

Table 16.1. A comparison of ant densities on bracken fronds at sites in the USA (Tempel 1983; Lawton and Heads 1984), South Africa (Rashbrook 1989), and the UK (Heads and Lawton 1984; Heads 1986; J. H. Lawton unpublished work). With the exception of New Mexico, densities were recorded during ant exclusion experiments.

Site	Proportion of fronds with ants		Mean ants/Frond	
	Over whole sampling period	Max. at any one sample	Over whole sampling period	Max. at any one sample
USA (New Jersey)				
Experimental station	0.16	0.35	—	—
State park	0.08	0.17	—	—
USA (New Mexico)				
Sierra blanca (open)	0.10	0.14	0.11	0.34
Sierra blanca (woodland)	0.20	0.65	1.10	3.70
South Africa (E. Cape)				
Faraway (1986)	0.62	0.80	1.72	2.40
Faraway (1987)	0.33	0.50	0.38	0.70
Glenthorpe HAD	0.27	0.40	0.54	0.95
Glenthorpe LAD	0.11	0.25	0.13	0.30
UK (N. Yorkshire)				
Skipwith common	—	—	2.50	3.14
Hazel head wood	—	—	0.71	2.14

Glenthorpe HAD and LAD refer to manipulated high and low ant density areas.

factors in addition to ant density must be taken into consideration when predicting the likely benefits of ants.

Extrafloral nectaries will be most valuable when both ant and herbivore densities are high (Keeler 1981a). Variable, but generally low levels of herbivory characterize the bracken system (Lawton and McNeill 1979; Caughley and Lawton 1981; Tempel 1983) and populations of many species of herbivore are very rare (Lawton et al. 1986; MacGarvin et al. 1986). Low herbivore densities may be another factor reducing the likelihood of bracken–ant mutualism.

Why does bracken have nectaries?

In the absence of detectable benefits from its nectaries, why does bracken offer a free source of food for no predictable gain? Excretory functions and attraction of non-ant predators seem unlikely explanations for the presence of nectaries on bracken (Heads and Lawton 1984). Conversely, while ants may not provide a continuous beneficial service, it could be argued that they may be effective against some herbivores for some of the time, or in as yet unstudied parts of the geographical range of bracken. Lawton and Heads (1984) put forward an additional hypothesis after an experimentally intro- duced, non-adapted moth species was found to suffer heavy mortalities from ants attracted by the nectar. They argued that the common bracken-specific insects either already possessed 'exaptations' (*sensu* Gould and Vrba 1982) or had evolved adaptations to minimize ant predation. In contrast, non- adapted, generalist species would fall prey to ants foraging on the plants. Thus the bracken–ant association was potentially mutualistic at times when non-adapted colonists were damaging the plant. Alternatively, in the same way that the aphid-tending ants on *Betula* species in Scandinavia switched to feeding on caterpillars during outbreaks of these defoliators (Laine and Niemelä 1980), bracken nectaries may ensure that ants are available to protect the plant at times when herbivore densities become exceptionally high. The nectaries could therefore function as a 'latent defence mechanism'.

For a mutualism to be detected, there has to be a sufficiently high frequency of interactions between ants and susceptible herbivores. These conditions do not appear to be met on bracken at most times and in most places. Given optimal defence theory (Rhoades 1979), a predicted con- sequence would be the loss of ant attractants in areas where benefits are no longer obtained, as reported in *Cecropia* (Janzen 1973; Rickson 1977). This may have happened in Hawaii, where bracken nectaries are reported to be non-functional (Keeler 1985). Loss of nectaries could be attributed to the absence of native ants and bracken herbivores on Hawaii (Lawton 1982).

A general absence of mutualism with ants has not, however, inhibited the success of bracken; it is one of the world's most widespread plants (Harper 1977) and a major economic weed (Taylor 1986; Lawton 1986, 1988).

17

Some associations between ants and euphorbs in tropical Australasia

D. A. Mackay and M. A. Whalen

Introduction

The Euphorbiaceae is a large and diverse plant family, widespread in both the Old and New World tropics (Webster 1967), many members of which have extrafloral nectaries (Elias 1983) or other adaptations that encourage ant visitation. Most studies to date on interactions between ants and Old World members of this family have been conducted on myrmecophytic species, such as *Macaranga triloba*, which provide specialized food bodies and hollow stem domatia for one or a few species of ants (Ridley 1910; Tho 1978; Rickson 1980; Maschwitz *et al.* 1984; Fiala *et al.* 1989, Chapter 18, this volume) and very little experimental work has been done on non-myrmecophytic associations in which plants provide food bodies and/or extrafloral nectar for ants, but not nest sites (Whalen and Mackay 1988; Fiala *et al.* 1989). In a study in Papua New Guinea of several euphorb tree species that were involved in the latter kind of ant association (Whalen and Mackay 1988), we found that plants were visited by several ant species and that the benefits to plants of ant visitation were variable, with only one out of three plant species showing significantly reduced herbivore loads and leaf damage levels when ants were experimentally excluded from branches. The benefits and costs of attracting ants to foliage presumably vary considerably depending on resource availability, the season, the location, and the density and composition of ant and herbivore faunas (e.g. Barton 1986; Horvitz and Schemske 1984; Inouye and Taylor 1979; Koptur and Lawton 1988) so we were interested in extending our previous study to examine associations in Australia between ants and euphorbs in some of the same genera that we had studied further north (*Macaranga* and *Homalanthus*).

Airy Shaw (1980*a*) considers that the Australian Euphorbiaceae (excluding the endemic Stenolobeae) is primarily composed of southern extensions of the Malesian euphorbiaceous flora. For example, the genus *Macaranga* has 75 species in Papua New Guinea and only six in Australia, *Homalanthus* is represented by 12 species in Papua New Guinea and three in Australia, and *Mallotus* has 19 species in Papua New Guinea and 12 in Australia (Airy Shaw 1980*a,b*). In addition to providing information on how variation in some of the above factors may affect ant–plant interactions, the study of such diffuse or facultative associations may provide information on how more species-specific myrmecophytic associations, such as that involving *Macaranga triloba*, may have evolved (McKey 1988; Fiala *et al.* 1989; Fiala *et al.* Chapter 18, this volume).

In this chapter we present the results of a short-term study of the ant and herbivore faunas found on the leaves of saplings of two euphorb tree species, *Macaranga subdentata* Benth. and *Homalanthus populifolius* Grah., in tropical northern Australia. The latter species is closely related and very similar in appearance to *Homolanthus novoguineensis* (Warb.) Schum. (Airy Shaw 1980*a*) which we studied in Papua New Guinea. These species were the two most common euphorbs with foliar nectaries in secondary forest at our study site but in the field they were not seen to provide nest sites or food bodies for attendant ants. Ants were commonly seen visiting the nectaries on the leaves of both species.

Our study was conducted in January–February 1988, in the late wet season at Paluma, near Townsville in Northern Queensland, Australia. The elevation of Paluma is about 890 m and the annual rainfall is approximately 2600 mm. During our study there was a flush of growth of new leaves on both tree species. *Macaranga subdentata* possesses 'petiolar' nectaries on the adaxial side of the leaf blade, near the insertion of the petiole, as do the species of *Macaranga* and *Mallotus* that we examined in Papua New Guinea. In *Macaranga subdentata* and in the Papua New Guinea *Macaranga* species, other nectaries are frequently present elsewhere on the lamina, particularly near the leaf tip, but by far the majority of observations of ants feeding at nectaries were made of ants feeding at the petiolar nectaries. Leaves of *Homalanthus populifolius* have one large, protruding nectary at the insertion of the petiole and up to two smaller nectaries nearby on the abaxial leaf surface. We refer to these nectaries collectively as 'petiolar' nectaries.

Ant, herbivore, and damage levels

To quantify the density and composition of the ant and herbivore faunas, five censuses of ten individuals of each tree species were conducted over 17 days in January–February. For logistical reasons, and because ant defence may be particularly important for saplings (Schupp 1986, Fiala *et al.* 1989), we restricted our study to saplings between 1 and 3 m high. The censusing protocol we employed was the same as that used in the earlier study. Censuses were performed during the day and were conducted in secondary forest, frequently along trails and roads where these plants were abundant. At each census, the top ten leaves (or fewer if less than ten leaves were present) on one branch of each sapling were examined for ants and herbivores. On the first census, the numbers of petiolar nectaries on the leaves were also recorded. Two plants of *Macaranga subdentata* and four of *Homalanthus populifolius* were attacked by leaf-rolling lepidopteran larvae during our censuses. Numbers of leaf-rollers (which were capable of causing considerable leaf damage) were only recorded in our herbivore counts if they could be seen without unrolling leaves, as that would have caused the larvae to drop off the plant and would have affected measurement of damage over the census period. Consequently, numbers of these larvae were underestimated in herbivore censuses, although the damage they caused was recorded. Initial damage levels of leaves were categorized on a scale of 0 to 8, with 0 representing no damage, 1 representing up to one-eighth of the leaf damaged, etc. The average leaf areas of both tree species were estimated from a sample of the top five leaves from each of six plants. Leaf outlines were traced onto paper and areas calculated using a digitizer.

For both tree species, there were significant correlations between leaf position and the proportion of petiolar nectaries that were functional (*Macaranga subdentata*, Spearman's $r = -0.63$, $n = 87$, $P < 0.001$; *Homalanthus populifolius*, $r = -0.35$, $n = 81$, $P < 0.002$) and between leaf position and ant abundance (*M. subdentata*, $r = -0.35$, $n = 461$, $P < 0.001$; *H. populifolius*, $r = -0.13$, $n = 402$, $P < 0.02$) such that more ants, and a higher proportion of functional nectaries, tended to be found on the younger leaves.

Macaranga subdentata was visited by ants in five genera, with *Iridomyrmex* being the most common, while *Homalanthus populifolius* was visited by ants in three genera, with *Paratrechina* being the most common (Fig. 17. 1). In casual observations at night, we did not see any ant species on the euphorbs that we did not also see in the daytime.

Figure 17. 2(a) shows the mean numbers of ants and insect herbivores per leaf on the top five leaves of our censused plants. Data from the top five leaves were used in order to standardize comparisons between species and because the foliar nectaries were usually active on these upper, younger leaves. Average leaf areas were estimated to be 60 cm² for both species and the mean numbers of ants and herbivores per leaf were divided by these averages to indicate their approximate relative densities (Fig. 17. 2(b)). The mean number of herbivores observed on *Homalanthus populifolius* was extremely low. We observed at least one ant on 22 and 12 per cent, respectively, of the total census observations of individual leaves of *Macaranga subdentata* and *H. populifolius*. The mean initial damage levels on leaves of *M. subdentata* and *H. populifolius* were 1.2 and 0.8, respectively, and the average changes in damage category by the last census were 0.7 and 0.9.

Ant exclusion experiment

To investigate the effects of ant visitation on herbivore numbers and levels of herbivory on *Macaranga subdentata* saplings, we excluded ants from the upper immature leaves of 15 pairs of saplings. The saplings in each pair were in close proximity to one another. The members of each pair were then

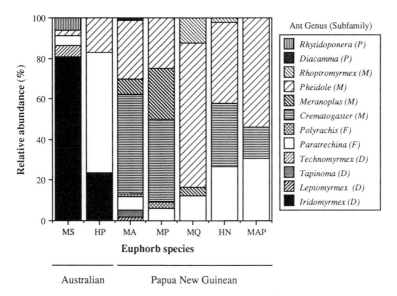

Fig. 17.1. Relative abundances of different ant genera on censused euphorb saplings at Paluma, Queensland and at Baiyer River, Papua New Guinea. Five daytime censuses were carried out on 10–12 individuals of each species during January–February 1988. (P) = Ponerinae, (M) = Myrmicinae, (F) = Formicinae, (D) = Dolichoderinae. MS = *Macaranga subdentata*, MA = *M. aleuritoides*, MP = *M. punctata*, MQ = *M. quadriglandulosa*, HP = *Homalanthus populifolius*, HN = *H. novoguineensis*, MAP = *Mallotus philippinensis*.

randomly assigned to control (ant access) or treatment (ant exclusion) groups. Ants were excluded from the top five immature leaves (or fewer if five immature leaves were not available) of treatment saplings by placing a layer of sticky Rentokil Bird Repel® over masking tape that was wrapped around the stem. Immature leaves were red to pale, reddish green in colour and soft in texture in comparison to mature leaves, which were tougher and darker green. Control saplings were simply banded with masking tape. The experimental protocol therefore followed that used by Whalen and Mackay (1988), except that herbivore numbers on experimental saplings were censused five times, rather than three times, during the experiment and the experiment was conducted over 20 rather than 15 days.

No significant difference was found in the mean number of herbivores per plant between ant-access and ant-exclusion groups (Fig. 17. 3) (MANOVA, $F = 0.04$, d.f. = 1 and 28, $P > 0.85$). In censuses of control saplings in this experiment, and in our censuses of non-experimental saplings, the main herbivores observed on this species were bug nymphs and leaf-hoppers. Mean initial damage levels on the leaves of ant-access and ant-exclusion

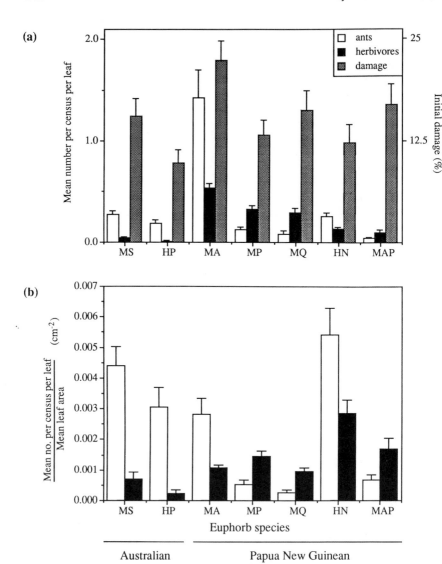

Fig. 17.2. (a) Mean initial damage levels and mean numbers of ants and herbivores per leaf census observation on euphorb saplings in Australia and Papua New Guinea. (b) Mean numbers of ants and herbivores per leaf census observation divided by mean leaf areas of euphorb species censused in Australia and Papua New Guinea. Data are for the top five leaves of saplings. Error bars are standard errors. Data for Papua New' Guinea euphorbs are from Whalen and Mackay (1988). MS = *Macaranga subdentata*, MA = *M. aleuritoides*, MP = *M. punctata*, MQ = *M. quadriglandulosa*, HP = *Homalanthus populifolius*, HN = *H. novoguineensis*, MAP = *Mallotus philippinensis*.

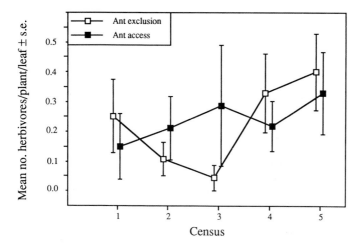

Fig. 17.3. Mean numbers of herbivores per leaf per plant (± s.e.) on ant-access and ant-exclusion saplings of *Macaranga subdentata* on five censuses in Queensland.

plants (0.95, 1.18 respectively) were not significantly different and the average increase in damage levels on ant-exclusion plants (0.56) was actually lower than that on ant access plants (0.94) by the end of the experiment, though not significantly so (Fig. 17. 4(a), Wilcoxon's test, one-tailed $P > 0.4$). Five ant-access plants and three ant-exclusion plants suffered damage from leaf-rolling larvae during the experiment. More new leaves were produced near the growing tip on ant-access plants than on ant-exclusion plants (Fig. 17. 4(b)) but this difference was not statistically significant (Wilcoxon's test, one-tailed $P = 0.063$).

Nectaries

We observed significant differences among censused individuals of both euphorb species in the mean numbers of petiolar nectaries present on their leaves. Numbers of petiolar nectaries on leaves of *Homalanthus populifolius* ranged from one to three (mean = 1.79, s.e. = 0.11), with all leaves having a central basal nectary but with the number of side nectaries varying from zero to two. All censused leaves on some plants of this species lacked side nectaries. The number of petiolar nectaries on leaves of *Macaranga subdentata* ranged from zero to ten (mean = 2.97, s.e. = 0.19) though no censused plant lacked nectaries on all its leaves.

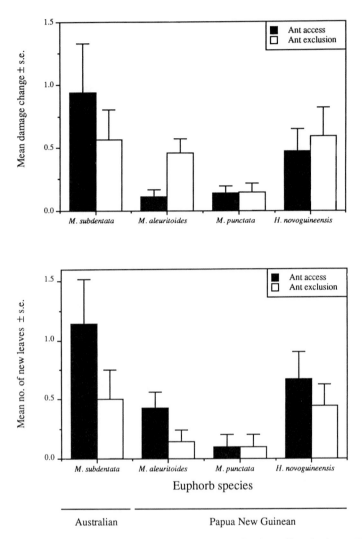

Fig. 17.4. Results of ant exclusion experiments on euphorb saplings in Australia and Papua New Guinea. (a) Mean damage changes per plant (± s.e.) on ant-access and ant-exclusion saplings. (b) Mean numbers of new leaves per plant (± s.e.) produced on ant-access and ant-exclusion saplings. Data for Papua New Guinea euphorbs are from Whalen and Mackay (1988).

Discussion

The results from this study are comparable with those obtained in a study of associations between ants and saplings of five species of euphorb trees in sub-montane Papua New Guinea (Whalen and Mackay 1988). The tree species examined in Papua New Guinea (*Macaranga aleuritoides* F. Muell., *M. punctata* Schum., *M. quadriglandulosa* Warb., *Mallotus philippinensis* (Lam.) Muell. Arg., and *Homalanthus novoguineensis*) were common in secondary forest habitats, had foliar nectaries, and, also, were not observed in the field to provide food bodies or nest sites for attendant ants. Figure 17. 2(a) shows the mean initial damage levels of leaves and the mean densities of ants and herbivores per leaf census observation on the top five leaves of the euphorbs that were censused in Papua New Guinea in 1986 and in Australia in 1988. Average ant densities and initial leaf damage levels per species did not differ significantly between Australian and Papua New Guinean species (Fig. 17. 2(a), Kruskal–Wallis non-parametric multiple contrasts (Zar 1984) using plant means as replicates, $S = 1.25$, $P > 0.05$ and $S = 1.453$, $P > 0.05$, respectively), while herbivore numbers tended to be lower on the Australian species (Fig. 17. 2(a), KW multiple contrast, $S = 3.784$, $P < 0.05$). The latter comparison is slightly biased, however, by the fact that leaf-rollers, although uncommon, were more frequently included in herbivore censuses in Papua New Guinea, as some species there could more easily be seen and counted without unrolling the leaves. Comparisons of damage levels do not suffer this bias. *Macaranga aleuritoides*, the species with the largest leaves, had the greatest densities of ants and herbivores per leaf and the highest levels of leaf damage. The most striking difference between the ant faunas was the absence of *Crematogaster* from Australian, and the absence of *Iridomyrmex* from Papua New Guinean euphorbs (Fig. 17. 1). The density of herbivores observed on the Australian *H. populifolius* was lower than on any other euphorb on a per leaf and on a per unit leaf area basis and this species also had the lowest levels of initial leaf damage (Fig. 17. 2).

In both study sites, nectary activity and ant numbers were higher on immature leaves than on mature leaves, trends that have been noted in several studies (e.g. O'Dowd 1979, Tempel 1983, Koptur 1984). Rates of herbivory are frequently higher on young leaves than on mature leaves (Coley 1983*a* and references therein) and biotic defences such as ants have been hypothesized to offer several advantages over physical or other defences for such relatively soft and palatable tissues (Beattie 1985; McKey 1988).

One aspect of variation in associations between ants and plants with extrafloral nectaries is variation in the number of nectaries that are produced. We compared numbers of petiolar nectaries within and among species of *Homalanthus* and *Macaranga*. There were significant differences among individuals of *Homalanthus populifolius* in Australia in the average number of petiolar nectaries. In contrast, all leaves examined on individuals of the closely related *H. novoguineensis* in Papua New Guinea had three petiolar nectaries—a protruding central nectary and a pair of side nectaries (Fig. 17. 5). There were significant differences in the average numbers of petiolar nectaries per leaf among individuals of all the *Macaranga* species examined. The Australian species *Macaranga subdentata* had significantly fewer nectaries per leaf (some leaves lacked nectaries altogether) than did the average *Macaranga* in Papua New Guinea (KW multiple contrast, $S = 3.538$, $P < 0.05$) and it also showed the greatest range of nectary numbers of the *Macaranga* species examined (Fig. 17. 5). Across all species, coefficients of variation for the number of petiolar nectaries ranged from 0 to 61 per cent and were highest for the two Australian species (Fig. 17. 5). In a myrmecophytic ant–plant system, McKey (1984) found coefficients of variation ranging from 13 to 34 per cent for the number of nectaries on basal, middle, and distal leaflets of *Leonardoxa africana* Aubrév. (Caesalpiniaceae).

Fiala *et al.* (1989, Chapter 18, this volume) examined associations between ants and *Macaranga* species in three low-elevation ($\leqslant 500$ m) sites in peninsular Malaysia. They found that, of 27 *Macaranga* species occurring

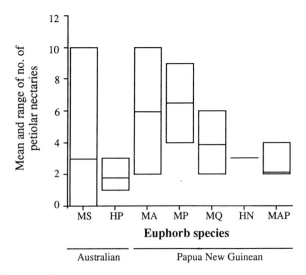

Fig. 17.5. Means and ranges of numbers of petiolar nectaries produced on leaves of euphorb saplings in Australia and Papua New Guinea. Statistics based on all leaves sampled in initial censuses. MS = *Macaranga subdentata*, MA = *M. aleuritoides*, MP = *M. punctata*, MQ = *M. quadriglandulosa*, HP = *Homalanthus populifolius*, HN = *H. novoguineensis*, MAP = *Mallotus philippinensis*. Coefficients of variation for the number of petiolar nectaries on leaves were 61, 28, 15, 23, 53, 0, and 18 per cent respectively.

mainly in pioneer habitats, nine were involved in myrmecophytic associations which involved the ant *Crematogaster borneensis* exclusively. For four myrmecophytic species, they found that naturally ant-occupied plants had significantly less leaf damage than did naturally ant-free plants. Fiala *et al.* (1989, Chapter 18, this volume) noted that non-myrmecophytic *Macaranga* species were visited by several species of ants and that the levels of leaf damage on three such species which they examined were quite variable; with two species showing damage levels that were similar to those on ant-free plants of a myrmecophytic species and one species showing levels that were lower and more similar to the levels found on an ant-tended myrmecophyte.

Our results from seven non-myrmecophytic euphorb species are similar in that all the species we examined apparently have opportunistic or facultative associations with several ant genera. However, initial leaf damage levels did not vary significantly among these seven species (Kruskal–Wallis $H = 10.03$, d.f. $= 6$, $P > 0.1$) or among the Macarangas alone (Kruskal–Wallis $H = 4.66$, d.f. $= 3$, $P > 0.1$). In contrast to Fiala's results from myrmecophytic species, we found that ant attendance reduced herbivory in only one of the four non-myrmecophytic euphorb species which we examined experimentally (Fig. 17. 4). This species, *Macaranga aleuritoides* in Papua New Guinea, was the only species that had an average of more than one ant per leaf and the ant density per leaf on this species was several times greater than that on any other species (Fig. 17. 2(a)). In view of the fidelity of the association of *Crematogaster borneensis* with myrmecophytic *Macaranga* species in Malaysia (see above) and of the effectiveness of these ants as defensive agents, it is interesting that *M. aleuritoides* was also the species on which ants in the genus *Crematogaster* were most common. In comparing the numbers of new leaves produced at the growing tips of our experimental saplings, we found that for three species there was a statistically non-significant trend for ant-access plants to produce more new leaves than ant-exclusion plants (Fig. 17. 4(b)) and this trend may have become more marked in a longer experiment. Data for the fourth species, *Macaranga punctata*, were not analysed statistically because too few new leaves were produced.

Ant-occupied plants of three of the myrmecophytic species examined by Fiala *et al.* (1989) had significantly fewer vines attached to them than did ant-free plants. Vines were uncommon on the saplings of the seven euphorb species we examined, and we did not observe ants clipping or pruning foliage encroaching on the saplings.

One possible reason for the apparent lack of effective ant defence in three of the non-myrmecophytic species that we examined experimentally may be that our study sites were located in sub-montane regions (900–1000 m elevation) where ant activity and the efficacy of ant defence may be reduced relative to more lowland regions (Fiala *et al.* 1989, Chapter 18, this volume). Koptur (1985) also found that ant activity and the effectiveness of ant defence on *Inga* (Mimosaceae) trees were lower in upland than in lowland populations.

Another reason may be the relatively low densities of ants and herbivores (less than one per leaf) found on these species. Barton (1986), in a study of variation in the effectiveness of ant defence of *Cassia fasciculata* Michx. (Caesalpinaceae), found that ants only had a significant effect on plant reproductive success in a site where both ant and herbivore numbers were high. In a study of bracken fern (*Pteridium aquilinum* (L.) Kuhn) in South Africa, Rashbrook *et al.* (Chapter 16, this volume), added and excluded ants from experimental fronds and found no evidence that ants were reducing either numbers of lepidopteran larvae or damage levels on fronds. They suggested that the lack of ant defence in their study and in several other studies of bracken fern may simply result from the low numbers of ants and herbivores commonly found on this plant species. The mean numbers of ants per leaf which we observed on these euphorb saplings (less than one ant per leaf) are similar to the densities found on bracken fronds in several studies.

Another possible reason why we failed to detect significant ant defence on three of four tree species examined experimentally, is that such species may not attract enough ants that are effective as defensive agents. Variation in the effectiveness of different ants as defensive agents and in the vulnerability of different herbivores have been demonstrated in a number of studies (e.g. Horvitz and Schemske 1984; Koptur 1984; Barton 1986; Koptur and Lawton 1988). Ants may also effectively defend more of our euphorb species at other sites that are connected by gene flow to the ones we worked in or, alternatively, ants may be effective in other seasons or years, or in other stages of the trees' life cycles (e.g. O'Dowd and Catchpole 1983; Koptur 1985; Barton 1986).

Selection on facultative ant–plant associations, such as those we have studied, might lead to the development of further adaptations to attract ants and/or the evolution of more specialized associations if damage caused by herbivores that are vulnerable to ant defence increases, or if ants that are particularly effective as defensive agents become more numerous. A correlation between the production of more valuable resources and increased specificity is known from several ant–plant systems (Schemske 1982, 1983, and references therein). In the genus *Macaranga*, studies of species that produce domatia have revealed highly specific associations with an ant species, *Crematogaster borneensis*, which is capable of acting as a defensive agent against herbivores and vines (Tho 1978; Fiala *et al.* 1989). The lower numbers of petiolar nectaries on euphorbs in Australia and the increased variance in nectary numbers relative to those found in Papua New Guinea suggests that selection for ant mutualisms involving these species may be relatively reduced in the Australian species. However, the failure of two of three species that we examined experimentally in Papua New Guinea to exhibit reduced herbivore levels or damage when ants were experimentally excluded must make this interpretation a rather tentative one.

Acknowledgements

We gratefully acknowledge Peter Daenke for technical assistance, Phil Ward for help with ant identifications, and Flinders University and the Ian Potter Foundation for financial support.

References to Part 3

Addicott, J. F. (1984). Mutualistic interactions in population and community processes. In *A new ecology* (ed. P. W. Price, C. N. Slobodchikoff, and W. S. Gaud), Ch. 16, pp. 438–55. John Wiley & Sons, New York.

Aide, T. M. (1988). Herbivory as a selective agent on the timing of leaf production in a tropical understorey community. *Nature*, **336**, 574–5.

Airy Shaw, H. K. (1980*a*). A partial synopsis of the Euphorbiaceae–Platylobeae of Australia (excluding *Phyllanthus*, *Euphorbia* and *Calycopeplus*(. *Kew Bulletin*, **35**, 577–700.

Airy Shaw, H. K. (1980*b*). The Euphorbiaceae of New Guinea. *Kew Bulletin Additional Series*, **8**, 1–243.

Altieri, M. A., von Schoonhoven, A., and Doll, J. (1977). The ecological role of weeds in insect pest management systems: a review illustrated by bean cropping systems. *PANS*, **23(2)**, 195–205.

Atsatt, P. R. (1981). Lycaenid butterflies and ants: selection for enemy-free space. *American Naturalist*, **118**, 638–54.

Barton, A. M. (1986). Spatial variation in the effect of ants on an extrafloral nectary plant. *Ecology*, **67(2)**, 495–504.

Bazzaz, F. A. and Pickett, S. T. A. (1980). The physiological ecology of tropical succession. *Annual Review of Ecology and Systematics*, **11**, 287–310.

Beattie, A. J. (1985). *The evolutionary ecology of ant–plant mutualisms*, 182 pp. Cambridge University Press, Cambridge.

Beckmann, R. L. and Stucky, J. M. (1981). Extrafloral nectaries and plant guarding in *Ipomoea pandurata* (L.) G. F. W. Mey (Convolvulaceae). *American Journal of Botany*, **68**, 72–9.

Bentley, B. L. (1976). Plants bearing extrafloral nectaries and the associated ant community: interhabitat differences in the reduction of herbivore damage. *Ecology*, **57**, 815–20.

Bentley, B. L. (1977*a*). Extrafloral nectaries and protection by pugnacious bodyguards. *Annual Review of Ecology and Systematics*, **8**, 407–27.

Bentley, B. L. (1977*b*). The protective function of ants visiting the extrafloral nectaries of *Bixa orellana* (Bixaceae). *Journal of Ecology*, **65**, 27–38.

Bentley, B. L. (1981). Ants, extrafloral nectaries, and the vine life-form: an interaction. *Tropical Ecology*, **22**, 127–33.

Bernays, E. A. and Woodhead, S. (1982). Plant phenols utilized as nutrients by a phytophagous insect. *Science*, **216**, 201–3.

Bishop, Y. M. M., Fienberg, S. E., and Holland, P. W. (1975). *Discrete multivariate analysis: theory and practice*. MIT Press, Cambridge.

Blom, P. E. and Clark, W. H. (1980). Observations of ants (Hymenoptera: Formicidae) visiting extrafloral nectaries of the barrel cactus *Ferocactus gracilis* (Cactaceae) in Baja California, Mexico. *Southwestern Naturalist*, **25**, 181–96.

Boecklen, W. J. (1984). The role of extrafloral nectaries in the herbivore defence of *Cassia fasciculata*. *Ecological Entomology*, **9(3)**, 243–9.

Bolton, B. (1976). The ant tribe Tetramoriini (Hymenoptera: Formicidae): constituent genera, review of smaller genera and revision of *Triglyphothrix* Forel. *Bulletin of the British Museum (Natural History), Entomology*, **34**, 281–378.

Briese, D. T. (1982). Damage to saltbush by the coccid *Pulvinaria maskelli* Olliff, and the role played by an attendant ant. *Journal of the Australian Entomological Society*, **21(4)**, 293–4.

Brokaw, N. V. L. (1985). Treefalls, regrowth, and community structure in tropical forests. In *The ecology of natural disturbance and patch dynamics* (ed. S. T. A. Pickett and P. S. White), pp. 53–69. Academic Press, Orlando.

Brokaw, N. V. L. (1987). Gap-phase regeneration of three pioneer tree species in a tropical forest. *Journal of Ecology*, **75**, 9–19.

Brown, W. L., Jr. (1973). A comparison of the Hylean and Congo–West African rain forest ant faunas. In *Tropical forest ecosystems in Africa and South America: a comparative review* (ed. B. J. Meggers, E. Ayensu, and W. D. Duckworth). Smithsonian Institution Press, Washington.

Buckley, R. C. (1982). Ant–plant interactions: a world review. In *Ant–plant interactions in Australia* (ed. R. C. Buckley), pp. 111–41. W. Junk, The Hague.

Carroll, C. R. and Janzen, D. H. (1973). Ecology of foraging by ants. *Annual Review of Ecology and Systematics*, **4**, 231–57.

Caughley, G. and Lawton, J. H. (1981). Plant–herbivore systems. In *Theoretical ecology. Principles and applications* (ed. R. M. May), pp. 132–66. Blackwell, Oxford.

Chapin, F. S. III. (1989). The cost of tundra plant structures: evaluation of concepts and currencies. *American Naturalist*, **133**, 1–19.

Clancy, K. M. and Price, P. W. (1986). Temporal variation in three-trophic-level interactions among willows, sawflies, and parasites. *Ecology*, **67(6)**, 1601–7.

Coley, P. D. (1983*a*). Herbivory and defensive characteristics of tree species in a lowland tropical forest. *Ecological Monographs*, **53**, 209–33.

Coley, P. D. (1983*b*). Intraspecific variation in herbivory on two tropical tree species. *Ecology*, **64(3)**, 426–33.

Coley, P. D. (1987). Interspecific variation in plant anti-herbivore properties: the role of habitat quality and rate of disturbance. *New Phytologist*, **106** (Suppl.), 251–63.

Coley, P. D. (1988). Effects of plant growth rate and leaf lifetime on amount and type of anti-herbivore defense. *Oecologia*, **74**, 531–6.

Coley, P. D. and Aide, T. M. (1990). A comparison of herbivory and plant defenses in temperate and tropical broad-leaved forests. In *Herbivory: tropical and temperate perspectives* (ed. P. W. Price, T. M. Lewinsohn, W. W. Benson, and G. W. Fernandes) pp. 25–49. John Wiley and Sons, New York.

Coley, P. D., Bryant, J. P., and Chapin, F. S. III. (1985). Resource availability and plant antiherbivore defense. *Science*, **230**, 895–9.

Compton, S. G. and Robertson, H. G. (1988). Complex interactions between mutualisms: ants tending homopterans protect fig seeds and pollinators. *Ecology*, **69**, 1302–5.

Compton, S. G., Lawton, J. H., and Rashbrook, V. K. (1989). Regional diversity, local community structure and vacant niches: the herbivorous arthropods of bracken in South Africa. *Ecological Entomology*, **14**, 365–73.

Cooper-Driver, G., Finch, S., Swain, T., and Bernays, E. (1977). Seasonal variation in secondary plant compounds in relation to the palatability of *Pteridium aquilinum*. *Biochemical Systematics and Ecology*, **5**, 177–83.

Crepps, W. F. (1975). Influence of specific non-crop vegetation on the insect fauna of small-scale agroecosystems. M.Sc. thesis. University of California, Davis.

Croat, T. B. (1978). *Flora of Barro Colorado Island.* Stanford University Press, Stanford.

Cronquist, A. (1981). *An integrated system of classification of flowering plants.* Columbia University Press, New York.

Darwin, F. (1877). On the glandular bodies of *Acacia sphaerocephala* and *Cecropia peltata* serving as food for ants. With an appendix on the nectar-glands of the common brake fern, *Pteris aquilina. Botanical Journal of the Linnean Society,* **15**, 398–409.

De Bach, P. (1964). *Biological control of insect pests and weeds,* 844 pp. Reinhold, New York.

Denno, R. F. and McClure, M. S. (1983). *Variable plants and herbivores in natural and managed systems.* Academic Press, New York.

Deuth, D. (1977). The function of extrafloral nectaries in *Aphelandra deppeana* Schl. & Cham. (Acanthaceae). *Brenesia,* **10/11,** 135–45.

DeVries, P. and Baker, I. (1989). Butterfly exploitation of an ant–plant mutualism: adding insult to herbivory. *Journal of the New York Entomological Society* (in press).

Douglas, M. M. (1983). Defense of bracken fern by arthropods attracted to axillary nectaries. *Psyche,* **90,** 313–20.

Duffey, S. S. (1986). Plant glandular trichomes: their partial role in defence against insects. In *Insects and the plant surface* (ed. B. E. Juniper and T. R. E. Southwood), pp. 151–72. Edward Arnold, London.

Elias, T. S. (1983). Extrafloral nectaries: their structure and distribution. In *The biology of nectaries* (ed. B. L. Bentley and T. S. Elias), pp. 174–203. Columbia University Press, New York.

Elias, T. S. and Gelband, H. (1975). Nectar: its production and functions in the trumpet creeper. *Science,* **189,** 289–91.

Feeny, P. P. (1976). Plant apparency and chemical defense. *Recent Advances in Phytochemistry,* **10,** 1–40.

Fiala, B., Maschwitz, U., Tho, T. Y., and Helbig, A. J. (1989). Studies of a South East Asian ant–plant association: protection of *Macaranga* trees by *Crematogaster borneensis. Oecologia,* **79,** 463–70.

Foster, R. B. and Brokaw, N. V. L. (1982). Structure and history of Barro Colorado Island. In *The ecology of a tropical forest. Seasonal rhythms and long term changes* (ed. E. G. Leigh, Jr., A. S. Rand, and D. M. Windsor), pp. 67–81. Smithsonian Institution Press, Washington, DC.

Fowler, S. V. and Lawton, J. H. (1985). Rapidly induced defences and talking trees: the devil's advocate position. *American Naturalist,* **126(2),** 181–95.

Fritz, R. S. (1982). An ant–treehopper mutualism: effects of *Formica subsericea* on the survival of *Vanduzea arquata. Ecological Entomology,* **7,** 267–76.

Gilbert, L. E. (1975). Ecological consequences of a coevolved mutualism between butterflies and plants. In *Coevolution of animals and plants* (ed. L. E. Gilbert and P. H. Raven), pp. 210–40. University of Texas Press, Austin.

Gilbert, L. E. and Smiley, J. T. (1978). Determinants of local diversity in phytophagous insects: host specialists in tropical environments. *Symposium of the Royal Entomological Society of London,* **9,** 89–104.

Goodland, R. (1971). A physiognomic analysis of the cerrado vegetation of central Brazil. *Journal of Ecology,* **59,** 411–19.

Gould, S. J. and Vrba, E. S. (1982). Exaptation—a missing term in the science of form. *Paleobiology,* **8,** 4–15.

Harley, J. L. and Smith, S. E. (1983). *Mycorrhizal symbiosis*. Academic Press, London.

Harper, J. H. (1977). *Population biology of plants*. Academic Press, London.

Hassell, M. P. (1978). *The dynamics of arthropod predator–prey systems*. Princeton University Press, New Jersey.

Hassell, M. P. (1980). Foraging strategies, population models and biological control: a case study. *Journal of Animal Ecology*, **49**, 603–28.

Hassell, M. P. (1982). Patterns of parasitism by insect parasitoids in patchy environments. *Ecological Entomology*, 7, 365–77.

Hassel, M. P. (1985). Insect natural enemies as regulating factors. *Journal of Animal Ecology*, **54**, 323–34.

Hassell, M. P. and Waage, J. K. (1984). Host–parasitoid population interactions. *Annual Review of Entomology*, **29**, 89–114.

Hawkins, C. P. and MacMahon, J. A. (1989). Guilds: the multiple meanings of a concept. *Annual Review of Entomology*, **34**, 423–52.

Heads, P. A. (1986). Bracken, ants and extrafloral nectaries. IV. Do wood ants (*Formica lugubris*) protect the plant against insect herbivores? *Journal of Animal Ecology*, **55**, 795–809.

Heads, P. A. and Lawton, J. H. (1984). Bracken, ants and extrafloral nectaries. II. The effect of ants on the insect herbivores of bracken. *Journal of Animal Ecology*, **53(3)**, 1015–31.

Heads, P. A. and Lawton, J. H. (1985). Bracken, ants and extrafloral nectaries. III. How insect herbivores avoid ant predation. *Ecological Entomology*, **10**, 29–42.

Hespenheide, H. A. (1985). Insect visitors to extrafloral nectaries of *Byttneria aculeata* (Sterculiaceae): relative importance and roles. *Ecological Entomology*, **10(2)**, 191–204.

Hill, M. G. and Blackmore, P. J. M. (1980). Interactions between ants and the coccid *Icerya seychellarum* on Aldabra Atoll. *Oecologia*, **45**, 360–5.

Horvitz, C. C. and Schemske, D. W. (1984). Effects of nectar-harvesting ants and an ant-tended herbivore on seed production of a neotropical herb. *Ecology*, **65(5)**, 1369–78.

Howe, H. F. (1984). Constraints on the evolution of mutualisms. *American Naturalist*, **123**, 764–77.

Hubbell, S. P. and Foster, R. B. (1986). Canopy gaps and the dynamics of a neotropical forest. In *Plant ecology* (ed. M. J. Crawley), pp. 77–96. Blackwell Scientific Publications, Oxford.

Huxley, C. R. (1986). Evolution of benevolent ant–plant relationships. In *Insects and the plant surface* (ed. B. E. Juniper and T. R. E. Southwood), pp. 257–82. Edward Arnold, London.

Inouye, D. W. and Taylor, O. R. (1979). A temperate region plant–ant–seed predator system: consequences of extrafloral nectar secretion by *Helianthella quinquenervis*. *Ecology*, **60**, 1–7.

Janzen, D. H. (1966). Coevolution of mutualism between ants and acacias in Central America. *Evolution*, **20**, 249–75.

Janzen, D. H. (1967). Interaction of the Bull's-Horn acacia (*Acacia cornigera* L.) with one of its ant inhabitants (*Pseudomyrmex ferruginea* F. Smith) in eastern Mexico. *Kansas University Science Bulletin*, **47**, 315–558.

Janzen, D. H. (1973). Dissolution of mutualism between *Cecropia* and its *Azteca* ants. *Biotropica*, **5**, 15–28.

Jeffree, C. E. (1986). The cuticle, epicuticular waxes and trichomes of plants, with reference to their structure, functions and evolution. In *Insects and the plant surface* (ed. B. E. Juniper and T. R. E. Southwood), pp. 23–64. Edward Arnold, London.

Jeffries, M. J. and Lawton, J. H. (1984). Enemy free space and the structure of ecological communities. *Biological Journal of the Linnean Society*, **23**, 269–86.

Jolivet, P. (1986). *Les fourmis et les plantes—un exemple de coevolution*, 245 pp. Société Nouvelle des Editions Boubee, Paris.

Keeler, K. H. (1979*a*). Distribution of plants with extrafloral nectaries and ants at two elevations in Jamaica. *Biotropica*, **11**, 152–4.

Keeler, K. H. (1979*b*). Species with extrafloral nectaries in a temperate flora (Nebraska). *Prairie Naturalist*, **11**, 33–7.

Keeler, K. H. (1980*a*). The extrafloral nectaries of *Ipomoea leptophylla* (Convolvulaceae). *American Journal of Botany*, **67**, 216–22.

Keeler, K. H. (1980*b*). Distribution of plants with extrafloral nectaries in temperate communities. *American Midland Naturalist*, **104**, 274–80.

Keeler, K. H. (1981*a*). Cover of plants with extrafloral nectaries in four northern California habitats. *Madrono*, **28**, 26–9.

Keeler, K. H. (1981*b*). Function of *Mentzelia nuda* (Loasaceae) postfloral nectaries in seed defense. *American Journal of Botany*, **68(2)**, 295–9.

Keeler, K. H. (1981*c*). A model of selection for facultative nonsymbiotic mutualism. *American Naturalist*, **118**, 488–98.

Keeler, K. H. (1985). Extrafloral nectaries on plants in communities without ants: Hawaii. *Oikos*, **44**, 407–14.

Keeler, K. H. (1989). Ant–plant interactions. In *Plant–animal interactions* (ed. W. G. Abrahamson), pp. 207–42. McGraw-Hill Book Co., New York.

Keeler, K. H. and Kaul, R. B. (1979). Morphology and distribution of petiolar nectaries in *Ipomoea* (Convolvulaceae). *American Journal of Botany*, **66**, 946–52.

Kelly, C. A. (1986). Extrafloral nectaries: ants, herbivores and fecundity in *Cassia fasciculata*. *Oecologia*, **69**, 600–5.

Kempf, W. W. (1973). A new *Zacryptocerus* from Brazil, with remarks on the generic classification of the tribe Cephalotini (Hymenoptera–Formicidae). *Studia Entomologica*, **16**, 449–62.

Kirk, A. A. (1977). The insect fauna of the weed *Pteridium aquilinum* (L.) Kuhn (Polypodiaceae) in Papua New Guinea: a potential source of biological control agents. *Journal of the Australian Entomological Society*, **16**, 403–9.

Kirk, A. A. (1982). Insects associated with bracken fern *Pteridium aquilinum* (Polypodiaceae) in Papua New Guinea and their possible use in biological control. *Acta Oecologica/Oecologia Applicata*, **3**, 343–59.

Koptur, S. (1979). Facultative mutualism between weedy vetches bearing extrafloral nectaries and weedy ants in California. *American Journal of Botany*, **66**, 1016–20.

Koptur, S. (1983). Flowering phenology and floral biology of *Inga* (Fabaceae: Mimosoideae). *Systematic Botany*, **8**, 354–68.

Koptur, S. (1984). Experimental evidence for defense of *Inga* (Mimosoideae) saplings by ants. *Ecology*, **65**, 1787–93.

Koptur, S. (1985). Alternative defenses against herbivores in *Inga* (Fabaceae: Mimosoideae) over an elevational gradient. *Ecology*, **66(5)**, 1639–50.

Koptur, S. (1989). Is extrafloral nectar production an inducible defense? In *Evolutionary ecology of plants* (ed. J. Bock and Y. Linhart), pp. 323–39. Westview Press, Boulder.

Koptur, S. and Lawton, J. H. (1988). Interactions among vetches bearing extrafloral nectaries, their biotic protection agents, and herbivores. *Ecology*, **69(1)**, 278–83.

Koptur, S., Smith, A. R., and Baker, I. (1982). Nectaries in some neotropical species of *Polypodium* (Polypodiaceae): preliminary observations and analyses. *Biotropica*, **14**, 108–13.

Kusnezov, N. (1957). Numbers of species of ants in faunae of different latitudes. *Evolution*, **11**, 298–9.

Laine, K. J. and Niemelä, P. (1980). The influence of ants on the survival of mountain birches during an *Oporinia autumnata* (Lep., Geometridae) outbreak. *Oecologia*, **47**, 39–42.

Law, R. (1985). Evolution in a mutualistic environment. In *The biology of mutualism: ecology and evolution* (ed. D. H. Boucher). Croom Helm, London.

Law, R. and Koptur, S. (1986). On the evolution of non-specific mutualism. *Biological Journal of the Linnean Society*, **27**, 251–67.

Lawton, J. H. (1976). The structure of the arthropod community on bracken. *Botanical Journal of the Linnean Society*, **73**, 187–216.

Lawton, J. H. (1978). Host–plant influences on insect diversity: the effects of space and time. In *Diversity of insect faunas* (ed. L. A. Mound and N. Waloff), pp. 105–25. Blackwell, Oxford.

Lawton, J. H. (1982). Vacant niches and unsaturated communities: a comparison of bracken herbivores at sites on two continents. *Journal of Animal Ecology*, **51**, 573–95.

Lawton, J. H. (1983). Plant architecture and the diversity of phytophagous insects. *Annual Review of Entomology*, **28**, 23–9.

Lawton, J. H. (1984). Non-competitive populations, non-convergent communities, and vacant niches: the herbivores of bracken. In *Ecological communities: conceptual issues and the evidence* (ed. D. R. Strong, D. Simberloff, and L. G. Abele), pp. 67–100. Princeton University Press, Princeton.

Lawton, J. H. (1986). Biological control of bracken: plans and possibilities. In *Bracken. Ecology, land use and control technology* (ed. R. T. Smith and J. A. Taylor), pp. 445–52. Parthenon Publishing, Carnforth, England.

Lawton, J. H. (1988). Biological control of bracken in Britain: constraints and opportunities. *Philosophical Transactions of the Royal Society of London B*, **318**, 335–55.

Lawton, J. H. and Heads, P. A. (1984). Bracken, ants and extrafloral nectaries. I. The components of the system. *Journal of Animal Ecology*, **53**, 995–1014.

Lawton, J. H. and McNeill, S. (1979). Between the devil and the deep blue sea: on the problem of being a herbivore. In *Population dynamics* (ed. R. M. Anderson, B. D. Turner, and L. R. Taylor), pp. 223–440. Blackwell Scientific, Oxford.

Lawton, J. H., MacGarvin, M., and Heads, P. A. (1986). The ecology of bracken-feeding insects: background for a biological control programme. In *Bracken. Ecology, land use and control technology* (ed. R. T. Smith and J. A. Taylor), pp. 285–92. Parthenon Press, Carnforth, England.

Leigh, E. G., Jr., Rand, A. S., and Windsor, D. M. (ed.) (1982). *The ecology of a tropical forest. Seasonal rhythms and long-term changes.*

Leius, K. (1967). Influence of wild flowers on parasitism of tent caterpillar and codling moth. *Canadian Entomologist*, **99**, 444–6.

Lersten, N. R. and Brubaker, C. L. (1987). Extrafloral nectaries in Leguminosae: review and original observations in *Erythrina* and *Munca* (Papilionoideae; Phaseoleae). *Bulletin of the Torrey Botanical Club*, **114**, 437–47.

Lloyd, F. E. (1901). The extra-nuptial nectaries in the common brake, *Pteridium aquilinum. Science*, **13**, 885–90.

MacGarvin, M., Lawton, J. H., and Heads, P. A. (1986). The herbivorous insect communities of open and woodland bracken: observations, experiments and habitat manipulations. *Oikos*, **47**, 135–48.

McKey, D. (1979). The distribution of secondary compounds within plants. In *Herbivores. Their interaction with secondary plant metabolites* (ed. G. A. Rosenthal and D. H. Janzen), pp. 55–133. Academic Press, New York.

McKey, D. (1984). Interaction of the ant–plant *Leonardoxa africana* (Caesalpiniaceae) with its obligate inhabitants in a rainforest in Cameroon. *Biotropica*, **16**, 81–99.

McKey, D. (1988). Promising new directions in the study of ant–plant mutualisms. In *Proceedings of the XIV International Botanical Congress* (ed. W. Greuter and B. Zimmer), pp. 335–55. Koeltz Konigstein/Taunus.

McKey, D. (1989). Interactions between ants and leguminous plants. In *Advances in legume biology* (ed. J. Zarucchi and C. Stirton). Missouri Botanical Gardens, St. Louis.

McLain, D. K. (1983). Ants, extrafloral nectaries and herbivory on the passion vine, *Passiflora incarnata. American Midland Naturalist*, **110**, 433–9.

Maschwitz, U., Schroth, M., Manel, M., and Yow Pong, T (1984). Lycaenids parasitizing symbiotic plant–ant partnerships. *Oecologia*, **64**, 78–80.

Messina, F. J. (1981). Plant protection as a consequence of an ant–membracid mutualism: interactions on goldenrod (*Solidago* sp.). *Ecology*, **62**, 1433–40.

Mooney, H. A. (1972). The carbon balance of plants. *Annual Review of Ecology and Systematics*, **3**, 315–46.

Mooney, H. A. and Gulmon, S. L. (1982). Constraints on leaf structure and function in reference to herbivory. *BioScience*, **32**, 198–206.

Morais, H. C. (1980). Estrutura de uma comunidade de formigas arborícolas em vegetação de campo cerrado. M.Sc. thesis, Universidade Estadual de Campinas, SP, Brazil.

Morais, H. C. and Benson, W. W. (1988). Recolonizaçao de vegetaçao de cerrado após queimada, por formigas arborícolas. *Revista Brasileira de Biologia*, **48**, 459–66.

Newbury, D. and Foresta, H. (1985). Herbivory and defense in pioneer, gap, and understory trees of tropical rain forest in French Guiana. *Biotropica*, **17(3)**, 238–44.

Nicholson, A. J. and Bailey, V. A. (1935). The balance of animal populations. Part I. *Proceedings of the Zoological Society of London*, **1935**, 551–98.

O'Dowd, D. J. (1979). Foliar nectar production and ant activity on a neotropical tree, *Ochroma pyramidale. Oecologia*, **43**, 233–48.

O'Dowd, D. J. (1980). Pearl bodies of a neotropical tree, *Ochroma pyramidale*: ecological implications. *American Journal of Botany*, **67**, 543–9.

O'Dowd, D. J. (1982). Pearl bodies as ant food: an ecological role for some leaf emergences of tropical plants. *Biotropica*, **14**, 40–9.

O'Dowd, D. J. and Catchpole, E. A. (1983). Ants and extrafloral nectaries: no evidence for plant protection in *Helichrysum* spp.—ant interactions. *Oecologia*, **59(2–3)**, 191–200.

Oliveira, P. S. (1988). Sobre a interação de formigas com o pequi do cerrado, *Caryocar brasiliense* Camb. (Caryocaraceae): O significado ecológico de nectários extraflorais. Ph.D. thesis, Universidade Estadual de Campinas, São Paulo, Brazil.

Oliveira, P. S. and Leitão-Filho, H. F. (1987). Extrafloral nectaries: their taxonomic distribution and abundance in the woody flora of cerrado vegetation in southeast Brazil. *Biotropica*, **19**, 140–8.

Oliveira, P. S. and Oliveira-Filho, A. T. (1990). Distribution of extrafloral nectaries in woody flora of tropical communities in Western Brazil. In *Herbivory: tropical and temperate comparisons* (ed. P. W. Price, W. W. Benson, T. M. Lewinsohn, and G. W. Fernandes). John Wiley and Sons, New York.

Oliveira, P. S., Silva, A. F. da, and Martins, A. B. (1987). Ant foraging on extrafloral nectaries of *Qualea grandiflora* (Vochysiaceae) in cerrado vegetation: ants as potential anti-herbivore agents. *Oecologia*, **74**, 228–30.

Page, C. N. (1982). Field observations on the nectaries of bracken, *Pteridium aquilinum*, in Britain. *Fern Gazette*, **12**, 243–5.

Pagel, M. D. and Harvey, P. H. (1988). Recent developments in the analysis of comparative data. *Quarterly Review of Biology*, **63**, 413–40.

Pemberton, R. W. (1988). The abundance of plants bearing extrafloral nectaries in colorado and mojave desert communities of southern California. *Madrono*, **35**, 238–46.

Pierce, N. E. and Mead, P. S. (1981). Parasitoids as selective agents in the symbiosis between lycaenid butterfly larvae and ants. *Science*, **211**, 1185–7.

Price, P. W., Bouton, C. E., Gross, P., McPheron, B. A., Thompson, J. N., and Weis, A. E. (1980). Interactions among three trophic levels: influence of plants on interactions between insect herbivores and natural enemies. *Annual Review of Ecology and Systematics*, **11**, 41–65.

Price, P. W. and Clancy, K. M. (1986). Interactions among three trophic levels: gall size and parasitoid attack. *Ecology*, **67(6)**, 1593–600.

Price, P. W., Westoby, M., Rice, B., Atsatt, P. R., Fritz, R. S., Thompson, J. N., and Mobley, K. (1986). Parasite mediation in ecological interactions. *Annual Review of Ecology and Systematics*, **17**, 487–505.

Putz, F. E. (1984). The natural history of lianas on Barro Colorado Island, Panama. *Ecology*, **65**, 1713–24.

Putz, F. E. and Windsor, D. M. (1987). Liana phenology on Barro Colorado Island, Panama. *Biotropica*, **19**, 334–41.

Rand, A. S. and Rand, W. M. (1982). Variation in rainfall on Barro Colorado Island. In *The ecology of a tropical forest. Seasonal rhythms and long-term changes* (ed. E. G. Leigh, Jr., A. S. Rand, and D. M. Windsor), pp. 47–59. Smithsonian Institution Press, Washington, D. C.

Rashbrook, V. K. (1989). Interactions between ants, herbivorous insects and bracken (*Pteridium aquilinum*), a fern with extrafloral nectaries. M.Sc. thesis. Rhodes University, South Africa.

Rhoades, D. F. (1979). Evolution of plant chemical defense against herbivores. In *Herbivores. Their interaction with secondary plant metabolites* (ed. G. A. Rosenthal and D. H. Janzen), pp. 3–54. Academic Press, London.

Rhoades, D. F. and Cates, R. G. (1976). Toward a general theory of plant anti-herbivore chemistry. In *Biochemical interaction between plants and insects* (ed. J. W. Wallace and R. L. Mansell), pp. 168–213. Plenum Publishing Co., New York.

Rickson, F. R. (1977). Progressive loss of ant-related traits of *Cecropia peltata* on selected Caribbean islands. *American Journal of Botany*, **64**, 585–92.

Rickson, F. R. (1980). Developmental anatomy and ultrastructure of the ant-food bodies (Beccarian bodies) of *Macaranga triloba* and *M. hypoleuca* (Euphorbiaceae). *American Journal of Botany*, **67**, 285–93.

Ridley, H. N. (1910). Symbiosis of ants and plants. *Annals of Botany*, **24**, 457–85.

Ridley, M. (1983). *The explanation of organic diversity*. Oxford University Press, Oxford.

Ridley, M. (1986). The number of males in a primate troop. *Animal Behaviour*, **66**, 1848–58.

Rodriguez, E., Healey, P. L., and Mehta, I. (ed.) (1984). *Biology and chemistry of trichomes*. Plenum Press, New York.

Schemske, D. W. (1980). The evolutionary significance of extra floral nectar production by *Costus woodsonii* (Zingiberaceae): an experimental analysis of ant protection. *Journal of Ecology*, **68**, 959–67.

Schemske, D. W. (1982). Ecological correlates of a neotropical mutualism: ant assemblages at *Costus* extrafloral nectaries. *Ecology*, **63**, 932–41.

Schemske, D. W. (1983). Limits to specialization and coevolution in plant–animal mutualisms. In *Coevolution* (ed. M. H. Nitecki) pp. 67–109. University of Chicago Press, Chicago.

Schemske, D. W. and Horvitz, C. C. (1988). Plant–animal interactions and fruit production in a neotropical herb: a path analysis. *Ecology*, **69(4)**, 1128–37.

Schupp, E. W. (1986). *Azteca* protection of *Cecropia*: ant occupation benefits juvenile trees. *Oecologia*, **70**, 379–85.

Simmonds, F. J. (1967). Possibilities of biological control of bracken *Pteridium aquilinum* (L.) Kuhn (Polypodiaceae). *Pest Articles and News Summaries (C)*, **13**, 200–3.

Smiley, J. T. (1985). *Heliconius* caterpillar mortality during establishment on plants with and without attending ants. *Ecology*, **66**, 845–9.

Smiley, J. T. (1986). Ant constancy at *Passiflora* extrafloral nectaries: effects on caterpillar survival. *Ecology*, **67(2)**, 516–21.

Smiley, J. T., Atsatt, P. R., and Pierce, N. E. (1988). Local distribution of the lycaenid butterfly, *Jalmenus evagoras*, in response to host ants and plants. *Oecologia*, **76**, 416–22.

Smith, L. L., Lanza, J., and Smith, G. C. (1990). Amino acid concentrations in extrafloral nectar of *Impatiens sultani* increase after simulated herbivory. *Ecology*, **71(1)**, 107–15.

Snelling, R. R. (1973). The ant genus *Conomyrma* in the United States. *Contributions in Science, Natural History Museum of Los Angeles County*, **238**, 1–6.

Sokal, R. R. and Rohlf, F. J. (1981). *Biometry*, 2nd edn. W. H. Freeman and Company, New York.

Southwood, T. R. E., Brown, V. K., and Reader, P. M. (1983). Continuity of vegetation in space and time: a comparison of insects' habitat templet in different successional stages. *Researches on Population Ecology*, **3** (Suppl.), 61–74.

Stephenson, A. G. (1982). The role of the extrafloral nectaries of *Catalpa speciosa* in limiting herbivory and increasing fruit production. *Ecology*, **63**, 663–9.

Strong, D. R., Lawton, J. H., and Southwood, T. R. E. (1984). *Insects on plants*. Blackwell Scientific Publications, Oxford.

Sudd, J. H. (1983). Distribution of foraging wood-ants in relation to the distribution of aphids. *Insectes Sociaux*, **30(3)**, 293–307.

Sudd, J. H. and Sudd, M. E. (1985). Seasonal changes in the response of wood-ants (*Formica lugubris*) to sucrose baits. *Ecological Entomology*, **10**, 89–97.

Takeda, S., Kinomura, K., and Sakurai, H. (1983). Effects of ant attendance on the honeydew excretion and larviposition of the cowpea aphid. *Applied Entomology and Zoology*, **17**, 133–5.

Taylor, F. (1977). Foraging behavior of ants: experiments with two species of myrmecine ants. *Behavioral Ecology and Sociobiology*, **2**, 147–67.

Taylor, J. (1986). The bracken problem: a local hazard and global issue. In *Bracken. Ecology, land use and control technology* (ed. R. T. Smith and J. A. Taylor), pp. 21–42. Parthenon Publishing, Carnforth, England.

Tempel, A. S. (1983). Bracken fern (*Pteridium aquilinum*) and nectar-feeding ants: a nonmutualistic interaction. *Ecology*, **64**, 1411–22.

Tho, Y. P. (1978). Living in harmony. *Nature Malaysiana*, **3**, 34–9.

Thompson, J. N. (1988). Variation in interspecific interactions. *Annual Review of Ecology and Systematics*, **19**, 65–87.

Tilman, D. (1978). Cherries, ants and tent caterpillars: timing of nectar production in relation to susceptibility of caterpillars to ant predation. *Ecology*, **59**, 686–92.

Waloff, N. and Richards, O. W. (1977). The effect of insect fauna on growth, mortality and natality of broom, *Sarothamnus scoparius*. *Journal of Applied Ecology*, **14**, 787–98.

Washburn, J. O. (1984). Mutualism between a cynipid gall wasp and ants. *Ecology*, **65**, 654–6.

Washburn, J. O. and Cornell, H. V. (1981). Parasitoids, patches, and phenology: their possible role in the local extinction of a cynipid gall wasp population. *Ecology*, **62(6)**, 1597–607.

Way, M. J. (1963). Mutualism between ants and honeydew producing Homoptera. *Annual Review of Entomology*, **8**, 307–44.

Webster, G. L. (1967). The genera of Euphorbiaceae in the southeastern United States. *Journal of the Arnold Arboretum*, **48**, 303–430.

Weis, A. G. and Abrahamson, W. G. (1985). Potential selective pressures by parasitoids on a plant–herbivore interaction. *Ecology*, **66(4)**, 1261–9.

Wells, H. and King, J. L. (1980). A general 'exact' test for N × M contingency tables. *Bulletin of the Southern California Academy of Sciences*, **79**, 65–77.

Whalen, M. A. and Mackay, D. A. (1988). Patterns of ant and herbivore activity on five understory Euphorbiaceous saplings in submontane Papua New Guinea. *Biotropica*, **20(4)**, 294–300.

Whitham, T. G. and Mopper, S. (1985). Chronic herbivory: impacts on architecture and sex expression of pinyon pine. *Science*, **228**, 1089–91.

Wilson, D. S. (1980). *The natural selection of populations and communities*. Benjamin Cummings, Menlo Park.

Wilson, E. O. (1976). Which are the most prevalent ant genera? *Studia Entomologica*, **19**, 187–200.

Winterbourne, M. J. (1987). The arthropod fauna of bracken (*Pteridium aquilinum*) on the Port Hills, South Island, New Zealand. *New Zealand Entomologist*, **10**, 99–104.

Wolf, P. G., Haufler, C. H., and Sheffield, E. (1988). Electrophoretic variation and mating system of the clonal weed *Pteridium aquilinum* (L. Kuhn) (bracken). *Evolution*, **42**, 1350–55.

Zar, J. H. (1984). *Biostatistical analysis*. 2nd edn, p. 201. Prentice-Hall, Englewood Cliffs, New York.

Part 4

Symbiosis between plants and ants

18

The association between *Macaranga* trees and ants in South-east Asia

Brigitte Fiala, Ulrich Maschwitz, and Tho Yow Pong

Introduction

In south-east Asia, myrmecophytic associations have so far been little investigated. Although the euphorb tree genus *Macaranga* has long been known to contain myrmecophytic ants (Smith 1903) existing information on its association is limited (e.g. Baker 1934; Ong 1978; Tho 1978; Rickson 1980). Nevertheless, the association has often been interpreted as an Asiatic equivalent to similar associations in Africa and America (Janzen 1969; Duviard and Segeren 1974; Buckley 1982). Our study was the first experimental investigation of the biological significance of the *Macaranga* association with ants. We were especially interested to find out what adaptations may have evolved convergently in South-east Asia.

Distribution and habitat

The genus *Macaranga* (Euphorbiaceae) is distributed from Africa to Polynesia and strongly centred in the Malesian region. Many species of this tree genus inhabit disturbed areas such as clearings, gaps, and forest edges. These habitats have enormously increased in extent over the last 100 years and the fast-growing *Macaranga* species have spread and become one of the most conspicuous trees in cleared areas (Whitmore 1967). In West Malaysia, where our study was carried out, 19 of the 27 *Macaranga* species occur in secondary forests.

West Malaysia is situated in the humid tropics with little seasonal variation in temperature. There is no pronounced dry season in the study area but two periods of heavy rainfall. The study areas are primarily covered with mixed dipterocarp forest that has been disturbed by forestry, road building, etc., and hence contains much secondary habitat.

Associated ants

Some of the most common species of *Macaranga* are closely associated with ants of the genus *Crematogaster* (Myrmicinae). In peninsular Malaysia a tight relationship exists between nine of the 27 *Macaranga* species and a small, non-stinging *Crematogaster*. Most samples from the investigated *Macaranga* plants probably belong to *C. borneensis* which seems to be very variable. Based on these samples, however, several additional, closely related *Crematogaster* species may be involved in the symbiosis in West Malaysia (J. T. Longino, personal communication). In *M. puncticulata* we also found a *Camponotus* (*Colobopsis*). The only myrmecophytic *Macaranga* species which occurs in Thailand (*M. griffithiana*) was also found inhabited by a (different) *Camponotus* (*Colobopsis*) sp. In Borneo other *Crematogaster* species seem to take part in the association; we found a *Crematogaster* species from a different subgenus living in *M. kingii* and D. W. Davidson (personal communication) found several *Crematogaster* species in Bornean *Macaranga*. So the type of community complexity which has been found in other associations may exist (see Chapters 19–21, this volume).

Life history

Young *Macaranga* plants start being colonized by ants when they are about 10 cm tall and occupation occurs all year round. The queen ant searches for an unoccupied plant, sheds her wings, and chews an entrance hole, which she then seals from inside. The importance of the plants for the ants is evident: they provide nesting space inside hollow stems and food in the form of food bodies. A carbohydrate-rich food source is obtained via scale insects cultivated inside the stem. *Crematogaster borneensis* were found to be totally dependent on the host plant: in laboratory tests the workers did not survive away from the plants and in their natural habitat they were never found anywhere else (Fiala and Maschwitz 1990).

Benefits to the plant

There is no indication that the ants contribute to the nutrient requirements of their host plant. We did not find evidence for a net nutrient gain by *Macaranga* plants from the association with ants (Fiala 1988). Only very small amounts of labelled nutrients were taken up by the plant from the hollow internodes. The interior wall of the stem does not show any special absorptive structures. In addition, the ants never leave the plants for foraging or feed on insects that they kill on the plant, so that no nutrient import takes place.

Workers of *Crematogaster borneensis*, however, do protect their host plant against herbivores (Fiala *et al.* 1989). These ants, although seemingly rather defenceless, are very aggressive and, with a mass recruiting system, are able to attack phytophagous insects. Ant-inhabited *Macaranga* plants had a significantly lower percentage of herbivore damage than ant-free specimens (Fig. 18. 1). More than 50 per cent loss of leaf area was observed only in ant-

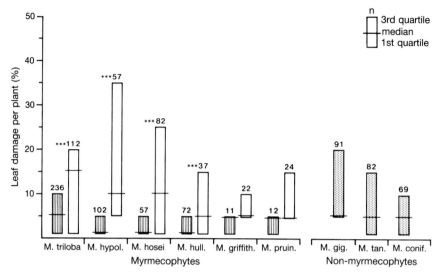

Fig. 18.1. Herbivore damage, as a percentage of leaf area, for ant-occupied (hatched columns) and ant-free (open columns) specimens of myrmecophytic and non-myrmecophytic *Macaranga* species. Differences between ant-occupied and ant-free plants according to Mann–Whitney U-Test: *** = $P < 0.001$. *Macaranga* species: hypol. = *hypoleuca*; hull. = *hulletti*; griffith. = *griffithiana*; pruin. = *pruinosa*; gig. = *gigantea*; tan. = *tanarius*; conif. = *conifera*.

free plants (e.g. *M. triloba*, Fig. 18.2). The ants also display a cleaning behaviour which results in removal of herbivores in the earliest developmental stages as eggs (Fig. 18.3). This contributes significantly to the reduction of herbivore damage.

Still more important is the ants' defence of the host plant against plant competitors, especially vines which are abundant in the well-lit habitats where *Macaranga* grows. The pruning of foreign plant material in contact with the host plant has not been found so far in other *Crematogaster* species. Both ant-free myrmecophytic *Macaranga* species and most of the principally uninhabited *Macaranga* species had a significantly higher degree of vine growth than plants with ants (Fig. 18.4).

The genus *Macaranga* comprises a full range of species, from those which are not regularly associated with ants to obligate myrmecophytes. This makes the genus especially suitable for interspecific comparison. Most of the myrmecophytic *Macaranga* species have a high percentage of occupation: plants > 1 m are usually colonized to more than 90 per cent. Three species out of the nine myrmecophytes do not seem to be fully adapted and can be seen as transitional. We will briefly introduce two of them: *M. hosei* and *M.*

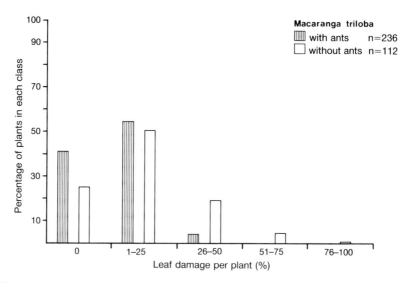

Fig. 18.2. Percentage of herbivore damage on ant-occupied (hatched) and ant-free (open) *Macaranga triloba*.

Fig. 18.3. Worker of *Crematogaster borneensis* removing a lepidopteran egg from the *Macaranga triloba* host plant.

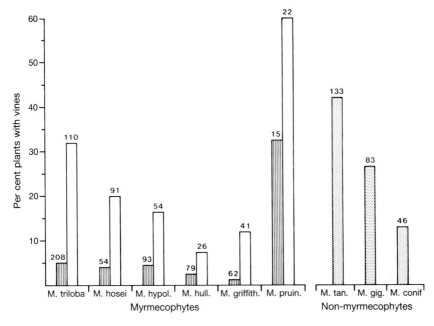

Fig. 18.4. Percentage of plants of myrmecophytic (left) and non-myrmecophytic (right) *Macaranga* species with vines. Hatched columns: ant-occupied; open columns: ant-free specimens. Differences between absolute frequencies according to Chi² analysis: *M. triloba* = $P < 0.001$; *M. hosei, M. hypoleuca* = $P < 0.05$; other species n.s. (not significant).

pruinosa, which are morphologically very similar, but differ in habitat. *M. pruinosa* grows in swampy places and *M. hosei* at drier sites. Their percentage of ant inhabitation is lower, and in *M. pruinosa* the effects of ant occupation are less clear cut: only parts of the plants (often the main stems) are inhabited by ants, which is reflected in a greater herbivore damage and vine growth on this species (Figs 18. 1 and 18. 4).

Most of the other myrmecophytic species have food bodies hidden under recurved stipules, as in *M. triloba*. In *M. hosei* and *M. pruinosa* food bodies are exposed on horizontal stipules. The stem interior of these species does not become hollow as in the other myrmecophytes but remains solid with a dry, soft pith. Although cavities can be excavated in the stem interior (Fig. 18. 5), it takes up to four hours for a founding queen to chew her way into the interior and she runs a high risk of predation or parasitoid attack during this time.

We also investigated nine of the *Macaranga* species known as non-myrmecophytes for their possible relationship with ants. Although there are

Table 18.1. Percentage of plants inhabited by *Crematogaster borneensis* for different *Macaranga* species in Peninsular Malaysia. Only plants over 1 m tall are recorded. *n* = sample size.

Species	*n*	Percentage of plants occupied by ants
M. hypoleuca Muell. Arg.	56	98
M. hullettii King ex Hook. f.	48	96
M. triloba Muell. Arg.	267	93
M. griffithiana Muell. Arg.	57	89
M. hosei King ex Hook	94	78
M. pruinosa Muell. Arg.	49	51

always ants on the plants (we found at least 24 species in 14 genera) we did not find evidence for any specific association or nesting of ants on these non-myrmecophytic plants. In order to explain this we looked for the presence or absence of morphological or other predispositions to a myrmecophytic way of life. Contrary to earlier assumptions most non-myrmecophytic *Macar-*

Fig. 18.5. Transverse section through stem of *Macaranga pruinosa*. Younger part of the stem (left side) excavated by *Crematogaster borneensis* ants (scale = 4 mm).

anga species do produce food bodies, some also have extrafloral nectaries. Therefore, food availability for potential ant partners does not appear to be a limiting factor. The lack of nesting space on or within the plants probably explains why ants are not present permanently. The stem morphology differs considerably from ant-inhabited species. Most non-myrmecophytes have a solid stem with a hard and wet pith and rings of resin ducts. These secrete large amounts of a sticky gum-like fluid when injured (Fig. 18. 6). This would be an effective barrier to ant colonization.

Fig. 18.6. Transverse section through stem of *Macaranga tanarius* Muell. Arg. Secretion of sticky fluid (scale = 4 mm).

As we have seen ant inhabitation offers some advantage for the plants. How do non-inhabited *Macaranga* species cope with herbivory and climber growth? Are there any principal differences between non-myrmecophytes and myrmecophytes? Preliminary evidence suggests that differences exist in habitat requirements and that competitive ability can be achieved by different types of growth. Some species occur exclusively at sites where plant competition and herbivore pressure are lower than in secondary habitats, e.g. in primary forest or at higher altitudes. Species without ants growing in well-lit lowland sites often have certain structures which probably favour them in the face of plant competition, e.g. very large leaves and a broad, roof-like canopy. Both attributes result in shading of the surrounding vegetation and will inhibit the growth of neighbouring plants. In particular, this is the case with *M. gigantea* and *M. tanarius*. Individuals of *M. gigantea* only 30 cm tall already have leaf widths of up to 40 cm.

Most myrmecophytic species have a much more open and branched canopy. They grow primarily at the interface between forest and openings or roads; these edges are rich in climbers. Here the ants have an important function in suppressing growth of climbers by biting them off. More work is under way on this comparative aspect of the different strategies of all Malaysian *Macaranga* species. This will also involve investigations of defence mechanisms against herbivores, to assess effectiveness of herbivore protection in non-myrmecophytic *Macaranga*. Thus, the relationships between ants and plants are not only curious examples of symbiotic mutualisms, but show patterns and complexity paralleling those in other ecological systems, as pointed out in Chapter 20.

In summary, it can be said that the ecology of ant-associated *Macaranga* trees is modified compared to uninhabited congeners and that these adaptations influence their competitive ability. A broader perspective on ant–plant associations may lead to new insights in colonization patterns in plants. In some features *Macaranga* resembles the pioneer tree genus *Cecropia* (Cecropiaceae) in Central and South America and may be regarded as an Asiatic analogue of this neotropical system. The *Macaranga–C. borneensis* association is equivalent to myrmecophytic systems in South America and Africa in its specificity and symbiotic character.

Acknowledgements

Dr Salleh Mohd Nor, the director of the Forest Research Institute Malaysia (FRIM), generously gave permission to work at the FRIM and its field stations. The Department of Zoology, University of Malaya, granted permission to use their field station in Ulu Gombak. We are indebted to Wulf Killmann for logistical and other support. Dr J. T. Longino kindly checked the *Crematogaster* identifications. Dr A. J. Helbig helped with the translation. Financial support by Deutsche Forschungsgemeinschaft, DFG, Deutscher Akademischer Austauschdienst, DAAD, and Freunde und Förderer der J.W. Goethe-Universität is gratefully acknowledged.

19

Azteca ants in Cecropia trees: taxonomy, colony structure, and behaviour

John T. Longino

Introduction

The most conspicuous ant–plant association in the wet neotropics is that of *Azteca* ants and *Cecropia* trees. In most low-elevation, wet regions of the neotropics, *Cecropia* is an ubiquitous and important invader of man-made clearings (Uhl *et al.* 1981; Brokaw 1987). The open, candelabra-shaped crowns of *Cecropia* often appear as the sole emergents over dense vine tangles in abandoned fields. The hollow internodes of these trees are almost always occupied by biting ants in the genus *Azteca*, each tree typically containing a single colony (Bequaert 1922; Wheeler 1942; Benson 1985).

In spite of being a common and species-rich ant–plant symbiosis, the *Azteca–Cecropia* association is only now receiving more than cursory investigation, with major research efforts underway in Costa Rica, Peru, and Brazil. Which *Azteca* species are obligate to *Cecropia* has been obscured by taxonomic uncertainty, past misidentifications, and inadequate collections (e.g. Wheeler 1942). It is important to provide a firm descriptive background to establish:

1. the data necessary for reconstruction of the evolutionary history of the relationship;
2. a basis for ecological studies of *Cecropia* and *Azteca*.

With the work of Harada (1982), Benson (1985), Harada and Benson (1988), Davidson and Fisher Chapter 20, this volume), and myself, involving surveys of *Cecropia*-inhabiting ant communities, a clearer picture is emerging.

Here I present a general description of *Azteca* and *Cecropia*, a discussion of the taxonomy of *Cecropia* ants (Table 19. 1), and a more detailed account of *Azteca* species in Costa Rica, contrasting their colony structure and behaviour (Table 19. 2), and their habitat and host use (Table 19. 3).

Cecropia (Cecropiaceae) and *Azteca* (Dolichoderinae)

Cecropia is a strictly neotropical genus (introduced elsewhere, see Ake Assi 1980; Putz and Holbrook 1988) related to the Urticaceae and Moraceae

Table 19.1. *Cecropia*-inhabiting *Azteca* and their status as obligate *Cecropia* inhabitants.

	Type locality	Status
constructor Emery (1896)	Costa Rica	1
var. *guianae* Wheeler (1942)	Guyana	2
muelleri Emery (1893)	Brazil	1
var. *brunni* Forel (1909)	Brazil	3
var. *janeirensis* Forel (1912)	Brazil	3
var. *nigella* Emery (1893)	Brazil	3
var. *nigridens* Forel (1908)	Brazil	3
var. *pallida* Stitz (1937)	Mexico	3
var. *wacketi* Emery in Forel (1908)	Brazil	3
subsp. *terminalis* Mann (1916)	Brazil	3
xanthochroa (Roger (1863))	Mexico	1
subsp. *costaricensis* Wheeler (1942)	Costa Rica	2
subsp. *isthmica* Wheeler (1942)	Panama	2
subsp. *australis* Wheeler (1942)	Bolivia	2
subsp. *salti* Wheeler (1942)	Colombia	2
coeruleipennis Emery (1893)	Costa Rica	1
alfari Emery (1893)	Costa Rica	1
ovaticeps Forel (1904)	Brazil	1

1: Known obligate *Cecropia* inhabitant. 2: Taxonomic status (i.e. valid name or junior synonym) unknown, usually known only from type specimens, but types collected from *Cecropia*. 3: Taxonomic status unknown; no biological data with type description.

Table 19.2. Nest structure and behaviour of obligate *Cecropia* ants in Costa Rica.

Azteca species	Nest organization	Aggression	Herbivory
A. constructor	Centralized in main stem	high	low
A. xanthochroa	Centralized in main stem	high	low
A. coeruleipennis	Polydomous in branch tips	?	?
A. alfari	Polydomous in branch tips	low	high
A. ovaticeps	Polydomous in branch tips	low	high

Table 19.3. Habitats and host species used by *Cecropia*-inhabiting *Azteca* in Costa Rica. + = present and common, r = present but rare, ? = present in habitat but not recorded from host, 0 = not present at locality.

Locality*	*Azteca* species†				
	con	xan	coe	alf	ova
La Selva					
second growth, saplings, and small trees	+	+	0	+	+
primary forest, mature *C. obtusifolia*	+	+	0	?	?
primary forest, mature *C. insignis*	?	?	0	?	+
Peñas Blancas					
sapling and mature *C. insignis*	+	+	0	0	r
Monteverde					
sapling *C. polyphlebia*‡	r	r	0	0	0
sapling and mature *C. obtusifolia*	+	+	r	r	0
Guacimal					
sapling and mature *C. peltata*	+	r	+	+	0

* Prov. Heredia, La Selva Biological Station, 50 m elevation; Prov. Alajuela, Río Peñas Blancas, 800 m; Prov. Puntarenas, Monteverde, 1300–1500 m; Prov. Puntarenas, Guacimal, 12 km SSW Monteverde, 500 m.
† con = *A. constructor*, xan = *A. xanthochroa*, coe = *A. coeruleipennis*, alf = *A. alfari*, ovat = *A. ovaticeps*.
‡ *Cecropia polyphlebia* Donn. Sm. is a non-myrmecophytic cloud forest species, the mature trees of which never contain *Azteca* colonies, but whose saplings exhibit myrmecophytic traits and sometimes attract colonizing queens.

(Berg 1978*a*). It exhibits a very regular modular structure. The basic unit of stem elongation is the internode, a cylinder bearing a single, long-petioled, palmately lobed leaf. Initially, the internode centre is filled with a soft pith. The pith is surrounded by a thin layer of extremely hard sclerenchyma, which is continuous across the two ends of the internode, and thus forms durable septa between the internodes (Bailey 1922*a*). The cambium, and hence all wood growth, is external to this capsule of horny tissue.

In most of the more than 100 *Cecropia* species, there is a suite of characters which relate to ant association. First, the internode pith splits and retracts during internode development, leaving a hollow internode with a thin layer of spongy pith on the walls. The hollow internodes form the domatia which ants inhabit. Secondly, each internode has a preformed thin spot, the prostoma, which lacks latex ducts (Bailey 1922*a*). The prostoma is often a well-defined oval depression in the wall, and is typically excavated and used as an entrance hole both by founding ant queens and workers of established colonies. Thirdly, the base of each petiole bears a trichilium, a pad of densely packed trichomes, from which sprout 1–2 mm long, glycogen-containing beads called Müllerian bodies (Rickson 1971). In Costa Rica, only the terminal two or three trichilia on leafy branches of mature trees produce Müllerian bodies; production ceases abruptly on more proximal leaves. Most ant inhabitants of *Cecropia* harvest Müllerian bodies as their primary food source.

Cecropia saplings, prior to the establishment of a single colony large enough to occupy all internodes, form stacks of potentially semi-autonomous nest sites for ants. Saplings frequently contain numerous incipient colonies in separate internodes (Longino 1989*a*; Davidson *et al.* 1990). The upper-most internodes typically contain live queens of the local species of *Azteca* that are obligate inhabitants of *Cecropia*. Internodes below those housing live queens routinely contain dead *Azteca* queens, evidence of colony failures. These failures occur inside sealed internodes, and so are not due to interactions with other colonies. Queens are sometimes killed by hymen-opteran parasitoids (*Conoaxima*: Eurytomidae, Wheeler 1942), which may be locally abundant (see Chapter 20). At some point during sapling growth colonies begin to survive, producing workers that reopen the prostoma and begin gathering Müllerian bodies from trichilia. How reduction in the number of colonies occurs, such that one *Azteca* colony comes to dominate the sapling, is poorly understood as yet, but is currently under investigation (D. Perlman, personal communication).

In the sapling stage, incipient and mature colonies of a variety of arboreal ants may be found inhabiting lower internodes. The species list is large and includes the genera *Gnamptogenys*, *Heteroponera*, *Pachycondyla*, *Pseudo-myrmex*, *Crematogaster*, *Solenopsis*, *Pheidole*, *Wasmannia*, *Zacryptocerus*, *Procryptocerus*, *Camponotus*, and *Myrmelachista*. These include species that are known to be nest-site generalists and some that may be specialized inhabitants of sapling *Cecropia*. However, only two non-*Azteca*, a *Pachy-condyla* and a *Camponotus* from Peru, are known to have persistent, dominant colonies in mature *Cecropia* (see Chapter 20).

Azteca is a diverse genus, with over 150 nominal taxa, which exhibits a variety of nesting habits (Emery 1893, 1912; Forel 1929). Many *Azteca* use carton, a cardboard-like construction material fabricated by the ants (Wheeler 1910*a*; Longino 1986). Some species build large carton nests which hang exposed from branches. Others build carton nests within dead branches or hollow trunks. Still others inhabit live stems, and exhibit varying degrees of host specificity. Among this group at least six species are obligate inhabitants of *Cecropia*. Harada and Benson (1988) also review the *Cecropia* ants, and discrepancies between their work and this need to be resolved.

Taxonomy of *Cecropia*-inhabiting *Azteca*

Worker size polymorphism and gradual changes accompanying colony ontogeny greatly compromise the utility of workers in species identification. In contrast, queens are monomorphic and exhibit greater structural diversity than workers, and so are often essential for the identification of a species. Several characters are of critical importance in the species-level taxonomy of *Azteca*. First, the abundance of erect setae on the scapes and tibiae varies between species, and this character is expressed similarly in workers and queens. Secondly, the shape and size of the queen head varies greatly, ranging from extremely elongate (twice as long as wide) to broadly heart-shaped and wider than long. Thirdly, colour, although rarely used by ant taxonomists, is stable for some *Azteca* species and may be a very useful field character.

The common *Cecropia*-inhabiting *Azteca* species can be unequivocally divided into two groups: those with and those without abundant setae on scapes and tibiae. The former group is the more diverse, containing *A. constructor*, the *A. muelleri* and *A. xanthochroa* species complexes, and *A. coeruleipennis*. The group with bare tibiae contains only two species, *A. alfari* and *A. ovaticeps* (Longino 1989*b*).

A few other *Azteca* have been reported as *Cecropia* ants, but so infrequently that they are either obligate to *Cecropia* and very rare, or they have a variety of nest sites and only occasionally occur in *Cecropia*. Colonies of the *A. schimperi* species complex have been found nesting on *Cecropia* trees (Benson 1985). Unlike the foregoing species, workers totally lack erect setae on the thorax, and they build external carton nests. Their status as obligate *Cecropia* ants is uncertain. *Azteca minor* Forel was collected once from a *Cecropia* in Brazil. I have seen what may be conspecific material from myrmecophytic melastomes from Amazonia and thus doubt that it is an obligate *Cecropia* ant. These species will not be discussed further here.

Azteca constructor occurs in northern South America and Central America. The queen is solid black, with the head about as long as it is wide (Fig. 19. 1(a)). The erect setae on the thoracic dorsum are long and extremely dense, a condition unique to this species. The workers are chocolate brown, and are often difficult to distinguish from workers of *A. xanthochroa*.

Azteca muelleri, the species originally observed by Müller (1876; 1880–1881), occurs in southern Brazil, but its morphological variability and range are unknown. Queens from the type locality (Santa Catarina state) have a head shape very similar to *A. constructor*, but the thoracic setae do not form as dense a brush, and the colour is a mottled orange-brown. There are seven infraspecific taxa, whose status and affinities are completely unknown (Table 19. 1).

The taxonomy of *A. xanthochroa* and its four infraspecific nominal taxa is in a state of flux. All of them live in *Cecropia* and have queens with body colour solid orange, or the ground colour orange with some darkening in the

centre of the head and on the extremities. Distributional disjunctions and
geographical variation confuse the species-level taxonomy. Queens from
Mexico, the type locality of *A. xanthochroa s.s.*, have heads that are
somewhat elongate, with sides slightly convex, and the petiolar apex bears 0–
4 weak setae. Moving south, the head generally becomes larger, more
elongate, and with very straight sides, until an extreme is reached in Costa
Rica, the type locality of the subspecies *costaricensis* (Fig. 19.1(b)). A
disjunction occurs across northern South America, *A. xanthochroa*-like
forms occurring again in the highlands of Peru and Bolivia, and extending

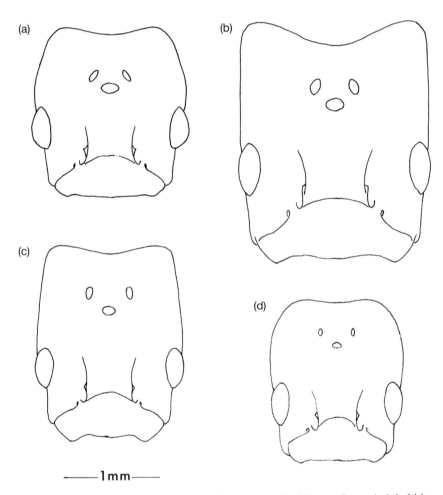

———1mm———

Fig. 19.1. Heads of queens of the major groups of obligate *Cecropia*-inhabiting
Azteca. (a) *Azteca constructor.* (b) *Azteca xanthochroa.* (c) *Azteca coeruleipennis.* (d)
Azteca alfari. All were drawn from Costa Rican specimens in the author's collection.

along the eastern slopes of the Andes and across the Amazon basin. Queens from the Andean highlands, the type locality of the subspecies *australis*, are nearly identical to queens from Mexico, but decrease substantially in size in the Amazonian lowlands (*A. xanthochroa* of Harada 1982). Thus, when queens from Costa Rica and the Amazon basin are placed side by side, they are strikingly different, yet they can be connected by a continuous series of intermediates.

In contrast to *A. xanthochroa s.s.*, several geographically isolated forms in northern South America have a conspicuous tuft of erect setae on the petiolar apex. These include the subspecies *isthmica* in Panama and northern Colombia, the subspecies *salti* from the Sierra Nevada de Santa Marta, Colombia, and an undescribed species from the coastal mountains of Venezuela. Subspecies *isthmica* is the smallest form in the *A. xanthochroa* species complex, with the head about as long as it is wide, and presents a strong contrast to the largest members of the complex which are found in neighbouring Costa Rica.

The *A. muelleri* and *A. xanthochroa* complexes are poorly known and probably are not distinct groups on either phenetic or phylogenetic grounds. At one site in the Venezuelan Andes I observed two new species of *Cecropia*-inhabiting *Azteca*: one was intermediate between *A. muelleri* and *A. xanthochroa*, the other was intermediate between *A. muelleri* and *A. constructor*. Harada (1982) reported subspecies *isthmica* from the Manaus area, but her description and measurements suggest a distinct species. Two sympatric *A. xanthochroa*-like species have been found in eastern Bolivia (P. S. Ward, personal communication), one of which is certainly new. These new species suggest that there is an unexplored diversity of related species throughout the South American tropics.

Azteca coeruleipennis, presumably of Mesoamerican origin, has a range from southern Mexico to the Pacific coast dry forest of Costa Rica. The setae are fine and difficult to see, which has resulted in frequent misidentification of this species as *A. alfari*. The workers are always concolorous yellow-orange, and the queen is a lustrous solid black. The name refers to the smokey blue reflections on the wings of alate queens. The queen head is somewhat rectangular and longer than it is wide (Fig. 19.1 (c)).

Workers and queens of the *A. alfari* species group can be distinguished from all other obligate *Cecropia* ants by the absence of erect setae on the scapes and tibiae. The queens are small relative to queens in the other species groups. The head is slightly longer than it is wide and the posterior border is evenly rounded to flat, never strongly cordate, or angulate (Fig. 19. (d)). Two species are recognized in this group, *A. alfari* and *A. ovaticeps* (Longino 1989*b*), which differ in the degree of pilosity. The two species are broadly sympatric, both geograpically and at the micro-habitat scale. This species group has the widest range of any obligate *Cecropia* inhabitant, occurring from southern Mexico to Argentina, and on several Caribbean islands.

Nest structure and behaviour of Costa Rican species of *Azteca*

Azteca constructor

A mature colony of *A. constructor* inhabiting a *C. obtusifolia* Bertol. tree near Monteverde was dissected (Fig. 19. 2). The colony centre was a spindle-shaped carton nest. To accommodate the nest, extensive excavation had occurred in the wood external to the hard, internodal capsules, the remains of which were incorporated into the carton. A major exit hole, 110 × 3 mm, was adjacent to the carton nest, and exit holes were common above and below the nest. The single carton nest contained the lone colony queen, all brood, all cached Müllerian bodies, abundant males, and abundant alate queens. No carton, brood, or sexuals were found elsewhere in the tree. The central nest communicated internally with every branch tip, the septa having been largely removed throughout, and passages maintained at branch junctions. All branch tips contained many exit holes, abundant workers, and Coccoidea. Eight branches from other *C. obtusifolia* trees containing *A. constructor* colonies in the area were all occupied throughout the 1–2 m of branch. All contained workers and Coccoidea only, with no carton, brood, or sexuals.

When the trunks of the trees occupied by *A. constructor* were tapped or shaken, workers poured from the central fissure and blackened the trunk near the internal carton nest. Workers also issued forth in numbers from all the branch tips. Workers responded similarly when branches were prodded with a pole. Even without disturbance, workers were often seen patrolling leaf surfaces (in part to tend Pseudococcidae, which were found tucked in the sinuses where the primary leaf veins radiate from the petiole). Leaves from the dissected tree and the eight branch samples were nearly devoid of other insects, and had almost no trace of herbivore damage.

Anastasio Alfaro claimed that *A. constructor* was the most aggressive of the Costa Rican *Cecropia* ants (in Emery 1896); I have also found this to be true. The behaviour and nest structure of *A. constructor* are similar to *A. muelleri* inhabiting *C. adenopus* in southern Brazil. Both Müller (1876; 1880–1881) and Eidmann (1945) observed that mature colonies of *A. muelleri* construct large, spindle-shaped carton nests in the boles of *Cecropia* trees, causing a swelling in the trunk which is visible from outside. Both Müller and Eidmann described *A. muelleri* as an extremely aggressive ant.

Azteca xanthochroa

A mature colony of *A. xanthochroa* inhabiting *C. obtusifolia* near Monteverde was dissected (Fig. 19. 3). The colony centre was a cylindrical carton nest filling two internodes in the centre of the trunk. To accommodate the nest a small amount of wood had been excavated exterior to one of the hard

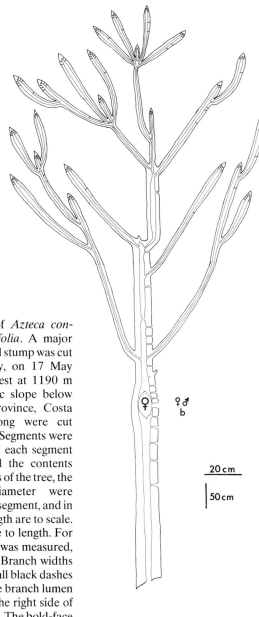

Fig. 19.2. Nest structure of *Azteca constructor* in *Cecropia obtusifolia*. A major stem from a coppice on an old stump was cut and dissected in its entirety, on 17 May 1989, in second growth forest at 1190 m elevation, on a steep Pacific slope below Monteverde, Puntarenas Province, Costa Rica. Segments 50 cm long were cut sequentially with a bow-saw. Segments were examined for exit holes, and each segment was split longitudinally and the contents examined. For the central axis of the tree, the internal and external diameter were measured at the base of each segment, and in the figure both width and length are to scale. Width is exaggerated relative to length. For lateral branches, length only was measured, thus widths are not to scale. Branch widths were mostly 5 cm or less. Small black dashes at branch tips, connecting the branch lumen to the exterior, and gaps in the right side of the central axis, are exit holes. The bold-face female symbol is where the physogastric colony queen was found. Smaller male and female symbols are where adult sexuals (males and alate queens) were found. 'b' is where brood was found.

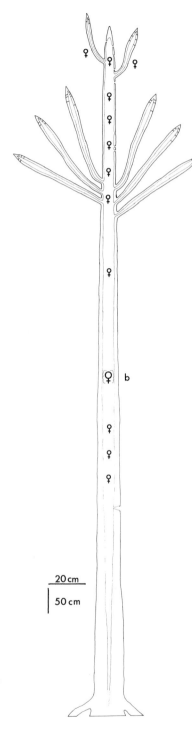

20 cm
50 cm

Fig. 19.3. Nest structure of *Azteca xanthochroa* in *Cecropia obtusifolia*. A free-standing stem was cut and dissected in its entirety, on 25 May 1989, within 20 m of the stem shown in Fig. 19.2. Methods and symbols are as in Fig. 19.2.

internodal capsules, but there was no external sign of the nest location. There were no exit holes near the carton nest. The only exit hole below the carton nest was a 4 cm long longitudinal fissure 4 m above the ground and 2 m below the carton nest. The next exit hole was over 4 m above the carton nest, near the branch tips. The single carton nest contained the lone colony queen, all brood, all cached Müllerian bodies, and abundant alate queens. Alate queens were also found scattered throughout the main trunk of the tree. The internodal septa were largely removed and the nest communicated internally with all branch tips, and all branch tips contained exit holes, workers, and Coccoidea, but no carton or brood.

Fifteen 1–2 m long branches were sampled from other *C. obtusifolia* trees containing *A. xanthochroa* colonies near Monteverde, and an additional three branches from La Selva Biologica Station, Heredia Province, Costa Rica. All were continuously occupied throughout, none contained brood, two contained very small amounts of carton, 13 contained sexuals (males, alate queens, or both), and all contained Coccoidea.

The colonies observed maintained a longitudinal fissure near the base of the tree, and major workers emerged from this fissure when the tree was tapped or shaken. Although they had large heads and strong mandibles, they had difficulty obtaining a grip on a flat surface such as a collector's hand. In contrast, smaller workers were fierce biters. When branches were prodded with a pole, workers issued from the branch tips vigorously and in large numbers. Workers patrolled leaves of undisturbed trees, the leaves were largely free of other insects, and there was minimal herbivore damage.

Alfaro (in Emery 1896) made similar observations of Costa Rican *A. xanthochroa* colonies. He observed that the nest entrance was a longitudinal fissure, and that workers emerged with their gasters elevated but did not bite the hand of the collector. The *A. xanthochroa*-like species in the eastern Amazon has a nest structure similar to that described here (D. W. Davidson, personal communication), and I have observed a similar nest structure in a related species from Venezuela.

Azteca coeruleipennis

A mature colony of *A. coeruleipennis* inhabiting a *C. peltata* L. tree near Monteverde was dissected (Fig. 19. 4). This tree was growing in a seasonally dry habitat, and, perhaps reflecting slow growth rates, the internodes were very compressed vertically and had small diameters relative to wood thickness. The colony was polydomous, with colony fragments occupying 23 separate branch tips and four separate cavities in the central axis. Brood and sexuals were evenly spread throughout the colony space. The colony queen was in a very protected region, surrounded by four large knots of dense wood, at a site where four branches had radiated from the central axis. Although adjacent portions of the queen's nest cavity contained abundant workers and small brood, the internode in which the queen was found and

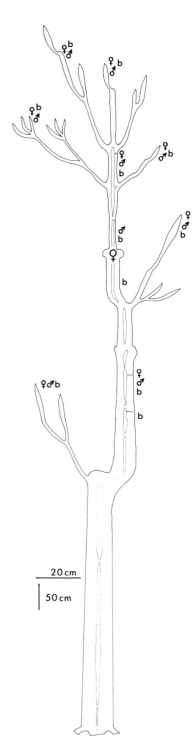

Fig. 19.4. Nest structure of *Azteca coeruleipennis* in *Cecropia peltata*. A freestanding stem was cut and dissected on 28 May 1989, in scrubby vegetation on a steep river bank, at 840 m elevation along the Rio Guacimal, on the Pacific slope below Monteverde. Methods and symbols are as in Fig. 19.2, except that lateral branches were not all cut and split, and so lumens are not drawn in the figure. All branches were inhabited and contained workers, brood, and sexuals. Not all of the 23 branch tips are shown in the figure because many broke in the felling and their positions were uncertain.

20 cm

50 cm

the adjacent four to five internodes were devoid of brood. In contrast to *A. xanthochroa* domatia, there was no trace of carton anywhere in the tree. In general, the septa were perforated with small holes, little larger than the workers themselves, and the lumens of lateral branches did not communicate with the central axis.

Eight branch tips were sampled for detailed dissection. Every branch contained workers, brood, adult males, alate queens, and Coccoidea. There were two species of Coccoidea, one of which was common in shoot apices, the other occurred more basally, 20–30 internodes down.

Azteca coeruleipennis was chary with its exit holes. Each of the four occupied nest spaces in the central axis had a single 2–3 mm diameter, round exit hole near the top of the cavity. Branch tips had very few exit holes, and these were often located 10–20 internodes below the active branch tip. This contrasts sharply with *A. constructor* and *A. xanthochroa*, which typically maintain 4–7 exit holes in the top ten internodes.

Workers responded to trunk tapping or shaking by descending to the trunk base, but their response was not as vigorous as *A. constructor* or *A. xanthochroa* colonies. The exit holes were small and inconspicuous; small yellow workers seemed to suddenly appear from nowhere, swarming over the crown and trunk. The leaves were free of other insects and herbivore damage was low. However, I have no data on patrolling, nor on between-tree or seasonal patterns of herbivory.

Azteca alfari *group*

A mature colony of *A. alfari* inhabiting *Cecropia insignis* Liebm. at La Selva Biological Station was dissected (Fig. 19. 5). The nest was polydomous, occupying only the terminal 1–3 m of most branch tips. The single colony queen was in a small carton nest in the lowermost occupied internode of the central axis of the tree. The carton nest was small, occupying only a portion of the internode, and it contained only a few workers and a small mass of eggs and first-instar larvae. Small carton nests, brood, males, and alate queens were spread diffusely throughout the entire colony space. Coccoidea were present in every occupied branch tip. The colony occupied 22 separate branch tips, none of which communicated internally. Internodal septa had 1–3 perforations which were little larger than the workers, in contrast to the extensive septum removal displayed by *A. constructor* and *A. xanthochroa*. Branch and trunk interiors basal to the inhabited regions generally exhibited evidence of former occupation (perforate septa, pith removed from internode walls), but there were scattered blocks of internodes that had clearly never been entered. Small colonies of two other ant species, a *Camponotus* and a *Pachycondyla*, were found basally in branch space formerly occupied by *Azteca*.

Sixteen branches were sampled from other *Cecropia* trees inhabited by *Azteca alfari* group colonies at La Selva (two *C. insignis* with *A. ovaticeps*,

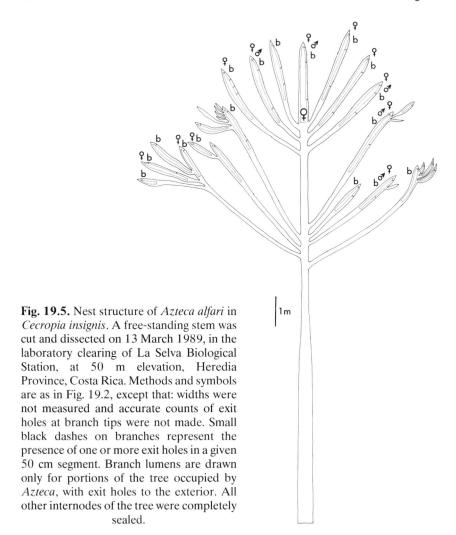

Fig. 19.5. Nest structure of *Azteca alfari* in *Cecropia insignis*. A free-standing stem was cut and dissected on 13 March 1989, in the laboratory clearing of La Selva Biological Station, at 50 m elevation, Heredia Province, Costa Rica. Methods and symbols are as in Fig. 19.2, except that: widths were not measured and accurate counts of exit holes at branch tips were not made. Small black dashes on branches represent the presence of one or more exit holes in a given 50 cm segment. Branch lumens are drawn only for portions of the tree occupied by *Azteca*, with exit holes to the exterior. All other internodes of the tree were completely sealed.

ten *C. obtusifolia* with *A. ovaticeps*, and four *C. obtusifolia* with *A. alfari*). Carton, brood, and Coccoidea were present in all of them; 13 contained sexuals. Most of the branches were long enough to observe a basal, unoccupied region and an apical, occupied region. One branch from a *C. obtusifolia* inhabited by *A. alfari* at Monteverde followed the same pattern, with an unoccupied base and workers, carton, brood, Coccoidea, and alate queens found in the inhabited apex.

When mature, branching *Cecropia* trees inhabited by *A. alfari* or *A. ovaticeps* were tapped or shaken, usually nothing happened. When branches were prodded with a pole, workers either did not emerge at all, or did so only in small numbers. If a few workers did emerge they were aggressive and bit, and when branches were cut large numbers of workers emerged and presented just as much of an irritation as *A. constructor* or *A. xanthochroa* workers.

I never observed workers of *A. alfari* group species patrolling leaf surfaces. Leaves from the La Selva branch samples supported a complex arthropod community, including nymphs and adults of Miridae and Tingidae (Hemiptera), lepidopteran larvae, spiders, and other species of foraging ants in the genera *Camponotus*, *Solenopsis*, and *Procryptocerus*. Six of the 16 sampled branches were heavily infested with a myrmecophilous caterpillar (*Theope* aff. *decorata* G. & S., Riodinidae, P. J. DeVries, personal communication) which was always tended by an abundant Diplorhoptrum *Solenopsis* species. Leaves from the 14 *C. obtusifolia* branch samples were heavily damaged, speckled throughout with small holes and necrotic spots, and some showed the characteristic feeding damage of *Theope* larvae near the petiolar insertion. The two *C. insignis* branches were less damaged.

The non-aggressive nature of Costa Rican *A. alfari* and *A. ovaticeps* should not be assumed in other parts of their ranges. P. S. Ward (personal communication) observed aggressive responses of *A. alfari* in southern Mexico, and D. W. Davidson (personal communication) observed *A. ovaticeps* near Manaus 'swarming over its host *Cecropia latiloba* Miq.'.

Inside *Cecropia* domatia inhabited by *Azteca xanthochroa* and *A. constructor*, there are regular mounds of brown, sticky, bran-like material. Müller (1880–1881) called these mounds 'knollen'. They have a uniform texture and are composed partly, if not entirely, from pith scraped from the internode walls. They are crawling with small fly larvae and nematodes, but there are no insect parts. Common inquilines are flies, especially a milichiid and a psychodid running freely and rapidly inside of the branches. Other inquilines (beetles, silverfish, cockroaches, Hymenoptera) seem to be extremely rare.

Phylogenetic considerations

Benson (1985) suggested that obligate *Cecropia* use arose many times in the genus *Azteca*, and the evidence presented here supports this conclusion. Structural and behavioural data strongly support an independent origin of *Cecropia* use in the *A. alfari* group, while the number of origins represented by the remaining *Cecropia* ants is unclear.

In many parts of the wet, lowland neotropics the canopy contains *Azteca* species that share the following characters:

1. external, pendent carton nests, in which all brood and sexuals are concentrated;

2. satellite carton shelters which house scale insects, Coccoidea;
3. queens with short, broad heads;
4. workers and queens with densely hairy tibiae and scapes; and
5. fierce defence of the main carton nest.

Azteca constructor and the *A. muelleri* complex share all these characters, except that they have moved into *Cecropia* trees and shelter their Coccoidea in branch tips. The construction of a bulging carton nest in the bole, with its requisite wood excavation and consequent deformation of the tree, and the maintenance of many large exit holes near the carton nest are probably retained characters which are not particularly adaptive to nesting in *Cecropia* trees.

In contrast to the builders of exposed carton nests, there is a large and little known set of *Azteca* species that nest in live stems of a variety of plants. Some nest in recognized ant plants such as *Cordia* (Boraginaceae), *Triplaris* (Polygonaceae), and *Tachigali* (Caesalpiniaceae), while others occupy a wide variety of canopy and understorey vegetation. These *Azteca* are often found in plants that do not have preformed domatia, and it is unknown whether the ants excavate the stems themselves or opportunistically follow other stem-boring insects. These species typically have small queens with narrow, elongate heads. They are polydomous with the colony spreading out in shoot tips, progressively moving into new growth and abandoning older stems. Part of the reason that they are so poorly known is that they are often cryptic, tending Coccoidea inside the live stems and not actively patrolling the plant surface. Workers and queens of these species often have greatly reduced pilosity, with no or few erect setae on scapes and tibiae. Some of these species have strong associations with particular myrmecophytes, for example several species are obligate to *Cordia alliodora* Cham. (Wheeler 1942; J. T. Longino, personal observation). The *A. alfari* group clearly belongs with these stem-nesting *Azteca*, on both behavioural and morphological grounds.

The affinities of *A. xanthochroa* and *A. coeruleipennis*, and whether they share close ancestry, are unknown. Based on queen head shape, male genital characters, and the centralized carton nest, species in the *A. xanthochroa* complex may be related to the *A. muelleri* complex and to *A. constructor*. There are no obvious species groups of *Azteca* with which *A. coeruleipennis* can be allied. Thus, for the non-*alfari* group of *Azteca* it is unknown whether the *Cecropia* specialists are one lineage within which the habit arose once, or multiple lineages with independent colonization of *Cecropia*.

Which ant behavioural traits are adaptations specific to association with *Cecropia*? The trait which all obligate *Cecropia* ants must share is host-specific searching by founding queens, and in a species such as *A. constructor* this might be the only specialized trait. The lack of exit holes in the vicinity of the colony queen, a character shared by *A. xanthochroa* and *A. coerulei-*

pennis, could be a specialized adaptation which takes advantage of *Cecropia* architecture to protect the queen from predators or parasites. The total lack of carton, and the location of the queen in a highly protected part of the tree, may also be specialized traits that have evolved in *A. coeruleipennis*. Thus, these two species may be the most specialized occupants of *Cecropia*, representing plant-ants which have undergone extensive modification in response to the *Cecropia*-nesting habit.

More detailed knowledge of how, and how many times, obligate use of *Cecropia* has arisen will require phylogenetic analysis, and the mapping of behaviour onto structure-based cladograms (see Chapter 22). Museum collections of full caste series, species-level taxonomy, and field observations of behaviour and nest structure for a range of *Azteca* species are essential.

Ecological consequences of *Azteca* behaviour

It has been observed that species of *Azteca* and species of *Cecropia* sort by habitat (Harada 1982; Benson 1985; Davidson *et al.* 1990; Longino 1989*a*, *b*), with the *A. alfari* group occupying open, disturbed habitats, and *A. constructor* and *A. xanthochroa* being more common in closed forest. A summary of these habitat relationships for Costa Rica is presented in Table 19. 3. Davidson *et al.* 1990 and Davidson and Fisher (Chapter 20, this volume) have proposed a mixture of ultimate and proximate causes for the observed habitat partitioning. The ultimate causes pertain to resource availability and herbivore defence. The suggested proximate causes are active habitat selection by flying queens and/or habitat-specific outcomes of competition between newly emergent workers and founding queens in saplings.

With such strong behavioural differences between *Azteca* species inhabiting *Cecropia*, the species of ant occupant probably influences tree growth and survivorship, providing an additional explanation for habitat partitioning. If *A. alfari* fails to protect its host tree, trees occupied by *A. alfari* may selectively die during succession. At La Selva Biological Station, Costa Rica, I have observed that:

1. at the founding stage, saplings of *C. obtusifolia* contain queens of a full range of species: *A. constructor*, *A. xanthochroa*, *A. alfari*, and *A. ovaticeps*;
2. small trees with one dominant colony contain a similar range of species; and
3. large *C. obtusifolia* trees in the forest canopy contain only *A. constructor* or *A. xanthochroa*.

If active habitat selection by queens or habitat-specific outcomes of competition are occurring, the effects are not strong. These observations suggest that either trees containing *A. alfari* are dying early or that there is a succession of

Azteca species in a tree after a dominant colony has emerged. There is no evidence for the latter.

The length of time a colony takes to reproduce could also enhance habitat partitioning. Sexuals of *A. alfari* and *A. ovaticeps* occur in branches of small trees, but sexuals of *A. xanthochroa* occur only in branches of large trees (J. T. Longino, personal observation). If *A. alfari* group species reproduce early, in small trees, and *A. xanthochroa* and *A. constructor* delay reproduction until their colonies and trees are larger, then *A. alfari* group species would be dominant in pastures and along river edges, where trees would be either cut or washed out before *A. xanthochroa* or *A. constructor* could reproduce.

A pattern of within-habitat partitioning, which none of the above hypotheses explains, occurs on Costa Rica's Atlantic slope. At La Selva Biological Station, a lowland site, canopy *C. obtusifolia* are dominated by *A. constructor* and *A. xanthochroa*, while canopy *C. insignis* are dominated by *A. ovaticeps*, an *A. alfari* group species (D. and D. Clark, personal communication). At higher elevations, where *C. insignis* occurs alone, *A. xanthochroa* and *A. constructor* dominate and *A. ovaticeps* is extremely rare or absent. The mechanism resulting in host partitioning of forest *Cecropia* is unknown. A related question is how *A. ovaticeps*, whose behaviour appears to be identical to *A. alfari*, can be both a common, early-reproducing 'parasite' of second-growth *C. obtusifolia* and the primary occupant of mature *C. insignis*, a long-lived canopy species.

The study of the community ecology of the *Azteca–Cecropia* association is in its infancy. The following questions should be addressed by further observation and experiment:

1. Do founding queens exhibit habitat-specific host choice?
2. Are there deterministic competitive relationships between *Azteca* species, and are they affected by host tree species or habitat?
3. For each *Cecropia* species, how is host-tree demography (and ultimately fitness) affected by species of ant occupant and habitat?

Acknowledgements

I thank those who have tolerated *Azteca* bites with me, in particular the students of OTS 88–2 and 89–1. I thank W. W. Benson, D. W. Davidson, D. Perlman, and D. and D. Clark for sharing their insights with me, and W. W. Benson, D. W. Davidson, C. R. Huxley, and P. S. Ward for comments on the manuscript. Taxonomic work has been supported by National Geographic Society grants 2900–84 and 4064–89 to the author, and NSF Biological Research Resources Program grant BSR–8800344 to the Los Angeles County Museum of Natural History (C. Hogue).

20

Symbiosis of ants with *Cecropia* as a function of light regime

Diane W. Davidson and Brian L. Fisher

Introduction

The relationships between ants and myrmecophytes are still viewed by many as unique and curious phenomena, unrelated to mainstream evolutionary ecology. However, as argued by McKey (1988), renewed interest in these relationships is leading to the discovery of both pattern and complexity resembling that found in other ecological systems. First, certain unifying principles of plant defence theory may be generalized to ant-protection mutualisms (McKey 1984, 1988). In addition, symbiotic ant–plant relationships are likely to be products of selection based on complex networks of direct and indirect interactions, the outcome of which can be modified by variation in the physical environment (e.g. Schemske and Horvitz 1988; Davidson and Epstein 1989; Davidson *et al.* 1990). Given these parallels with other ecological systems, symbiotic ant–plant associations may have a unique role to play in elucidating the determinants of evolutionary specialization. Both myrmecophytes and plant-ants vary in their degree of specialization, and this variation can be quantified explicitly by experiment, and related to present-day selection environments.

To characterize factors driving evolutionary specialization in these relationships, it is necessary to describe the environmental context in which the associations evolve. The framework developed here draws principally on studies of myrmecophytic *Cecropia* in western Amazonia (Davidson *et al.* 1990). However, some of the patterns noted may apply both to Central American *Cecropia* (Longino 1989*a*, Chapter 19, this volume) and to other ant–plant symbioses (Davidson *et al.* 1990).

Myrmecophytic *Cecropia* and their ants

Cecropia (Cecropiaceae) is the most widespread and important genus of early successional trees in neotropical forests, and most of the more than 100 species host symbiotic ants (C. C. Berg, personal communication). All myrmecophytic *Cecropia* provide similar benefits to ants. Early in sapling growth, hollow stems expand and internodes develop with prostomata, or small areas of unvascularized tissue, where queens enter stems without rupturing phloem and flooding internodes with mucilage. Colonies develop within these stems, and emerging workers collect glycogen-rich Müllerian bodies at the bases of petioles on hairy platforms termed trichilia (Rickson 1971, 1973, 1976). The production of pearl bodies on lower leaf surfaces is variable among *Cecropia* species.

Although early naturalists noted a diversity of ant genera dwelling in *Cecropia* stems (e.g. Wheeler 1942), recent ecological studies have emphasized associations with *Azteca* ants (Dolichoderinae). *Azteca* species predominate on fast-growing *Cecropia* in open, sunny environments. Trees growing slowly in shaded habitats are often occupied by specialized *Cecropia* ants in other genera, including *Pachycondyla* (Ponerinae), *Camponotus* (Formicinae), and *Crematogaster* (Myrmicinae) (Davidson *et al.* 1989; 1990; D. W. Davidson and B. L. Fisher, unpublished work). *Cecropia* seedlings receive significant protection from at least some ant associates which repel insect herbivores (Schupp 1986) and/or remove encircling vines (Janzen 1969; Davidson *et al.* 1988).

Cecropia in south-eastern Peru

Our studies have focused on the *Cecropia* of tropical moist forests of southeastern (Madre de Dios) Peru (Davidson *et al.* (in press)). Three pairs of closely related species appear to be included among six *Cecropia* species which are common at Reserva Tambopata and the Estación Biológica de Cocha Cashu (including borders of the Rio Manu from Cocha Cashu to Tayacome). *Cecropia membranacea* and *Cecropia tessmannii* (a name used in ornithological literature (Koepcke 1972) but not yet confirmed) are difficult to distinguish by vegetative morphology. Once classified as subspecies, they differ conspicuously only in developmental and myrmecophytic traits (see below), but are quite distinct from other species. *Cecropia engleriana* and *Cecropia* sp. A differ strikingly from one another as seedlings and early saplings but are difficult to distinguish by vegetative characters later in development. Seeds from individual maternal parents of both species occasionally give rise to seedlings that are morphological intermediates. Either the two species may hybridize, or their separate gene pools may not yet have eliminated particular gene combinations that lead to similar morphology. The last two species, *Cecropia polystachya* and *Cecropia ficifolia* also bear a striking resemblance to one another in vegetative phenotypes, but are very distinct from the other species pairs. Since the characteristic habitats of each of the six *Cecropia* species differ markedly from those of postulated close relatives, morphological similarities within pairs are unlikely to reflect evolutionary convergence. Both the Peruvian *Cecropia* and their symbiotic ants exhibit strong habitat specificity, and the distributions of at least some ant associates are related more obviously to habitat than to host species (Davidson *et al.* 1990). Apparently a very general feature of the symbiosis, the habitat specificity of both *Cecropia* and its specialized ants occur principally with respect to difference in the light regime (Benson 1985; Harada and Benson 1988; Longino 1989*a*). To explain the pattern, we consider how light intensity may affect the allocation of plant resources to growth and defence and, consequently, the resources available to ants.

Patterns of plant investment in defence

In all ecological systems, variation in the physical environment sets the pattern of resource availability at successive trophic levels. Ecological and evolutionary responses of myrmecophytes to variation in their own resource environments should then determine rates of resource supply to ants in symbiotic ant-protection mutualisms. Plant responses can be investigated at two levels. First, environmental variation often correlates predictably with proportionate investment in growth versus defence. Secondly, for any given level of defence, relative allocation to biotic versus chemical (or physical) defences may also exhibit a pattern in relation to the environment.

Total defence investment

Circumstantial evidence suggests that both biotic and chemical defence of myrmecophytic *Cecropia* may be costly, and may compromise the capacity for rapid growth. First, *Cecropia* populations growing without their specialized ants tend to lose the ability to produce trichilia (Janzen 1973; Rickson 1977; Putz and Holbrook 1988). Secondly, intraspecific variation in leaf tannin levels is negatively correlated with rates of leaf production in Central American *Cecropia peltata* L. (Coley 1986). Thirdly, interspecific differences in the defence investment of myrmecophytic *Cecropia* in western Amazonia are correlated with habitat in a manner predicted from tradeoffs between growth and defence (Davidson *et al.* 1990). In south-eastern Peru, *C. membranacea*, *C. engleriana*, and *C. polystachya* grow regularly at high light intensities in large disturbances along major watercourses, and in clearings associated with human habitation (Table 20. 1). Here, competition from fast-growing pioneer species of similar size places a premium on rapid growth, and the smallest *Cecropia* in these habitats usually die before reproducing (J. Terborgh, unpublished work).

The apparent closest relatives of light-demanding species establish mainly within small forest light gaps, in the shade of much larger and older neighbours. For these species, diversion of limiting carbon from defence to growth might confer little advantage but jeopardize the persistence required to take advantage of later canopy openings. Not surprisingly, shade-tolerant gap species develop their trichilia at earlier leaf nodes (Fig. 20. 1) and at smaller sizes than do closely related congeners of more open environments. Moreover, in the three pairwise comparisons, preliminary data indicate greater dry

Table 20.1. Percentage contributions of six species to total *Cecropia* across habitats in south-eastern Peru. Values for relatively large plants and small seedlings (in parentheses)*.

Cecropia	% Contribution of species to *Cecropia* in habitats			
	Forest gaps	Human clearings	Stream banks	River banks
Light-demanding species				
1. *membranacea* Trec.	0	0	98	85
	(3)	(19)	(62)	(66)
2. *engleriana* Snethlage	0	48	0	0
	(0)	(39)	(8)	(3)
3. *polystachya* Trec.	1	52	2	15
	(2)	(32)	(28)	(28)
Shade-tolerant species				
1. *tessmannii* Mildbr.	36	0	0	0
	(36)	(5)	(0)	(3)
2. species A	31	0	0	0
	(34)	(5)	(2)	(0)
3. *ficifolia* Warb.	32	0	0	0
	(25)	(0)	(0)	(0)
Sample size	94	21	45	13 = 173
	(67)	(57)	(50)	(67) (241)

* *Cecropia engleriana* also occurs frequently along streams and rivers but was uncommon where we worked. Like-numbered light-demanding and shade-tolerant species are thought to be close relatives.

weight production of Müllerian bodies in shade-tolerant species (Table 20. 2). These same species tend to have a greater percentage of trichilia active at a given time (Table 20. 2, two of three comparisons only, see below). If alternative chemical defences were strongly reduced with the acquisition of biotic defence (P. D. Coley, personal communication for *C. peltata*), then shade-tolerant species would have proportionately greater total defence investment than their light-dependent counterparts after the onset of myrmecophytism. Both the slower intrinsic growth rates of shade-tolerant species (Fig. 20. 2) and a delay in trichilia development at low light intensities (intraspecific comparisons, Fig. 20. 3) also support the hypothesis that these carbon-based defences are costly.

Patterns in the biotic defence of *Cecropia* are remarkably similar to those exhibited by chemically defended plants under different light regimes

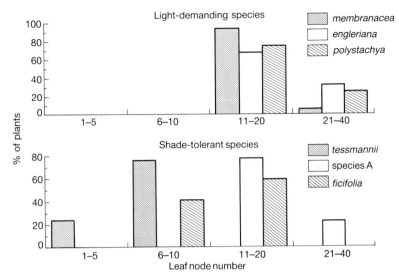

Fig. 20.1. Distributions of leaf node numbers for the first leaf with trichilia, for six Peruvian species of *Cecropia* grown in tropical greenhouses in Utah. Relationships for closely related species pairs were: *C. tessmannii* < *C. membranacea* ($P < 0.00001$); *Cecropia* sp. A. < *C. engleriana* ($P = 0.06$), and *C. ficifolia* < *C. polystachya* ($P < 0.00001$). Mann–Whitney U-Tests were two-tailed.

(Janzen 1974*b*; Bryant *et al.* 1983; Coley *et al.* 1985). In intraspecific comparisons of carbon-based chemical defences, secondary compounds occur at higher concentrations in environments of high light intensity (e.g. Woodhead 1981; Waterman *et al.* 1984; Larsson *et al.* 1986; Müller *et al.* 1987; Mole *et al.* 1988), or where plants are limited by nutrients other than carbon (e.g. nitrogen, Bryant *et al.* 1985). This trend is commonly reversed in interspecific comparisons, probably because greater defence investment is an adaptation to slower growth rates and longer leaf lifetimes (Coley 1983; Coley *et al.* 1985; Bryant *et al.* 1985; Bazzaz *et al.* 1987; Coley and Aide 1990). However, the pattern may vary with plant growth form. Some understorey shrubs grow at low light levels throughout their lives and exhibit lower defence investment at reduced light in interspecific comparisons (see Baldwin and Schultz (1988) for Melastomataceae). It is not yet possible to compare total defence investment by myrmecophytes across such disparate lifeforms.

Investment in biotic versus chemical defence

If defence is costly, plants should avoid investing in redundant defence systems, but deploy alternative systems in ways contingent on their relative

Table 20.2. Production of Müllerian bodies by trichilia in three pairs of Peruvian *Cecropia* species. Plants were grown in a temperate greenhouse, but were not standardized for history or size.

Cecropia species[a]	Mass of MBs[b] (mg d^{-1})	% Active trichilia[c]
tessmannii (ST)	7.3	50
	6.7	57
membranacea (LD)	5.3	86
	4.3	76
	—	64
species A (ST)[d]	7.2	82
	10.1	89
engleriana (LD)	1.2	69
	2.7	60
ficifolia (ST)	24.4	92
polystachya (LD)	2.2	67
	1.6	56

[a] ST = shade tolerant and LD = light-demanding member of species pair.
[b] 24-hour production of Müllerian bodies (MBs) by trichilia of the two youngest leaves. Food bodies are produced in irregular pulses, but production peaks at dusk in all species. Irregularity may stimulate relatively constant attention by ants.
[c] Percentage of trichilia actively producing Müllerian bodies on plants whose leaves all have trichilia. Production declines regularly with leaf age.
[d] Leaf size of *Cecropia* species A is approximately one-third that of other species; production of Müllerian bodies, per unit leaf area, is greatest in this species.

cost effectiveness in different circumstances. Chemical defences have been reduced or lost in some myrmecophytes (Janzen 1966; Rehr *et al.* 1973; Siegler and Ebinger 1987; Jolivet Chapter 26, this volume), but other ant-plants are protected by secondary compounds early in development, when ant defence is lacking or unreliable. Young seedlings with few total resources may grow too slowly or unevenly to maintain ant colonies. Not surprisingly, many ant-plants delay expression of myrmecophytic traits until later in seedling or sapling development (Janzen 1975; Davidson *et al.* 1990; D. W. Davidson, personal observation). Like *Barteria fistulosa* (Passifloraceae) (Waterman *et al.* 1984), myrmecophytic *Cecropia* apparently have chemical protection in their leaves prior to producing ant-attractive traits (Coley 1986; P. D. Coley, personal communication for *C. peltata*). Each of the six Peruvian *Cecropia* species has measurable leaf concentrations of condensed tannins, although we do not yet have data on hydrolyzable tannins, nor on developmental and interspecific trends in total tannins.

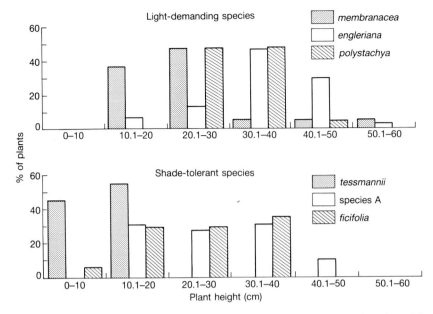

Fig. 20.2. Distribution of plant sizes for six Peruvian *Cecropia* species after eight months of cultivation from seed in the University of Utah tropical greenhouse (July 1987–February 1988). The exception, *C. tessmannii*, was grown from January 1987–October 1987, and its growth was related to that of the other species indirectly, by comparisons with *C. engleriana* cultivated independently during both growth trials. Larger individuals of *C. membranacea* are probably more representative of the species' potential for rapid growth. A significant fraction of its seedlings were stunted at the outset for unexplained reasons. Growth rates of related species differed significantly in two-tailed Mann–Whitney U-Tests: *C. membranacea* > *C. tessmannii* ($P < 0.0001$); *C. engleriana* > *Cecropia* sp. A. ($P \approx 0.002$), and *C. polystachya* > *C. ficifolia* ($P < 0.05$). Sample sizes ranged from 17 to 30.

Weighing the disparities in the synthesis and maintenance costs of biotic and chemical defences, McKey (1984, 1988) concluded that habitat-correlated variation in plant growth patterns should determine relative investment in the two defence systems. Food and domatia for ants must be synthesized continuously and should have higher maintenance costs than chemical defences like tannins and lignins. However, the non-toxic con-stituents of food rewards may be reclaimable and more easily shunted from senescent leaves to new growth. Ant defences then take the place of phenological defences in plants with continuous leaf production and rapid leaf turnover. In contrast, because chemical defences such as tannins and lignins have relatively high synthesis costs but low maintenance costs, their net benefit should increase with the lifespan of the plant part. These

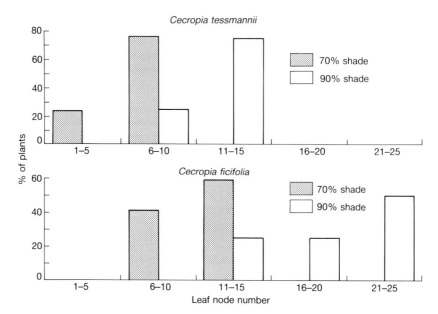

Fig. 20.3. Distributions of leaf node numbers for the first leaf with trichilia in two shade-tolerant *Cecropia* species grown in the University of Utah tropical greenhouse. A 70 per cent shade cloth covered the greenhouse bays to shield tropical plants from intense sunlight. Additional shading significantly delayed trichilia development in both species (in two-tailed Mann–Whitney-U-Tests, $P = 0.002$ (*C. tessmannii*) and $P = 0.003$ (*C. ficifolia*). Conceivably, selection for the timing of trichilia development could be uniform across shade-tolerant and light-demanding species, with shade-tolerant saplings showing earlier onset of myrmecophytism only at comparatively high light levels in the greenhouse (comparing Fig. 20.1 with Fig. 20.3). However, two factors suggest the interpretation that selection has favoured earlier myrmecophytism in shade-tolerant hosts. First, the growth form of greenhouse seedlings at 70 per cent shade is entirely normal, while that of 90 per cent shaded plants is unusually elongate with narrow internodes. This suggests that field seedlings grow at light levels higher than 90 per cent shade. Nevertheless, even comparing shade-tolerant species grown at low light with light-demanding species grown at high light, shade-tolerant and light-demanding species differ in the same direction as in the comparisons in Fig. 20.1.

non-reclaimable chemical defences are more typical of plants with long-lived leaves. Even specialized ant-plants with continuous leaf production may protect mature leaves chemically, if long leaf lifetimes make the continuing costs of ant defence uneconomic (McKey (1984) for *Leonardoxa africana*).

Leaf lifetimes and tannin concentrations must be measured under natural conditions in the field, and we do not yet have such data. However, in many non-myrmecophytes, leaf lifetimes vary inversely with intrinsic growth rate (e.g. Chabot and Hicks 1982; Mooney and Gulmon 1982). If the extremely slow growth rate of *C. tessmannii* (Fig. 20. 2) is indicative of especially long leaf lifetimes, mature leaves would be defended more economically by tannins than by ants. Mature leaves of greenhouse *C. tessmannii* have moderate to high concentrations of condensed tannins. Interestingly, this species had the lowest percentage of leaves with active trichilia, despite the relatively high percentages of ant-defended leaves in the other two shade-tolerant species (Table 20. 2). This species is also unique in lacking pearl bodies on leaf under-surfaces. In contrast, *Cecropia* sp. A, the species with the greatest investment in biotic defences per unit leaf area (Table 20. 2), had the lowest levels of condensed tannins (proanthocyanin assay in greenhouse saplings with trichilia) measured for mature leaves of any of the six species. Comparatively high tannin concentrations might characterize *Cecropia* species that often house parasitic ants, i.e. ants that are relatively poor host plant defenders (Janzen 1975; McKey 1984; Davidson *et al.* 1990). *Cecropia ficifolia* may have the highest frequency of occupation by parasitic ants (*Camponotus balzani*, Table 20. 3, but see below), and levels of condensed tannins were highest in leaves of this species.

Rates of resource supply to established ant colonies

Rates of resource supply to young ant colonies depend not just on the host plant's proportionate investment in biotic defence, but on plant size and rate of leaf production at colonization. Somewhat counter-intuitively, then, light-demanding *Cecropia* may represent better resources than shade-tolerant species at the stage when ant workers first emerge on the plant surface. Rates of resource provisioning should be greater in light-demanding species because of a combination of factors:

1. higher light intensities that promote rapid growth in the characteristic habitats;
2. faster intrinsic growth rates (positively correlated with new leaf production); and
3. larger sizes at colonization, together with size-dependency in growth rates.

Rates of Müllerian body production decline with leaf age in all the Peruvian *Cecropia* and, on average, active trichilia may be newer and more productive on fast-growing, light-demanding species.

Figure 20. 4(a) and (b) summarizes the hypothesized relationship between leaf lifetimes (presumed negatively correlated with intrinsic growth rate), defence investment, and rate of resource provisioning to ants. These hypotheses are preliminary because none of our measurements of resource supply were made under natural conditions, where light-demanding and shade-tolerant species grow in very different resource environments (including

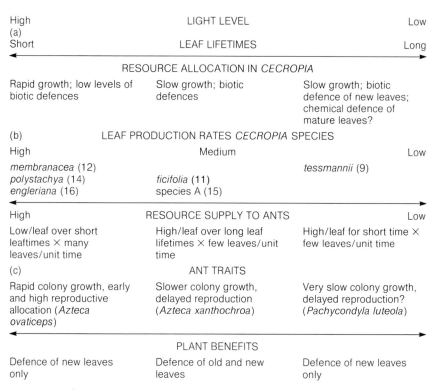

Fig. 20.4. (a) Hypothesized relative investment by *Cecropia* in growth versus defence, in relation to average leaf lifespan (assumed proportional to growth rates). (b) Hypothesized relationship between leaf production rates of particular *Cecropia* species and rates of resource supply to ants. In parentheses are numbers of leaves and leaf scars at the end of the greenhouse growth experiment; the numbers are equated with leaf production rates. In the field, leaf production rates should be proportionately much greater for light-demanding species (left-hand column) than for related shade-tolerant species (centre and right-hand columns), because of far greater light intensities in the typical habitat. (c) Ant traits and plant benefits correlated with patterns of resource provisioning to ants in (b).

red/far red light ratios which are known to affect plant development, Lee 1986).

Specialized ants of *Cecropia*

Early tropical naturalists recorded, without interpretation, much variability in the species composition of ant–plant associations (e.g. Bequaert 1922; Wheeler 1942; summarized in Huxley 1986). Relationships of most myrme-cophytes and plant-ants are oligophilic, with taxonomic or functional ant–

plant guilds, regularly associated with small guilds of specialized congeneric or unrelated ants. Four ants of *Cecropia* in south-eastern Peru (Table 20. 3) can be classified as specialists because:

1. they occur on no other myrmecophyte genus;
2. queens colonize hosts principally at prostomata;
3. they recognize Müllerian bodies as food; and
4. except for *Camponotus balzani*, they reject standard ant baits like tuna and cheese.

Competition between specialized ants

Many ant-plants have a modular structure which allows colonization of leaf or stem domatia by several foundresses of one or more species. Multiple colonization should lead to intense intra- and interspecific competition among incipient colonies for dominance of the host plant. Competition for hosts is usually resolved with the emergence of the first young colony, whose

Table 20.3. Percentages of four specialized *Cecropia* ants on six *Cecropia* spp. from south-eastern Peru. Values for larger plants with established colonies and for seedlings with only foundresses or incipient colonies (in parentheses).

Cecropia	*n**	*Azteca ovaticeps*	*Azteca xanthochroa*	*Camponotus balzani*	*Pachycondyla luteola*
Light-demanding species					
1. *membranacea*	53	49	21	30	0
	91	(48)	(18)	(15)	(20)
2. *engleriana*	6	83	17	0	0
	41	(10)	(76)	(15)	(0)
3. *polystachya*	7	57	29	14	0
	64	(12)	(77)	(11)	(0)
Shade-tolerant species					
1. *tessmannii*	31	0	0	10	90
	49	(0)	(2)	(12)	(86)
2. species A	17	0	100	0	0
	43	(0)	(100)	(0)	(0)
3. *ficifolia*	24	0	48	52	0
	18	(0)	(83)	(17)	(0)

*n = Total number of trees sampled per *Cecropia* species (or, for seedlings, total number of foundresses or incipient colonies censused per *Cecropia* species). Data from combined censuses at Reserva Tambopata and Estación Biológica de Cocha Cashu, Peru.

workers kill all remaining colonies. Thus, Davidson *et al.* (1989) demonstrated negative associations between alternative ant taxa for populations of six myrmecophytes (in six genera) tested after worker emergence, but none of four hosts tested before colonies were active on plant surfaces. *Cecropia tessmannii* with established colonies housed either *Pachycondyla luteola* or *Camponotus balzani* but not both. Similar negative associations among ants are evident from Longino's (1989) data on Central American *Cecropia*.

Included niches in relation to a variable physical environment

The species of ants associated with particular *Cecropia* species often change abruptly across habitat boundaries, within as well as between host plant species (Benson 1985; Longino 1989; Davidson *et al.* 1990). This pattern suggests that the relative success of different ant species could differ in response to rates of resource supply, governed by resource availability to hosts (Davidson *et al.* 1990). For example, in south-eastern Peru several *Cecropia* species can house established colonies of either *Azteca ovaticeps* (see Chapter 19), *Azteca xanthochroa*, or *Camponotus balzani* (Table 20. 3). While the former species predominates in the most favourable plant resource environments along banks of major rivers (high light and alluvial soils), the latter two species dominate numerically at reduced light intensity inside the forest edge, or on poorer soils along steep banks of minor streams, and in large clearings on high terraces. Together with *A. ovaticeps'* apparent restriction to relatively fast-growing *Cecropia* species (Table 20. 3), the pattern suggests that this ant's niche may be included in that of the other two species, and is dependent on rapid rates of resource supply. Davidson *et al.* (1990) have called attention to the commonness of similar included niche phenomena in relation to light environments of other terrestrial and epiphytic myrmecophytes (Huxley 1978; Janzen 1983; McKey 1984; Benson 1985; Jebb 1985; Harada and Benson 1988; Longino 1989*a*). As is often the case in other systems of included niches (reviewed in Colwell and Fuentes 1975), species dependent on abundant resources often survive by superior colonizing or competitive ability, while others persist by virtue of greater tolerance of less favourable environments.

Developmental differences between ants

The early resolution of competition has at least two important implications. First, there is a premium on rapid colony development if the requisite resources are available. This is suggested by the observation that colony (brood) size was greater on average for *Azteca ovaticeps* than for other specialized *Cecropia* ants six weeks after colonization of *C. membranacea* growing in full sun (D. W. Davidson and P. Herrera unpublished work). On slow-growing *C. membranacea* in the greenhouse, rapidly expanding colonies of *A. ovaticeps* destroyed new leaves by excavating immature Müllerian bodies from trichilia, and swarmed off the host plant in search of additional hosts.

Secondly, the resources determining the competitive outcome may be those available to foundresses, rather than to established colonies. The flight muscles of foundresses are often catabolized to provide resources for the first worker brood (Janet 1907; reviewed in Wilson 1971). With poorly developed flight muscles (relative to *Azteca xanthochroa*), *A. ovaticeps* queens may be especially dependent on the nutritional parenchyma lining internal walls of *Cecropia* internodes (Fig. 20. 5(a)).

Parenchymal thickness depends on host species and light intensity, the lining being best developed in host species growing regularly at high light intensity. *Azteca ovaticeps* queens transferred to saplings of *C. tessmannii*, which lacks the parenchyma, tended to die or desert at disproportionately high rates, or to produce smaller broods over a six-week period, than did queens transferred to *C. membranacea*, a host species with well-developed

Fig. 20.5. (a) Thick parenchyma lines the internal walls of *Cecropia membranacea* stems in plants growing at high light intensities. Patches of this material have been removed by ant larvae. (b) *Pachycondyla luteola* queen investigates the prostoma of *Cecropia tessmannii*. The prostoma of this host is unique for its large size, outward swelling, and long, urticating hairs which exclude small *Azteca* queens.

parenchyma (D. W. Davidson and P. Herrera, unpublished work). In the *Cecropia* specialist *Pachycondyla luteola* (and the newly discovered *Pachycondyla* sp. nov. on Panamanian *C. hispidissima*), egg production by foundresses apparently depends on nutrition from unusually large Müllerian bodies, which queens themselves collect and store inside colonized internodes of the typical host, *C. tessmannii*. Of the six common host species, *P. luteola* queens colonize only two apparently close relatives (*C. membranacea* and *C. tessmannii*) with Müllerian bodies > 0.09 mg (Davidson *et al.* 1990). Queens transferred to *C. membranacea* deserted or died at higher rates than those transferred to *C. tessmanii*. When foundresses transferred to the typical host were sealed inside their internodes by corking prostomata, most of them either died or escaped by cutting new exit holes within two days of transfer. Queen dependency on Müllerian bodies may explain why the only two *Pachycondyla* species presently confirmed as *Cecropia* specialists are specific to hosts whose Müllerian body production is skewed to new leaves (Table 20. 2), safely accessible to queens occupying the newest internodes.

Later in colony life histories, interspecific differences in relative allocation to reproductive versus worker castes may also vary in relation to the pattern of resource provisioning by hosts. Thus, the ability of *Azteca ovaticeps* to dominate numerically on fast-growing hosts may also depend on its disproportionately large investment in colonizing reproductives. Unlike *A. xanthochroa*, which occupies and defends the entirety of its tree including the trunk, *A. ovaticeps* maintains smaller worker populations restricted to branch tips (see Chapter 19). Moreover, preliminary observations suggest that reproductives may constitute a greater fraction of total colony biomass (J. T. Longino, personal communication; D. W. Davidson, personal observation). The proportionately greater reproductive allocation of this species might have been predicted from the normally brief lifespan of its most common host, *Cecropia membranacea*, which is replaced in riverine succession at an average age of approximately ten years (Terborgh 1983). In contrast, *C. tessmannii* is a dominant canopy species on the edges of swamps and aguajals, and probably reaches its full stature (c. 35 m) only after several to many decades.

Figure 20. 4(c) summarizes our interpretation of correlations between ant traits and patterns of resource provisioning by hosts. The hypothesized implications for plant defence are also shown.

Competition between specialized and unspecialized ants

Competition may also influence the composition of ant–plant symbioses at later stages in their life histories. An example outside the *Cecropia* system is that of *Pheidole minutula*, which can take over branches of *Clidemia heterophylla* Steud. (Melastomataceae) from *Crematogaster cf. victima* when branches of their separate shrubs come into contact (Davidson *et al.* 1989).

The behaviour of attacking vines and other vegetation which is in contact with hosts may be an evolutionary response to competition from unspecialized, behaviourally dominant competing or predatory ants (Davidson *et al.* 1988). Pruning of vegetation is especially well developed in some pseudomyrmecine ants (e.g. Janzen 1966, 1972; Davidson *et al.* 1988) which create areas that are barren of vegetation around the host tree. Although none of the four specialized Peruvian *Cecropia* ants (Table 20. 3) makes distinct clearings around its hosts, all but *Camponotus balzani* prune vines to some extent (Davidson *et al.* 1988, 1990). Pruning is facultative in *P. luteola*, occurring only when vines are occupied by predaceous or competing ants (D. W. Davidson and P. Herrera, unpublished work). Perhaps because of pruning, there is no indication that unspecialized ants ever take over *Cecropia* hosts from specialized ants once colonies of the latter are established. Nevertheless, some ant species may be specialists on mature trees that have lost their previous residents (e.g. Benson (1985) for *Azteca schimperi*).

Many unspecialized ant species are occasional residents of *Cecropia* (Longino 1989*a*), usually inhabiting the lower trunks or seedling stems, and typically foraging off the host. In Peru, unspecialized species of *Camponotus*, *Crematogaster*, *Pachycondyla*, and *Solenopsis* are most common in the lower internodes of trees housing *Azteca ovaticeps* in their upper stems. These occasional *Cecropia* residents usually nest in a variety of other plant stems, with colonies often fragmented over different hosts. For reasons not yet clear, none of them appears to recognize Müllerian bodies as food. The necessity for foraging off the host plant may prevent these species from becoming *Cecropia* specialists.

Parasitoids might affect the competitive outcome

On both ecological and evolutionary time-scales, predation and parasitism may interact with competition to determine the distribution of alternative ant species in particular hosts and habitats. The evolutionary case would be especially difficult to detect (Bernays and Graham 1988). Since established ant colonies tend to be immune to mortality from predation and because host-plant ownership is resolved early in establishment, the most significant effects of predation and parasitism should be those affecting queens and young colonies. At Cocha Cashu, mortality due to parasitoids is highest for the most frequent colonists of *Cecropia membranacea*, the most abundant and conspicuous myrmecophyte in this region. In October 1988, beach seedlings were visited daily in order to date colonization of particular internodes. Six weeks after colonization, internodes were opened to reveal their contents. Based on conservative estimates, 21 per cent of *Azteca ovaticeps* queens were found to have been killed by parasitoids, most frequently *Conoaxima* sp. (Eurytomidae, Chalcidoidea). Members of this parasitoid genus apparently specialize in attacking *Azteca* foundresses (R. R. Snelling, personal communication; Longino 1989*a*). Parasitoids may locate queens by searching for host plants; female *Conaxima* were observed to visit various seedlings, where they inspected newly sealed prostomata.

The brood of incipient *Pachycondyla luteola* colonies may be particularly easy targets for parasitoids because host prostomata are especially large, and are left open for queens to collect Müllerian bodies. As many as nine perilampid wasps (Perilampida, Chalcidoidea) can emerge from individual pupae of *P. luteola*. Present evidence is too preliminary to evaluate whether these and other parasitoids influence the distribution of specialized plant-ants in particular hosts and habitats.

Variation in host-plant protection by alternative ants

If alternative ants differ in the protection afforded to their hosts, then some of the ants may be parasites of relationships between these hosts and other ants. Among the four specialized *Cecropia* ants of Amazonian Peru, *Azteca xanthochroa* and *Pachycondyla luteola* are the most aggressive, racing down the trunk in response to the slightest disturbance. *Camponotus balzani* probably offers the least effective defence; this ant is extremely timid, principally nocturnal, and seldom ventures away from trichilia to patrol leaves. *Camponotus* may even directly injure its hosts by tending scale insects within stems. This practice would buffer variation in rates of resource provisioning and, perhaps, explain how this ant can inhabit so many host species (Table 20. 3). In contrast to Central American *Azteca*, the Peruvian species do not regularly tend Homoptera inside *Cecropia* stems.

Before classifying *Camponotus balzani* and/or *Azteca ovaticeps* as parasites, one must take into account other defences of host plants. Hosts with either short leaf lifetimes or chemical protection of mature leaves may require biotic defence only for young leaves. If this were so, then even relatively timid ants that foraged only at branch tips might proffer adequate defence. Additionally, if host fitness were most sensitive to herbivory at the seedling and sapling stages, then colony development rates may be as, or more, important than differences in ant behaviour in determining the benefits that different ants afford their hosts. Finally, parasitic ants might be most likely to invade ant–plant associations (and even to become specialized plant-ants) on hosts that provision ants in a way that permits timid ants to monopolize the food resource. Thus, both *P. luteola* and *Pachycondyla sp. nov.* (B. L. Fisher, personal observation) are specialized on hosts whose Müllerian body production is skewed toward the youngest leaves. *Pachycondyla* queens open partitions between internodes to colonize new internodes as these are produced. Trichilia are easily accessible to queens (and later workers) foraging through prostomata at the very tips of branches. If hosts do not require protection of older leaves, host evolution is less likely to oppose parasites and may even lead to increasing accommodation of these new arrivals. Arguments like these may help to resolve the paradox of *Cecropia hispidissima* traits that seem to favour *Pachycondyla sp. nov.*, despite the reluctance of these timid ants to leave their stems except when foraging at the newest trichilia (B. L. Fisher, personal observation).

Habitat variation in herbivore pressures

Herbivore pressure doubtless varies with geography and habitat. Thus, although leaf-cutting ants are a major threat to *Cecropia* in many areas, they are rare and unimportant in the frequently flooded forests of south-eastern Peru. At Reserva Tambopata, herbivores of young *Cecropia* lacking trichilia included Coleoptera, Diptera, Hemiptera, Homoptera, and Orthoptera. Most abundant and devastating in large clearings with numerous *Cecropia* seedlings are *Coelomera* beetles (Coleoptera, Chrysomelidae) which are specialists on *Cecropia* (Andrade 1984; Jolivet 1987*b*, Chapter 26, this volume). These beetles were practically absent from isolated *Cecropia* saplings in small forest light gaps. Preliminary evidence suggests that *Coelomera* numbers vary with habitat rather than with (correlated) host species. The few shade-tolerant species sampled within large disturbances had beetle densities comparable to those on the light-demanding species. Herbivore pressure from *Coelomera* and other *Cecropia* specialists (Table 20. 4) may be low on plants in small isolated light gaps because the smaller seedling populations are more difficult to find, and are not sufficiently large or dense to support breeding populations of the herbivores. The relatively high investment by some shade-tolerant species in biotic defence (Table 20. 2) is even more noteworthy, given the comparatively low herbivore densities in their characteristic habitats.

Inferring determinants of evolutionary specialization

Our understanding of the factors leading to evolutionary specialization will depend on knowing which plants and ants have evolved traits that increase the species specificity of their relationships. Such traits are difficult to detect because ecological and evolutionary effects are confounded. Habitat specificity in the outcome of ecological processes (e.g. ant competition and parasitoid infection) may produce specific and repeatable patterns of ant association with particular habitats and hosts in the absence of evolved specificity. Experimental manipulations will be required to distinguish such species sorting (Jordano 1987) from special adaptation (Davidson *et al.* 1989).

Although existing data can suggest the distributions of evolutionary specialization leading to specificity, they cannot distinguish whether such specializations arose before or after symbiotic associations were common. The absence of *Azteca ovaticeps* foundresses from slow-growing, shade-tolerant

Table 20.4. Herbivory on natural populations of *Cecropia* seedlings lacking established ant colonies (at Reserva Tambopata, south eastern Peru).

Cecropia species	$n_P(n_L)$	All herbivores		Most common herbivores	
		H/P	*H/L*	*H/P*	*H/L*
Large disturbances (human clearings)					
C. membranacea	3 (24)	11.7	1.5	4.0	0.5[a]
C. engleriana	16 (175)	8.2	0.8	3.5	0.3[a]
C. polystachya	7 (47)	5.1	0.8	2.1	0.3[a]
C. tessmannii	2 (15)	6.5	0.9	3.1	0.5[a]
C. species A	1 (4)	2.0	0.5	2.0	0.5[a]
Forest gaps					
C. tessmannii	21 (153)	0.3	0.0	0.1	0.0[b]
C. species A	11 (85)	0.4	0.1	0.3	0.0[c]
C. ficifolia	11 (75)	0.6	0.1	0.4	0.1[d]

* n_P = number of plants sampled per species; n_L = number of leaves sampled per species: *H/P* = mean number of herbivores per plant; *H/L* = mean number of herbivores per leaf. Herbivores varied markedly in type and number on individual seedlings and no statistical analysis is attempted. Rigid habitat associations of plants restricted our sample sizes of plants in uncharacteristic habitats.
[a-d] Most common herbivores of various *Cecropia* species: a = *Coelomera* sp. (Coleoptera, Chrysomelidae), accounting in large disturbances for 35 per cent of herbivores of *C. membranacea*, 43 per cent of those on *C. engleriana*, 42 per cent of those on *C. polystachya*, 54 per cent of those on *C. tessmannii*, and 100 per cent of those on *C.* species A. In forest light gaps, b = any of 6 herbivores represented by a single individual *C. tessmannii*; C = Coleoptera sp. 3, accounting for 75 per cent of herbivores on *C.* species A; d = Coleoptera sp. 4 (Curculionidae), comprising 57 per cent of herbivores on *C. ficifolia*.

Cecropia (Table 20. 3) implies either host or habitat specificity. This species (probably a species group, Longino 1989*a*, *b*) became a *Cecropia* specialist independently of at least three other major radiations of *Azteca* onto *Cecropia* (Benson 1985; Longino Chapter 19, this volume). Habitat-specific searching could have arisen prior to specialization on *Cecropia* if this species were dependent on the high productivity of early successional habitats. Similarly, rapid colony development, early and high reproductive allocation, and short queen lifespans may have been pre-adaptions for the present lifestyle as a colonizing species. However, both habitat and host-restricted searching and traits related to the occupation of fast-growing hosts could have originated as responses to differentially high fitness on *Cecropia* with high rates of resource provisioning. Preferences may also have evolved for hosts whose favourable resource environments allowed rapid growth and high leaf turnover to compensate for herbivory. In *Pachycondyla luteola*, host specificity appears more likely than habitat specificity, because the species colonizes only two, probably closely related, species of *Cecropia* in different environments. Nevertheless, to determine confidently whether either *Azteca ovaticeps* or *P. luteola* have specialized on particular habitats or hosts, colonization must be observed on hosts transferred experimentally to uncharacteristic habitats. Observations of queens transferred to atypical hosts would show whether ants have evolved preferences for certain hosts, or whether the unused host species are adapted to exclude particular ants.

Indications that host trees may specialize on particular ants are strongest for slow-growing, shade-tolerant species, which should have the most to lose by housing ants that provide inferior protection. First, in *Cecropia tessmannii*, long urticating hairs on an enlarged and outwardly swollen prostoma (Fig. 20. 5(b)) exclude small *Azteca* queens, which apparently never colonize these plants. At least, *A. xanthochroa* foundresses search in habitats where *C. tessmanii* seedlings volunteer, and experimentally transferred *A. ovaticeps* queens attempt unsuccessfully to enter these prostomata (D. W. Davidson and P. Herrera, unpublished work). Although *A. xanthochroa* colonies appear to afford excellent protection to their hosts, low rates of resource provisioning by this especially slow-growing host may frequently lead to the mortality of incipient colonies and, thus, a delay in the host's acquisition of biotic defences. Secondly, the very narrow and short internodes of *Cecropia* sp. A may exclude large queens of both *Camponotus balzani* and *Pachycondyla luteola*. These queens appear not to colonize *Cecropia* sp. A despite their presence in other shade-tolerant hosts (Table 20. 3). Especially low leaf tannin concentrations (see above) and disproportionately high reliance on biotic defence (Table 20. 3) may make this species particularly susceptible to herbivory when housing *C. balzani*.

There are other indications that *Cecropia* species have specialized to particular ants. Müllerian bodies of *C. tessmannii* are unusually large, averaging approximately three times the dry weight of those on *Cecropia* species that typically house small *Azteca* ants, and almost twice as large as those of *C. membranacea* (Davidson *et al.* 1990). Comparatively large *P. luteola* workers will collect the smaller food bodies when these are supplied artificially from other plants, but queens regularly reject them. *Pachycondyla sp. nov.* is even larger in size and the Müllerian bodies of *C. hispidissima* may be larger and harder than those of any other *Cecropia* species (as well as being purple!). Subtle differences in Müllerian bodies are almost certainly more apparent to ants than to human observers. In the greenhouse, we successfully colonized *C. ficifolia* with *A. ovaticeps*, an ant which never occurred on this host in the field (Table 20. 3). Here, in contrast to their frenzied activity on other hosts, workers were lethargic, standing motionless on or near trichilia with mandibles often open, as in an 'attack posture'.

Although the workers did collect some Müllerian bodies, many remained pendant on the trichilia and eventually fell.

Finally, Berg (1978*b*) suggested that the tiny stem hairs of *Cecropia* function as 'ladders', giving sturdy footing to attendant ants. In this context, it is interesting that dense stem hairs are much longer (2 mm) on *C. tessmannii* and *C. hispidissima*, which regularly house the large *Pachycondyla*.

Defence-mediated speciation in *Cecropia*, an hypothesis

Ant-plants wear their defences on the outside, and patterns in defence investment are more easily observed than in chemically defended plants. One remarkable pattern stands out for the western Amazonian *Cecropia*. Defensive investment changes with habitat in similar ways for the three pairs of probable close relatives. If phylogenetic evidence supports the hypothesis that closest relatives inhabit different light environments, then plants may have speciated across habitat boundaries, perhaps by mechanisms related to defence investment. *Cecropia* trees produce numerous small seeds, dispersed widely by mobile vertebrates (birds, bats, and monkeys; reviewed in Holthuijzen and Boerboom 1982). Many of these seeds regularly reach atypical environments and genetic variation in anti-herbivore defence systems would be tested each year in habitats differing in herbivore pressure and availability of plant resources. Variation in defence investment could occasionally permit saplings to survive in otherwise unsuitable habitats, especially if isolated plants experienced lower herbivore pressure (e.g. Feeny 1970).

At maturity these *Cecropia* might be reproductively isolated by differences in flowering season; intraspecific variation in reproductive season of one *Piper* species is known to be influenced by light regime (Marquis 1988). Among the six *Cecropia* species, light-dependent *C. membranacea* and *C. engleriana* flower and fruit in the wet season but each of their probable close relatives reproduces in the dry season, when the photosynthate needed to fill fruits may be less limiting to these often shaded plants. Although *C. polystachya* and *C. ficifolia* both fruit in the wet season, fruiting seasons are slightly displaced, but flowering times are not yet known. The hypothesis of sympatric (but not syntopic) speciation in relation to evolutionary tradeoffs between defence and growth need not be restricted to myrmecophytes, but could generalize to other tropical plants with small and widely distributed seeds. This hypothesis is consistent with Gentry's (1989) view that habitat specialization (principally in vegetative traits) has been of great importance to speciation in Amazonian rainforest plants.

Cecropia as a model for studying ant–plant symbioses

Locally high species diversities of both *Cecropia* and its specialized ants make this system particularly complex, though perhaps not uniquely so given superficial convergences with the *Macaranga–Crematogaster* association in Asia (see Chapter 18). With complexity comes opportunity to move from a purely descriptive to an explanatory phase in the study of ant–plant symbioses. Patterns in the allocation of plant defence investment may be useful in predicting rates and patterns of resource supply to symbiotic ants. In turn, interspecific differences in resource supply may determine both the identities of ant associates and adaptations of both ants and plants to one another. Where variation in the species composition of ant–plant associations is possible, given the evolved attributes of plants and ants, associations may be determined by the sometimes habitat-specific outcome of competition among specialized ants for hosts. The competitive outcome might occasionally be modified by foundress mortality from parasitoids.

The selection environments of symbiotic ants and plants are multivariate and must be characterized fully before we can answer the kinds of questions posed by A. J. Beattie in Chapter 37. Relationships may be mutualistic or not, depending on the protection proffered by alternative and competing ants, the growth rates, leaf lifetimes, and alternative defences of various congeneric hosts, and subtle changes in plant resources across habitat boundaries. If the selection environment is not well characterized, even the most straightforward predictions set out for expensive and time consuming tests of micro-evolutionary hypotheses may be completely wrong. The evaluation of a pattern within these highly context-dependent associations may also provide the best means of identifying mechanisms that both promote and oppose the diversification of symbiotic ant and plant species.

Acknowledgements

Preliminary work was funded by a National Geographic Society Expedition Grant (to D.W.D.), and D.W.D. financed later field studies. The University of Utah kindly provided tropical greenhouse space. We thank Mary Alyce Kobler, Sandy Adondakis, Alan Perry, P. Lozada, and Patricia Herrera for assistance. Jack Longino identified *Azteca* and C. C. Berg, the *Cecropia*.

21

Phylogenetic analysis of the evolution of a mutualism: *Leonardoxa* (Caesalpiniaceae) and its associated ants

Doyle McKey

Two notorious disciplines in biology seem to be lurching toward each other, groping for mutual understanding and even fusion into a more powerful version of evolutionary biology. The two disciplines are cladistics and adaptationist theory.... The two fields obviously need each other. The current estrangement between adaptationist theory and 'pure' cladistics (which tends to focus on character patterns rather than on evolutionary processes) finds its metaphor in a Greek explanation of sex. Plato believed that each male and female once formed a perfect hermaphroditic whole, but through some misunderstanding (nb!) were sundered into different, individually sterile genders. As such they are condemned to wander, miserably seeking their complementary half. I would ... advise cladists and adaptationists to give up, join up, and procreate.

Coddington (1985)

Introduction

The study presented in this chapter illustrates an approach to analysing the evolution of ant–plant mutualisms which combines the methods of evolutionary ecology and those of cladistics. The first are used to determine the functional significance of ant and plant characters involved in interactions, and cladistic methods are used to reconstruct the evolutionary history of organisms and characters. By providing information on the number of times a trait evolved, the direction of character transformation, and the sequence in which characters were assembled in a lineage, cladistic analysis suggests which ecological comparisons and experimental designs will be useful in distinguishing among alternative explanations of the causes of character change (Donoghue 1989). Cladistic analysis of character correlations in two lineages of associated organisms can inform studies of co-evolution (Mitter and Brooks 1983). Phylogenetic analysis of *Leonardoxa* suggests that the principal novel feature of myrmecophytic *Leonardoxa*, the swollen-twig myrmecodomatium, had its origin as a pleiotropic consequence of selection on another trait, and was then 'co-opted' and elaborated by selection acting on its new function in interactions with ant mutualists. Phylogenetic analysis of associations between *Leonardoxa* and ants also suggests that co-evolution has occurred, in the strict sense that plants and ants have evolved specifically and reciprocally in response to each other (Futuyma 1986; 'co-accommodation' of Mitter and Brooks (1983)), but firm conclusions about co-evolution must await cladistic treatment of the ants associated with these ant-plants.

Historical explanation in evolutionary ecology

Only since the emergence of phylogenetic systematics as the dominant school of taxonomic thought have evolutionists begun to perceive that methods exist for producing testable estimates of the history of organisms and their characters (Brooks 1985). Donoghue (1989) has summarized what phylogenies can add to the study of evolutionary processes.

First, phylogenetic analysis, because it is designed to detect convergence, provides an estimate of the number of times a trait has evolved, enabling distinction between similarities that are homologous and can be 'explained' by common descent, and those that have evolved more than once, in a parallel or convergent fashion. While either selective forces (some form of balancing or stabilizing selection) or simply 'inertia' may cause a trait that is inherited from a common ancestor to be retained in all the members of a lineage, selection is the only force likely to lead to repeated independent evolution of a trait (Ridley 1983; Coddington 1988). Secondly, phylogenetic analysis gives information about the direction of character evolution, indicating which states of a character are old and which are new ones that may be new adaptations (Wanntorp 1983). Thirdly, phylogenetic analysis allows us to infer the order in which different characters evolved in a lineage, permitting tests of hypotheses which (implicitly or explicitly) require a particular sequence in the evolution of two or more characters (e.g. Sillén-Tullberg 1988; Donoghue 1989). Although useful, the approach does have its limits. For example, using cladistic methods alone, it is not possible to resolve the sequence of events of character evolution that are not separated by a branching event (Donoghue 1989). The more rapid character evolution has been relative to speciation, the lower the resolution of sequences will be.

Perhaps the most important frontier in the application of cladistics to the study of evolutionary processes is in clarifying the adaptive nature of homologies (Coddington 1985). Convergent similarities find only one strong explanation, adaptation by natural selection. On the other hand, adaptational explanations for homologies, in terms of maintenance by stabilizing or directional selection, face the strong competing alternative of simple heritage or 'phylogenetic inertia' (Coddington 1985). According to Coddington (1988), 'the view that features may have been adaptive at some point but are now maintained by something other than selection is probably the most widely accepted explanation of synapomorphy among biologists today.' (A synapomorphy is a shared derived character: Wiley 1981.) Must we concede that the most common pattern in evolutionary history, the maintenance of characters inherited from ancestors, is intractable to adaptational analysis? Not if it can be shown that new adaptations confer an advantage only when they are combined with old adaptations inherited from an ancestor. Examining the order of character assembly in a lineage may be of use, allowing us to trace the evolution of functional and developmental interdependencies. The hypothesis that traits inherited from ancestors are maintained by selection acting on such functional interdependencies (Coddington 1988; Donoghue 1989) may be difficult to distinguish from the alternative hypothesis of 'phylogenetic inertia'. However, it leads to corollary hypotheses that are bolder, more precise, and more easily falsifiable than is the concept of inertia.

The path from constructing a cladogram to using it to test adaptational hypotheses is neither simple nor well trodden. Methods for transforming adaptational scenarios for phylogenies from 'fairy tales' (Eldredge 1979) into a set of testable propositions are only now being developed (Coddington 1988). While palaeontologists such as Eldredge may rightly despair of testability in scenarios, those working with extant organisms can devise experiments, guided by cladistic analysis, to test at least some classes of adaptational hypotheses. Testability '... derives from the willingness to suppose continuity of cause, a kind of uniformitarianism, from the origin of the trait to present times' (Coddington 1988).

Phylogenetic analysis of co-evolution

The study of co-evolution is potentially fertile ground for the application of the phylogenetic approach (Mitter and Brooks 1983; McKey 1988). Yet phylogenetic analyses have so far concentrated on only one aspect of co-evolution, termed 'association by descent'. The goal of this research has been to determine the extent to which the phylogenies of associated organisms are congruent, reflecting association by descent as opposed to host-switching or some other form of 'colonization'. Most studies compare cladograms of hosts and host-specific parasites. These studies have so far provided little evidence to distinguish between two competing explanations of association by descent: (1) co-evolutionary interactions and (2) independent (non-interactive) responses to a common cause such as a shared history of geographical isolating events (Mitter and Brooks 1983).

Relatively little attention has been paid to another aspect of co-evolution, termed 'co-accommodation', or co-evolution *sensu stricto*, the reciprocal evolutionary responses that may occur in lineages of interacting organisms (Mitter and Brooks 1983). Testing for association by descent in two lineages involves construction and comparison of two cladograms. Testing for co-accommodation requires both systematic and ecological data on the two associated lineages. The hypothesis that interacting species influence each other's evolution 'can be tested by systematics by finding some form of temporal relatedness of evolutionary events (character changes or divergences) between asssociated lineages' (Mitter and Brooks 1983). However, distinguishing co-evolution from non-interactive causes for such a pattern requires independent evidence that traits evolving in a temporally related fashion in the two lineages are also functionally correlated: new traits in each lineage must have a selective advantage when paired with new traits in the other lineage. Studies providing the necessary combination of systematic and ecological information are rare (Mitter and Brooks 1983). The result is that few studies have gone beyond comparing cladograms of associated organisms to examining evolutionary sequences of characters important in biotic interactions for evidence of co-accommodation.

Testing for co-accommodation

In mutualistic associations with some degree of specificity, traits of the two organisms sometimes seem to be functionally matched, in the sense that a trait in one organism confers an advantage only in combination with the appropriate trait of its associate, and vice versa. Ramirez (1974) discusses morphological and physiological traits of agaonid figwasps that suggest reciprocal adjustment with syconial features of associated taxa of *Ficus*. Similarly, Janzen (1966) lists corresponding traits of ants and acacias believed to have been produced by co-evolution. Cladistic analysis provides methods for distinguishing among three possible explanations of such cases.

1. One associate evolved an adaptation functionally matched to a trait that was already present in the other lineage before the association evolved. In hindsight, we may term the trait in the second lineage a pre-adaptation. Plant-ants often trim vines from their host plants, and this trait has been presumed to have arisen in co-evolution with ant-plants. It now seems likely that this behaviour preceded the origin of specific associations and functioned originally to prevent invasion of a plant by competitively superior ants (Davidson *et al.* 1988). Figure 21. 1(a) shows the cladistic structure that would suggest such an explanation.

2. The traits may have been present in *both* lineages before the association appeared (Fig. 21. 1(b)). The appearance of co-adaptation is due to ecological fitting of two pre-adapted species. Davidson and Epstein (1989) postulate that dispersal of seeds of ant-garden epiphytes by ants is based on rather general chemical-ecology traits of both ants and plants which were already in place before specific associations arose.

3. In each of the associated taxa, functionally correlated traits evolved during or after the origin of the association. The cladistic structure shown in Fig. 21. 1(c) would rule out explanations 1 and 2, and would strongly suggest the mutual evolution of adaptations that we term co-accommodation. In effect, this cladistic test for co-accommodation is a partial test, for the special case of associated lineages, of the assumption that the selective advantage currently maintaining a trait is the same as that which fixed it

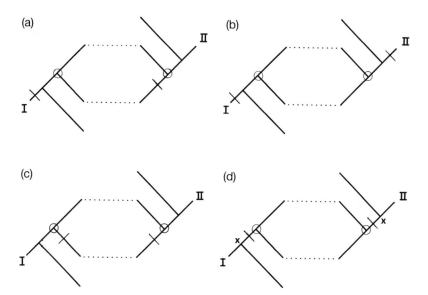

Fig. 21.1. Sequences in evolution of functionally related characters in two lineages (I and II). Dotted lines connect pairs of associated species. Bars on cladogram internodes represent new character states of functionally related characters in the two lineages. Circled nodes represent co-temporal nodes. In these examples, circled nodes mark the origin of the association. (a) Sequence suggesting adaptation in one taxon of lineage II to a character ('pre-adaptation') already existing in lineage I prior to the origin of the association. (b) Sequence suggesting ecological fitting of 'pre-adapted' taxa of two lineages; in both lineages, characters of current functional importance in the context of the association were already present before the association was formed. (c) Sequence suggesting reciprocal evolutionary response or co-accommodation. (d) Example of a sequence that neither confirms nor rejects co-accommodation. Ambiguity is caused by limits to temporal resolution of information about character-state change provided by cladograms. The pattern indicates that the origin of association between the two lineages, and the character-state changes in each lineage, all took place between the nodes marked 'x' and the circled co-temporal internodes, but provides no evidence about the order of these events along this branch.

originally. This is Coddington's assumption of continuity of cause from the origin of a trait to the present, and it would appear to be violated by confirmed instances of explanations 1 or 2. Once traits of the associates survive the cladistic test for co-accommodation, we can then ask if there is any other kind of evidence that these traits owe their origin to selective advantages different from those maintaining them today.

Examining cladograms of two associated lineages for evidence of reciprocal evolutionary interactions (as in explanation 3 above) requires several steps. First, we must identify characters that are known, or suspected, to be functionally related in the two lineages and may thus be co-evolved. Secondly, in each lineage polarity must be established for different states of single characters and evolutionary sequences (Donoghue 1989) must be established for characters suggested to be functionally interdependent. Thirdly, the two cladograms, or relevant portions of them, must be 'calibrated' to establish co-temporality of branching events in the two lineages. Calibration of the two cladograms lets us superimpose sequences established for functionally related characters in the two lineages, to produce an overall co-evolutionary sequence. The co-evolutionary sequence can then be examined for evidence of co-accommodation. These procedures can be considered an extension of methods of mapping the phylogeny of one lineage onto that of an associated lineage (e.g. Brooks and Mitter 1984 and references therein) by hybridizing them with methods of analysing evolutionary sequences (Donoghue 1989). (For lack of a more appropriate alternative, I use the neologism 'co-temporal' to describe evolutionary events in two lineages that are inferred to have happened at the same time, to the limit of the resolving power of the methods used.)

Determining polarity in character transformation and establishing evolutionary sequences are general problems in cladistic analysis. The remaining two steps, calibrating cladograms and determining co-evolutionary sequences, present complex problems unique to the analysis of co-evolution. Congruence in cladograms of two lineages suggests that the lineages are associated by descent. Such a pattern of association, however, may have been produced by either of two quite distinct histories. First, there may have been strict, continuous, mutual descent of taxa in the associated lineages. In this case, speciation events in the two lineages are assumed to have been at least roughly co-temporal, and biogeography (Brooks and Mitter 1984) and molecular clocks (Hafner and Nadler 1988) can sometimes provide independent evidence for this. An alternative explanation of congruence in cladograms, posed originally for plants and associated herbivores, is the 'sequential colonization' (Brooks and Mitter 1984) or 'reciprocal radiation' (Mitter and Brooks 1983) model that involves alternating periods of diversification of each lineage (Mitter and Brooks 1983). In this model, plants evolve a new anti-herbivore defence and undergo an adaptive radiation, producing descendant taxa with the new defence. Later, a group of herbivores evolves an adaptation against this defence and undergoes a radiation on the group of plants that possess it. Herbivores track plant resources that contain defences to which they are adapted and upon which they have come to depend as behavioural cues. Because host phylogeny is a significant, albeit imperfect, predictor of the occurrence of these substances, this non-interactive resource tracking results in limited congruence between cladograms of the two lineages (Mitter and Brooks 1983).

The sequential colonization model is plausible in host-parasite systems because it is easy to see how the host traits are maintained until, at some later time, the association is established. The defence traits of the hosts are presumably being maintained by selection, and leading to diversification of hosts, precisely *because* they destroy or weaken associations with parasites. (More puzzling is why the host defence traits persist once the parasites have evolved adaptations against them.) In contrast, selection maintains and reinforces mutualistic interactions. In species-specific mutualistic associations it is difficult to see how traits important only to the interaction would be maintained if the continuity of the association were broken. Continuous mutual descent thus seems to be the only plausible explanation of congruence in cladograms of species-specific mutualistic associates. If this is so, corresponding nodes and internodes of congruent portions of their cladograms should be at least roughly co-temporal.

Once congruent portions of the cladograms have been calibrated, the next step is to superimpose evolutionary sequences in the two lineages into a single sequence of co-evolutionary interplay. Character state changes appearing on corresponding internodes of the two cladograms represent evolutionary events that happened at the same time (to the limit of our resolving power) in the two lineages. In many co-evolving lineages, character evolution probably proceeds too rapidly, relative to speciation, to produce highly resolvable sequences. Cladistic analysis may often allow us to distinguish between co-accommodation and alternative explanations of functionally related traits in two lineages, but may rarely tell us which of the two associated lineages first underwent evolutionary change and which responded.

Leonardoxa and its ant associates

Leonardoxa is a genus of trees and shrubs confined to forests of Central Africa. Five species are recognized, two of them, *L. bequaertii* (de Wildeman) Aubréville and *L. romii* (de Wildeman) Aubréville, occurring in the Zaire basin, and three closely related species (*L. africana* (Baillon) Aubréville, *L. gracilicaulis* McKey, and *L. letouzeyi* McKey) which occur allopatrically within the Lower Guinea coastal forest (Aubréville 1968; McKey (in press)).

Like many caesalpinioid legumes of the tribes Detarieae and Amherstieae (McKey 1989), all species of *Leonardoxa* possess foliar nectary glands (McKey (in press)). The two Zaire basin species have not been studied in the field but they are presumed to interact facultatively with ants attracted to nectaries. No other ant-related structures are apparent in these two species. The three species of the *L. africana* complex vary dramatically in their interactions with ants. *Leonardoxa gracilicaulis*, found from south central Cameroon to Gabon, is like the two Zaire basin species in having foliar nectaries that attract ants but lacking any other ant-related structures. A variety of arboricolous ants usually nest in the canopy of *L. gracilicaulis*. The other two species of the *L. africana* complex are myrmecophytes which possess structures used by ants as nest sites, in addition to foliar nectaries. These myrmecodomatia are swollen internodes with thick pith that either dries up or is excavated by ants. Ants gain entrance into the young internode via the prostoma, a small unlignified spot at the apex of the internode.

Cladistic analysis of Leonardoxa

To produce a hypothesis for the phylogeny of *Leonardoxa*, herbarium specimens, field observation of living plants, and information from the literature were used to compile a set of 19 morphological characters for the five species of *Leonardoxa* and for an outgroup (Group III of *Cynometra*: Léonard 1951). Of the 19 characters, 7 are reproductive, 6 involve vegetative characters apparently unrelated to interaction with ants, and 6 involve characters directly related to interaction with ants. Table 21. 1 describes the characters used and presents the distribution of character states across taxa. Full discussion of the analysis, including choice of outgroup, character selection, and character argumentation, will be presented elsewhere (D. McKey, unpublished work).

The computer program Phylogenetic Analysis Using Parsimony (PAUP: Swofford 1985) was used to search for the most parsimonious tree(s). In the first analysis, ant-related characters (14–19) were specifically excluded in order to avoid circular reasoning by inferring cladistic structure from characters involved in the hypothesis of adaptation (Coddington 1988). This yielded the single most parsimonious tree shown in Fig. 21. 2. Next, the analysis using only the six ant-related characters was run. The purpose of this was to examine whether ant-related characters gave the same estimate of evolutionary history, as would be expected if shared similarities in these traits are homologous rather than due to homoplasy (similar characters that have evolved independently: Wiley 1981). This data set gave three equally parsimonious trees (Fig. 21. 3). All three are consistent with the tree generated by the first analysis in recognizing the *L. africana* complex as a monophyletic group and in agreeing on the relationships inferred within this group. Taken together they fail to resolve the relationships between *L. romii*, *L. bequaertii*, and the *L. africana* complex. These three trees collapse to the strict consensus tree shown in Fig. 21. 4, which is compatible with the tree shown in Fig. 21. 2. Since information from ant-related characters is consistent with relationships inferred from other characters, all 19 characters were used to produce the single most parsimonious tree shown in Fig. 21. 5. This tree has a consistency index (Kluge and Farris 1969) of 0.962, with only two instances of homoplasy:

1. a reversal in number of leaflet pairs in the *L. letouzeyi*/*L. africana* clade; and
2. independent evolution of cauliflory in this clade and in *L. bequaertii.*

Doyle McKey

Table 21.1. Character matrix for the phylogenetic analysis of *Leonardoxa*.

	OG	L.r.	L.b.	L.g.	L.l.	L.a.
Reproductive traits						
1. Inflorescence type: panicle (0)/raceme (1)	0	1	1	1	1	1
2. Sepals: reflexed at anthesis (0)/extended at anthesis (1)	0	1	0	0	0	0
3. Petal length: < 10 mm (0)/> 10 mm (1)	0	1	1	1	1	1
4. Corolla colour: yellow or white (0)/violet to red (1)	0	0	1	1	1	1
5. Ovary insertion in hypanthium: central (0)/lateral or adnate to hypanthium (1)	0	1	1	1	1	1
6. Inflorescence position: terminal or axillary (0)/cauliflorous (1)	0	1	1	1	1	1
7. Number of sepals: 4–5 (0)/4 (1)	0	1	1	1	1	1
Vegetative traits not directly related to ants						
8. Modal number of leaflet pairs: > 3 (0)/2 (1)	0	0	1	1	0	0
9. Relative sizes of leaflets: proximal < distal (0)/proximal < middle > distal (1)	0	0	0	0	1	1
10. Leaflet size: largest leaflet < 15 cm long (0)/15–25 cm (1)/ >25 cm (2)	0	0	0	1	2	1
11. Rachis: adaxially grooved (0)/not grooved or only slightly grooved distally (1)	0	0	0	1	1	1

	OG	L.r.	L.b.	L.g.	L.l.	L.a.
12. Number leaf-bearing internodes produced with each flush of young growth: >4 (0)/2–4 (1)/1 (2)	0	0	0	1	1	2
13. Leaflet apex: not emarginate (0)/emarginate (1)	0	0	1	0	0	0
Vegetative traits directly related to ants						
14. Size of nectary glands: very small (0)/moderately sized (1)/large (2)	0	0	0	1	2	3
15. Vascularization of glands: no veins directed to glands (0); veins directed to glands, but not branched (1); veins branched, with large gaps between veins (2); veins branched, with small gaps between veins (3)	0	1	2	2	3	3
16. Position of glands on leaflets: scattered over basal one-third of leaflet (0)/restricted to extreme base of leaflet (1)	0	0	0	1	1	1
17. Number of basal glands on proximal leaflet pair: 0–2 (0)/2–5 (1)	—	—	0	0	0	1
18. Relative number of glands on different leaflet pairs: most numerous on basal leaflets (0)/most numerous on distal leaflets (1)	0	1	0	0	0	0
19. Myrmecodomatia: none (0)/appear late in development [beginning in second or third flush of growth in a proleptic shoot] (1)/appear early in development [in first leaf-bearing internode of a proleptic shoot] (2)	0	0	0	0	1	2

OG = Outgroup. L.r. = *Leonardoxa romii*. L.b. = *L. bequaertii*. L.g. = *L. gracilicaulis*. L.l. = *L. letouzeyi*. L.a. = *L. africana*.

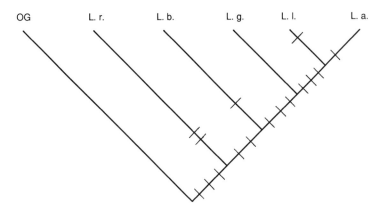

Fig. 21.2. Cladogram of *Leonardoxa* based on characters not directly related to inter-actions with ants (characters 1–13, Table 21.1). Bars on the internodes of the cladogram indicate character-state changes. OG = outgroup; L.r. = *Leonardoxa romii*; L.b. = *L. bequaertii*; L.g. = *L. gracilicaulis*; L.l. = *L. letouzeyi*; L.a. = *L. africana*.

Six synapomorphies, involving reproductive characters and vegetative structures both related and unrelated to ants, define the *L. africana* complex as a monophyletic group. Within this group, six synapomorphies, again involving a diversity of characters, delimit the *L. letouzeyi/L. africana* clade from its sister group, *L. gracilicaulis*.

Relationships of ants associated with Leonardoxa

Each of the two species of myrmecophytes in *Leonardoxa* is associated with a different species of closely related formicine ant, currently placed in two monotypic genera as the sole African members of the tribe Myrmelachistini Forel (Wheeler 1922). *Leonardoxa letouzeyi* is associated with *Aphomomyrmex afer*, a widespread but rarely collected ant which also occurs in many areas where *Leonardoxa* ant-plants are absent. *Leonardoxa africana* is associated with *Petalomyrmex phylax*, a morphologically highly specialized ant described just over 10 years ago (Snelling 1979*a*) which is restricted to *L. africana*, whose young leaves it protects from insect herbivores (McKey 1984). The *L. africana/P. phylax* mutualism is 'parasitized' by *Cataulacus mckeyi* (Snelling 1979*b*), a myrmicine ant that provides little or no protection and excludes *P. phylax* from plants that it occupies (McKey 1984).

If the tribe Myrmelachistini is accepted as natural (i.e. monophyletic), information on its biology and geographical distribution, summarized in Table 21. 2, suggests an interesting picture. First, as the group is represented in the Asian, African, and New World tropics, it may be an old group in which diversification followed the breakup of the southern continents.

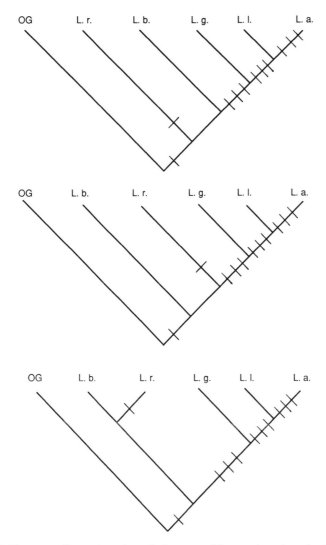

Fig. 21.3. Three equally parsimonious cladograms of *Leonardoxa*, based only on ant-related characters (characters 14–19, Table 21.1). Symbols as in Fig. 21.2.

Secondly, most of its members occupy cavities inside living plants, tending Homoptera (coccids and pseudococcids) within the cavities. This way of life was apparently established before the group diversified. Thirdly, even though particular species may be restricted to one or a few host plants, the group is collectively associated with a range of plants, as are the Pseudomyrmecinae (see Chapter 22). These three features all suggest a long and

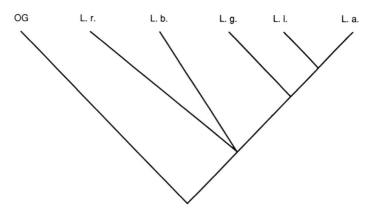

Fig. 21.4. Strict consensus tree of the cladograms shown in Fig. 21.3. Symbols as in Fig. 21.2.

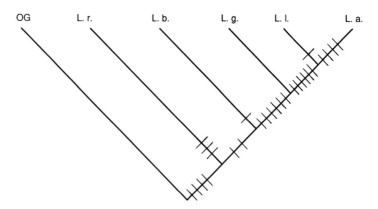

Fig. 21.5. Cladogram of *Leonardoxa*, based on all 19 characters. Symbols as in Fig. 21.2.

varied history of interaction with plants that provide nesting sites for ants, substrates for associated Homoptera, and sometimes foliar nectar.

Evolutionary sequences in ant and plant characters related to interactions

The origin and elaboration of an ant-related adaptation: myrmecodomatia

How evolutionary novelties arise is one of the most challenging questions in evolutionary biology. By clarifying phylogenetic relationships, direction of character transformation, and evolutionary sequences, cladistic analysis helps us estimate the morphological and ecological context in which a new adaptation first appeared. Ideally, we would like to compare ancestral taxa just before the origin of the trait with those soon after its appearance. Using parsimony methods, we can estimate character states of these hypothetical ancestral taxa and determine which extant taxa come closest to providing the material for this critical comparison. Since novel morphological features are rarely truly new (Futuyma 1986), this comparison may identify alternative, primitive character states that are homologous predecessors of the 'new' adaptation. Once identified, these plesiomorphic (= primitive: Wiley 1981) character states (which we may in hindsight term 'pre-adaptations', or exaptations: Gould and Vrba (1982)) can be studied to test hypotheses about how they were modified by selection acting on a new function. That we can generate testable corollary hypotheses about putative pre-adaptations saves their identification from being a completely *post hoc* exercise. Furthermore, we may find some surprising regularities in the idiosyncratic tinkering of evolution (Jacob 1977). Similar adaptations which have evolved independently in different lineages may be built upon similar raw material. If analysis of one lineage successfully predicts similar pre-adaptations in others, the approach has survived a test.

Myrmecodomatia (the structures modified to function as nest cavities for ants) are the most striking novel features of myrmecophytes. Cladistic analysis points to the taxa that are critical in any attempt to determine the origin of this adaptation. Figure 21. 5 shows that *L. gracilicaulis* is the sister taxon of the clade in which the trait evolved. Furthermore, since cladistic analysis has so far revealed no autapomorphies (new characters evolved from their plesiomorphic homologues in a single species: Wiley 1981) of *L. gracilicaulis*, this species may not be very different from the common ancestor it shared with the myrmecophyte clade. Morphological and ecological study of this species may therefore cast light on the origin of symbiotic associations with ants in this genus.

In *Leonardoxa*, the myrmecodomatia are strongly swollen young twigs filled with soft pith. Comparisons of myrmecophytes with non-myrmecophytic species, principally *L. gracilicaulis*, point to a key pre-adaptation for this feature in the genus: an allometric relationship between leaf size and stem size that appears to be quite general in flowering plants. This relationship is itself adaptive, but in a functional context unrelated to ant–plant interactions. If my analysis is correct, the evolution of myrmecophytes in this genus may be the highly indirect consequence of a habitat shift that led to selection for larger leaves. Larger twigs 'hitch-hiked' along with larger leaves. Thicker twigs have thicker pith. Before twig-nesting ants exerted any selection on twig characteristics, *Leonardoxa* were already producing twigs with extensive pith that often dried up spontaneously or could be easily excavated to produce a cavity. Selection acting to increase the size of nest cavities available

Table 21.2. Geographical distribution and associations with plants and homopterans of the ant genera included in the tribe Myrmelachistini Forel (Formicinae).

Genus	Geographic distribution	Nest sites	Association with Homoptera	Sources
Pseudaphomomyrmex	Philippines	?	?	Wheeler (1920)
Cladomyrma	Borneo, Sumatra	4 spp. in hollow stems: 2 spp. in *Saraca thaipingensis* (Leguminosae: (Caesalpinioideae); 1 sp. in *Millettia nieuwenhuisii* Papilionoideae); 1 sp. in *Crypteronia griffithii* (Crypteroniaceae) 'swollen internodes of a shrub'	+	U. Maschwitz *et al.*, (personal communication) Wheeler (1910*b*)

Aphomomyrmex	Tropical Africa	Swollen internodes of *Leonardoxa letouzeyi* (other hosts?)	+	Snelling (1979*a*), McKey (in press)
Petalomyrmex	Tropical Africa	Swollen internodes of *Leonardoxa africana*	—	Snelling (1979*a*), McKey (1984)
Myrmelachista	Neotropics	Hollow stems of *Ocotea* (Lauraceae) and other plants; hollow petioles and stems of several ant-plants; leaf-pouch domatia of *Duroia* (Rubiaceae)	+	Stout (1979), J. T. Longino (personal communication), Wheeler (1942), D. W. Davidson (personal communication)
Brachymyrmex	Nearctic, neotropics	Terrestrial; or in domatia of several ant-plants: *Cordia* (Boraginaceae); *Tachigali* (Leguminosae: Caesalpinioideae); *Triplaris* (Polygonaceae)	+	Wheeler (1942), Alayo (1974)

to protective ants merely elaborated a previously existing structure that originated in another context. After presenting the argument for *Leonardoxa*, I also show that myrmecodomatia in other ant-plant lineages may have originated in a similar context.

Evolutionary increase in leaf size in the Leonardoxa africana complex

The *Leonardoxa africana* complex is characterized by having leaves larger than in other members of the genus and larger than those of taxa used as outgroups, suggesting that leaf size has increased during the evolution of this group. This increase mostly reflects an increase in the size of individual leaflets, with leaflet number remaining comparable.

Although determining the cause of this evolutionary increase in leaf size is not an objective of this chapter, ecogeographical patterns, coupled with cladistic analysis, do offer clues. The members of the *L.africana* complex inhabit forests where the rainfall, atmospheric humidity, and cloudiness are all greater than in the Zaire basin (Pearce and Smith 1984; Griffith 1987). Furthermore, the two Zaire basin species appear to be canopy members in riparian forest, while at least two members of the *L. africana* complex, *L. africana* and *L. letouzeyi*, are typically understorey trees (McKey (in press)). Givnish (1987) reviews the considerable evidence that in diverse plant lineages effective leaf size (the width of a leaf or its lobes or leaflets) increases along gradients of increasing rainfall and humidity, and decreases with increasing irradiance. These patterns appear to be due to effects of leaf size on conductance of the leaf boundary layer, which in turn affects leaf heat exchange, CO_2 uptake, and loss of water vapour. Evolution of larger leaves in the *L. africana* complex is consistent with a widespread pattern and may thus be interpreted as an adaptation accompanying a shift to more humid and shaded habitats.

Correlation between leaf size and stem size

Whatever the reason for the evolution of larger leaves in the *L. africana* complex, it seems to have had an important consequence: stems also became larger. One of 'Corner's rules' (Hallé *et al.* 1978) posits a correlation between the thickness of a plant stem (before secondary growth) and the size of appendages (e.g. leaves) borne by it. This 'rule' was first tested by White (1983), who found a strong allometric relationship between leaf area and twig cross-sectional area in 48 species of deciduous tree of eastern North America. White (1983) argued that this relationship could be explained in terms of requirements for two functions of stems, support and transport.

I examined herbarium specimens of *Leonardoxa* to determine whether a similar relationship between leaf size and stem size exists in this genus. In order to obtain comparable estimates in both myrmecophytes and non-myrmecophytes, I measured the stem at the *base* of the internode subtending the leaf; the stem swellings in the myrmecophytes are pronounced only in the apical two-thirds of the internode. Only the two youngest mature internodes of a twig were measured, to avoid the confounding effects of secondary growth.

Figure 21.6 shows that leaf area and twig cross-sectional area are highly correlated among the five species of *Leonardoxa*. The large-leafed species of the *L. africana* complex, including the non-myrmecophytic *L. gracilicaulis*, have thicker stems than the smaller-leaved Zaire basin species. Because the divergence of *L. gracilicaulis* and the rest of the complex preceded association with twig-nesting ants and the evolution of myrmecodomatia, we can infer that twigs were becoming thicker in this lineage before these associations were established. Cladistic structure of the group is thus consistent with the postulated evolutionary sequence of pre-adaptation and adaptation.

Another morphological feature of *Leonardoxa* has enhanced the underlying correlation between leaf size and stem size, leading to even greater thickening of internode apices as leaves have increased in size. At the base of the rachis of the compound leaf is found a thickened pulvinus, a structure that functions in leaf movement. Presumably because of the requirements of this thickened structure for attachment and support, the apex of the internode is thicker than the base, in non-myrmecophytic as well as in myrmecophytic species of *Leonardoxa*.

The crucial consequence of thicker twigs, most marked near the internode apex, was that the pith also became thicker, especially near the internode apex. In *L. gracilicaulis* the extensive pith often dries up leaving a small

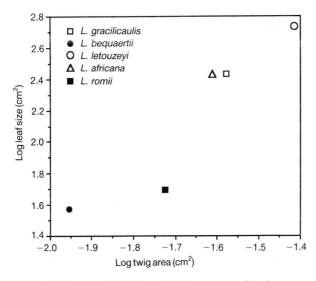

Fig. 21.6. Relationship between leaf size and twig cross-sectional area among the five species of *Leonardoxa*. Leaf areas measured with a Hi-Pad digitizer (Houston Instruments) and twig diameters with Vernier calipers.

cavity, largest near the tip of the internode (Fig. 21. 7). If arboricolous ants that provided some protection to the plant somehow began to enter these cavities (or excavate the pith to create cavities) and use them for nesting or tending Homoptera, selection might favour enlargement of these structures.

If *L. gracilicaulis* already possesses these crucial pre-adaptations, why has it not evolved myrmecodomatia? Ecological data provide a clue. This plant interacts with numerous ant species which are attracted by its foliar nectaries. In Cameroon, one of its most frequent associates is the myrmicine *Tetramorium aculeatum*, which builds felt nests on the undersides of leaves (McKey (in press)). This ant is an aggressive predator and probably provides some protection to plants in which it nests. It is also a dominant species in the arboreal ant mosaic (Leston 1973; Majer 1976), excluding access of other ants to resources present on its nest trees. Twig-nesting ants thus have little

Fig. 21.7. Longitudinal section of a young twig of *L. gracilicaulis*, from Mbalmayo, Cameroon, showing the small (c. 4 mm diameter) cavity (C) formed by spontaneous drying of the thickened pith at the internode apex. P = pulvinus at the base of the rachis of the pinnately compound leaf; S = continuation of the stem. The ant shown in the figure was foraging on the plant surface; ants have not been noted to enter these cavities.

opportunity to occupy *L. gracilicaulis*, so there is no selection to enlarge the structures in which they might nest.

The myrmecophytes occur in much wetter forest near the coast and around Mount Cameroon, in an area near the Bight of Biafra characterized by a long and monsoon-like rainy season (Moby Etia 1980). This area is not only the wettest part of Africa but also the only part of the African forest zone with a single, long rainy season. During the southern winter, while areas to the east and west experience a short dry season, the coast of Cameroon undergoes a 'pluvial paroxysm' (Maley 1989). The exposed leaf-surface nests of *Tetramorium aculeatum* are sometimes destroyed by wind and rains (Majer 1976). In the climate along the coast of Cameroon, these nests would be subject to much more frequent destruction. The ants would be less able to maintain large colonies and monopolize resources. With the balance of ant–ant competition for plant resources thus tipped in favour of ants with nest sites protected from rains, association with twig-nesting ants became sufficiently frequent and durable to generate selection for larger internodes.

These hypotheses, suggested by field observations, remain to be tested in field experiments. The cardinal point is that tests can be devised. Testing hypotheses that explain how an adaptation is maintained is a common approach in evolutionary ecology. When cladistic relationships within a group are known, we can also test hypotheses about how an adaptation originated.

Did similar pre-adaptations precede domatia in other myrmecophytes?

An evolutionary increase in leaf and stem size may have similarly provided the hollow stems for myrmecodomatia in two groups of pioneer trees, Asian *Macaranga* (Euphorbiaceae) and neotropical *Cecropia* (Moraceae) (see Chapters 18–20). Increased leaf size and sparse branching are architectural traits found in fast-growing rainforest pioneer trees, because this combination allows the tree to economize on the woody frame necessary to support a given surface area of leaves (White 1983). (Although leaf size increases, effective leaf size is held low in these plants of sunny, low-humidity sites by division of the leaf into lobes or leaflets.) White (1983) noted that the pith of large-stemmed, large-leaved pioneer trees is often thick, soft, chambered, or hollow, and suggested that this hollow beam construction further reduced costs of the tree's woody frame. *Cecropia* appears in his list of examples, and many species of the Old World genus *Macaranga* would fit as well. Examining cladistic structure and evolutionary sequences in size of leaves and stems, and in ant associations, would enable a partial test of the hypothesis that in these lineages, as in *Leonardoxa*, selection in a context unrelated to interaction with ants may have fortuitously produced structures that could then be shaped into domatia when the appropriate selective pressures arose (see also McKey 1989).

Coevolution of *Leonardoxa* and myrmelachistine formicine ants

Association by descent or colonization?

Did ants of this lineage independently colonize *Leonardoxa* twice, or are *Petalomyrmex phylax* and *Aphomomyrmex afer* descendants of a common ancestor that also occupied *Leonardoxa*? In the approach followed here, the answer to this question hinges on resolution of cladistic structure of the associated taxa. The two plants occupied by these ants are sister species. If the ants are also shown to be sister taxa within the tribe Myrmelachistini, then the two cladograms are congruent. In this case, association by descent is the more parsimonious explanation, because it allows us to posit a single origin of the association, rather than two independent origins. If, on the other hand, these two ants are shown not to be sister taxa, we would have to conclude either that each association with *Leonardoxa* evolved independently, or that one or more members of an ant lineage originally associated with *Leonardoxa* switched to a new host. (This last assumption seems unlikely, since the other members of the lineage occur in other continents, where *Leonardoxa* is absent.)

Evidence for co-accommodation between Leonardoxa *and ant associates*

Match between shape of the prostoma and shape of the ant associate

In each of the two myrmecophytic species of *Leonardoxa* the prostoma is closely matched to the ant associate and its entrance hole not only in size, but also in shape (McKey (in press)). In twig-nesting ants in general, correspondence in size of entrance holes and of ant dimensions is important in excluding intruders, and several taxa have independently evolved entrance-blocking behaviour (phragmosis) and associated morphological adaptations that function to 'plug up' entrance holes (Baroni Urbani 1989). The shield-shaped heads of *Cataulacus* and *Paracryptocerus* (both Myrmicinae) and the flat-topped heads of *Camponotus* (*Colobopsis*) (Formicinae) are notable examples (Wilson 1971).

In *L. letouzeyi* the prostoma is round. Workers and alates of this plant's ant associate, *Aphomomyrmex afer*, are round in cross-section, like most ants, and their entrance holes are also round. In *L. africana* the prostoma is narrowly elliptical, and the entrance hole is a narrow slit (McKey 1984). The workers of the obligate associate of *L. africana*, *Petalomyrmex phylax*, are round in cross-section but being very small they encounter no problem in passing through the entrance slit. The alates are much larger and are strongly flattened dorso-ventrally, a specialized trait to which this monotypic genus owes its name (Snelling 1979a). Alate females of this species are very closely matched in size and shape to the prostoma of *L. africana* and to the entrance slit cut by *P. phylax*. In this pair of associates, the match in shape and size between the prostoma and the ants functions as a lock-and-key system that reduces the impact of an ant 'parasite' of the mutualism, *Cataulacus mckeyi* (Snelling 1979*b*; McKey 1984).

Is this shape matching due to co-accommodation? In testing for co-accommodation, the first step is to establish whether the traits of interest evolved since the association originated. The prostoma is a feature that first appeared in the common ancestor of the sister species *L. letouzeyi* and *L. africana*, and it is clearly interpretable as an adaptation to twig-nesting ants. Because only these two species possess a prostoma, and there is no hint of a precursor in taxonomic outgroups, cladistic methods are not informative about the shape of the prostoma in the common ancestor. Three evolutionary sequences are possible (Fig. 21. 8):

1. the round prostoma is primitive, the elliptical prostoma derived (Fig. 21. 8(a));
2. the elliptical prostoma is primitive, the round prostoma derived (Fig. 21. 8 (b)); and
3. each shape is a separate autapomorphy, derived from a prostoma of unspecified shape (Fig. 21. 8(c)).

Unless the prostoma or a precursor is identified in additional taxa, cladistic methods do not allow us to decide among these alternatives.

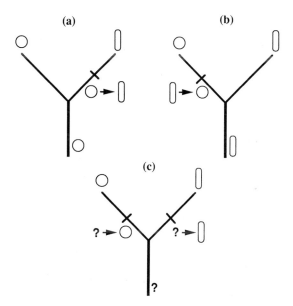

Fig. 21.8. Three possible character transformation series in the evolution of prostoma shape in *Leonardoxa*.

With regard to ant shape, even though cladistic relationships are un-resolved it is clear that *A. afer* is of conventional shape; in fact, workers of *A. afer* collected from *L. letouzeyi* in Cameroon are not distinguishable morphologically from those collected in parts of Africa where *Leonardoxa* ant-plants are absent (Snelling 1979*a*). Because none of the obvious out-groups related to Myrmelachistini deviates from this conventional shape, the round cross-sectional shape of *A. afer* may be safely considered as plesio-morphic, regardless of what the cladistic relationships of this species within the tribe are discovered to be. Analysis of the sequence of events in this association suggests that evolution of the round prostoma of *L. letouzeyi* was simply an adaptation of it (or its ancestor) to a trait already present in *A. afer* prior to the association, and typical of most other ants as well. Regardless of which prostoma shape is primitive, this pair of associates offers no evidence for co-accommodation in this set of traits.

In the pair *L. africana/P. phylax*, on the other hand, narrowly elliptical prostoma and flattened alates provide some evidence for co-accommoda-tion, but not enough to confirm it. The two traits are clearly functionally matched, and outgroup comparison confirms that dorso-ventral flattening of *P. phylax* alates is an apomorphic trait. The uncertainty comes from our inability to establish polarity in prostoma shape. In two of the three possible sequences (Fig. 21. 8), the elliptical prostoma is apomorphic (derived). Co-accommodation would be ruled out only if we suppose that the elliptical prostoma were primitive.

If additional evidence allows us to decide that the elliptical prostoma of *L. africana* is apomorphic, like the flattening of *Petalomyrmex* alates, there remains an additional question: did these autapomorphies evolve before the association was established, or after? This question is more easily answered if these two ants are shown to be sister species. If they are, we infer that their common ancestor was associated with the common ancestor of *L. letouzeyi* and *L. africana*, and that the matching autapomorphies of *L. africana* and *P. phylax* evolved after divergence from these associated ancestors. This cladistic structure would rule out the two alternatives to co-accommodation. If, on the other hand, the ants are shown not to be sister taxa, we must infer that they independently colonized *Leonardoxa*. In this case, we cannot calibrate the origin of *L. africana* with the origin of *P. phylax* and the origin of their association. We have no way of knowing whether the functionally matched autapomorphies of these two species appeared before, during, or after the origin of the association. Thus we cannot distinguish among the three alternative explanations for the appearance of co-adapted traits. Also, in the particular case considered here, if *A. after* and *P. phylax* are not sister species, other problems are raised. The closest living relative of *P. phylax* would be a myrmelachistine ant on another continent, implying a much longer temporal gap between extant sister taxa, with the greater likelihood that extinction of intermediates has obscured the patterns upon which we depend for the reconstruction of evolutionary history.

Match between nectary size and number in plants and tending of homopterans by associated ants

The foliar nectaries of *L. africana* are larger, more highly vascularized, and more numerous than in any other species of the genus (McKey (in press)). Cladistic analysis clearly shows these traits to be autapomorphies of this species. Foliar nectaries of *L. letouzeyi* are not very different in size and number from those of non-myrmecophytic members of the genus. The differences in nectary size and number between the two myrmecophytes seem to be functionally matched with different states of a character of their associated ants: formation of associations with coccids or other homopterans. The associate of *L. letouzeyi*, *A. afer*, tends large numbers of coccids in the hollow internodes of its host, while the associate of *L. africana*, *P. phylax*, does not tend coccids or other homopterans (McKey (in press)). This pattern suggests that nectar and homopteran secretions are alternative trophic bases for ant colonies.

In the ants, because cladistic relationships are as yet unresolved, polarity is not completely clear. *Petalomyrmex* is the only genus in the tribe Myrmelachistini in which associations with homopterans are unknown, and such associations appear to be widespread in formicines; therefore the absence of associations with homopterans is probably an apomorphic loss in *Petalomyrmex*. If so, this autapomorphy and its functionally matching autapomorphy in *L. africana*, the evolution of larger, more numerous nectaries, may constitute a case of co-accommodation.

Stabilizing selection, phylogenetic inertia, and the functional interdependency of old and new adaptations in *Leonardoxa*

The preceding examples concern single characters of ants and plants that are functionally related. The next step is to examine the interdependencies that mould the different traits of each of these organisms into a functionally integrated whole. Analysis of evolutionary sequences in *Leonardoxa* is beginning to provide examples of how new adaptations may confer an advantage only when combined with older adaptations, supporting the contention of Coddington (1988) and Donoghue (1989) that synapomorphies may be maintained by selection acting on such interdependencies. Such interdependencies are part of the burden that leads to phylogenetic inertia. In *Leonardoxa*, *L. africana* is the species in which traits related to ant association are most specialized. The features of this species important to its interaction with *P. phylax*, for example, the characteristics of domatia, are a composite of traits of varying age that are now inseparable in functional terms. But there are indications of even farther-reaching functional interdependence. In contrast to related species, *L. africana* produces young leaves in small, more or less continuously produced flushes, rather than in large, intermittently produced bursts. This new autapomorphy may represent the replacement of phenological defence by biotic defence (McKey 1988, and in press), and is thus functionally inseparable from older ant-related adaptations that began accumulating earlier in this lineage. Future field work, guided by phylogenetic analysis, will be aimed at providing ecological tests for this and other hypotheses about the origin and maintenance of adaptations in *Leonardoxa* and their associated ants.

Acknowledgements

Field work was carried out with the co-operation of the Ministry of Higher Education and Scientific Research of the Republic of Cameroon, and was made possible by financial assistance from the Zoological Institute of the University of Basel (Switzerland), the Swiss Society for the Study of Nature, the University of Miami, and the National Geographic Society. Funding for attending the symposium was provided by the organizers and by the Department of Biology, University of Miami. I thank the Directors of the following herbaria for loans of plant specimens and/or permission to work in their institutions: Royal Botanic Gardens, Kew; Missouri Botanical Garden; United States National Herbarium; Jardin Botanique de l'Etat, Brussels; Botanisches Institut, University of Basel; Jardin Botanique et Conservatoire, Geneva; National Herbarium of Cameroon; and Botanisches Institut, University of Zürich. The study has benefited from discussions with D. Davidson, C. Horvitz, and M. Hossaert, and from the tutelage of J. Slowinski on cladistics. U. Maschwitz and colleagues supplied unpublished information about *Cladomyrma*. R. Calvo, M. Hossaert, J. Slowinski, and an anonymous reviewer provided useful comments on the manuscript. C. Huxley is also thanked for her comments and for her great patience. This study is Contribution No. 359 of the Program in Tropical Biology, Ecology, and Behavior, University of Miami.

22

Phylogenetic analysis of pseudomyrmecine ants associated with domatia-bearing plants

Philip S. Ward

Introduction

The close relationship between domatia-bearing plants and certain ants has long held the attention of naturalists and raised questions about the nature and origin of such interactions (Belt 1874; Müller 1876; Ule 1906; Spruce 1908; Bequaert 1922; Wheeler 1942; recent reviews in Benson 1985; Huxley 1986; Jolivet 1986). Tropical arboreal ants of the subfamily Pseudomyrmecinae are especially prone to involvement in intimate associations with plants, and their behaviour in defence of the plant is often quite apparent, at least to sting-vulnerable observers. Indeed, studies on pseudomyrmecines inhabiting swollen-thorn acacias in Central America and *Barteria* (Passifloraceae) in Africa have provided some of the best evidence of the mutualistic nature of such ant–plant interactions (Janzen 1966, 1972).

Despite these ecological studies, we remain rather ignorant about the origin and evolution of associations between ants and domatia-bearing plants. This may be partly because we have focused too narrowly on a few specific cases, and ignored the wider ecological and historical context in which they are enmeshed. There is increasing recognition of the need for broad based comparative studies of ant–plant mutualisms (Davidson *et al.*, 1989; McKey 1989; Longino Chapter 19, this volume; Davidson and Fisher Chapter 20, this volume); yet the cloud of taxonomic uncertainty hanging over the ant–plant interactants continues to impede research on a large geographical and phylogenetic scale.

In this chapter I adopt a taxonomic–historical approach to the problem of pseudomyrmecine–plant interactions, by probing the phylogenetic relationships of the obligate plant-ants and their non-specialist relatives. The resulting inferences about phylogeny provide a framework upon which the known ant–plant associations and other comparative natural history can be mapped. This admittedly myrmecocentric approach permits us to consider such questions as:

1. How many times have domatia-inhabiting ants evolved in the Pseudomyrmecinae?

2. Have obligate plant-ants expanded or altered their range of plant associations over evolutionary time?
3. Is there any correspondence between the phylogenetic histories of the ants and their plant partners?
4. What are the pre-conditions for, and consequences of, the evolution of obligate associations with plants?

The Pseudomyrmecinae are a diverse group of ants, containing an estimated 250–300 species, of which a considerable number are poorly known or undescribed. The results of the phylogenetic analyses presented here must be considered tentative and coarse-scaled. Details of relationships within certain difficult sections (e.g. the *Pseudomyrmex sericeus* group) cannot be clarified at this stage. Nevertheless, the results do reveal certain general patterns which are unlikely to be altered by changes in taxonomic detail.

Phylogenetic relationships among the Pseudomyrmecinae

Ongoing systematic research on the Pseudomyrmecinae (Ward 1989*b*, 1990) supports the recognition of three genera: *Pseudomyrmex* (New World; 150–200 species), *Tetraponera* (Palaeotropical; approximately 100 species), and *Myrcidris* (Neotropical, monotypic). I also recognize several major species groups within each of the two principal genera. The African 'satellite' genera *Viticicola* (for *V. tessmanni*) and *Pachysima* (for *P. aethiops* and *P. latifrons*) have recently been formally put into synonymy with the genus *Tetraponera* (Ward 1990). For a first assessment of phylogenetic relationships 50 pseudomyrmecine species were chosen, representing all three genera and all major species groups, and a systematic survey of character variation among these taxa was undertaken. This resulted in a set of 125 morphological characters (both binary and discrete multi-state), of which 66 are male-based and 59 are derived from queen and worker morphology. The characters involve aspects of shape, size, colour, sculpture, and pilosity, of various structural features including male genitalia; biological attributes of the ants (e.g. plant associations) were specifically excluded. The full data matrix (50 taxa by 125 characters) is given in Appendix 22. 1. Undescribed species of pseudomyrmecines are referred to by code numbers (e.g. *Pseudomyrmex* sp. PSW-02).

David Swofford's (1985) PAUP program (version 2.4.1) was employed to search for the most parsimonious tree(s) consistent with this data set, using the options MULPARS, SWAP = GLOBAL, and three different ADDSEQ conditions. *Myrcidris* was used as an outgroup to root the tree since recent studies (Ward 1990) indicate that *Pseudomyrmex* and *Tetraponera* are more closely related to one another than to this third genus. When the data set produced a large number of equally parsimonious trees, a strict consensus tree (Rohlf 1982) of all such minimum-length trees was calculated.

The result of these calculations (Fig. 22. 1) depicts the inferred phylogenetic relationships between the 50 representative species of Pseudomyrmecinae. This strict consensus tree shows only the groupings common to all of the 154 equally parsimonious trees (length 543, consistency index 0.343) generated by PAUP. Despite the lack of resolution in parts of the tree, there is enough overall structure to indicate that obligate plant-ants have arisen a number of times throughout the subfamily and are not confined to one or two lineages.

Nesting behaviour and plant associations of pseudomyrmecine ants

Most species of pseudomyrmecine ants nest in dead stems of woody and herbaceous plants, usually in cavities which have been at least partially excavated by other insects. Although they may exhibit habitat preferences, these species are nest-site generalists, inhabiting dead hollow twigs and branches opportunistically. A minority of pseudomyrmecines live only in specialized live plant cavities (domatia), which are intrinsically hollow or which can be easily excavated by the ants. Of the approximately 230 species of Pseudomyrmecinae (described and undescribed) known to me, at least 37 (16 per cent) appear to be obligate plant-ants (Table 22. 1). Undoubtedly, additional species of pseudomyrmecines have yet to be discovered so that the total number of species probably lies in the region of 250–300; but the percentage of known plant specialists is not likely to change markedly. Before focusing on these obligate plant-ants, however, it is worth noting that some generalist pseudomyrmecines will also nest facultatively in domatia (Table 22. 1) or in insect-bored cavities in *live* parts of non-myrmecophytes (Table 22. 2). Some of the latter cases may provide insight into the origins of more intimate ant–plant associations. In fact, where details of natural history are sparse, it may be difficult to determine the status of the plant and its relationship to the ants. More information is needed, for example, on the interaction between *Stereospermum* (Bignoniaceae) and *Tetraponera penzigi* in Africa (Penzig 1894), and on the association between *Acacia caven* and various ants, in South America (Wheeler 1942).

Information on plant associates and other relevant natural history was compiled from the primary literature, museum records, colleagues, and the results of my own field work in the neotropics, Madagascar, South-east Asia, and Australia. The known obligate plant associations have been plotted on the terminal branches of the cladogram (Fig. 22. 1).

A review of the pseudomyrmecine plant-ants on a case by case basis reveals some interesting patterns and permits us to estimate the minimum number of times that domatia-inhabiting ants have originated in this subfamily. Starting at the top of the tree (Fig. 22. 1), and referring to the

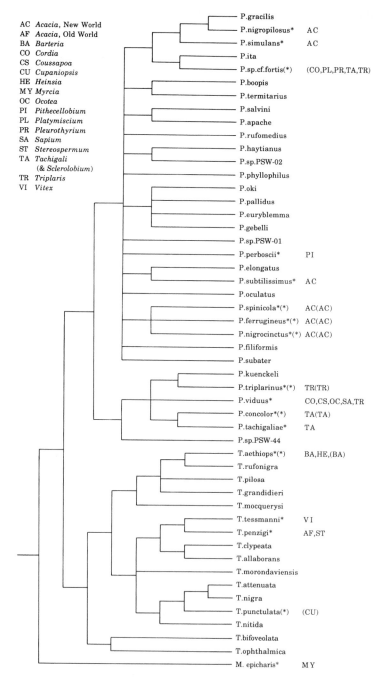

AC *Acacia*, New World
AF *Acacia*, Old World
BA *Barteria*
CO *Cordia*
CS *Coussapoa*
CU *Cupaniopsis*
HE *Heinsia*
MY *Myrcia*
OC *Ocotea*
PI *Pithecellobium*
PL *Platymiscium*
PR *Pleurothyrium*
SA *Sapium*
ST *Stereospermum*
TA *Tachigali*
    (& *Sclerolobium*)
TR *Triplaris*
VI *Vitex*

P.gracilis
P.nigropilosus*   AC
P.simulans*   AC
P.ita
P.sp.cf.fortis(*)   (CO,PL,PR,TA,TR)
P.boopis
P.termitarius
P.salvini
P.apache
P.rufomedius
P.haytianus
P.sp.PSW-02
P.phyllophilus
P.oki
P.pallidus
P.euryblemma
P.gebelli
P.sp.PSW-01
P.perboscii*   PI
P.elongatus
P.subtilissimus*   AC
P.oculatus
P.spinicola*(*)   AC(AC)
P.ferrugineus*(*)   AC(AC)
P.nigrocinctus*(*)   AC(AC)
P.filiformis
P.subater
P.kuenckeli
P.triplarinus*(*)   TR(TR)
P.viduus*   CO,CS,OC,SA,TR
P.concolor*(*)   TA(TA)
P.tachigaliae*   TA
P.sp.PSW-44
T.aethiops*(*)   BA,HE,(BA)
T.rufonigra
T.pilosa
T.grandidieri
T.mocquerysi
T.tessmanni*   VI
T.penzigi*   AF,ST
T.clypeata
T.allaborans
T.morondaviensis
T.attenuata
T.nigra
T.punctulata(*)   (CU)
T.nitida
T.bifoveolata
T.ophthalmica
M. epicharis*   MY

Fig. 22.1. Estimated phylogenetic relationships between 50 representative species of pseudomyrmecine ants. This is a strict consensus tree of 154 equally parsimonious trees generated by PAUP. (Data set given in Appendix 22.1.) Obligate plant-ants are designated by an asterisk, and their plant associates are given. Information in parentheses refers to other (one or more) closely related species of obligate plant-ants not included in the cladistic analysis.

information in Table 22.1, we find the following. In the *Pseudomyrmex gracilis* group (Kempf 1958; Ward 1989*b*), represented in the cladistic analysis by *P. gracilis*, *P. nigropilosus*, and *P. simulans*, there are two obligate plant-ants (*P. nigropilosus*, *P. simulans*), both associated with swollen-thorn acacias of Central America. Since these two are not sister species and since most of the remaining members of the *P. gracilis* group (20 or more species) are generalist twig-nesters, *P. nigropilosus* and *P. simulans* probably represent two independent origins of specialization on acacia. Moreover, *P. nigropilosus* is a non-protective 'parasite' of the ant–acacia system (Janzen 1975).

Next comes the taxonomically difficult *Pseudomyrmex sericeus* group, represented on the cladogram by *P. ita* and *P.*sp.cf.*fortis*. Although the *P. sericeus* group as a whole is distinctive, and almost certainly monophyletic (diagnosis in Ward 1989*b*), the relationships and species limits within the group are quite unclear. As a result it is difficult to hazard a guess about how many times plant-ants have arisen. There appear to be eight or more species of obligate plant-ants in this group (out of about 15 species in total), showing a tendency to be associated with *Tachigali* and *Sclerolobium* (Caesalpiniaceae), but also recorded from domatia of *Cordia* (Boraginaceae), *Triplaris* (Polygonaceae), *Pleurothyrium* (Lauraceae), and *Platymiscium* (Papilionaceae). At least two species (*P. pictus*, *P. rubiginosus*) occupy more than one plant genus. This looks like a group in which plant switching and secondary colonization of ant-plants have occurred several times. Most, if not all, of the species are aggressive plant defenders, however, and do not appear to be opportunistic parasites like *P. nigropilosus*.

Pseudomyrmex perboscii is a taxonomically isolated and distinctive species, which (at least in northern Colombia and Venezuela) occupies live branches of the mimosoid legume, *Pithecellobium saman* (Benth.) Lyons (= *Albizia saman*) (Mimosaceae) (Ward 1989*b*). Since the ant ranges from Mexico to Bolivia and Brazil, it probably inhabits other ant-plants. Ducke's (1925) record of a *Pseudomyrmex* sp. from *Inga cinnamomea* Spruce, ex Benth. (Mimosaceae) might refer to this species.

Pseudomyrmex subtilissimus is a small timid species, known only from a few collections on swollen-thorn acacias, where it is apparently able to co-exist with aggressive acacia-ants (*P. ferrugineus* group). The latter group, represented on the cladogram by *P. spinicola*, *P. ferrugineus*, and *P. nigrocinctus*, is well defined and undoubtedly monophyletic. Its distribution is co-extensive with that of the swollen-thorn acacias (Mexico to northern Colombia). A brief diagnosis of the group and of the commoner species is given in Ward (1989*b*). Janzen's (1967) detailed experimental study of the ant–acacia interaction (using *P. ferrugineus* and *Acacia cornigera*) demonstrated that the ants provide essential protection against herbivores and competing plants. In many parts of Central America several species of acacia and acacia-ant co-occur, with no specificity of pairing. Moreover, the

Table 22.1. Pseudomyrmecine ants recorded from live plant domatia. Principal sources: Bequaert (1922), Wheeler (1942), Janzen (1966, 1972), Hocking (1970), Benson (1985), Huxley (1986), McKey (1990), personal observations. Plants associated with *Pseudomyrmex* and *Myrcidris* are New World taxa; those associated with *Tetraponera* are from the Ethiopian (ET) and Australian (AU) regions. st = swollen-thorn *Acacia* of Central America and northern South America.

Genus/species group	Species	Plant(s)	Obligate plant-ant?	Coccoidea in cavities?
Pseudomyrmex *ferrugineus* group	*ferrugineus* and 8 other spp.	*Acacia* (st)	Y	N
P. gracilis group	*nigropilosus*	*Acacia* (st)	Y	N
	simulans	*Acacia* (st)	Y	N
	gracilis complex (2+ spp.)	*Acacia* (st)	N	N
	gracilis complex (1 sp.?)	*Cordia*, *Triplaris*	N	Y
	sp. PSW-35	*Tachigali*	?	Y
P. oculatus group	*elongatus*	*Cecropia*, *Cordia*, *Pithecellobium*, and *Triplaris*	N	?
P. pallens group	*urbanus*	*Cordia*	N	Y
	sp. PSW-03	*Cecropia*, *Inga*	?	?

Group	Species	Plant association		
P. sericeus group	ita	Acacia (st)	N	N
	sericeus complex (2+ spp.)	Triplaris	?	Y
	sericeus complex (2+ spp.)	Cordia	Y	Y
	sericeus complex (4+ spp.)	Tachigali, Sclerolobium, and Platymiscium	Y	Y
	sericeus complex (1 sp.)	Pterocarpus	?	?
	pictus	Tachigali, Platymiscium	Y	Y
	rubiginosus	Triplaris, Pleurothyrium	Y	Y
P. subtilissimus group	subtilissimus	Acacia (st)	Y	N
P. tenuis group	boopis	Cordia	N	?
	tenuis	Patima	N	N
P. viduus group	viduus	Triplaris, Sapium, Coussapoa, Cordia, and Ocotea	Y	Y
	triplarinus complex (4+ spp.)	Triplaris	Y	Y
	concolor complex (5+ spp.)	Tachigali	Y	Y
	kuenckeli	Acacia (st)	N	N
Pseudomyrmex	perboscii	Pithecellobium	Y	Y
Tetraponera	aethiops, latifrons	Barteria, Heinsia (ET)	Y	Y
	tessmanni	Vitex (ET)	Y	Y
	penzigi	Acacia, Stereospermum (ET)	Y[a]	Y[b]
	sp. PSW-77	Cupaniopsis (AU)	Y	Y
Myrcidris	epicharis	Myrcia	Y	Y

[a] Possibly facultative.
[b] In *Stereospermum*; coccoid association in *Acacia* less certain.

Table 22.2. Pseudomyrmecine species recorded nesting in live stems or thorns, apparently excavated by other insects. Most, if not all, of these species are generalist twig-nesters.

Species		Plant(s)	Coccoidea in cavities?	Source
Pseudomyrmex	*acanthobius*	*Acacia caven*	?	Wheeler (1942)
	boopis	*Lantana*	N	Ward (unpublished work)
	gracilis (*s.l.*)	*Acacia caven*	?	Museum specimens
	tenuis	*Witheringia*	N	Longino (personal communication)
	urbanus	*Tecoma*	Y	Santschi (1936), Ward (1989*b*)
	sp. cf. *fortis*	*Avicennia germinans*	Y	Ward (unpublished work)
Tetraponera	*grandidieri*	*Ixora, Leea, Rhus*	N	Ward (unpublished work)
	hysterica	*Albizia*	N	Ward (unpublished work)
	punctulata	*Eucalyptus, Avicennia*	Y	Ward (unpublished work)
	rufonigra	*Sonneratia*	N	Ward (unpublished work)

swollen-thorn acacias apparently represent several independent lines of evolution towards a myrmecophytic habit (Janzen 1966, 1974a). Thus, here there is not expected to be any clear pattern of co-speciation between the ants and plants.

The *Pseudomyrmex viduus* group (previously known as the *P. latinodus* group; see Ward 1989b) contains an intriguing set of species, of which five representative taxa were chosen for cladistic analysis (*P. kuenckeli*, *P. triplarinus*, *P. concolor*, *P. tachigaliae*, and *P. viduus*). Of these, *P. kuenckeli* is not an obligate plant-ant, and the cladogram suggests that it has secondarily reverted to more generalized nesting habits. If confirmed by an in-depth phylogenetic analysis of the *P. viduus* group this would represent the only known case of such a reversal. The remaining members (10 or more species) of the *P. viduus* group are all obligate plant-ants, with the *P. concolor* complex specialized on *Tachigali* (Caesalpiniaceae) and the *P. triplarinus* complex confined to *Triplaris* (Polygonaceae). Perhaps most interesting of all is *P. viduus*, an obligate plant-ant whose recorded hosts include plants from five genera in as many families: *Cordia* (Boraginaceae), *Coussapoa* (Moraceae), *Ocotea* (Lauraceae), *Sapium* (Euphorbiaceae), and *Triplaris* (Polygonaceae). It is unclear whether such flexibility is ancestral or derived. In the majority of cladograms produced by PAUP, *P. viduus* is positioned as an outgroup to the remaining members of its species group. This would suggest that association with *Triplaris* (if not additional plants) is part of the ground plan of the *P. viduus* group, with one lineage subsequently remaining on *Triplaris*, and another switching to *Tachigali*.

In the Old World genus *Tetraponera*, *T. aethiops* and its sister species, *T. latifrons*, inhabit and protect small trees in the genus *Barteria* (Passifloraceae) in equatorial West Africa (Janzen 1972). *Tetraponera aethiops* has also been recorded from live stems of the rubiaceous plant, *Heinsia myrmoecia* (Stitz 1910; Schnell and Grout de Beaufort 1966).

Two other Afrotropical species, *Tetraponera tessmanni* (associated with *Vitex*) and *T. penzigi* (associated predominantly with whistling-thorn *Acacia*), appear as sister groups in the cladistic analysis; however they probably represent independent origins of the domatia-inhabiting trait, since there are a number of species more closely related to *T. penzigi* (e.g. *T. liengmei*, *T. gerdae*) which are apparently generalist twig-nesters. Even the status of *T. penzigi* as an obligate plant-ant is uncertain (Bequaert 1922). It is not the most common ant in the whistling-thorn acacias, that role being taken by *Crematogaster* (Hocking 1970), and it has been recorded from live stems of an unrelated plant, *Stereospermum dentatum* (Bignoniaceae) (Penzig 1894).

An undescribed species of *Tetraponera* (*T.*sp.PSW-77), allied to *T. punctulata*, occupies live terminal branches of *Cupaniopsis anacardioides* (Sapindaceae) in south-east Queensland (Ward, unpublished work). The plant has a much wider geographical range than the ant, indicating that this represents an ant mutualism in the early stages of development (see further discussion below).

Finally, *Myrcidris* contains a single described species *M. epicharis* (Ward 1990), discovered by W. W. Benson, which lives in swollen terminal stems of *Myrcia* sp. (Myrtaceae), an Amazonian understorey tree. This curious ant, possesses some morphological features apparently ancestral to all other pseudomyrmecines and, at the same time, a number of unique and strongly derived (autapomorphous) traits, suggesting that the mutualism may be quite old.

In summary, obligate domatia-inhabiting ants appear to have evolved at least 12 times in the Pseudomyrmecinae. Three groups of plant-ants, all from the New World, have undergone considerable diversification: the *P. ferrugineus* group, on swollen-thorn acacias; the *P. viduus* group, principally on *Triplaris* and *Tachigali*; and a section of the *P. sericeus* group, on *Cordia*, *Tachigali*, *Triplaris*, and other plants. None of these three groups is particularly closely related to the other (Fig. 22. 1). The remaining pseudomyrmecine plant-ants occur in small, independent groups of one or two species. From the plant point of view, the majority of pseudomyrmecines associated with *Acacia*, *Tachigali*, and *Triplaris* form monophyletic groups (the *P. ferrugineus* group, *P. concolor* complex, and *P. triplarinus* complex, respectively), while the remaining species in each assemblage are (with the exception of *P. viduus*) not closely related to these three majority groups, and presumably represent secondary colonizers. In the case of *Tachigali*, the 'secondary group' comprises several species in the *P. sericeus* group, whose diversity and geograpical range suggest that they may have been involved with the *Tachigali* mutualism since its early stages.

These patterns emerge from a pairing of the data in Table 22. 1 with the results of a cladistic analysis of 50 representative taxa (Fig. 22. 1). Given the preliminary nature of this analysis, it is worth inquiring whether the observed patterns continue to hold up with other, slightly less parsimonious trees. I examined about 440 trees of slightly greater length (544 to 549 steps) than the set of most parsimonious trees (543 steps), by sub-optimizing PAUP's performance (using the SWAP and ADDSEQ options). Although differing in details, these trees showed substantial agreement with the set of most parsimonious trees with respect to the major groupings shown in Fig. 22. 1. Perhaps the most noteworthy difference was that the *P. viduus* group appeared as paraphyletic or polyphyletic in a minority of cases. None of the slightly less parsimonious trees altered the major conclusions of:

1. multiple origination of plant-ants;
2. monophyly of the majority of species associated with each of the major ant-plants;
3. the general lack of a close relationship between these majority groups and the remaining group of ants found on the same plants.

The observation of considerable host-plant diversity within some groups of plant-ants also remains valid; and of course the same observation continues to apply to certain species (such as *P. viduus*), regardless of the outcome of phylogenetic inference.

Discussion

Patterns of association

These findings confirm the view that pseudomyrmecine ants have repeatedly developed obligate associations with domatia-bearing plants. The number of origins is not likely to be a simple function of the number of times that the myrmecophytes have evolved, however. This is because there is considerable evidence of plant switching and secondary colonization of pre-existing ant-plants. This is suggested by the occurrence of two or more pseudomyrmecine lineages on *Acacia*, *Triplaris*, and *Tachigali*; by the diversity of plants associated with some groups of plant-ants (*P. sericeus* group, *P. viduus* group); and by the existence of *intraspecific* variation in host-plant choice. Thus, about one-quarter of the 37 obligate plant-ants listed in Table 22. 1 occupy more than one plant genus, sometimes (e.g. *P. viduus*, *P. rubiginosus*, *T. aethiops*, *T. penzigi*) in widely different plant families. These observations belie the notion of a close correspondence between ant and plant phylogenies. This is true even for the close association between acacia-ants (*P. ferrugineus* group) and their plants, ostensibly a classic example of co-evolution in the strict sense, and yet one which is diffuse enough to preclude a clear pattern of co-speciation (Janzen 1966, 1974*a*). Diffuse co-evolution is even more likely for the pseudomyrmecine–*Tachigali* system, and that involving *Triplaris*, in which there has evidently been a period of co-participation by the *P. viduus* and *P. sericeus* groups. In some regions of South America ants of the genus *Azteca* are also part of the contemporary guilds on *Triplaris* and *Tachigali*.

Of course, where the present-day assemblage of ants associated with a particular myrmecophyte is taxonomically diverse, it may be difficult (if not impossible) to ascertain with certainty which lineage(s) was involved in the original development of the interaction. In the case of New World *Cordia* and African *Acacia*, ants other than pseudomyrmecines (*Azteca* and *Crematogaster*, respectively) are the predominant inhabitants; but whether the pseudomyrmecines are secondary arrivals or early associates which have been competitively pushed aside cannot be easily determined. Some evidence could be adduced from the geographical ranges of the ants and plants, the morphological fit between ants and domatia, and the behaviour of colony-founding queens; but without a much more thorough knowledge of the historical biogeography of the ant and plant the evidence would remain circumstantial in nature.

Origins of pseudomyrmecine–plant relationships

What factors have predisposed pseudomyrmecines to become obligate plant-ants? Janzen (1966) lists a number of potentially important traits of which arboreal nesting, a tendency to glean foliage for small pieces of organic matter, and the possession of a well-developed sting, seem particularly significant. The pseudomyrmecine sting sheath and lancets are both barbed (the latter are not smooth, as claimed by Janzen (1966)). Although most generalist twig-nesters sting reluctantly and only in defence, a few Old World species can be aggressive (e.g. *T. rufonigra*, *T. difficilis*, *T. morondaviensis*).

A closer look at the nesting habits of pseudomyrmecines provides further insight into the evolution of specialist plant-ants. As noted previously, a number of generalist pseudomyrmecines have been recorded nesting opportunistically in plant domatia (Table 22. 1), illustrating the potential for secondary colonization. A clue to the primary origin of ant–myrmecophyte associations lies in the tendency of some pseudomyrmecine ants to occupy the *live* stems or thorns of unspecialized plants, in cavities excavated by beetle or lepidopteran larvae. Examples are given in Table 22. 2; these involve 10 species of Pseudomyrmecinae, most of which are known to nest in dead stems of woody plants also. This phenomenon has probably been under reported, both for pseudomyrmecines and for other ants (J. T. Longino, personal communication). Of particular interest are two cases (involving *Pseudomyrmex* sp.cf. *fortis* and *Tetraponera punctulata*) in which the ants have been observed tending scales (Coccidae) inside the insect-bored cavities. Colonies of *Pseudomyrmex* sp.cf.*fortis*, which occupy live, excavated branches of *Avicennia* in northern Colombia and tend scales of the genus *Cryptostigma*, have more aggressive workers than those of a sympatric relative, *P. ita*, nesting in dead twigs. A member of the *P. sericeus* complex, *P.* sp.cf.*fortis*, closely resembles species which have been collected in *Cordia* and *Triplaris*. The Australian species, *Tetraponera punctulata*, typically occupies dead branches but it will occasionally nest in live stems of *Avicennia* and *Eucalyptus* with coccids (?*Coccus* sp., and an undescribed species in, or near, *Cryptostigma*). *Tetraponera punctulata* is closely related to the obligate plant-ant, *T.*sp.PSW-77, which keeps *Cryptostigma* in hollow stems of *Cupaniopsis anacardioides*. A related phenomenon has been reported from Malaysia: an undescribed species of *Tetraponera* in the *T. attenuata* complex (*T.*sp.PSW-80) inhabits live hollow internodes of a bamboo, *Gigantochloa ligulata* (Poaceae), and maintains scale insects in the nest (D. Kovac, personal communication).

Nearly all pseudomyrmecines which are obligate plant-ants tend scale insects (principally Coccidae and Pseudococcidae) in their domatia (see Table 22. 1). The only exceptions are, (i) species living in swollen-thorn acacias of Central America, where alternative trophic rewards are available in the form of Beltian bodies and extrafloral nectar, and (ii) *T. tessmanni*, a

Vitex (Verbenaceae) inhabitant, for which there is an indication that workers obtain nutrients directly from plant tissue (Bailey 1922*b*). There is evidence that ants which keep scales in their domatia not only harvest the honeydew but also periodically consume the scales themselves, thus obtaining both carbohydrates and protein (Bailey 1923; Wheeler 1942; Janzen 1972; Carroll and Janzen 1973; Schremmer 1984).

It seems probable that scale insects (Coccoidea), especially those prone to reside in protected crevices or cavities, have been crucial to the inception of most pseudomyrmecine–plant mutualisms, because they provide the trophic resource which draws the ants into a closer relationship with a potential myrmecophyte. If ant activity does not provide a consistent benefit to the plant, the relationship may develop no further than that permitted by the presence of twig-boring insect larvae. However, if ant occupancy confers a net gain in fitness on the plant, despite the cost of maintaining scales, and if the plant possesses the necessary structural features, it may be able to encourage ant inhabitation by developing, or elaborating upon, pithy swellings (McKey 1989). This would allow the ants to escape dependency on twig-boring insects, and (by pre-emption) lower the frequency of the latter. Such a dependency on the activity of insect larvae which bore live or dead stems is, incidentally, very common among generalist pseudomyrmecines and other twig-nesting ants. It is conceivable that the removal of this dependency in the early stages of a pseudomyrmecine–plant relationship does not even require special development in the plant. Consider the *Tetraponera–Cupaniopsis–Cryptostigma* system mentioned previously. The ant, whose workers and queens excavate the pithy terminal stems in which brood and coccids are kept, appears restricted to a limited area in south-east Queensland, yet the plants here are not manifestly different in stem morphology from those found without ants throughout the plant's wide range in eastern and northern Australia. It is not at all obvious that the plant is 'in control' and using homopterans to lure the ants into a close partnership (cf. Janzen 1979; Beattie 1985). Rather, we may be observing the consequences of selection on ants, particularly colony-founding queens, to circumvent reliance on stem-borers while establishing nest sites suitable for coccids. Of course, such a system is ripe for mutually beneficial elaboration by both the ant and plant, and for exploitation by other species of ants and plants in the same community. The association between *Pseudomyrmex perboscii* and *Pithecellobium saman* may represent a slightly later stage in this process, involving additional species of plants in parts of the ant's range.

The evolutionary scenario developed above is applicable to other ants which keep scales in domatia. In New Guinea rainforests, for example, where myrmecophytism appears to be relatively recent, ants of the genera *Iridomyrmex*, *Crematogaster*, and *Podomyrma* can be found in association with *Cryptostigma* scales, both in understorey myrmecophytes, such as *Myristica* (Myristicaceae), and in insect-bored twigs of unspecialized plants (P. S.

Ward, unpublished work). J. T. Longino (personal communication) has made similar observations on ant–coccoid associations involving the New World ant genera *Azteca*, *Myrmelachista*, and *Zacryptocerus*, and has reached similar conclusions about their importance in the development of ant–plant relationships. Benson (1985) suggested a role for homopterans (scales and membracids) in the evolution of so-called secondary domatia (leaf pouches). The argument put forward here is that Coccoidea have been equally important in the origin of many primary domatia (fistulose stems, swollen petioles, etc.), and that the process involves interactions between four classes of participants: ants, plants, scales, and stem-borers (Fig. 22. 2). A major unknown factor is the specificity of the scale insects to particular plants or ants.

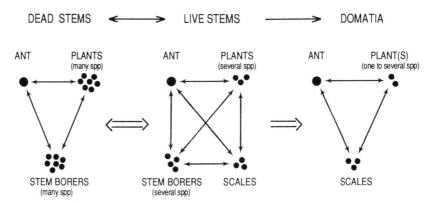

Fig. 22.2. Postulated transition in nesting behaviour of pseudomyrmecine ants, from occupancy of dead twigs to inhabitation of domatia, illustrating the involvement of scale insects (Coccoidea) and stem-boring insects. The process is depicted for a single species or lineage of ants.

Characteristics of obligate plant-ants

Finally, what are the morphological and behavioural consequences of ant–plant specialization in the Pseudomyrmecinae? Domatia-inhabiting species tend to display a series of morphological traits including small eyes, reduced palpal segmentation, short, broad petiolar segments, and hypertrophied metapleural glands (the last two features are not exhibited by the acacia-ants). The significance of some of these traits is obscure; but the enlarged metapleural glands probably reflect the increased importance of antibiotic control of pathogens in the nest environment of plant domatia. This accords with reports (e.g. Bailey 1922; Wheeler 1942) of accumulations of debris, faecal matter, and fungi in the domatia of some rainforest plants; and it is consistent with the absence of notably enlarged metapleural glands in acacia-ants, whose nest conditions appear to be cleaner and drier.

Behavioural convergences of pseudomyrmecine plant-ants include extended foraging activity on the plant surface, increased aggressiveness, and a tendency to prune alien vegetation encroaching on the ant-plant (Janzen 1966, 1972; Davidson *et al.* 1988). Polygyny and large colony size have been cited as specialized traits of acacia-ants (Janzen 1966), but they are neither peculiar to, nor characteristic of, pseudomyrmecine plant-ants, also being found in some generalist twig-nesting species (Ward 1989a, unpublished work), although not in the extreme form shown by one or two species of acacia-ants.

For the most part pseudomyrmecines which have become ant-plant specialists appear locked into this life history mode. One possible exception is *P. kuenckeli* which may be derived from a *Triplaris*-inhabiting lineage (see above), but this needs confirmation. While such reversals are rare or absent, some plant-ants retain considerable flexibility by inhabiting a range of myrmecophytes. Studies on these species, particularly on geographical variation in plant use and on competitive interactions with other ants in the same myrmecophyte guilds, promise to provide insight into the factors influencing specialization in plant-ants.

Conclusions

Phylogenetic analysis of the Pseudomyrmecinae indicates that obligate plant-ants, which comprise almost 20 per cent of the estimated 250–300 extant species in the subfamily, have arisen at least twelve times. The lack of species-specific pairing, the existence of two or more ant clades on the major ant-plants (*Acacia*, *Triplaris*, and *Tachigali*), and the diversity of ant-plants associated with some groups (and even single species) of ants, all point to a pattern of 'diffuse co-evolution' in pseudomyrmecine–plant associations. There is evidence of both secondary colonization of ant-plants by twig-nesting generalists and of plant switching by obligate plant-ants. The latter process may involve the adoption of new host plants in quite different families, judging by the intraspecific variation in plant use exhibited by some ants. In other words, both ants and plants have been able to exploit pre-existing mutualisms.

The arboreal nesting habits and foraging behaviour of pseudomyrmecines have pre-adapted them to develop close relationships with plants. In many cases their propensity to tend and defend scales (Coccoidea) has probably aided the transition from occasional inhabitation of live, insect-bored stems to more specialized interactions with plants. Studies of incipient ant–myrmecophyte associations (e.g. that involving *Tetraponera* and *Cupaniopsis*) and of obligate plant-ants, such as *Pseudomyrmex viduus*, which maintain a considerable latitude of host plant choice, should prove to be particularly illuminating of the selective forces propelling or arresting such ant–plant mutualisms.

Appendix 22.1

	1	11	21	31	41	51
P.gracilis	0101000141	0200000000	1000100001	1002002112	0310000012	1001010200
P.nigropilosus	0101000141	0200000000	1000100001	1002002111	0310000012	0001013200
P.simulans	0101000141	0200000000	1000100001	1000002111	0310000012	0001010000
P.ita	0101000151	0200000000	1000100001	1010001101	0320000003	0012020000
P.sp.cf.fortis	0101000151	0210000000	1000100001	1010001111	0310000003	0012020000
P.boopis	0101110131	0200000000	1000100001	1010002112	0310000012	0011012000
P.termitarius	0101110131	0200000000	1000100001	1010002112	0310000012	0001012000
P.salvini	0101110131	0200000000	1000100001	1010002111	0310000012	0001013300
P.rufomedius	0101110131	0200000000	1000100001	1000002012	0200200112	0001013001
P.apache	0100110131	0200000000	1000100001	1010001111	0210000012	0001013300
P.haytianus	2100110131	0201000000	1100100001	1010001111	0310000002	0001011200
P.phyllophilus	0101110121	0200000000	1000100001	1000001111	0310000012	0001012000
P.sp.PSW-02	2100110131	0221000000	1100100001	1000002112	0310000012	0001010000
P.oki	1100110121	0211000000	1100100001	1000001101	0310000003	1001012201
P.sp.PSW-01	1100110131	0211000001	1100110001	1010001101	0320000003	1001010200
P.pallidus	1100110121	0211000000	1100100001	1010001111	0310000002	0001013301
P.euryblemma	2100110121	0211000000	1100100001	1010001111	0310000003	1001012301
P.perboscii	1101110231	0200000000	1000100001	1010002101	0210010013	1001013400
P.gebelli	1101110121	0211000000	1100100001	1010002112	0310000012	0001013401
P.kuenckeli	0101110131	0200000000	1000000001	1002002013	0201210111	1013012400
P.triplarinus	1101110131	0200000100	1000000001	1002002012	0201210102	0012012500
P.viduus	1101110131	0211000000	1000100001	1002001011	0210110003	0012012500
P.concolor	1101110121	0221000000	1000000001	1002001011	0201210002	0012023500
P.tachigaliae	1111010141	0211000610	1100000001	1000001011	0200100002	0012012500
P.spinicola	0101110131	0221000000	1100000001	1000001121	0210010012	0001012200
P.ferrugineus	0101010131	0211000000	1000100001	1000001111	0210100002	0001012000
P.nigrocinctus	0101110131	0211000000	1000100001	1000001121	0210100002	0001013200
P.filiformis	1100110131	0210000000	1000100001	1000001111	0310010002	1001013400
P.subater	1100110121	0211000000	1000000001	1000001111	0310000002	0011012500
P.sp.PSW-44	1100110121	0211000000	1000100001	1002001011	0210110012	0012012500
P.elongatus	1001110131	0201000000	1000100001	1000000111	0320000003	1012012000
P.oculatus	0100110121	0201000000	1000100001	1001001111	0310000012	1012012000
P.subtilissimus	0101110121	0201000000	1000110001	1010000110	0320000003	2001013000
T.aethiops	2010291112	1000110200	0001000000	0010003012	1102220211	0002120300
T.mocquerysi	1021991103	1000140000	0000000101	0001102011	1211221112	0001010100
T.pilosa	2010291112	1000100200	0001000000	0010002011	1211121112	0001010100
T.rufonigra	0109111023	1000100210	0001000000	0002003011	1101221211	0001103200
T.grandidieri	0019291012	1000130000	0000000001	0010002021	1101222211	0101003110
T.bifoveolata	1119191013	1200100000	0000000101	1000011011	1212221002	0001013100
T.ophthalmica	0109191013	1200100000	0000000101	1010010021	1211221002	0000003100
T.tessmanni	1000111122	1031000010	0000000211	1010100001	1002220202	0112113400
T.penzigi	1009291012	1100100010	0000000211	0010101011	1102320102	0012110100
T.clypeata	1021991203	1100100000	0010000211	0010100011	1101221202	0100012310
T.morondaviensis	1109191013	1000101000	0000000201	0000101021	1101221202	0001012300
T.allaborans	2021991202	1100120000	0010000201	1010001011	1201221102	0100000410
T.attenuata	1010291112	1100110310	0000000201	0101001021	1101221211	0200000500
T.nigra	1010291112	1100110410	0000000101	0101001021	1101221212	0201010300
T.nitida	2000191113	1000110000	0000000200	0110001001	1211220002	0101010400
T.punctulata	1009291013	1100110510	0000000201	0100001011	1201221103	0001010500
M. epicharis*	2009290010	0211000000	0000001000	0000000001	1201220003	0011013400

```
61         71          81           91           101          111          121
1111111100 000010100   0000000000   0010000010   0021000000   0000001100   01100
1111111100 000010100   0000000000   0010000010   0021000000   0000001100   01100
1111111100 000010100   0000000000   0010000010   0021000000   0000001100   01100
1011111000 000010110   0000000000   0000000010   0021000010   0000000100   01100
1011111000 000010110   0000000000   0000000010   0021000010   0000000100   01100
1011011000 000010100   0000000000   0000000010   0021000100   0000000000   01100
1011011000 000010100   0000010000   0000000010   0021000100   0000000000   01100
1011111100 001010100   0100010000   0000000010   0021000000   0100000100   01100
1111101100 001010101   0000000000   0000000010   0021010010   0000000100   09100
1011111100 000010100   0100010000   0000000010   0021000000   0000000100   01100
1011111100 000010100   0000010000   0100000020   0021000999   9999901000   09100
1011111010 001010101   0000000000   0000000010   0021000010   0000001100   01100
1011111200 000010100   0000010000   0100000020   0021000001   0000001001   09100
1011111100 001010100   0010000000   0000000000   0021000000   0000001000   01100
1011111000 001010100   0000010000   0100000010   0021000000   0000001000   01100
1011111100 002010100   0010000000   0000000100   0021000000   0000001000   01100
1011111100 012010100   0010000000   0000000101   0021000000   0000001000   01100
1011111300 001010100   0000000100   0000000000   0021000010   0000001001   01100
1011111100 001010100   0010000000   0000000000   0021000000   0010001000   01100
1011111100 001010100   0000010000   0000000010   0021000000   0000000200   01100
1011111100 001010100   0000020000   0000000000   0011010001   0000000101   01100
1011111100 001010120   0000020000   0000000000   0021000001   0000000101   01100
1011111200 001010100   0000010000   0100000010   0021000000   0001001000   01101
1011111100 001110100   0000010000   0100000010   0021000000   0000001001   01101
1011111200 001010100   0000000000   0100000020   0021000001   0000001021   01100
1011111200 001010100   0000001000   0000000020   0021000001   0000001020   02100
1011111100 001010100   0000000000   0100000020   0021000001   0000001010   01100
1011111100 000010100   0000000000   0000000010   0021000000   0100000000   01100
1011111100 001010100   0000010000   0000000010   0021000000   0000000100   01100
1010111000 001010100   1000000001   0000000010   0021000000   0000000000   01100
1011111000 000010100   0000000010   0000000010   0021000000   0000000100   01100
1011111100 001010100   0000000010   0000000010   0021000001   0000000100   01100
1011111000 000010100   0000000010   0001000010   0021000000   1000000100   09100
1011111100 000010100   0000000000   0000001000   0011100000   0000001000   00100
1011111100 002011100   0000000000   0000000000   0021100000   0000000100   00100
1001111000 000010000   0000100000   0000010000   0021100000   0000001000   00100
1011111100 000012100   0000000000   0000111000   0012101000   0000001000   00100
1011110000 000010100   0000000000   0000000000   0021100000   0000100101   00100
1011111000 000010100   0000000000   0000001000   1022100000   0000001000   19100
1010111000 000010100   0000000000   0000001000   1022100000   0000001000   10100
1011111000 000010000   0000000000   0000011000   0022100000   0000100001   00110
1010111000 000010100   0001000000   0000001000   0022101000   0000101001   00110
1010111000 000010100   0000010000   0000001000   0122100000   0000101001   09110
1011111000 000011100   0000000000   1000011000   0012101000   0000101000   09100
1010111001 000010100   0000000000   0000001000   0122100000   0000101001   00110
1011111000 000012100   0000010000   0000111000   0022101000   0000200100   00100
1011111100 000012100   0000010000   0000111000   0012101000   0000300100   09100
1011111100 100012100   0000010000   0000111000   0022101000   0000300100   00100
1011111100 100012100   0000010000   0000111000   0022101000   0000300100   00100
0011111000 001000100   0000000000   0000000000   0001000000   0000010000   00000
```

Acknowledgements

I am grateful to Jack Longino and Woody Benson for sharing their field knowledge; to Penny Gullan for identifying coccids; to Tony Irvine and George Schatz for plant identifications; and to Jack Longino, Steve Shattuck, Susan Harrison, Diane Davidson, Penny Gullan, Pete Cranston, Barry Bolton, and Camilla Huxley for helpful discussions and comments on this paper. The work was supported by NSF grants BSR 8507865 and INT 8714076.

23

Myrmecotrophy: origins, operation, and importance

David H. Benzing

Introduction

Tropical forests harbour a poorly understood but potentially important mutualism between arboreal ants and epiphytic vascular flora. Partners are diverse and the benefits to ant and plant vary, indicating that myrmecotrophy (the feeding of plants by ants) arose repeatedly under different conditions. Plants fed by ants belong to two groups which are distinguished by the type of organ that comes into contact with useful ant products. These are:

1. ant-house epiphytes which provide domatia for zoobionts in leaf, stem, root, or rhizome cavities (Figs 23. 1–3) (Huxley 1980); and
2. the roots of ant-garden epiphytes which enter arboreal ant nests (Fig. 23. 1) (Davidson 1988).

Food and perhaps special metabolites, in addition to nest chambers or carton reinforcement, are the rewards for the ants. Myrmecotrophs, in turn, may receive protection in addition to nutrients and seed dispersal.

The overview of myrmecotrophy presented here covers: participating ant and plant taxa; the specificity of symbiotic combinations; ant and plant modifications for mutualism; benefits and liabilities associated with ant and plant use; and the importance of the relationship. Suggestions for future inquiries and several testable hypotheses are offered. Except for the few instances where the definition is broadened, application of the term 'myrmecotroph' is limited to the currently recognized ant-house and ant-garden flora.

Definitions and boundaries

Myrmecotrophy is not a discrete or precisely defined phenomenon, nor is it sharply differentiated from other ant–plant relationships. Consequently, labels have been difficult to apply and disagreements have occurred. Myrmecochores (plants dispersed by ants (perhaps 35 per cent of all herbaceous and many woody species) Beattie 1985) have not been considered to be ant-fed, but the possibility of profit to terrestrial plants from ant

Fig. 23.1. (a) *Hydnophytum formicarum* Jack illustrating the intact plant and its chambered hypocotyl. (b) A well-developed ant-garden dominated by bromeliads and gesneriads in fertile humid forest at Río Palenque, Ecuador. (c) A partially dissected ant-garden filled with bromeliad roots at Río Palenque. (d) A small carton nest supporting epiphyte seedlings in humid forest in south-western Venezuela.

waste and nest material cannot be discounted in all cases. Vegetation confined to arboreal ant-gardens has a narrower, less arguable resource base, yet Ule's original attributions (1901, 1905, 1906) were vigorously contested by another prominent authority (Wheeler 1921*a*). Longino (1986) viewed the neotropical ant-garden as simply the most conspicuous expression of a much wider use of carton (a complex maché of varied composition) by arboreal flora. Some of the same, or closely related, ant-house occupants (e.g. members of *Azteca*, *Camponotus*, *Crematogaster*) also manufacture nests in which seedlings that eventually provide domatia may be farmed (e.g. Hydnophytinae, *Iridomyrmex* in Sarawak; Janzen 1974).

The epiphytes (e.g. certain Hydnophytinae) which offer the most conspicuous and costly ant housing were presumed to be ant-fed many years ago (Beccari 1884–1886; Yapp 1902; Miehe 1911): but the trophic implications, if any, of contact between ant colonies and much additional, less specialized vegetation are still unclear (e.g. many bromeliads, a Malaysian palm; Rickson and Rickson 1986). Another set of potential myrmecotrophs are the tree myrmecophytes (e.g. *Acacia*, *Cordia*, *Plectronia*; Beattie 1989) that house often aggressive ants in swollen petioles, leaf pouches, stipules, or hollow stems. The importance of ants to plant nutrition is influenced by the properties of nest cavity walls, the housekeeping habits of the ants, and, in particular, by the availability of nutrients from more conventional sources. Utilization of ant-provided nutrients by certain *Cecropia* species is suggested by modest root development compared to non-myrmecophytic congeners (Luizão and Fortunato de Carvalho 1981). Uptake of nutrients from melastome leaf sacs is suggested by K. Jaffé (unpublished work).

One criterion is paramount: myrmecotrophs must routinely attract ants that can provide significant nutrition over part of the plant life cycle. Neither domatia or production of food for ants are necessary qualifications. Plant cavities may intercept rainfall to maintain a dependable water supply (e.g. many tank bromeliads, *Hydnophytum kajewskii* Merrill & Perry and certain other relatively unspecialized Hydnophytinae; Jebb 1985) rather than to promote feeding by ants. Extrafloral nectar is a routine enticement for ants that afford only protection to the plants.

Ant-gardens

A variety of vascular epiphytes grow on ant-made substrates in New and Old World tropical forests (Docters van Leeuwen 1929*a*,*b*; Horich 1977). Ule (1901, 1905, 1906) recognized mutualisms in South American associations, but demonstration of myrmecochory and the occurrence of a specific flora on carton required additional studies (Kleinfeldt 1978; Madison 1979; Davidson 1988). The ants belong to a variety of advanced taxa (Table 23. 1), but they all share high calorie diets (mostly from tended Homoptera) and the ability to produce nests of carton. Ant-garden plants are also diverse and

Table 23.1. Ant genera containing species that regularly create ant-gardens or inhabit domatia of myrmecotrophs or similar plants.

Genus	Plant association	Reference
Azteca *Camponotus* *Crematogaster* *Monacis* *Solenopsis*	Create carton substrates for an extensive Amazonian ant-garden flora	Davidson 1988 Wilson 1987 Weber 1943
Azteca *Crematogaster* *Hypoclinea* *Odontomachus* *Paratrechina*	Nest in hollow pseudobulbs of *Caularthron bilamellatum* (Orchidaceae) in Panama	Fisher and Zimmerman 1988
Anechotus *Azteca* *Camponotus* *Crematogaster* *Cyphomyrmex* *Hypoclinea* *Hypoponera* *Leptogenys* *Megalomyrmex* *Monomorium* *Odontomachus* *Pachycondyla* *Pheidole* *Smithistruma* *Solenopsis* *Tetramorium* *Wasmannia* *Zacryptocerus*	Nest among leaf base cavities of *Aechmea bracteata*	A. Dejean and I. Olmstead unpublished
Azteca *Camponotus* *Crematogaster* *Cryptocerus* *Dolichoderus* *Monomorium* *Myrmelachista* *Pheidole* *Solenopsis*	Nest within bulbs of *Tillandsia butzii* and *T. caput-medusae*	Benzing 1970
Iridomyrmex	Domatia provided by: *Dischidia*, Hydnophytinae, *Lecanopteris*, *Pachycentria*	Rickson 1979, Weir and Kiew 1986 Jebb 1985 Huxley 1978 Janzen 1974*c*

include ferns (Polypodiaceae) and angiosperms (Table 23. 2) which, except for myrmecochory and possession of close, less specialized, epiphytic relatives, offer no hint to explain their shared propensity for substrates created by arboreal ants. Eleven of 16 flowering plant families with > 50 epiphytic members (Kress 1986) contain taxa that regularly root in carton and (in at least four instances) produce ant-inhabited domatia. Occasional species grow on carton constructed inside the domatia of ant-house plants. For example, *Dischidia nummularia* R. Br. utilizes pouch leaves of *D. rafflesiana* Wall. and various Hydnophytinae (see Chapters 25 and 26). *Pachycentria* (Melastomataceae) so consistently occupied the tubers of *Hydnophytum formicarum* Jack at a site in Sarawak that Janzen (1974c) considered it to be an ant-house parasite.

Amazonian ant-garden ants are often parabiotic, with two or three species sharing a single nest (Weber 1943; Davidson 1988). Body size tends to be graded as large, small, and tiny in, for example, *Camponotus*, *Crematogaster*, and *Solenopsis* respectively (Weber 1943). Perhaps this effectively partitions common living space or avoids other competition, although the diminutive species may be parasitic. Several factors discussed below influence the composition of the usually mixed garden floras.

Carton, like several types of rooting media in tree crowns, can be quite nutritive for plants and much richer than subjacent earth soil (Table 23. 3) if it contains vertebrate faeces or decayed foliage. Ants which use more inert fibre produce inferior seed beds. Adding to the complexity of ant-garden substrates are honeydew and the ants' potentially microbiocidal secretions (Maschwitz and Hölldobler 1970). Myrmecochores could further enhance nest quality for ants if volatiles associated with their seeds suppressed the growth of pathogens (Davidson 1988). However, mycelia of at least one fungus (*Cladosporium myrmecophilum*) regularly permeate active carton with no obvious impact on animal or plant inhabitants.

Effects of plants on ants can be positive or negative. Rather than improving carton by reinforcement, some ant-garden plants may clog up brood chambers and slow down temperature-dependent larval development through shading (Longino 1986). Extensive rooting eventually drives large ants from nests dominated by certain robust *Anthurium* and bromeliads. The benefits vary: extrafloral secretions are common (e.g. on *Codonanthe uleana* Fritsch, *Codonanthopsis ulei* Mansf. (= *Codonanthe ulei* (Mansfield) H. E. Moore), and *Philodendron myrmecophyllum* Engler; Madison 1979); pearl bodies develop on *Ficus paraensis* Miq. Ants tend Homoptera on cultivated flora (Kleinfeldt 1978), but additional foraging occurs well beyond the garden (up to 10 m for *Camponotus femoratus* Fab. in Peru (Davidson 1988). Favoured homopteran feeding sites tend to be ephemeral; for example, the young fruit and pedicels of *Ficus* and flower buds of the garden-dwelling bucket orchid *Coryanthes*.

Table 23.2. Families of vascular plants with > 50 epiphytic species and the occurrence of ant-garden and ant-house members (numbers according to Kress 1986).

Family	Number of epiphytic species	Genera containing ant-fed species	Ant-garden species	Ant-house species
Polypodiaceae	1 023	*Lecanopteris*	−	+
		Solenopteris	−	+
Piperaceae	710	*Peperomia*	+	−
Moraceae	522	*Ficus*	+	−
Cactaceae	150	*Epiphyllum*	+	−
Marcgraviaceae	89	*Marcgravia*	+	−
Clusiaceae	85	None		
Ericaceae	672	None		
Melastomataceae	648	*Pachycentria*	+	−?
Araliaceae	78	None		
Asclepiadaceae	137	*Dischidia**	+	+
		Hoya*	+	−?
Solanaceae	56	*Markea*	+	−?
		Ectatomma	+	−
Gesneriaceae	560	*Codonanthe*	+	−
		Codonanthopsis	+	−
		Aeschynanthus	+	−
Rubiaceae	85	*Myrmecodia*	−	+
		Hydnophytum	−	+
		Squamellaria	−	+
		Anthorrhiza	−	+
		Myrmephytum	−	+
Cyclanthaceae	86	None		
Araceae	1 349	*Anthurium*	+	−
		Philodendron	+	−
Bromeliaceae	1 144	*Tillandsia**	−	+
		Aechmea*	+	+?
		Streptocalyx*	+	+?
		Araeococcus*	+	−
		Neoregelia	+	−
Orchidaceae	13 951	*Coryanthes*	+	−
		Dendrobium	+	−
		Schomburgkia	−	+
		Caularthron	−	+
		Epidendrum	+	−
		Acriopsis	+	−

* The concept of domatium is sometimes difficult to apply but ants nest within, or closely against, foliage.

Table 23.3. Chemical characteristics of mineral soil and epiphytic substrates (one sample each) in wet forest at Río Palenque, Ecuador (altitude approximately 350 m).

Description of material	pH	% Base saturation	Total cation exchange capacity (meq/100 g dry wt)	(meq/100 g dry wt)				(ppm in dry sample)		
				K	Ca	Mg	Na	N	P	K
Outer bark of large *Theobroma* branches with associated debris and non-vascular plants	6.2	79	124	20.0	50	25.5	2.6	3.0	0.34	0.67
Outer bark of *Theobroma* twigs	6.7	86	137	18.7	67	31.5	0.3	2.2	0.22	0.71
Rotten wood of *Theobroma*	7.1	90	163	4.6	112	30.1	0.3	1.5	0.09	0.18
Fern root ball	5.2	56	135	7.5	57	11.1	0.4	1.8	0.10	0.25
Carton of ant-garden	6.3	78	115	20.1	56	12.2	1.9	2.9	0.39	0.79
Earth soil beneath *Theobroma*	6.3	55	31	0.5	14	2.5	0.2	0.3	trace	trace

Plant dispersal and fidelity to carton

Evidence of plant specialization for association with plant-feeding ants comes from several quarters. Three Amazonian epiphytes were almost completely restricted to carton at Cocha Cashu, Peru (Davidson 1988). Species very rarely found except in ant substrates were *Peperomia macrostachya* (Vahl) A. Dietr., *Ficus paraensis*, *Anthurium gracile* (Rudge) Lindl., *Vanilla planifolia* Andrews, and a *Codonanthopsis*. Nest size, exposure, and ant identity influenced plant occurrence. *Anthurium gracile* was over-represented on small ant-gardens, whereas *Codonanthe uleana* succeeded regardless of nest size. Six of ten regular nest epiphytes at Cocha Cashu were shade tolerant. All of the relatively common ant-garden species, except a *Neoregelia*, were more likely to co-occur with other plant taxa than to live alone on carton. The decline of most myrmecotrophs on deteriorating carton suggests that the plants have no other alternatives. Bromeliads, probably because they draw resources from materials trapped in leaf axils, appeared to be least vulnerable to carton erosion. All but two ant-garden species were less successful in *Azteca* gardens versus those tended by *Camponotus femoratus* and its usual nest associate *Crematogaster linata parabiotica* (Forel). Substrate quality was probably responsible for the differences. *Azteca* carton was brittle and contained less organic matter compared with the rooting media provided by the parabiotic ants.

Ants responding to food and chemicals, and possibly other cues, maintain plant populations by depositing diaspores in their nests (Fig. 23. 1; Ule 1901, 1905, 1906; Docters van Leeuwen 1929*a,b*; Madison 1979; Davidson and Epstein 1989). Fleshy arils are the attractant offered by *Codonanthe*, the most widely distributed neotropical ant-garden genus. *Peperomia macrostachya* fruit bears a large, sticky, basal oil gland. All ant-cultivated bromeliads (e.g. *Aechmea* and *Streptocalyx*) and *Anthurium* have fleshy fruit from which attending ants mine seeds. Some Australasian *Aeschynanthus*, *Dendrobium*, *Dischidia*, and *Lecanopteris* species regularly root in carton, apparently after seeds or sporangia featuring oil-filled cells in walls or accessory appendages are incorporated by the ants. Docters van Leeuwen (1929*a,b*) demonstrated selective germination of several taxa on carton rather than on unmodified bark.

Field experiments confirmed the importance of olfaction in western Amazonian ant-garden establishment (Davidson and Epstein 1989). Porcelain 'seeds' impregnated with *ortho*-vanillyl alcohol attracted *Crematogaster linata parabiotica* and *Camponotus femoratus*; limonene had a similar effect on an *Azteca* species. In more extensive bioassays of about 60 substituted phenyl derivatives, ants were most responsive to 6-substituted phenyl derivatives; propagules produced by nine of ten common ant-garden epiphytes belonging to eight families in eastern Peru contained this ester (Seidel 1988). Eight members of the same ant-garden guild contained benzothiazole in fruit and/or seeds. Seeds of *Ficus paraensis* remained attractive to ants even after passage through the guts of captive bats, perhaps because eliaosomes remained intact on defecated seeds, as noted by S. C. Kaufman and D. B. McKey (unpublished) for *F. microcarpa* Bahl following ingestion by birds.

Circumstantial evidence that pupal mimicry promotes ingress to ant nests merits testing. Seeds of *Codonanthe* differ from all others produced by neotropical gesneriads and are, like those of *Anthurium gracile* (Madison 1979), reportedly strikingly similar to ant pupae. Seeds of *Aechmea mertensii* Schult. f. and *A. brevicollis* L. B. Smith, in addition to the fruit of some *Peperomia*, share the same size and shape, but not colour, of appropriate ant brood.

Ant-house epiphytes

Ant-house myrmecotrophs are mainly small herbs or compact succulent shrubs (Figs. 23. 1 to 23. 3). They belong to fewer families (Table 23. 2) than do ant-garden flora, but there are more species (although many may not be regularly ant-inhabited; see Chapter 24). *Hydnophytum* and *Myrmecodia* are sizeable genera, others are much smaller (e.g. *Squamellaria* and *Solenopteris*), or, if larger, contain only a few members modified for ant occupancy (e.g. *Tillandsia* and *Hoya*). Myrmecotrophs that produce domatia evolved separately several times in tropical America and Asia; those most specialized for ant habitation are Australasian.

The zoobionts also belong to a variety of unrelated genera (Table 23. 1). Oligophily rather than polyphily is usual in combinations involving the more specialized ant-house groups, with different members of a small group of taxa tending to dominate plant-provided housing in certain kinds of habitat (see discussion below and Davidson *et al.* (1990)). Plants with less elaborate nest cavities, or those primarily serving other purposes (e.g. water/litter impoundment in tank bromeliads), tend to harbour a more diverse, less faithful ant fauna (e.g. Hydnophytinae versus *Aechmea bracteata* (SW.) Griseb.; Table 23. 1).

A sizeable, predominantly opportunistic ant fauna inhabits epiphytes in tropical America (Table 23. 1). Nine ant species were utilizing *Schomburgkia tibicinis* Batem. in populations surveyed by Rico-Gray and Thien (1989). Seventeen species representing 11 genera in four subfamilies occupied bulbs of *Tillandsia bulbosa* Hooker in a census of the forest reserve at Quintana Roo, Mexico (Olmsted and Dejean 1987). Taxa assigned to 18 genera were encountered among the ants billeted in *Aechmea bracteata* rosettes in the same region (Table 23. 1), although some species were considered secondary occupants. In another study, species belonging to nine genera were nesting in leaf bases of *Tillandsia butzii* Mez and *T. caput-medusae* E. Morren collected at several Mexican and Costa Rican locations (Table 23. 1).

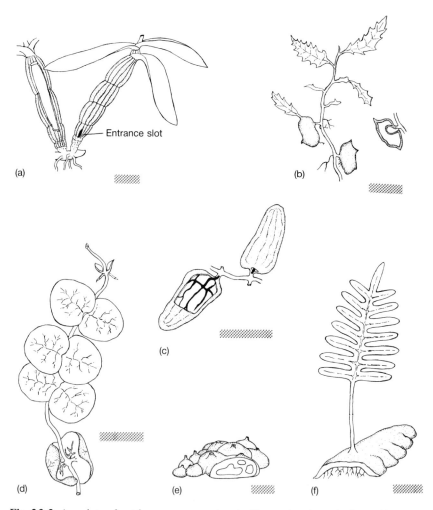

Fig. 23.2. A variety of ant-house myrmecotrophs illustrating domatia (bar = 5 cm). (a) *Schomburkia*; (b) *Solenopteris brunei*; (c) *Dischidia rafflesiana*; (d) *Dischidia coryllis*; (e) *Lecanopteris carnosa*; (f) *Lecanopteris mirabilis*.

Inhabitants of examined Hydnophytinae were mostly found to be three species of *Iridomyrmex* (Janzen 1974c; Huxley 1978; Rickson 1979; Jebb 1985) and nesting also occurred in tree myrmecophytes and on the ground. The Hydnophytinae featured the most elaborate of all domatia, suggesting long association with ant colonies. Galleries resulting from necrosis of parenchyma isolated by internal layers of cork tissue are differentiated into warted portions for ant refuse and plant absorption in advanced species, and

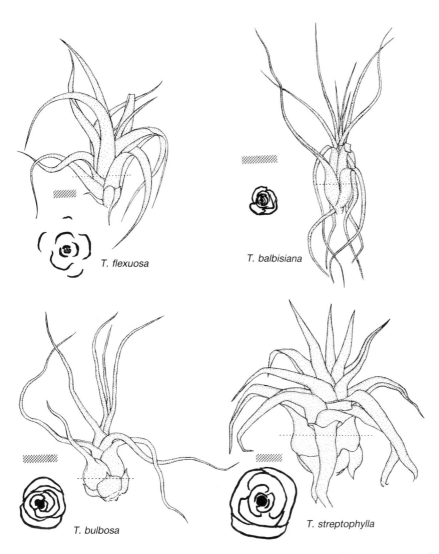

Fig. 23.3. Four epiphytic bromeliads with different volumes and arrangements of interfoliar cavities and the frequency of occupation of those chambers (after Olmsted and Dejean 1987; bar = 5 cm; dotted lines indicate where bulbs were cut to provide cross sections). Note that the one species without ants, *Tillandsia flexuosa*, is the only species without enclosed leaf bases.

into smoother-walled regions for brood and possibly ventilation or insulation (Spanner 1939; Huxley 1978; Jebb Chapter 24, this volume).

Ant-inhabited Hydnophytinae offer additional advantages for symbionts. Tuber surfaces are often ridged and sometimes equipped with stout spines derived from roots (see Chapter 24), perhaps detering ant predators that could tear open unprotected domatia. Shielded leaf axils harbouring the inflorescences (*Myrmecodia*) may protect ants foraging for nectar on discs of the developing fruits. The ripe fruits are exserted and dispersed by the same birds that disperse mistletoes (Wallace 1989). Honeydew is available from Homoptera farmed on adjacent host tissue or on the myrmecotroph. Here, as in ant-garden symbioses, ant colonies may be large (> 15 000 workers in a single tuber) and forage 50 m or more from the nest (Janzen 1974c). Usually, several ant-house epiphytes are inhabited by a single colony.

Most ant-house myrmecotrophs offer transitory food rewards in the form of extrafloral nectar (e.g. *Schomburgkia*) and edible fruit or seed appendages (e.g. Hydnophytinae), but some apparently offer none (e.g. *Tillandsia*). Ants supposedly mine sugar-laden tissue in *Solenopteris* rhizome tubers to hollow them out for occupation (Fig. 23. 2) (Gomez 1974). Nutritive appendages on sori occur on at least one Australasian *Lecanopteris* (Holttum 1954). Orchidaceae are exceptional, with their adjustable reward system which can provide short-term support for temporary ant guardians or more continuous food for plant feeders. A survey of 23 Panamanian orchids revealed that just one of the 16 species that attract ants with extrafloral nectar (*Caularthron bilamellatum* (Reichb. f.) R. E. Schult) did so with unusually long-lived glands on leafy stems in addition to the more common ephemeral EFNs on inflorescences (Fisher and Zimmerman 1988). *Caularthron* also produces hollow pseudobulbs with natural entrances, and so is better equipped than the others to profit from longer term and more intimate contact with ants. In turn, plant occupants exhibited an uneven capacity to supplement housing with calories. Isotopic analysis revealed that different ant species obtained 11–48 per cent of their carbon from host orchids (Fisher *et al.* 1990).

A small colony of similarly myrmecotrophic *Schomburgkia tibicinis* was in use by five ant taxa in south-western Mexico, with varied results for both partners (Rico-Gray and Thien 1989). In addition, hollow pseudobulbs (Fig. 23. 2) capable of absorbing solutes from radiolabelled ant carcasses were probably able to extract nutrients from nests regardless of the builder's identity. Other ant qualities, however, influenced plant fitness unevenly, i.e. the outcome of symbiosis was contingent on specific ant behaviour and diet. A relatively timid ant, *Crematogaster brevispinosa* Mayr, dominated about two thirds of all inflorescences, but it failed to prevent destruction of numerous flowers by a common curculionid beetle. Additional damage inflicted by farmed mealybugs reduced plant fecundity substantially below that of orchids patrolled by larger, more aggressive ants (*Camponotus planatus* Roger, *C. rectangularis* Emery, *C. abdominalis* (Fabricius), and *Ectatomma tuberculatum* (Oliver)).

Plant nutrition

Isotopic tracers have been used to monitor nutrient flux from ant to plant in various myrmecotrophs: *Tillandsia* (Benzing 1970); *Myrmecodia* (Huxley 1978); *Hydnophytum* (Rickson 1979); *Schomburgkia* (Rico-Gray and Thien 1989); and *Lecanopteris* Gay (1990). Ants that fed upon labelled substrate or prey were either killed and placed on domatia (Rico-Gray and Thien 1989) or allowed to continue normal nest-related activity for several weeks (Huxley 1978; Rickson 1979). Benzing (1970) applied $^{45}CaCl_2$ on moist cotton swabs to leaf base chambers of *Tillandsia caput-medusae* in order to determine if a phloem-immobile ion could be acquired from symbionts. Huxley (1978) used ^{32}P and ^{35}S as tracers, Gay (1990) injected several radioisotopes into fern rhizome cavities, and ants that were fed heavy nitrogen, ^{15}N, urea, and glycine were also employed. Rickson (1979) and Rico-Gray and Thien (1989), utilized organic ^{14}C which could have entered plant tissue after being respired by ants or microbes, rather than moving as a solute.

Radioactivity in harvested plants was broadly distributed, with the greatest concentration in young tissue. Neither the points of entry nor the absorption kinetics were determined in these labelling experiments, although the trichomes covering *Tillandsia* leaves possessed absorptive capacity (Benzing *et al.* 1976). Isotopes preferentially penetrated to tissues below warted rather than below smooth chambers of *Myrmecodia* (Huxley 1978), a pattern confirmed by Rickson (1979).

Most plant surfaces are at least modestly permeable to solutes, but epidermal specializations, such as those of carnivorous plants (Juniper *et al.* 1989), might also improve the capacity of a myrmecotroph to utilize similar animal products. Except for certain Hydnophytinae, however, no such modifications have been reported and the putatively absorptive warts of *Myrmecodia* and its relatives could perform additional or alternative functions to aid plant nutrition. Secretions could dissolve nutrients prior to uptake, humidify ant nests, or encourage the diverse resident biota, including fungi and other microflora, to process complex organic materials for plant use.

Scant attention has been devoted to the fate of nitrogen in myrmecotrophy even though shortages of this element, its occurrence in diverse chemical forms, and peculiarities of mobility, and disposal of by-products required to maintain acid–base balance have driven plant evolution in multiple directions, including mutualisms with diverse partners (Raven 1988). Little is known about the identity of combined nitrogen present in ant-inhabited domatia or cartons, or if the metabolism of myrmecotrophs is tailored to particular molecular species. The presence of ammonia and a tentative identification of urea have been attributed to nest contents of certain Hydnophytinae (Janzen 1974c). Water-soluble materials extracted from *Tillandsia caput-medusae* bulb chambers contained considerable Kjeldahl nitrogen but supported little bromeliad seedling growth in an aseptic bio-assay (Benzing 1970; however, *Aechmea bracteata* was the seed source). Additional investigation is needed to determine the plant's return on expenditure for ant feeding versus other sources of nutrients in tree crowns.

Ecological and other constraints

Restricted geographical distribution suggests that elaborate plant modi-fication to attract foundresses or to house fragmenting colonies is econo-mical only under special conditions. Domatia for plant-feeding ants are supposedly most cost effective where photosynthesis and turnover of plant organs are slow due to aridity, shade, and scarcity of key ions (Janzen 1974c; Thompson 1981). Thompson's use of an economic construct to explain why ant-house myrmecotrophs are virtually all epiphytic needs to be expanded; more variables must be included, such as local demand for ant housing and the status of alternative defences where vegetation is protected by ants (Beattie 1989). Another improvement would be clearer concepts of habitat fertility, plant nutrition, and economics than have been applied in the past.

The forest canopy is a patchy space in terms of plant nutrients. Epiphytes anchored to naked bark occupy 'sterile' micro-sites, but a tree crown in its totality is nutrient-rich by the standards of most earth soils (Benzing 1990, Chapter 4, Table 3). However, most of the essential elements present are concentrated in the biomass (Fig. 23. 4) and require processing prior to use by non-parasitic plants. Even oligotrophic forest accumulates such scarce nutrients as nitrogen and phosphorus to levels sufficient to sustain animals which, in the case of suitable ants (Table 23. 1), may serve arboreal vegetation as nutrient processors and collectors. The plant secured to scattered bark with a hold-fast root system can tap sources of nutrients if ants can be attracted to nest in domatia or among roots. Attracting ants might be less costly than more direct access to such scattered nutrients.

Terrestrial ants are also abundant enough to feed co-occurring vegetation. Failure to do so probably concerns features (Fig. 23. 4) that distinguish the biology and habitats of ground-dwelling ants and plants from those of their arboreal counterparts (but see Luizão and Fortunato de Carvalho 1981). Plant economy provides some insight on one important difference. Root systems of trees and shrubs which must support woody crowns and absorb dilute ions from large volumes of soil often constitute one-half or more of the plant's body. Even if the requirement for nutrients were met by ants, the requirement for moisture and anchorage would remain, and render costs associated with myrmecotrophy redundant. Absence of ant-feeding among

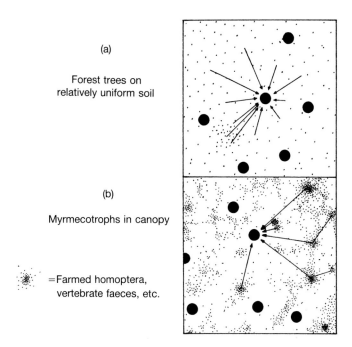

(a)

Forest trees on
relatively uniform soil

(b)

Myrmecotrophs in canopy

=Farmed homoptera,
vertebrate faeces, etc.

Fig. 23.4. A graphic model depicting availability of key nutrients to trees (circles in (a)) and, following accumulation in the canopy, to arboreal ants and subsequently to the epiphytic flora (circles in (b)). Stippling indicates the location and concentration of nutrients. Arrows represent root foraging by one tree and movements of ants returning from feeding sites to one myrmecotroph. Ant movement is influenced by location of food sources and competing ant colonies. Direction and extent of root growth is determined by dispersion of competing trees and nutrients in earth soil.

terrestrial herbs with myrmecotroph-like habits calls for a different explanation.

The belief that conditions are most suited to myrmecotrophy in forest over sterile soil (Janzen 1974c; Kleinfeldt 1978; Madison 1979) is challenged by certain Amazonian ant-gardens and some Hydnophytinae. Ant-gardens on trees over young alluvium in south-eastern Peru are constructed by aggressive ants and support diverse plant taxa (Davidson and Epstein 1989). Over older, impoverished terra firma nearby, gardens are smaller and contain one or a few myrmecotrophs tended by single-queen ant colonies with lower energy requirements. Ant-garden abundance is generally highest in open, relatively disturbed sites, indicating that exposure equals or exceeds nutrient supply in determining the success (particularly the vigour and complexity) of these mutualisms. Much ant-garden flora is heliophilic and so are the ants; although less directly. Regenerating woodland probably favours ant-garden development because colonizing trees, compared to those comprising primary forest, more often produce extrafloral nectar and support the Homoptera that carton-building ants seem to require in considerable abundance (Janzen 1974c; Kleinfeldt 1978; Davidson and Epstein 1989; Schupp and Feener Chapter 13, this volume).

Ant and plant characteristics and habitats ensure that some myrmecophytic partnerships grant little nutritional advantage to the phytobiont. Epiphytic myrmecophytes differ qualitatively from soil-rooted *Acacia* and *Cecropia* which shelter ants in what are probably non-absorptive thorns and stems (but see Luizão and Fortunato de Carvalho 1981 and Beattie 1989). Bull's-horn *Acacia* feeds and houses massive and costly populations of savage ants, a pattern which is ideal for defence but unsuited for plant nutrition. Unlike wider ranging, plant-feeding ants, those protecting myrmecophytes are too well fed by the tree to seek further nutrition, thereby gathering nutrients from beyond the tree's roots. Moreover, ant-trees are not noted for occurring on infertile soil where ant inputs would be most beneficial.

Numerous factors probably influence where and how certain trophic mutualisms operate and account for their generally conditional nature. Regional differences in the need for ant housing may help explain why epiphyte occupation rates are about the same (90 per cent) in relatively unspecialized domatia (e.g. *Schomburgkia* and *Tillandsia*; Figs 23. 2–3) and in more elaborate, and presumably more costly, housing (Hydnophytinae). Perhaps plants must provide greater or different incentives to secure the same level of ant-provided benefits in different locations, or the more elaborately modified myrmecotrophs may receive greater or different service from mutualists than plants which secure equal occupancy rates at lower cost. A species that exhibits site-specific differences in its associations with ants offers interesting possibilities for research. *Tillandsia flexuosa* Sw. shoots are usually free of nests in south Florida (D. H. Benzing, personal observation) and southern Mexico, but occupancy was frequent enough to convince Griffiths *et al.* (1989) that this taxon is myrmecophytic on a northern Venezuelan island. Placement of artificial domatia in trees would help to identify housing needs and design preferences for arboreal ant billets. Fitted with the exchange resins now used to investigate soil fertility (see, for example, Lajtha 1988), these devices could also provide insight on nutrient inputs from recruited colonies.

Ant-house and ant-garden myrmecotrophs have faced different constraints on promotion of ant aggressiveness and growth rates of plant and ant colonies. The former group must allocate substantial resources to nest-cavity walls, the latter is freer to produce the foliage necessary for manufacture of a diet that can satisfy the most aggressive and reproductive mutualists. Ant colonies may provide a parallel: after production of workers and sexual brood, the largest amounts of energy, time, and material go into nest (particularly carton) construction (Sudd and Franks 1987).

Importance of myrmecotrophy

Ant-house myrmecotrophs and neotropical ant-gardens can be locally abundant. A 225-m^2 quadrant sampled by Janzen (1974c) in Malaysia contained 494 *Hydnophytum*, 31 *Myrmecodia*, 20 *Dischidia rafflesiana*, and 2 *Lecanopteris* specimens. A tally of 79 woody plants over one metre tall showed that *Hydnophytum* was accommodated by 44 of these trees, *Myrmecodia* by 13, *Dischidia* by 20, and *Lecanopteris* by 2. At least one representative of each of the 11 local tree species supported one to several epiphytes. Ant habitation approached 95 per cent of available domiciles.

Thirty-four nests, each nourishing a sizeable garden, were recorded on a 100-m^2 plot in an Amazonian caatinga (Madison 1979). Citrus trees in a grove in northern Trinidad each contained 3–10 discrete ant-gardens (personal observation). Ant-garden aggregations, or what were probably the products of fragmented, polygynous colonies of *Camponotus femoratus*, occurred on 16–39 per cent of the trees in five Peruvian forest types (Davidson 1988). Colonization was highest in seasonally flooded locations.

A better index of ant-garden importance than colony frequency is the density of foraging workers. Ants other than parabiotic species discovered fewer aerial cheese baits in communities hosting abundant ant gardens than in control sites at Cocha Cashu (Davidson 1988). Ant-garden ants dominated the arboreal fauna at Tambopata, Peru (Wilson 1987). Tree crowns fogged with insecticide in four types of lowland forest yielded enormous numbers of ant taxa (43 species representing 26 genera in one tree crown alone, more than the entire ant fauna of the British Isles). *Crematogaster linata parabiotica*, here the nest companion of *Camponotus femoratus* and *Monacis debilis* (Emery), occurred in almost half of the 513 sampled crowns. Wilson (1987) attributed the great success of ant-garden species to their production of capacious nests, aided by symbiotic epiphytes.

Durability may also determine importance. Kleinfeldt (1978) discovered that subunits of fragmented Costa Rican *Crematogaster longispina* Emery colonies, cultivating the myrmecochore *Codonanthe crassifolia* (Focke) Morton, turned over rapidly through displacement of the original garden builders by a more aggressive but less industrious *Solenopsis* species. Within months, neglected cartons deteriorated and previously covered roots became exposed. Davidson (1988) reported 118 abandoned nests compared to 758 housing-active ant colonies along a forest trail at Cocha Cashu, Peru. Mineral cycling could be substantially enhanced at both locations. Less obvious manifestations of this process may have even broader implications. Carton-producing ants mix and rework substrates for extensive epiphytic flora even beyond that cultivated in ant-gardens, much as lumbricoids and similar detritivores benefit earth-rooted vegetation. Considerable insight could result from examination of this earthworm–ant analogy (see Woodell and King Chapter 34 this volume).

Evolution of ant–epiphyte associations

Benefits to both mutualists vary too much to propose one evolutionary pathway or a single selective advantage. However, ubiquity in canopy habitats, diet, eusociality, and year-round activity has ensured that ants, above all other insects, could evolve specialized mutualisms with plants (Davidson and Epstein 1989). Especially critical for ant–plant pairing is the workers' discharge of dangerous extranidal tasks, an arrangement that grants the parent colony and its insular queen(s) considerable immunity from predators and a potential for long life (often as long as, or longer than, that of the plants they utilize and serve).

Comparisons of plant morphology, ant behaviour, and patterns of residence suggest that co-evolution marks some alliances whereas other combinations have wrought little or no change in either symbiont. Plants can be opportunistic or obligate: in Peru's Manu National Park, several ferns, aroids, orchids, and peperomias root on bark or carton, while ten other epiphytes are restricted to carton, as documented above. Ants also show a variety of habits. One of the most abundant Amazonian ant-garden ants (*Camponotus femoratus*) seldom, if ever, occurs without its phytobionts (Davidson 1988). On the other hand, opportunists associate with trash-basket ferns (e.g. *Asplenium* and *Platycerium*) and tank bromeliads. Several primitive ant taxa also inhabit rotting wood and other plant chambers (e.g. *Pachycondyla* and *Odontomachus*). Horich (1977) described combinations of neotropical ants and epiphytic orchids ranging from sporadic to consistent. Ancestral ant traits involving behaviour, life history, and diet, as well as qualities of plant habit and seed, probably preceded specific kinds of myrmecotrophy. Ecological sorting of appropriate partners actually united compatible (pre-adapted) pairs which, in some instances, subsequently changed because of contact with symbionts.

Vagrant ant species demonstrate a preference for plant cavities; they occupy 10–15 per cent of the modest interfoliar leaf base chambers provided by *Tillandsia paucifolia* Sesse & Moc. in south Florida (Benzing and Renfrow 1971). Correlation between ant occupancy and plant habit was illustrated in *Tillandsia* native to Quintana Roo, Mexico (Olmstead and Dejean 1987). Four co-occurring epiphytes (Fig. 23. 3), listed in order of water-tightness and volume of bulb chambers (ant presence denoted in parentheses), were: *Tillandsia flexuosa* (0 per cent); *T. balbisiana* Schultes f. (42 per cent); *T. bulbosa* (41 per cent); and *T. streptophylla* Scheidweiler ex Morren (53 per cent). Of the three species occupied by ants, only *T. bulbosa* is reported to be a myrmecophyte.

Arboreal ants that manufacture large carton nests incorporating soil or faeces, or that collect seeds with oils which maintain fungus-free nests, are better candidates for cultivating plants than are producers of less nutritive and absorbent substrates or smaller domiciles. Such phenomena as the sprouting of *Ficus paraensis* seeds in carton laid down over fruit to shield ant-tended Homoptera, and use of seed for purposes other than planting, could help foster regular ant-garden status (Davidson and Epstein 1989). Several arboreal ant species exhibit appropriate behaviour but fail to establish ant-gardens. Seeds sown in thin brittle carton constructed by Peruvian *Hypodinea bidens* (L.) consistently failed to germinate (Davidson 1988) and seeds cultivated by often parabiotic ants nesting alone (e.g. *Azteca* sp. and *Crematogaster linata parabiotica*) produced seedlings that never matured.

Ant-garden plants that produce seeds laced with methyl-6-methylsalicylate, benzothiazole, and other bioactive, substituted phenyl derivatives probably acquired these traits through more conventional myrmecochory. *Peperomia macrostachya* and *Ficus paraensis* both have congeneric myrmecochorous but uncultivated relatives. Some terrestrial gesneriads, like epiphytic *Codonanthe* species, are dispersed by ants. Increased chemical synthesis with some fine-tuning may be a result of the ant-garden partnership. Congeners of some ant-garden ants (e.g. *Camponotus sericeiventris* (Guèrin)) are repelled by the same seeds that ant-garden relatives avidly collect (Davidson and Epstein 1989).

Habitual poor housekeeping and foraging to fill domatia beyond the needs of occupying ants may be other potential pre-adaptive features of the plant-feeding ant (Janzen 1974*c*; but see Davidson and Epstein 1989). The plant-feeder ant, *Iridomyrmex cordatus* (Fr. Smith), packs nutrient-rich refuse into absorptive chambers of *Hydnophytum* and *Myrmecodia* tubers in contrast to the habit of ejecting debris ostensibly practised by some guardian ants.

Ecological characteristics are shared by advanced ant-garden and certain Australasian ant-house mutualisms, suggesting that accelerated growth of plant and ant colonies and greater competitiveness in the zoobionts are evolutionary products of both types of symbiosis. The most refined and obligate partnerships in each case are found in exposed sites, probably, in part, because strong irradiation allows the plants to produce food and living space rapidly. Correlations between site quality and specialized ant and plant characteristics in the Amazonian ant-garden syndrome have already been mentioned. The Hydnophytinae illustrate a similar progression towards advanced ant-house status, culminating in this case in at least four different cavity arrangements (Jebb 1985, Chapter 24, this volume).

Non-myrmecotrophic species in the Hydnophytinae (mostly species of *Hydnophytum* and one of *Anthorrhiza*) are shade tolerant and slow growing, and maintain rather low ratios of leaf to total biomass (Jebb 1985). Related ant-house forms are heliophilic, mature faster, and produce proportionately larger amounts of green tissue. Likewise, the plant–ant guild attending the plant subtribe contains both generalists and specialists featuring an included niche pattern that may characterize other ant–plant associations (Davidson *et al.* 1990, Chapter 20, this volume). Generalist ant taxa, with their less demanding resource requirements and broadest niches, inhabit accommodating plants at all locations. The specialist's more stringent needs mandate better plant growing conditions and, accordingly, nesting possibilities are but a fraction of those available to less fastidious occupants.

Colonies on high-quality sites (the specialists) are typically those with multiple queens, high activity, and vigorous growth, aided sometimes by fragmentation. Shaded plants normally harbour more docile ants that produce slower growing single-queen colonies with less exacting housing requirements (e.g. *Iridomyrmex scrutator* (det. R. Taylor) as compared with *I. cordatus*). Occasionally, a generalist taxon occupies a tuber in a well-illuminated site if a foundress managed to fend off late-arriving queens of the normally dominant species (Davidson and Epstein 1989).

Conclusions

Davidson *et al.* (1990) have suggested that species-sorting mechanisms, involving both insect and plant qualities and habitat type, account for the composition of ant–epiphyte associations and influence their outcome. Co-adaptation has also occurred, but varies with the partnership, from significant to zero; plant and ant benefits depend on local conditions and the symbionts' identities. Study of the evolution and operation of ant-fed plant systems should capitalize on the variety of interactions which range from opportunistic to obligate and from commensalism and parasitism to unequivocal mutualism. Five subjects should receive special attention:

1. ant behaviour at the level of the individual worker and that resulting from the particular colony hierarchy;
2. use of plant products as ant metabolites or fungistatic agents to protect ant brood;
3. ant colony energetics and housing requirements;
4. increased plant fitness due to ant-enhanced photosynthesis and defence; and
5. the nature of ant-provided nutrients and their assimilation by plants.

An explanation for the relegation of myrmecotrophs to epiphytic habitats will emerge with a clearer picture of arboreal ant biology and the disposition of plant nutrients on the ground and in the canopies of tropical forests.

Ants promote plant nutrition beyond that of the recognized myrmecotrophs. Important questions concern how much additional vegetation is benefited and how broad are the consequences for other biota. How much ant activity results from plant products made possible by nutrients that these insects process or directly provide to phytobionts? Ant abundance, variety, and biomass in tropical forests (Wilson 1987) indicate that these insects are more directly dependent on the base of the food chain than formerly believed (see Chapter 35). Such considerations are not trivial given the context (i.e. the most complex and diverse of all terrestrial ecosystems). Epiphytes represent more than a third of the total species in some woodland floras (Gentry and Dodson 1987); they account for half or more of the photosynthetic tissue (Benzing 1984); and they significantly affect nutrient cycling (Nadkarni 1986). Moreover, much of the fauna in the humid tropical forest is canopy based and, hence variously dependent on plants fed by, and in other ways affected by, the multitudes of often dominant ants living there.

24

Cavity structure and function in the tuberous Rubiaceae

Matthew Jebb

Introduction

The epiphytic rubiaceous ant-plants are currently classified into a subtribe of the Hydnophytinae comprising six closely related genera: *Hydnophytum* Jack, *Myrmecodia* Jack, *Myrmephytum* Becc., *Myrmedoma* Becc., *Squamellaria* Becc., and *Anthorrhiza* Huxley & Jebb (Huxley 1981; Jebb 1985; Huxley and Jebb 1990). The group is confined to the Far-eastern tropics, and is most diverse in New Guinea (Fig. 24. 1). In all species of these genera the hypocotyl of the seedling swells to form a tuber which always becomes chambered and is usually inhabited by ants (Forbes 1880; Treub 1883, 1888). Parts of the chambers are smooth and impervious while areas with 'warts' are absorptive (Miehe 1911; Spanner 1939; Janzen 1974*c*; Huxley 1978). Comparatively little has been done on the morphology and architecture of the cavities (Beccari 1884–1886; Spanner 1938; Huxley 1976, 1978). The taxonomic and evolutionary implications of the cavity structure, gross plant morphology, and ant occupation are outlined here. The lack of material for *Myrmephytum* and *Myrmedoma* has meant that the tuber structure of these genera remains unknown, although predictions about their organization are made in the light of current knowledge.

Hydnophytum is the largest genus, with 46 species (Jebb 1985), and is the most widely distributed (Fig. 24. 1). *Myrmecodia* is only slightly restricted within this area, reaching Malaysia to the west and the Solomon Islands to the east; 25 species are recognized. The remaining genera are small and restricted in range. *Myrmephytum*, with four species, is confined to the Philippines, northern Sulawesi, and the westernmost tip of New Guinea; the single species of *Myrmedoma* should probably be included in *Myrmephytum* (Huxley 1981). *Squamellaria*, with three species, is confined to the Fijian Islands. Lastly, *Anthorrhiza* consists of eight species all of which are confined to south-eastern New Guinea (Huxley 1981; Jebb 1985).

When Jack (1823) named *Hydnophytum* = 'truffle plant' and *Myrmecodia* = 'ant-house', he drew attention to the markedly different vegetative morphology exhibited by the two genera. *Hydnophytum* (Fig. 24.2(a)) typically has a smooth tuber, with numerous slender, freely branching stems;

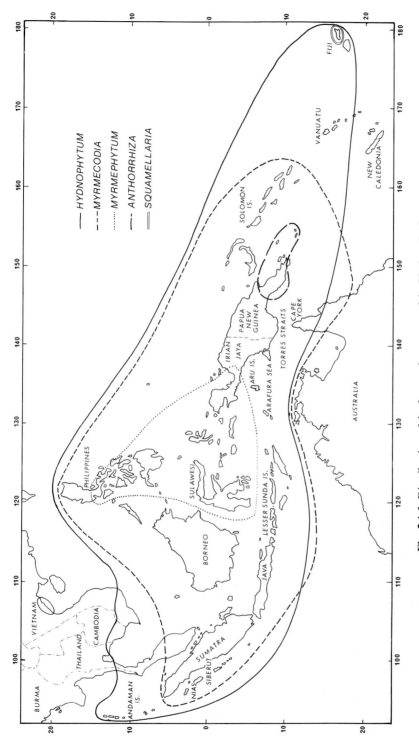

Fig. 24.1. Distribution of the four major genera of the Hydnophytinae.

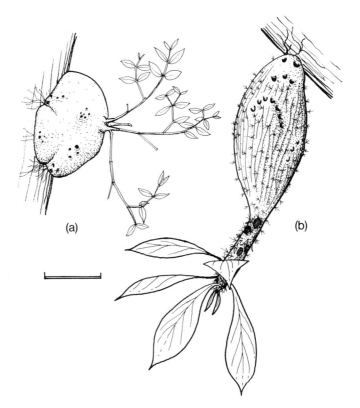

Fig. 24.2. Typical external appearance of (a) *Hydnophytum* and (b) *Myrmecodia*.
Scale bar = 10 cm.

sessile or pedunculate inflorescences; and small, leathery or sclerophyllous leaves. *Myrmecodia* (Fig. 24. 2(b)) is densely spiny; has few, thickened, condensed, and sparsely branched stems; large sunken inflorescences; and large mesomorphic leaves. Species of the two genera often grow together in a range of habitats including mangrove, savannah, ever-wet forest, and alpine grassland. In Ambon, Rumphius (1750) believed the two genera to represent the nests of black ants (*Nidus formicarum niger* (= *Hydnophytum*)) and red ants (*Nidus formicarum ruber* (= *Myrmecodia*)). In New Guinea, *Hydnophytum* is indeed more regularly occupied by one of two species in the *Iridomyrmex scrutator* group (black ants), and *Myrmecodia* in the lowland savannah is regularly occupied by *I. cordatus* Fr. Smith (a red ant) (Huxley 1976, 1978; Jebb 1985).

Beccari (1884–1886) distinguished *Squamellaria* from *Hydnophytum* by the presence of small barbate plates in the corolla tube, and the 'bony' endosperm. The tubers of this genus, then unknown, demonstrate important structural features which confirm the genus as a valid and distinct taxon (Jebb 1985). *Anthorrhiza* exhibits a range of external morphology from *Hydnophytum*-like to *Myrmecodia*-like (Jebb 1985) (Fig. 24. 3). There is also a morphological series in the tuber cavities of this genus which corresponds to the external appearance of the plant. Tuber content, in particular regular ant occupation, is also correlated with this morphological series, and *Anthorrhiza* therefore provides an insight into the morphological distinction between *Myrmecodia* and *Hydnophytum*. From a study of tuber cavity structure and its contents, it is apparent that the two morphological types in

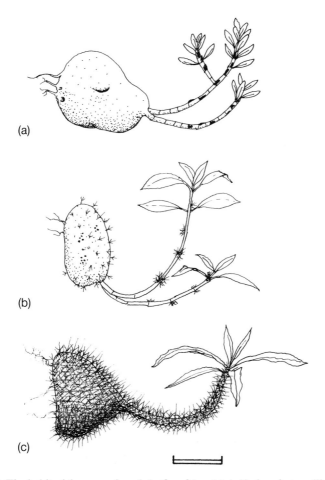

(a)

(b)

(c)

Fig. 24.3. The habit of three species of *Anthorrhiza*. (a) A *Hydnophytum*-like species, lacking spines (*A. areolala*). (b) An intermediate species (*A. recurrispina*). (c) A *Myrmecodia*-like species (*A. caerulea*), which is densely spinous. Scale bar = 10 cm.

the Hydnophytinae (Fig. 24. 2) are functionally distinct in relation to the association with ants.

The tuber cavities

The tuber varies in shape, the majority are globose to sub-globose, while many *Myrmecodia* species have cylindrical tubers. The tuber grows throughout the life of the plant, and a series of cavities are formed, *de novo*, of phellogens, each of which encloses a volume of parenchymatous tissue. The proportion of the total volume occupied by the cavities varies from species to species; some have few, small cavities in a mainly fleshy tuber, others have extensive cavities separated by thin walls (Huxley 1978). The future cavities are sealed off from the remaining cells by suberization and, being connected to the exterior or to earlier cavities, the enclosed tissue shrivels, leaving a hollow space (Treub 1883; Miehe 1911; Huxley 1978).

In the majority of species, the tuber cavities have two characteristic wall surfaces: smooth surfaces, which are impervious to water and are usually dry when a tuber is cut open, and rough or 'warted' surfaces, which are covered by numerous small, pale-coloured swellings, approximately 0.5 to 2 mm in diameter (Treub 1883; Beccari 1884–1886; Janzen 1974c; Huxley 1976, 1978). These 'warts' were likened to lenticels by early authors, who also pointed out, however, that they lack air spaces (Treub 1883; Beccari 1884–1886). The 'warts' can absorb water, dyes (Miehe 1911), and nutrients (Huxley 1978; Rickson 1979), although cracked surfaces of these cavities appear to be equally absorptive (Huxley 1978).

When the tuber of a plant which is occupied by ants of the genus *Iridomyrmex* is cut open, ant brood is found in the smooth-walled chambers, while debris consisting of dead insects and decayed amorphous matter is found on the warted surfaces (Beccari 1884–1886; Miehe 1911; Janzen 1974; Huxley 1978). Fungi and inquilines occur in this debris (see Chapter 26). Brood, however, is moved by the workers to different clean chambers depending on wind, sunlight, etc. The proportion of smooth and warted surfaces in the tubers appears to be maintained throughout the life of the plant. This is important for the two functions of housing ant brood and absorbing nutrients for both to develop together. This constant proportion can be maintained in two ways. Either each new cavity may be either smooth or warted, or each cavity may comprise both warted and smooth chambers.

The tuber cavity structure of *Myrmecodia*, *Squamellaria*, and to some extent *Anthorrhiza*, is characteristic for the genus. The tuber cavities of *Hydnophytum*, on the other hand, show greater diversity.

Hydnophytum *tuber structure*

Broadly speaking there are two types of cavity organization in *Hydnophytum*. Those species in which the tuber cavities are of two major forms (sub-globose tubers with a diffuse form of tuber growth), and those species in which all the cavities are essentially similar repeat units with a regular differentiated structure. The 22 species of *Hydnophytum* for which tuber structure has been studied fall into four more or less distinct types which in turn fall into the above two categories.

Moseleyanum-*type tubers* (Fig. 24.4(a))

The tuber cavities of this tuber type show a gradation from narrow, planar cavities near the tuber surface (the *open-type* cavities) to larger bulbous cavities which lie towards the tuber centre (the *closed-type* cavities).

The *open-type* cavities are planar to tubular branched cavities, they lie parallel to the tuber surface, and open by several lipped, or unlipped, entrance holes which vary in size from 2 to 6 mm. Generally they have smooth surfaces, but may be warted towards their extremities. These cavities are only occasionally interconnected, and overlie one another in older tubers.

The *closed-type* cavities open to the base of the tuber, and sometimes elsewhere, often by only one or few, large (10–40 mm) entrance holes. These cavities are large, bulbous-ended, and have densely warted wall surfaces.

The tuber grows over its whole surface, and it seems that the *closed-type* cavities are formed first and enlarge in step with the tuber during growth, while new *open-type* cavities are inserted amongst, and peripheral to, earlier *open-type* cavities. The outer surface of the tuber and deep-lying cavities often have a scabrous, or areolate appearance, suggesting that the epidermal layers have been torn and fragmented during growth.

In lowland species these tubers are usually occupied by ants, while in montane species, the tuber may be home to spiders, beetles, myriopods, etc. (Jebb 1985).

Spiral-type tubers (Fig. 24. 4(b))

In the montane forests of New Guinea there are several species of *Hydnophytum* in which the tuber tissue is thin and hard. The cavities comprise a large proportion of the tuber volume. Warts are entirely lacking. One or more large spiral cavities occupy the centre of the tuber. These open by large funnel-like entrance holes. The spirals of separate cavities tend to run in opposite directions, as well as lying in different planes to one another. The remaining cavities are narrow, tubular, little branched, and blind-ended. They open to the outside by small, often lipped, entrance holes. The two cavity types are rarely interconnected.

These tubers may be occupied by a wide range of arthropods. In two species the cavities normally contain rainwater, and a small tree frog, *Cophixalus riparius* Zweifel, appears to breed in the cavities of *H. myrtifolium* Merr. & Perry (Jebb 1985).

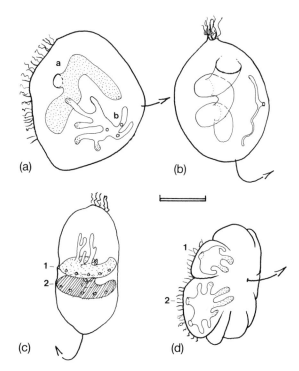

Fig. 24.4. Diagrams of the four major tuber cavity types in *Hydnophytum*. Dots indicate distribution of warts within the cavities. Two cavities are shown in each tuber. The arrow represents the stem(s). Scale bar = 10 cm. (a) *Hydnophytum moseleyanum*, showing Moseleyanum-type cavities, with deep-lying, densely warted *closed-type* cavities, (a), and superficial, largely smooth-walled *open-type* cavities, (b). (b) *Hydnophytum myrtifolium*, showing *spiral-type* tuber cavities. (c) *Hydnophytum guppyanum*, showing *apical-type* tuber cavities, two separate cavities are shown, 1, 2, 2 is hatched. (d) *Hydnophytum* sp. with *lateral-type* tuber cavities. Two separate cavities are shown, 1, 2.

Apical-type tubers (Fig. 24. 4(c))

A pair of closely related species of *Hydnophytum*, from the northern Solomon Islands, have cylindrical tubers which grow apically. Each cavity is essentially the same and is differentiated into chambers. New cavities are produced apically to the preceding cavities, being inserted at the tuber apex between existing cavities and the base of the stem.

The cavities are bilaterally symmetrical and form repeat units, each cavity opens by means of a pair of large, prominently lipped entrance holes in *H. kajewskii* Merr. & Perry, and by a ring of entrance holes in *H. guppyanum* Becc. Each cavity consists of a single, large, planar chamber, which lies perpendicular to the long axis of the tuber, and a basal system of narrow smooth-walled tubular tunnels, which ramify amongst earlier cavities or immediately below the tuber surface. The surfaces of the large planar chamber are rough with small clusters of warts. These have the appearance of short roots with root caps (Beccari 1884–1886).

Hydnophytum kajewskii contains both rainwater and cockroaches, while *H. guppyanum* is regularly occupied by *Iridomyrmex cordatus*.

Lateral-type tuber (Fig. 24. 4(d))

Only one, undescribed, species of *Hydnophytum* is known to exhibit wholly lateral growth of the tuber (Jebb 1985). The cavities are all essentially similar and are added laterally and, as a result, the tuber develops a flattened and lobed form. The cavities are not interconnected, and each opens by one or more entrance holes at the base of the tuber. The broad, flattened chamber may or may not be densely warted, and gives rise to several narrow tubular extensions, which ramify towards the base of the stem. In some cavities the planar chamber is smooth walled and the extremities alone are warted. This species is occupied by a high-altitude form of *Iridomyrmex scrutator* (Jebb 1985).

Myrmecodia *tuber structure*

The tubers of *Myrmecodia* are spheroid when young, becoming fusiform to cylindrical with growth. In the majority of species the surface is spiny, the spines being simple to stellate and sparse to dense. The stem or stems arise from the apex of the tuber, and only rarely do subsidiary stems arise laterally, although branching of the tuber is not unknown (Jebb 1985).

Unlike the tuber cavities of *Hydnophytum*, those of a given species of *Myrmecodia* are all more or less similar. The cavities are added sequentially and spirally at the tuber apex. Each cavity is differentiated into several chambers, each with distinct shapes, surface types, and colours.

All species of *Myrmecodia* for which adequate material is available possess a broadly similar organization of chambers within each cavity. Each cavity is composed of three distinct chambers; a '*superficial chamber*', several '*warted tunnels*', and '*inner smooth-walled chambers*' (Fig. 24. 5) (Huxley, 1976).

The *superficial chamber* is an oval, planar chamber, lying immediately below the tuber surface and may open by entrance holes and pores, and may connect to the *superficial chamber* of adjacent cavities. This chamber is apically warted and basally smooth-walled. From its apical end the superficial chamber gives rise to tubular, *warted tunnels*, and from its inner wall gives rise, via circular entrances, to the *inner smooth-walled chamber*. The warted tunnels are tubular, blind-ended, and branched; in some species they have smooth-walled extremities. The warted tunnels are dark brown or blackish due to the deposition of detritus by the ants, and the growth of fungal

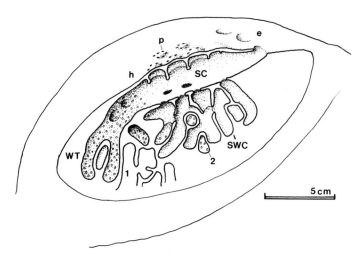

Fig. 24.5. Diagram of a single cavity in the tuber of *Myrmecodia schlechteri*. Apex of tuber lies at the left. SC = superficial chamber; WT = warted tunnels; SWC = smooth-walled chambers; h = honeycombing; e = entrance holes; p = pores; 1 = smooth-walled chamber of subsequent cavity; 2 = warted tunnel of preceeding cavity.

mycelia (Huxley 1976). The extremities of these tunnels are usually packed with moist plugs of decomposing debris (Janzen 1974c; Huxley 1976, 1978). The *inner smooth-walled chambers* ramify amongst the *warted tunnels* of the previous cavity, but the two do not interconnect. These chambers are broad and planar, opening by numerous entrances to the overlying superficial chamber. The walls of the inner smooth-walled chambers are smooth, dry, and pale in colour, but the apical parts may be more deeply pigmented, and even warted, suggesting that during formation of a new cavity the most apical parts are determined as warted, and the proximity of the apical parts of the inner smooth-walled chamber to the warted tunnels results in their some-times being influenced by this factor (Jebb 1985).

In some species in which the tuber hangs pendulously from the host tree, the inner smooth-walled chambers form horizontal shelves. In these species the superficial chambers are honeycombed (Huxley 1976) and the tuber surface overlying them is perforated by pores, which are too small to admit ants (Fig. 24. 5).

The form of these chambers develops gradually through the first four or five cavities (Treub 1888; Huxley 1976). The first cavity grows only slightly after its formation, reaching 4 to 5 cm in length. By the time the fifth cavity forms, all the features of the mature *Myrmecodia* cavity are usually present. At this stage one of the major divisions in the genus becomes apparent. In some species the superficial chambers are interconnected, while in others they are not, and open only to the outside via their own entrance holes (Huxley 1976). Three artificial groupings of *Myrmecodia* tubers are recognized on the basis of interconnections of the cavities and the location of entrance holes (Jebb 1985).

In a few species entrance holes are present at the tuber apex, e.g. *M. platytyrea* Becc., and in others the most apical cavities connect to tunnels within the stem, which open at the inflorescences (in the alveoli). Those species with interconnected superficial chambers are confined to New Guinea and northern Australia.

Squamellaria *tuber structure* (Fig. 24. 6)

Squamellaria resembles *Myrmecodia* in having a cylindrical tuber to which essentially similar cavities are added at the tuber apex, singly and sequentially. Each cavity is bilaterally symmetrical and is inserted in the same orientation. The tuber is somewhat flattened from side to side, exhibiting a degree of bilateral symmetry. The surface lacks ridges and pores and is smooth except for numerous small swellings, some of which bear slender, flexible, almost hair-like spines up to 1.5 cm in length. Entrance holes are arranged in circles concentric around the tuber axis, each circle belonging to a single cavity.

In a mature tuber each cavity is composed of several chamber types, with characteristic shapes and wall surfaces, the arrangement being uniform throughout the genus. The cavities comprise a *girdle chamber*, and several *superficial smooth-walled chambers*, and *deep-lying smooth-walled tunnels*.

The large *girdle chamber* saddles the tuber and the lowermost arms of this chamber are branched and bulbous-ended, and the walls here are densely

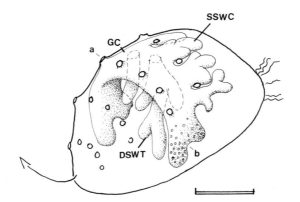

Fig. 24.6. Diagram of a typical tuber of *Squamellaria imberbis*, with rings of entrance holes, a. A single cavity is shown as a solid within a transparent tuber. GC = girdle chamber; DSWT = deep-lying smooth-walled tunnels; SSWC = superficial smooth-walled chambers; b = densely warted ends of girdle chamber. Scale bar = 10 cm.

warted (Fig. 24. 6). The outer wall of the girdle chamber opens along its apical edge by a ring of conical entrance holes. The outer wall of this chamber ramifies outwards, between the *superficial smooth-walled chambers* of the adjacent, younger cavity. The walls of the girdle chamber are smooth and rugose, the rugae being 1.5 to 2 mm across. From the basal edge of this chamber a number of superficial smooth-walled chambers arise. These lie above the girdle chamber of the immediately preceding cavity. They are from 0.4 to 1.5 cm across, branched, and ramify amongst the numerous outward extensions of the girdle chamber of the preceding cavity. The last chamber type is the *deep-lying smooth-walled tunnels*, which lie towards the centre of the tuber. These chambers arise from the inner wall and towards the rear of the girdle chambers.

The girdle chambers of successive cavities are interconneted. Along their basal edge small circular holes give access to the preceding cavity's girdle chamber, while along its front margin similar holes connect to the later cavity. In addition there may be a few holes connecting non-adjacent cavities, namely via the basal edge of the superficial smooth-walled cavities, which may connect with the apical edge of the next but one preceding cavity's girdle chamber.

As in the other genera the first cavity consists of a simple, bulbous, and densely warted chamber, reaching up to 7 cm in length. As in *Myrmecodia*, there is a gradual increase in complexity, from the first to the fourth cavity, after which the structure of individual cavities remains much the same. The differentiation of each cavity into several chambers is a striking parallel to the situation in *Myrmecodia*, however, the cavities of the two genera being unalike, and there are no obvious homologous features.

The cavity structure of the *Squamellaria* tuber shows interesting parallels to that of the two apically growing species of *Hydnophytum*: *H. guppyanum* and *H. kajewskii*. Each has a large chamber which opens to the outer surface of the tuber and encircles, or fills, much of the tuber space. From this chamber deep-lying or peripheral tunnel-like chambers arise (Fig. 24. 3(c) and 24. 6). Moreover, the tubers are bilaterally symmetrical in all of these species.

Anthorrhiza *tuber structure*

Anthorrhiza is of great interest in understanding evolutionary changes in these ant-plants. Externally, the species show a marked morphological series: from those which closely resemble *Hydnophytum* (Fig. 24. 3(a)), through intermediates (Fig. 24. 3(b)), to those which have a *Myrmecodia*-like morphology (Fig. 24. 3(c)).

This external morphology corresponds to their internal cavity structure; the externally, more *Myrmecodia*-like species show more segregation of cavity wall types. Unlike the tubers of *Myrmecodia* and *Squamellaria*, however, this segregation is not achieved by chamber types within a cavity unit, but through segregation of the warted and smooth surfaces into separate cavities, as in the tubers of Moseleyanum-type *Hydnophytum*.

Anthorrhiza areolata lacks warts entirely, and like the spiral-type *Hydnophytum* species it traps rainwater and houses a range of opportunistic arthropods, in particular spiders and cockroaches. Externally it resembles *Hydnophytum*, with a smooth spineless tuber, slender stems, and thick xeromorphic leaves (Fig. 24. 3(a)).

The majority of *Anthorrhiza* species have tuber cavities similar to the Moseleyanum-type *Hydnophytum* tubers, with large, deep-lying, and warted cavities which open to the tuber base, and more finely built superficial, tubular, branched, and smooth-walled cavities which open by numerous entrance holes to the tuber surface. The superficial cavities are narrow and tubular, warted at their extremities, and little interconnected. The deep-lying cavities are little branched and have globose endings which are usually densely warted. Externally these species are intermediate in morphology, with scattered spines and slender stems (Fig. 24. 3(b)).

In *A. caerulea* the cavities are more differentiated. The deep-lying warted cavity is solitary. Its opening onto the tuber base is long and narrow, and may reach up to half the length, or more, of the diameter of the tuber. This opening enters a narrow, planar chamber from which a number of bulbous chambers arise, each of which is entered by a somewhat restricted neck. Apically it gives rise to a flattened chamber which lies perpendicular to the tuber axis (Fig. 24. 7(a)). Superficial to this central warted cavity, are smooth-walled

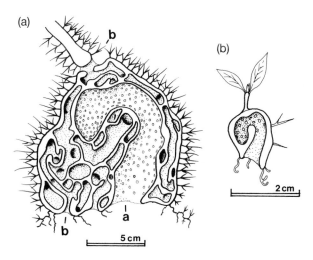

Fig. 24.7. Tuber cavities of a *Myrmecodia*-like species, *Anthorrhiza caerulea*, in the (a) mature and (b) juvenile tuber. The warted surfaces are confined to a large central cavity, a, which is derived from the first cavity of the tuber, (b). Smooth-walled cavities surround this central cavity, opening to the base and apex of the tuber, b.

cavities, which are narrow, branched, dark walled, and much interconnected. These cavities open both to the base and the apex of the tuber, but do not connect to the warted cavity. Isolated chambers may be pale walled and warted. Externally the tuber is markedly conical in form, the stem is thickened and contracted, and both are densely beset with stellately branched spines (Fig. 24. 3(c)).

Unlike mature tubers of *Myrmecodia*, *Squamellaria*, or most *Hydnophytum*, the first warted cavity of some *Anthorrhiza* does not remain as a small basal feature in the mature tuber but grows into a large central warted cavity. In *A. caerulea* no other warted cavities develop (Fig. 24. 7(a)), while in other species there are several large warted cavities.

Vascular connections between stem and roots

The growth of a tuber has implications for the continuity of vascular connections from root to stem. The water-conducting system of plants (xylem) is normally operative when the cells are dead, and the cytoplasm and other cell contents are absent. Therefore, functional xylem cannot be extended in the middle but only at the ends, where living, but as yet non-functional xylem initials can differentiate. Certain plants, such as grasses, or parts of plants, such as seeds, have interstitial meristems but, on the whole, these are rare and short-lived.

In the tubers of *Myrmecodia* new cavities are added apically, being inserted between the apex (youngest part) of the tuber and the base (oldest part) of the stem (Fig. 24. 8(b)). If vascular connections are to be maintained then an interstitial meristem must be active at the tuber/stem junction.

In those species with entrance holes around the base of the stem, or in which the stems are hollow, growth of the tuber does not destroy these connections. Tuber size appears to be limited in these species of *Myrmecodia* probably due to the presence of these apical complexities, which may make growth here disruptive (Jebb 1985).

In *Squamellaria* the tuber is also cylindrical, and new cavities are added apically (Fig. 24. 8(c)). Unlike *Myrmecodia*, however, the stems are not confined to the tuber apex. In *S. imberbis* (A. Gray) Becc. the stems are scattered over the apical end of the tuber, being densest towards the apex. In *S. major* A. C. Smith, however, one stem appears to be related to each successive cavity unit. It may be that in *Squamellaria*, apical growth of the tuber causes stems formerly near the apex to be 'left behind' on the tuber surface. Therefore, vascular connections between the stems and roots would not be disrupted by the increase in tuber size (Fig. 24. 8(c)).

Tuber growth of the *Myrmecodia*-like *Anthorrhiza caerulea* is achieved by internal enlargement of the central, warted cavity and the insertion of new, superficial, planar chambers amongst those already present. The conical shape of the tuber in this species suggests a possible basal direction of growth

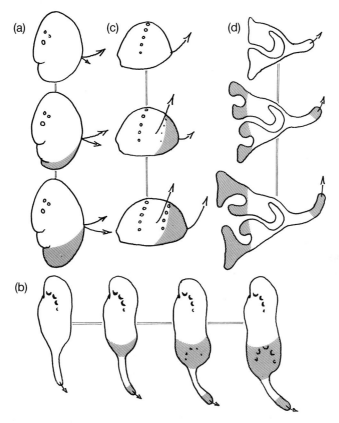

Fig. 24.8. Diagram of tuber growth in the four major genera of the Hydnophytinae. (a) *Hydnophytum*; (b) *Myrmecodia*; (c) *Squamellaria*, and (d) *Anthorrhiza* species 8, simplified tuber section to show growth of first cavity. Cross-hatching indicates new tissue. The arrows represent the stems. See text for explanation.

(Fig. 24. 8(d)). The shape of the warted cavity certainly supports this idea (Fig. 24. 7(a), (b)), and the occurrence of new cavities at the base of the tubers also lends weight to this suggestion. The apical entrance holes of this species are joined in an almost continuous ring around the stem (Fig. 24. 7(a)), and apical tuber growth would disrupt this arrangement.

In two other species of *Anthorrhiza* a series of young tubers was available for study and these, likewise, indicated that cavities formed early in the life of the tuber continued to expand; although there was no suggestion that the tubers of these species tended to grow at their basal periphery only.

The mature tuber structure as an ant-house

In *Myrmecodia*, *Squamellaria*, and *Hydnophytum guppyanum* and *H. kajewskii*, the first cavity is present in the mature tuber as a small, hooked structure, which does not grow significantly. In other *Hydnophytum* tubers the fate of the first cavity is less clear, mainly because growth of the tubers in many species is irregular and it is not apparent where the first cavity lies in the mature tuber. In Moseleyanum-type and spiral-type tubers, it may be that the large deep-lying, closed-type cavities are derived from the early cavities in a similar manner to those of *Anthorrhiza*.

The uniform and repeat cavities of *Myrmecodia*, *Squamellaria*, and certain *Hydnophytum* spp. appear to be the main key to their success in producing an efficient 'ant-house'. New cavities are added to the tuber with minimal disruption of earlier cavities. Since each cavity is a 'repeat unit', it contains the same essential structure, and thus the same ratio of different chambers, with their corresponding wall type, is maintained throughout the life of the tuber. This would allow a growing colony to increase its number of brood at the same rate as new areas are made available for debris and defecation.

The three 'ant-house' genera appear to have solved the problem of cavity growth with regard to the maintenance of tuber infrastructure in different ways; or at least to have circumvented the problem by a different system of growth. Most species of *Myrmecodia* appear to have evolved an efficient interstitial meristem at the apex of the tuber, which enables the stem to be carried forwards in front of new cavities; *Squamellaria* has evolved a system whereby it largely avoids the problem of extending the vascular connections between stem and root; while at least one species of *Anthorrhiza*, which shows some of the most *Myrmecodia*-like features in the genus, appears to grow 'backwards', again avoiding a young tuber/old stem complication, forming a large warted cavity, and developing a conical form.

Phylogeny as evidenced by the tubers

Evidence from general morphological and tuber cavity characters indicates that *Hydnophytum* is a paraphyletic, ancestral, and diverse genus, as opposed to the 'derived' monophyletic genera *Myrmecodia*, *Squamellaria*, and *Anthorrhiza*. The range of tuber structure is both greater and more varied in *Hydnophytum* than in the others. Although tuber cavity structure is an important, generic character in these other genera, there are reasons for maintaining *Hydnophytum* as a single, paraphyletic genus. The alternative systematic arrangement would be to sink all the genera into a single taxon (*Hydnophytum*) and raise the present genera to sub-generic rank, a cumbersome nomenclatural exercise. In the current arrangement, the three successive, parallel lines of evolution in the 'ant-house' genera are explicit in the taxonomy.

Ant occupancy

The taxonomy of *Iridomyrmex* is inadequate to determine whether taxa restricted to ant-plants should be recognized at specific level (e.g. *I. myrmecodiae* Emery). Information does suggest, however, that the incidence of certain species of ant in the tubers of, say, *Myrmecodia* is more an indication of co-ecological preference rather than of co-evolution. *Iridomyrmex cordatus* is normally present in the Hydnophytinae in mangroves, savannahs, and disturbed habitats up to 3000 m, while the *I. scrutator* group is present in Hydnophytinae in humid closed forests, with one species in lowland forest and another in montane forest. A number of other genera have been reported, especially from disturbed areas, but none of them regularly (Huxley 1978). Some species of *Hydnophytum* and of *Anthorrhiza* are not normally occupied by ants.

Conclusion

Combining information on inhabitation by ants with an analysis of the complex growth of the tubers of the Hydnophytinae suggests that ant specificity is associated with a tuber structure in which the ratio of smooth to warted chambers is maintained during the growth of the plant. This allows a steady increase in the size of the ant colony and in the plants' requirements for nutrients. The ratio is maintained either by producing a series of similar, repeat cavities, each with a fixed ratio of smooth to warted chambers, or by continuous growth of the warted cavity with addition of new smooth-walled cavities (as in *Anthorrhiza*). A repeat cavity system is seen in *Myrmecodia*, and *Squamellaria*, and is well developed in two species of *Hydnophytum*. In *Squamellaria* the arrangement of the repeat cavities resembles that in the two species of *Hydnophytum* but is more complicated. However, the repeat cavities of *Myrmecodia* are different in arrangement and form, suggesting an independent line of evolution. It is likely that *Myrmedoma* represents another independent line of evolution of maintenance of the ratio, but this needs to be tested. These independently developed modes of maintaining the ratio of smooth to warted cavities suggest a strong selection pressure for this feature.

The position of the tuber domatium between the stem and roots of the plant poses anatomical problems for sustaining vascular connections during growth of the plant. This has been solved in different ways as well: basal growth in some *Anthorrhiza*; leaving one shoot associated with each cavity in one species of *Squamellaria*; and lateral growth in some *Hydnophytum*. The morphogenetic complexities generated by symbiosis with ants in this group vie with floral adaptations as some of the most remarkable structures in the plant kingdom.

25

Parasitism of ant–plant mutualisms and the novel case of *Piper*

D. K. Letourneau

Parasites of diffuse associations

An organism can be considered a parasite of a mutualism if it exploits resources or the services of mutualists, such that they are unavailable to the organisms that are capable of providing reciprocal benefits. Diffusely associated mutualists such as plants with extrafloral nectaries and their visiting coterie of ants and other insects have a large potential for parasitism. Visitors that do not protect the plant but imbibe nectar and thereby compete for nutrients with the most effective mutualists may reduce the efficiency with which the plant is protected. Exploitation of nectar by herbivores and scavengers (e.g. Agnew *et al.* 1982) is clearly parasitic. Among ant and wasp visitors, differences in plant-guarding capabilities resulting from differences in size, territorial behaviour, and frequency of visits can translate into differences in herbivory levels (Beckmann and Stucky 1981; Horvitz and Schemske 1984; Oliveira *et al.* 1987). Koptur (1984) also attributed much of the variation in herbivory of *Inga* spp. (Mimosaceae) to the 'protective ability' of different nectary-visiting ants. The determination of such ability would be based upon morphology (size, offensive structures), physiology (e.g. sting vs. spray defence), and behaviour characteristics (recruitment rate, speed, strength, and appropriateness of response to key herbivores). Complications involve the tradeoffs in these factors, the evolutionary context, and the variable herbivore community. For example, a large, stinging ant that readily attacks grasshoppers would appear to be much more valuable to its host plant than a small, slow ant that imbibes nectar but is reluctant to sting (Seigler *et al.* 1982; Oliveira and Brandão Chapter 14, this volume). However, if the most detrimental plant feeder were a seed borer that is only vulnerable to ants as an egg, then the behaviour of the small ant may, in fact, embody a greater 'protective ability'. Similarly, Koptur and Lawton (1988) and Hespenheide (1985) suggested that parasitoids, not ants, are the important mutualists of some plants with extrafloral nectaries.

Comparable complexities emerge in attempting to define parasites of seed dispersal mutualisms. Ants that collect elaiosome-laden seeds but destroy the seed may be considered parasites, and species that regularly deposit seeds in locations where they cannot grow are other candidates (Janzen 1985). The variable quality of protection received per unit of nectar invested, or the variable probability of successful seed dispersal per unit of food provided, are complicating factors in cost–benefit models of facultative ant–plant mutualisms (Keeler 1981; Koptur 1984). The potential loss of fitness of ants and plants in diffuse associations due to resource use by non-mutualists has been recognized but has not been evaluated empirically or experimentally.

Parasitism of symbiotic ant–plant associations

Relatively unambiguous examples of parasitism have been discovered among the associates of symbiotic ant–plant mutualists. A review of six cases of parasitism that cross several taxa and span three trophic levels is presented here: omnivorous ants, opportunistic plants, a herbivorous beetle, and coleopteran predators of plant-ants.

The ant *Pseudomyrmex nigropilosus* Emery acts as a parasite of the ant-acacia mutualism in Central America (Janzen 1975). The ants nest in the thorns of bull's-horn acacia and harvest Beltian bodies and extrafloral nectar from the trees. However, unlike the obligate acacia-ants, also in the genus *Pseudomyrmex*, *P. nigropilosus* does not protect the host plant from encroaching vines or herbivores (see Ward Chapter 22 this volume). The ants are diurnal, leaving the foliage unattended during the night, and even when present they do not disturb other organisms that come into contact with the tree. The negative effect on the plant becomes apparent with time, as herbivores exploit its foliage and neighbouring vegetation shades the tree. The mutualistic ants are not as likely to colonize a sapling already occupied by the parasitic ants as they are an unoccupied acacia, and if they do, the growth of their colony is retarded because of competition for Beltian bodies by *P. nigropilosus*. The life history and colony structure of *P. nigropilosus* reflects its opportunistic way of life. The colony produces reproductives very quickly and maintains fewer workers per gram of brood than do the obligate acacia-ants.

A similar parasitic relationship occurs between *Cataulacus mckeyi* Snelling (Formicidae: Myrmicinae) and the mutualism between *Leonardoxa africana* (Baill.) Aubrev. (Caesalpiniaceae) and *Petalomyrmex phylax* Snelling (Formicinae) in the lowland rain forests of Cameroon (McKey 1984). *Cataulacus* exclude the mutualistic ants from branches or shoots of the plant and, compared to *P. phylax*, allow greater exploitation by herbivores. The cumulative damage to young leaves and shoot tips of parasitized *L. africana* results in suppressed growth and increased mortality of saplings.

The case of *Cataulacus* differs from the acacia example in the permanence of parasite occupation (acacia saplings occupied by *P. nigropilosus* rarely survive more than one year). Contributing factors are the lack of shoot take-over by *Petalomyrmex phylax*, the lower level of plant competition in the understorey versus light environments, and the weak patrolling behaviour of *Cataulacus*. Consistent with these trends, colony production of alates is proportionately greater in *P. nigropilosus* colonies than in *Cataulacus*.

Kleinfeldt (1978, 1986) described a situation in which plants inhabiting ant-gardens are parasitized by ants. In ant-gardens, the ants construct carton nests in trees and incorporate seeds of certain epiphytic myrmecochores (see Chapter 23). Several species of ant act as parasites of ant-gardens; for instance, *Solenopsis picta* Emery which invades and takes over the nests of the carton builder *Crematogaster longispina* and imbibes nectar but never repairs or extends the carton nest, which eventually deteriorates.

In the Far East, ant-house epiphytes have been parasitized by other plants which utilize nutrients contributed by resident ants without providing shelter for the colony. Janzen (1974c) described the epiphytic climber *Dischidia gaudichaudii* Dcne. (= *D. nummularia* R. Br.) (Asclepiadaceae) and the small shrub *Pachycentria tuberculata* (Melastomataceae) as parasitic on *Hydnophytum* and *Myrmecodia* (Rubiaceae) in Borneo (see also Weir and Kiew 1986). The seeds of the parasite *P. tuberculata* are gathered by ants and placed in the debris deposits in the tuber chambers of *Hydnophytum formicarum* Jack. They germinate inside the ant-plant and grow out of the entrance holes of the *H. formicarum* tubers. It is not known to what degree these plants lower the fitness of the ant-plant by removing nutrients from the debris dumps, but the relationship is potentially harmful.

Herbivorous beetles are known to exploit the resources of mutualists. Many *Cecropia* trees house protective ants but also serve as a dwelling for parasitic *Coelomerus* beetles (see Jolivet Chapter 26 this volume).

A novel parasite of Piper

A sixth example of parasitism has been discovered in the form of a specialized ant predator which exploits the relationship between *Piper* ant-plants and their *Pheidole bicornis* Forel inhabitants in Costa Rican forests (Letourneau 1990). Two species of *Phyllobaenus* beetles (Cleridae) have been observed on occasion in the hollow petiole chambers of *Piper ceno-cladum* C.DC. and *P. obliquum* R. & P. over the last five years at La Selva Biological Station (Heredia, Costa Rica) and at Finca Loma Linda (Cañas Gordas, Coto Brus), respectively. The larvae are usually associated with piles of dead ants and seemed to be the most common cause of colony demise in *Piper* plants.

Predatory behaviour of *Phyllobaenus B* was observed by rearing them in excised petiole sheaths in petri dishes. The beetle larva typically backs into the cone-shaped end of the chamber with mandibles protruding. As ants approach the larva, it cracks their head capsules or crushes the pronotum and tosses the bodies aside. After all oncoming adults have been killed or disabled, the beetle larva proceeds to consume the ant brood. The beetle larvae do not feed on adult ants.

These fascinating beetles show evidence of their novel ability to go beyond their role as ant predators and to parasitize the interaction. First, the relationship of the ants and their *Piper* hosts must be described (Risch *et al.* 1977). Initially, the ants reside in hollow chambers formed by the recurved, sheathing leaf bases. As the colony grows, the workers hollow out the stem and the entire plant becomes a domicile. The plant provides not only shelter but also nutrients in the form of hundreds of single-celled, opalescent bodies, rich in lipids and proteins, produced on the adaxial surface of the sheathing leaf bases. *Pheidole bicornis* ants collect and consume the food bodies. The ants may provide dual benefaction for the plant: they deposit frass on the inside of the hollowed stem which might release nutrients for the plant (Janzen 1974*c*; Risch *et al.* 1977) and they reduce herbivory on the plant by removing small larvae and eggs from the leaves (Risch 1982; Letourneau 1983; D. K. Letourneau and F. Arias G., unpublished work).

The unique feature of this interaction is that food production by *Piper* is induced by its ant inhabitants (Risch and Rickson 1981). The production of food bodies nearly ceased when *P. bicornis* were removed from the plant, and food body production commenced again when the ants reinvaded. There is circumstantial evidence that only this ant species is able to stimulate food body production by *Piper*. In examinations of nearly a thousand plants in four *Piper* species, I have observed many instances in which single petioles were inhabited by ants other than *P. bicornis*, but these petioles never contained food bodies.

In most places in Costa Rica, plants without ant colonies are rare. However, I recently discovered an unusual population of *Piper obliquum* in Carara Biological Reserve, Puntarenas, in which only one plant was occupied by an ant colony. The other plants had no external signs of previous occupation by *P. bicornis* ants, and none had hollow stems. According to previous studies, the hollow petiole chambers on these vacant plants should not have contained food bodies.

To test this assumption, a systematic survey at Carara Biological Reserve of plants appearing to be unoccupied by *P. bicornis* was conducted, by carefully opening each sheathing leaf base enough to see its contents with a light. Only one petiole on one tree contained *P. bicornis* ants and food bodies. Yet all the plants that lacked ant colonies had substantial food-body production (21 of 56 petioles on the remaining seven plants each contained hundreds of food bodies). In every case, the petioles that had produced food bodies also contained a *Phyllobaenus A* beetle larva. The presence of *Phyllobaenus A* in trees without ant colonies, however, was puzzling since *Phyllobaenus* beetles are ant predators. In addition to the larvae, nine beetle eggs were deposited, one or two per petiole, and four pupae and one teneral adult were closed into the ends of hollow petioles by a partition presumably made by larval secretions. Thus, the *Piper* plants at Carara seemed to produce ant-food even when ants were absent, and *Phyllobaenus* predators seemed to be subsisting where there was no prey.

Could the predators have evolved the ability to stimulate food-body production, and thus subsist directly on plant products? Seven *Phyllobaenus* larvae at Carara were removed from their chambers and added to smooth, empty petioles on plants without ants. Eighteen days later, six of the newly occupied chambers had hundreds of food bodies. In the seventh, the larva was absent and the petiole contained a small colony of ants (*Monomorium*) but no food bodies. The beetle larvae exploited the mutualistic association between ant and plant by causing the plant to produce food for them in the absence of ants.

Additional field observations at La Selva and rearing of *Phyllobaenus* larvae from Carara and Loma Linda suggest that these predators spin a partition as first instars and induce food-body production for nourishment until they reach the final instar. Then they break the partition and attack the ant inhabitants. Although I have not quantified their populations in all eight forest sites in Costa Rica, a survey of *P. obliquum* in Cañas Gordas revealed only one larva in 105 petiole chambers (< 1 per cent) of 15 plants. In 278 petioles on 33 *Piper cenocladum* plants, surveyed in 1985, only seven (2.5 per cent) contained a *Phyllobaenus* larva. In contrast, these larvae were common in the population at Carara (29 per cent of petioles occupied by a larva).

If the parasitic larvae are maintaining a low occupancy rate of *Piper* ant-plants by *P. bicornis* ants at Carara Reserve, they may be indirectly affecting another aspect of the plant's productivity. In early experiments I found that average herbivory levels of unoccupied *Piper* ant-plants were greater than those of plants containing *P. bicornis* ants (Letourneau 1983). One might expect the *Piper* population at Carara, in which plants are likely to house beetles but are unlikely to contain ants, to have relatively high levels of herbivore damage. Herbivory on the Carara population of *P. obliquum* was compared with herbivore damage at two other sites (on *P. cenocladum* at La Selva and *P. obliquum* at Cañas Gordas, Puntarenas) by estimating the leaf area missing from each leaf to obtain a mean herbivory index per plant for the four most recently produced, mature leaves. At Carara, all eight plants at the study site were used. At La Selva and Cañas Gordas, eight plants were chosen randomly from larger samples of 15 and 25 plants, respectively. Although this was not a definitive experiment since it compares herbivory at different sites, each presumably with a different herbivore load, the data were striking. Herbivore damage to plants at Carara ($x = 29.7\% \pm 15.1$ s.d.) was significantly greater and approximately four times that of plants at La Selva ($x = 7.6\% \pm 3.5$ s.d.) and Cañas Gordas ($x = 7.9\% \pm 5.8$ s.d.) (ANOVA, d.f. = 2, $F = 12.3$, $P = 0.0004$; Tukey's mean separation test, $\alpha = 0.05$). Thus, *Phyllobaenus* predators not only exploit the ants, but also the plants, as food, and may secondarily reduce plant fitness by disrupting plant defence against other herbivores.

Mechanisms to avoid parasitism

Some special attributes of plant-ants and ant-plants, both physical and behavioural, have been presented as puzzles or have been explained as having a primary function of reciprocal benefits when instead they may have originated as mechanisms to avoid parasitism of the system! Janzen (1985) suggests that many of the fine-tuned aspects of the morphology and behaviour of mutualists have in fact evolved to discourage parasitism of the relationship. For example, it is well known that plant-ants often cut vine tendrils and prune potentially competing vegetation that grows near the host plant. The net result of this behaviour can be an extreme reduction in shading, increased plant vigour, and protection from fire (Janzen 1967). The maintenance of plant health benefits the ant inhabitants, and this eventual benefit might explain the pruning behaviour of mutualistic ants. Davidson *et al.* (1988) conducted experiments which support an alternative explanation for the origin of this behaviour: the pruning of encroaching vegetation by plant-ants limits accessibility of the foliage to enemy ants and reduces the frequency of invasion of some ant-plants by non-mutualistic ant species. They also showed that pruning behaviour is more common in ant species with stings rather than those with the more effective chemical spray defence. Some of these invading ants harvest plant-produced food (Davidson *et al.* 1988), which suggests that pruning behaviour can protect against both predators of ants, plants, and parasites of the mutualism.

In a similar manner, an anti-parasite function may be accorded to the facultative production of food bodies by *Piper* ant-plants. It has been suggested that energy saving governs the cessation of food-body production by uninhabited plants (Risch and Rickson 1981). An additional effect of having food-body production directly induced by a specific ant species is to reduce the suitability of the plant as a host to other ant species. The extent to which these mechanisms exclude parasites seems to differ in different systems. The specificity of entrance-hole shape excludes the parasitic ant *Cataulacus* from *Leonardoxa* shoots colonized by the mutualist ant *Petalomyrmex*, but this also works the other way around, allowing long-term occupancy by the parasite when it invades first (McKey 1984). The custodial behaviour of *Pheidole* may be the factor that limits parasite loads in forests with high ant occupancy rates, since *Phyllobaenus* beetles oviposit on the exterior of the plant (Letourneau, unpublished work).

Role of parasites in the dynamics of two-species mutualisms

Simple mathematical models that predicted instability of obligate mutualisms were accepted without much question in the early 1970s, when these associations were deemed of little importance in natural communities (Williamson 1972; May 1973). Theoretical ecologists have since added complexity to these models to accommodate the prominence of mutualistic associations in both the tropics and temperate zones. Time delays (May 1974), non-linearities (May 1976; Vandermeer and Boucher 1978), re-interpretation of stability criteria (Addicott 1981), and the impingement of competitors and predators (Heithaus *et al.* 1980) have been proposed to allow the achievement of stable equilibria in these models. Parasites of obligate mutualisms may, however, also enhance the stability of these associations when they are embedded in complex communities.

Acknowledgements

This research was supported by the University of California at Santa Cruz through Faculty Research Grants from the Division of Social Sciences and the Academic Senate, the National Geographic Research and Education Foundation, and University Research Expeditions Grants through the University of California at Berkeley. I wish to express my thanks to F. Arias G. for his interest, involvement, and insights into the *Piper–Pheidole* research. I am grateful to R. Gomez M., L. Gomez M., L. Lobos, and P. Lockwood for excellent technical assistance, W. Barr and E. O. Wilson for describing the beetles and identifying ants, respectively, and D. Cole, the Organization for Tropical Studies, and the Costa Rican National Park Service, for the use of their land and their support.

26

Ants, plants, and beetles; a triangular relationship

Pierre Jolivet

Introduction

Although many invertebrates are successfully removed or deterred by the ants of ant-plants, there are also many phytophagous insects which have evolved ways of appeasing or repelling the ants (Jolivet 1987*a*). Different solutions are used by beetles and also by a range of other arthropods. The biocoenoses of myrmecophytic plants are usually less diverse and abundant than those of non-myrmecophytic congeners, but consist of more specialized insects (Table 26. 1). The most extraordinary case of adaptation to an ant-tree is that of the beetle *Coelomera* (Galerucinae) to *Cecropia* in tropical America, and there are other cases where beetles mimic ant species. These beetles feed with impunity on ant eggs and on plant food bodies.

Tropical America: *Cecropia* trees

The 106 species of *Cecropia* (Cecropiaceae) are dioecious trees and, like other members of the family, have a rather restricted latex system (see Chapters 19 and 20). *Cecropia* are pioneer trees and grow well in forest clearings. Most *Cecropia* are myrmecophytic, the hollow internodes being inhabited by *Azteca* ants which enter via weak points (prostomata) in the stems. The ants feed on Müllerian bodies produced in petiolar pads (trichilia). However, a few species are devoid of ants and there are various reasons for this absence. Some species appear to have lost ant adaptations because of an absence of *Azteca* or absence of pressure from *Atta*, leaf-cutting, ants (Andrade and Carauta 1982). This has occurred in species on islands (e.g. *Cecropia peltata* L., *C. schreberiana* Miq., *C. pittieri* B.L. Robinson) and at high altitude (*C. palmatisecta* Cuatrec., *C. suyantepuiana* Cuatrec., *C. santanderensis* Cuatrec.). Relics of trichilia and Müllerian bodies can sometimes be seen in these species. In other species the internodes remain filled with pith (*C. monostachya* C.C. Berg) or the trees lack prostoma and trichilia (e.g. *C. hololeuca* Miq. from the Atlantic forest and *C. sciadophylla* Mart. from Amazonia).

 Cecropia, both with and without ants, has a very rich associated fauna.

Table 26.1. Myrmecophytes, ants, and their associated beetles.

Plant genus	Ant genus	Beetle genus	Beetle family	Relationship with the ants
Latin America				
Cecropiaceae				
Cecropia	Azteca	Rhabdopterus	Eumolpinae	antagonistic
(Andrade 1981)	(Dolichoderinae)	Coelomera	Galerucinae	antagonistic
(Jolivet 1987b)		Dircema	Galerucinae	antagonistic
(Jolivet 1988a, b; and in press)		Cacoscelis	Alticinae	indifferent
(Baird 1967)		Sceloenopla	Hispinae	indifferent
		Metaxycera	Hispinae	indifferent
		Epitragus	Tenebrionidae	antagonistic
		Ophtalmoborus	Curculionidae	indifferent
SE Asia				
Rubiaceae				
Myrmecodia	Iridomyrmex	Oribius gen.	Curculionidae	indifferent
(Jolivet 1973)	(Dolichoderinae)		Staphylinidae	indifferent
East Africa				
Mimosoidae				
Acacia	Crematogaster	Hockingia	Clytrinae	symphilic
(Hocking 1970)	(Mrymicinae)	Isnus	Cryptocephalinae	symphilic
		Syagrus	Eumolpinae	?synoeketic

However, only the species living on ant-trees have devised ways of resisting or escaping *Azteca* agression. On *Cecropia lyratiloba* Miq. alone, Andrade (1981) recorded 20 species of Coleoptera, one Diptera, three Hemiptera, eight Homoptera, 20 Hymenoptera (including 18 ants), one Isoptera, and eight Lepidoptera. Taking into account the hundred or so *Cecropia* species, the fauna is diverse and extremely abundant. All these insects, like many neotropical birds, spiders and bees, feed on the Müllerian bodies.

The beetle *Coelomera* with 33 species, is a specialist of *Cecropia*, feeding also on non-myrmecophytic species (Figs. 26. 1(a), and 26. 2(a), (b), (c), (d)). Adults and larvae feed on the leaves, Müllerian bodies, and pearl bodies on the leaves. In several species the female *Coelomera* excavates the prostoma and lays eggs inside the internode; as does the *Azteca* queen, only the beetle is much slower at opening the hole. Other species lay eggs under the leaf-lobe tips where patrolling ants are rare. The protection in that case is, however, less perfect and some of the external eggs of the oothecae are eaten by bugs. The ants are appeased by secretions of specialized hairs on the beetles. Adult *Coelomera* show reflex bleeding, and the larvae show enteric discharge. The larvae rest during the night under the leaves, and show ring defence behaviour, cycloalexy, a very efficient defence against ants (Vasconcellos and Jolivet 1988; Jolivet 1989) (Fig. 26. 2(c), (d)). The biology of a closely related genus, *Dircema*, which also feeds on *Cecropia* in Amazonia, is little known but may be similar to that of *Coelomera* (Jolivet *et al.* 1988).

Other beetles show remarkable adaptations to *Cecropia* (Andrade 1981) and also to the related *Pourouma*. The Hispinae (*Sceloenopla* and *Metaxycera*) nibble at the ventral side of the leaves and the female inserts a small ootheca into the mesophyll. The larvae mine into the leaf-lobes. The tenebrionid, *Epitragus*, feeds nocturnally on pollen and Müllerian bodies. The curculionid, *Ophtalmoborus*, lives in the male inflorescences, feeds on pollen, and oviposits inside the flowers. The Hispinae and the weevil avoid ant-patrolled zones and they do not pollinate the flowers, but cause little damage. Several anthridid weevils also feed on Müllerian bodies. The ants are either indifferent to these beetles, or the beetles have a repellent strategy. The Alticinae (*Cacoscelis*) are less adapted to the plant and probably escape by jumping or flying. However, in certain cases beetles provide the ants with sweet secretions; for instance a longicorn on *Cecropia* in Manaus, Amazonia, which, according to Wheeler and Bequaert (1929), was surrounded by many ants that appeared to be licking its anal region.

Non-myrmecophytic *Cecropia* harbour more phytophagous insects, or at least more individuals, than myrmecophytic species do. For instance, I have described *Coelomera helenae* Jolivet (Fig. 26. 1(a)) from *Cecropia hololeuca* in the Atlantic forest in Brazil, and another galerucine (*Gynandrobrotica fenestrata* (Baly)) and an eumolpine (*Nodonota* sp.) feed on the leaves of *Cecropia santanderensis* in the Andes of Merida, together with many other insects (Orthoptera, Coccinellidae, Curculionidae, Hemiptera,

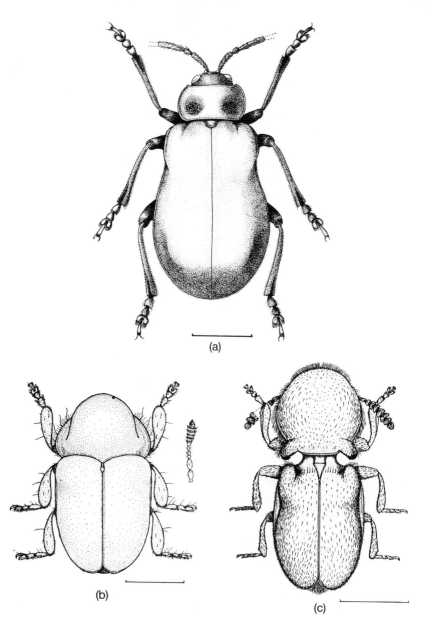

Fig. 26.1. (a) *Coelomera helenae* Jolivet (Coleoptera, Galerucinae) living on the leaves of a myrmecophobic *Cecropia* (*C. hololeuca*) in Brazil (Minas Gerais). Adult and larvae are lighter than in the other species. The eggs are deposited under the folioles. Scale bar = 5 mm (after Jolivet 1987*b*). (b) *Isnus petasus* Selman (Coleoptera, Cryptocephalinae), symphilic with *Crematogaster nigriceps* Emery (Myrmicinae) in the stipular thorns of *Acacia drepanolobium* (Harms) Sjostedt, Tanzania. The long curved setae on the edges of the pronotum are probably secretory. Scale bar = 0.6 mm. (c) *Hockingia curiosa* Selman (Coleoptera, Clytrinae), symphilic with *Crematogaster* sp. (Myrmicinae) in the stipular thorns of *Acacia drepanolobium* in Tanzania. The species mimic of *Crematogaster* has tufts of yellowish secretory setae at the tips of the pronotal tubercles. Scale bar = 1.0 mm. ((b) and (c) after Selman 1962).

Fig. 26.2. (a) *Coelomera lanio* Dalman (Coleoptera, Galerucinae), female laying eggs under a leaf-lobe of *Cecropia adenopus* Mart. *s. lat.*, Brazil, Minas Gerais. (b) Eggs of *Coelomera raquia* Bechyne laid inside an internodes of *Cecropia* sp., Brazil, Goias. (c) Cycloalexy among two groups of larvae (second instar) of *Coelomera raquia* on *Cecropia* leaves. (d) Cycloalexy among a group of larvae (third instar) of *Coelomera lanio* under a *Cecropia* leaf. (e) *Acacia seyal* Del. var. *fistula* (Schweine), stipular thorns inhabited by *Crematogaster* ants and leaf-beetle symphytes, Tanzania. (f) *Myrmecodia schlechteri* Valeton. The holes in the leaves are due to weevils (*Oribius*) which are not attacked by *Iridomyrmex* ants. Four fruits have been placed on the leaf.

Homoptera, etc.). The scolytid, *Scolytodes atratus* Col., breeds exclusively in petioles of recently fallen *Cecropia* leaves in Costa Rica and is not in contact with the ants (Wood 1983).

Africa: whistling-thorn *Acacia* trees (Mimosaceae)

Some 15 East African *Acacia* species have large stipular thorns inside which ants (*Crematogaster*) live in symphily with several beetles; e.g. *Hockingia*, a clytrine, and *Isnus*, a cryptocephaline (Selman 1962) (Fig. 26. 1(b), (c)). The Clytrinae, and to a certain extent the Cryptocephalinae, normally have a myrmecophilous tendency at the larval stage (Jolivet 1952). *Hockingia* is a mimic of *Crematogaster*. Several species of *Syagrus* (Eumolpinae) which live as synoeketes are also tolerated. Both *Isnus* and *Hockingia* have developed yellow setae which appear to have secretions attractive to ants; this does not seem to be true for *Syagrus*.

It is not clear whether the leaf-beetle larvae develop from eggs within the thorns or are transported there by the ants. In either case the hole is too small for the adult to escape. Normally, eumolpine larvae are root feeders but cryptocephaline larvae are found developing at the foot of the plant and clytrine larvae inside the ant-nest. The clytrine eggs, and sometimes young larvae, are transported by the ants into the nest. Many other insects (Buprestidae, Curculionidae, etc.) also share the hollow thorns with the ants (Hocking 1970). The case of symphily is interesting as it shows a close association between ant and beetle, where the beetle, like lycaenid larvae, scale insects, and aphids, produces sugary secretions which appease the ants.

Arboreal invertebrates in epiphytes

Ant-gardens in South and Central America, and groups of epiphytes in SE Asia, are the natural habitat of a community of invertebrates, especially insects, spiders, and myriapods. The biocoenosis of the ant-gardens in America is poorly known, as is that of Madagascar. A list of Coleoptera associated with epiphytes in Vietnam is given by Kabakov (1967) and many of these beetles have developed special relationships with ants. Coleoptera and ants are the most typical components of the epiphytic fauna.

Myrmecodia, *Hydnophytum*, and the related genera *Squamellaria*, *Myrmedoma*, and *Myrmephytum* (Rubiaceae), are tuberous epiphytes which mostly harbour *Iridomyrmex* ants (see Chapter 24). *Iridomyrmex* are aggressive but, curiously, tolerate various herbivorous insects such as the polyphagous weevil *Oribius* (Fig. 26. 2(f)). As soon as the leaf is touched the ants come out and attack. The weevil feeding on the leaves is not disturbed at all; it may be protected by its toxicity. Several undescribed staphylinids live in the warted chambers where they are not in close contact with the ants. These Staphylinidae do not show any special adaptation to their hosts. Nematodes,

Uropoda, dipteran larvae, etc. are found also among the ant excreta and cadavers.

Other beetles which resist ants

Table 26. 2 lists beetles found on the foliage of several ant-trees in America and Asia (Jolivet 1988a). Unfortunately, no data are available on the relationships involved. On myrmecophytic Verbenaceae (*Vitex* and *Clerodendrum*) many beetles are tolerated by the ants. *Phyllocharis* (Chrysomelinae) and *Hoplasomoides* (Galerucinae) feed on the leaves of *Clerodendrum* and the ants seem totally indifferent to them, probably because of their toxicity (Jolivet 1985). *Agriotes* (Elateridae) feeds on extrafloral nectar of *Clerodendrum fragrans* (Vent.) Wild. in Vietnam, and palpate the ants with their front legs to obtain food by trophallaxis (Jolivet 1983). Francis Darwin (1877) reported Elateridae feeding at nectaries of bracken (*Pteridium aquilinum* (L.) Kuhn) in England. (It would be interesting if trophallaxis occurs here also.) It is, however, surprising that *Formica* ants prey on adults and larvae of the Colorado beetle (Godzinska 1986) as they are coated with a mixture of defensive secretions which repels other ants. The relative quietness of the leaf-beetles on ant-plants may be a reaction with selective advantage to the ant species involved.

In the myrmecophytic tree *Tachigalia* of the neotropics, social beetles are responsible for the initial entering and excavation of the leaf rachis (Wheeler 1921b). Only when the trees mature are the beetles replaced by *Pseudomyrmex*. It is also noteworthy that this tree is hapaxanthic, flowering once at maturity and then dying.

Conclusion

The ants of myrmecophytes sometimes appear inefficient in attacking phytophagous insects. However, the main 'role' of the ants seems to be in repelling leaf-cutter ants, *Atta* (Jolivet 1990), at which they are generally successful. All myrmecophytes, even *Triplaris* (Polygonaceae) inhabited by the ferocious *Pseudomyrmex* (see Chapter 22), have several insects feeding on their leaves, food bodies, internal pith, extrafloral nectar, or on the inflorescence nectar (in *Cecropia*). This fauna includes beetles, Lepidoptera, and others, including spiders. Observations on *Cecropia* (in America), *Myrmecodia* (in Asia), and on *Acacia* (in East Africa), show that the reactions of ants to 'their' beetles range from hostility through indifference, to symphily. The beetles, adults, and larvae, have different ways of dealing with 'their' ants: toxicity, reflex bleeding, enteric or buccal discharge, behaviour (cycloalexy), avoidance (miners, borers, egg protection), or commensalism. They are successful, but it is notable that phytophagous insects seem to exclude each other on

Table 26.2. Some leaf-beetles of ant-plants and their ants.

Plant genus	Ant genus	Beetle genus and species	Beetle family	Geographical area
Palmae: *Korthalsia*	*Camponotus* (Formicinae)	*Lema karimui*	Criocerinae	New Guinea
		Octodonta subparallela *O. korthalsiae* *O. maffinensis*	Hispinae	New Guinea
		Palmispa parallela *P. korthalsivora*	Hispinae	
		Aspidispa albertisi *A. bicolor* *A. korthalsiae* *A. subviridipennis*	Hispinae	
Calamus	*Camponotus* (Formicinae)	*Octodonta subparallela* *O. depressa*	Hispinae	New Guinea Vietnam
		Palmispa parallela *Aspidispa pinange* *A. wilsoni* *A. striata*	Hispinae Hispinae	New Guinea New Guinea

Plant	Ant	Beetle	Subfamily	Location
Daemonorops	*Camponotus* (Formicidae)	*A. papuana* *A. calami* *A. lata* *A. rotanica* *A. palmella* *A. sedlaceki* *Aspidispa daemonoropa* *A. papuana*	Hispinae	New Guinea
Gesneriaceae: *Codonanthe*	*Crematogaster* (Myrmicinae)	gen. sp.	Alticinae	South America
Euphorbiaceae: *Macaranga*	*Crematogaster* (Myrmicinae)	*Coenobius macarangae*	Cryptocephalinae	Palau
		Cadmus macarangae	Cryptocephalinae	New Guinea
		Phytorus lineolatus	Eumolpinae	Mariannes
		Demotina dissimilis	Eumolpinae	Fiji
		D. evansi	Eumolpinae	Fiji
		Trichochrysea similis	Eumolpinae	South China
		Aphthona strigosa	Alticinae	Vietnam
Ehretiaceae: *Cordia*	*Azteca* (Dolichoderinae)	gen. sp.	Alticinae	South America

ant-plants except on very young trees; while on non-myrmecophytic species
(e.g. of *Cecropia*) they abound.

It is most regrettable that so little is known of the invertebrates associated
with ant-plants. More information may reveal adaptations as intricate as
those of *Coelomera* on *Cecropia*.

Acknowledgements

I wish to thank Dr M. Cox from the Department of Entomology, British
Museum (Natural History) who identified insects from *Cecropia santanderensis*.

References to Part 4

Addicott, J. H. (1981). Stability properties of 2-species models of mutualism: simulation studies. *Oecologia*, **49**, 42–9.

Agnew, C. W., Sterling, W. L., and Dean, D. A. (1982). Influence of cotton nectar on red imported fire ants and other predators. *Environmental Entomology*, **11**, 629–34.

Ake Assi, L. (1980). *Cecropia peltata* Linné (Moracées): ses origines, introduction et expansion dans l'est de la Côte d'Ivoire. *Institut Fondamental d'Afrique Noire, Bulletin, Serie A. Sciences Naturelles*, **42**, 96–102.

Alayo, D. P. (1974). Introducción al estudio de los himenópteros de Cuba. Superfamilia Formicoidea. *Academia de Ciencias de Cuba, Instituto de Zoologia, Serie Biológica*, **53**,

Andrade, J. C. (1981). Biologia de *Cecropia lyratiloba* Miq. var. *nana* Andr. & Car. (Moraceae) na restinga do recreio dos Bandeirantes. M.Sc. thesis, Rio de Janeiro.

Andrade, J. C. (1984). Observaçoes preliminares sobre a eco-etalogia de quatro Coleopteros que dependem da Embauba na restinga do recreio dos Bandeirantes, Rio de Janeiro. *Revista Brasileira Entomologia*, **28(1)**, 99–108.

Andrade, J. C. and Carauta, J. P. P. (1982). The *Cecropia–Azteca* association: a case of mutualism? *Biotropica*, **14(1)**, 15.

Aubréville, A. (1968). *Leonardoxa* Aubréville, genre nouveau de césalpinoidées guinéo-congolaises. *Adansonia Série 2*, **8**, 177–9.

Bailey, I. W. (1922*a*). Notes on neotropical ant-plants I. *Cecropia angulata*, sp. nov. *Botanical Gazette*, **74**, 369–91.

Bailey, I. W. (1922*b*). The anatomy of certain plants from the Belgian Congo, with special reference to myrmecophytism. *Bulletin of the American Museum of Natural History*, **45**, 585–621.

Bailey, I. W. (1923). Notes on neotropical ant-plants. II. *Tachigalia paniculata* Aubl. *Botanical Gazette*, **75**, 27–41.

Baird, J. N. (1967). Observations on the *Azteca–Cecropia* interaction (Hym. Formicidae). M.A. thesis, University of Kansas.

Baker, J. A. (1934). Notes on the biology of *Macaranga* spp. *Gardens Bulletin Straits Settlement*, **8**, 63–8.

Baldwin, I. T. and Schultz, J. C. (1988). Phylogeny and the patterns of leaf phenolics in gap- and forest-adapted *Piper* and *Miconia* understorey shrubs. *Oecologia*, **75**, 105–9.

Baroni Urbani, C. (1989). Phylogeny and behavioural evolution in ants, with a discussion of the role of behaviour in evolutionary processes. *Ethology, Ecology and Evolution*, **1**, 137–68.

Bazzaz, F. A., Chiariello, N. R., Coley, P. D., and Pitelka, L. F. (1987). Allocating resources to reproduction and defense. *Biotropica*, **37**, 58–67.

Beattie, A. J. (1985). *The evolutionary ecology of ant–plant mutualisms*, 182 pp. Cambridge University Press, Cambridge.

Beattie, A. J. (1989). Myrmecotrophy: plants fed by ants. *Trends in Ecology and Evolution*, **4**, 172–5.

Beccari, O. (1884–1886). Piante ospitatrici. In *Malesia*, Vol. 2. Istituto Sordo Muti, Genoa.

Beckmann, R. L., Jr., and Stucky, J. M. (1981). Extrafloral nectaries and plant guarding in *Ipomoea pandurata* (L.) (Convolvulaceae). *American Journal of Botany*, **68**, 421–6.

Belt, T. (1874). *The naturalist in Nicaragua*. John Murray, London.

Benson, W. W. (1985). Amazon ant plants. In *Amazonia* (ed. G. T. Prance and T. E. Lovejoy), pp. 239–66. Pergamon, Oxford.

Benzing, D. H. (1970). An investigation of two bromeliad myrmecophytes: *Tillandsia butzii* Mez, *T. caput-medusae* E. Morren and their ants. *Bulletin of the Torrey Botanical Club*, **97**, 109–15.

Benzing, D. H. (1984). Epiphytic vegetation: a profile and suggestions for future inquiries. In *Physiological ecology of plants of the wet tropics* (ed. E. Medina, H. A. Mooney, and C. Vázquez-Yánes), pp. 155–72. Junk, The Hague.

Benzing, D. H. (1990). *Vascular epiphytes: general biology and related biota*, 353 pp. Cambridge University Press, Cambridge.

Benzing, D. H. and Renfrow, A. (1971). The biology of the atmospheric bromeliad *Tillandsia circinnata* Schlecht. I. The nutrient status of populations in South Florida. *American Journal of Botany*, **58**, 867–73.

Benzing, D. H., Henderson, K., Kessel, B., and Sulak, J. (1976). The absorptive capacities of bromeliad trichomes. *American Journal of Botany*, **63**, 1009–14.

Bequaert, J. (1922). Ants in their diverse relations to the plant world. *Bulletin of the American Museum of Natural History*, **45**, 333–621.

Berg, C. C. (1978*a*). Cecropiaceae, a new family of Urticales. *Taxon*, **27**, 39–44.

Berg, C. C. (1978*b*). Espécies de *Cecropia* da Amazonia brasileira. *Acta Amazonica*, **8**, 149–82.

Bernays, E. and Graham, M. (1988). On the evolution of host specificity in phytophagous arthropods. *Ecology*, **69**, 886–92.

Brokaw, N. V. L. (1987). Gap-phase regeneration of three pioneer tree species in a tropical forest. *Journal of Ecology*, **75**, 9–19.

Brooks, D. R. (1985). Historical ecology: a new approach to studying the evolution of ecological associations. *Annals of the Missouri Botanical Garden*, **72**, 660–80.

Brooks, D. R. and Mitter, C. (1984). Analytical approaches to studying coevolution. In *Fungus–insect relationships. Perspectives in ecology and evolution* (ed. Q. Wheeler and M. Blackwell). Columbia University Press, New York.

Bryant, J. P., Chapin, F. S., and Klein, D. R. (1983). Carbon/nutrient balance of boreal plants in relation to vertebrate herbivory. *Oikos*, **40**, 357–68.

Bryant, J. P., Chapin, F. S., III, Reichardt, P., and Clausen, T. P. (1985). Adaptation to resource availability as a determinant of chemical defense strategies in woody plants. In *Chemically mediated interactions between plants and other organisms. Advances in Phytochemistry* (ed. G. A. Cooper-Driver and T. Swain), Vol. 19, pp. 219–37.

Buckley, R. C. (ed.) (1982). *Ant-plant interactions in Australia*, 162 pp. W. Junk, The Hague.

Carroll, C. R. and Janzen, D. H. (1973). Ecology of foraging by ants. *Annual Review of Ecology and Systematics*, **4**, 231–57.

Chabot, B. F. and Hicks, D. J. (1982). The ecology of leaf life spans. *Annual Review of Ecology and Systematics*, **13**, 229–59.

Coddington, J. A. (1985). Review of the explanation of organic diversity: the comparative method and adaptations for mating. *Journal of Cladistics*, **1**, 102–7.

Coddington, J. A. (1988). Cladistic tests of adaptational hypotheses. *Cladistics*, **4**, 3–22.

Coley, P. D. (1983). Herbivory and defensive characteristics of tree species in a lowland tropical forest. *Ecological Monographs*, **53**, 209–33.

Coley, P. D. (1986). Costs and benefits of defense by tannins in a neotropical tree. *Ecology*, **70**, 238–41.

Coley, P. D. and Aide, T. M. (1990). A comparison of herbivory and plant defenses in temperate and tropical broad-leaved forests. In *Herbivory: tropical and temperate perspectives*. (ed. P. W. Price), pp. 25–49. John Wiley & Sons, New York.

Coley, P. D., Bryant, J. P., and Chapin, F. S., III. (1985). Resource availability and plant antiherbivore defense. *Science*, **230**, 895–9.

Colwell, R. K. and Fuentes, E. R. (1975). Experimental studies of the niche. *Annual Review of Ecology and Systematics*, **6**, 281–310.

Darwin, F. (1877). On the glandular bodies of *Acacia sphaerocephala* and *Cecropia peltata*, serving as food bodies. *Journal of the Linnean Society, Botany*, **15**, 398–409.

Davidson, D. W. (1988). Ecological studies of neotropical ant-gardens. *Ecology*, **69**, 1138–52.

Davidson, D. W. and Epstein, W. W. (1989). Epiphytic associations with ants. In *Vascular plants as epiphytes* (ed. U. Lüttge), pp. 200–33. Verlag, Berlin.

Davidson, D. W., Foster, R. B., Snelling, R. R., and Lozada, P. W. (1990). Variable composition of some tropical ant-plant symbioses. In *Herbivory: tropical and temperate perspectives* (ed. P. W. Price), pp. 145–62. John Wiley & Sons, New York.

Davidson, D. W., Longino, J. T., and Snelling, R. R. (1988). Pruning of host plant neighbors by ants: an experimental approach. *Ecology*, **69**, 801–8.

Davidson, D. W., Snelling, R. R., and Longino, J. T. (1989). Competition among ants for myrmecophytes and the significance of plant trichomes. *Biotropica*, **21**, 64–73.

Docters van Leeuwen, W. M. (1929*a*). Miereneepiphyten. *De Tropische Natuur*, **18**, 57–65, 131–9.

Docters van Leeuwen, W. M. (1929*b*). Kurze Mitteilung über Ameisen-Epiphyten aus Java. *Mitteilungen der Deutschen botanischen Gesellschaft*, **47**, 90–9.

Donoghue, M. J. (1989). Phylogenies and the analysis of evolutionary sequences with examples from seed plants. *Evolution*, **43**, 1137–57.

Ducke, A. (1925). As leguminosas do estado do Para. *Archivos do Jardim Botanico do Rio de Janeiro*, **4**, 211–341.

Duviard, D. and Segeren, P. (1974). La colonization d'un myrmécophyte, le parasolier, par *Crematogaster* spp. (Myrmicinae) en Côte d'Ivoire forestière. *Insectes Sociaux*, **21**, 191–212.

Eidmann, H. (1945). Zur Kenntnis der Ökologie von *Azteca muelleri* Em. (Hym. Formicidae), ein Beitrag zum Problem der Myrmecophyten. *Zoologische Jahrbücher, Abteilung für Systematik, Ökologie und Geographie der Tiere*, **77**, 1–48.

Eldredge, N. (1979). Cladism and Common sense. In *Phylogenetic analysis and paleontology* (ed. J. Cracraft and N. Eldredge). Columbia University Press, New York.

Emery, C. (1893). Studio monografico sul genere *Azteca* Forel. *Memorie della R. Royal Accademia delle Scienze dell'Istituto di Bologna*, **(5)3**, 119–52.

Emery, C. (1896). Alcune forme nouve del genere *Azteca* For. e note biologiche. *Bollettino dei Musei di Zoologia ed Anatomia comparata della R. Università di Torino*, **11(230)**, 1–7.

Emery, C. (1912). Subfam. Dolichoderinae. *Genera insectorum* Fasc., **137**, 1–50.

Feeny, P. P. (1970). Plant apparency and chemical defense. In *Biochemical interactions between plants and insects. Recent Advances in Phytochemistry* (ed. J. Wallace and R. L. Mansell), Vol. 10. Plenum Press, New York.

Fiala, B. (1988). Biologie, Funktion und Evolution eines malaysischen Myrmekophytiesystems: Die Assoziation von *Crematogaster borneensis* (Form.: Myrmicinae) mit Bäumen der Gattung *Macaranga* (Euphorbiaceae). Dissertation, Universität Frankfurt.

Fiala, B. and Maschwitz, U. (1990). Studies on the South East Asian ant–plant association *C. borneensis–Macaranga*: adaptations of the ant-partner. *Insectes Sociaux*, **37**, 212–31.

Fiala, B., Maschwitz, U., Tho, Y. P., and Helbig, A. J. (1989). Studies of a South East Asian ant–plant association: protection of *Macaranga* trees by *Crematogaster borneensis. Oecologia*, **79**, 463–70.

Fisher, B. L., Sternberg, L. de S. L., Price, D. (1990). Variation in the use of extrafloral nectar by ants. *Oecologia*, **83**, 263–6.

Fisher, B. L. and Zimmerman, J. K. (1988). Ant–orchid associations in the Barrow Colorado National Monument, Panama. *Lindleyana*, **3**, 12–16.

Forbes, H. O. (1880). Notes from Java. *Nature*, **22**, 148.

Forel, A. (1908). Ameisen aus Sao Paulo (Brasilien), Paraguay etc. gesammelt von Prof. Hern. v. Ihering, Dr. Lutz, Dr. Fiebrig, etc. *Verhandlungen der Zoologische-botanischen Gesellschaft in Wien*, **58**, 340–418.

Forel, A. (1929). *The social world of the ants* (translated by C. K. Ogden). Albert & Charles Boni, New York.

Futuyma, D. J. (1986). *Evolutionary biology*, 2nd edn. Sinauer Associates, Sunderland, Massachusetts.

Gay, H. J. (1990). The taxonomy and ecology of the Far Eastern epiphytic ant-fern *Lecanopteris*. D.Phil. thesis, Oxford University.

Gentry, A. H. (1989). Speciation in tropical forests. In *Tropical forests: botanical dynamics, speciation and diversity* (ed. S. Nielsen, H. Balslev, and L. Holm-Nielsen). Academic Press, London.

Gentry, A. H. and Dodson, C. H. (1987). Diversity and biogeography of neotropical vascular epiphytes. *Annals of the Missouri Botanical Gardens*, **74**, 205–33.

Givnish, T. J. (1987). Comparative studies of leaf form: assessing the relative roles of selective pressures and phylogenetic constraints. *New Phytologist*, **106** (Suppl.), 131–60.

Godzinska, E. J. (1986). Ant predation on Colorado beetle (*Leptinotarsa decemlineata* Say). *Zeitschrift für angewandte Entomologie*, **102(1)**, 1–10.

Gomez, L. (1974). Biology of the potato-fern, *Solanopteris brunei. Brenesia*, **4**, 37–61.

Gould, S. J. and Vrba, E. (1982). Exaptation—a missing term in the science of form. *Paleobiology*, **8**, 4–15.

Griffith, J. F. (1987). Climate of Africa. In *The encyclopedia of climatology* (ed. J. E. Oliver and R. W. Fairbridge). Van Nostrand Reinhold, New York.

Griffiths, H., Smith, J. A. C., Lüttge, U., Popp, M., Cram, W. J., Diaz, M., Lee, H. S. J., Medina, E., Schäfer, C., and Stimmel, K-H. (1989). Ecophysiology of xerophytic and halophytic vegetation of a coastal alluvial plain in northern Venezuela. IV. *Tillandsia flexuosa* S.W. and *Schomburgkia humboldtiana* Reich., epiphytic CAM plants. *New Phytologist*, **111**, 273–82.

Hafner, M. S. and Nadler, S. A. (1988). Phylogenetic trees support the coevolution of parasites and their hosts. *Nature*, **332**, 258–9.

Hallé, F., Oldeman, R. A. A., and Tomlinson, P. B. (1978). *Tropical trees and forests: an architectural analysis*. Springer-Verlag, Berlin, Heidelberg, and New York.

Harada, A. Y. (1982). Contribuição ao conhecimento do gênero *Azteca* Forel 1878 ... *Cecropia* Loefling, 1758. Masters thesis, Instituto Nacional de Pesquisas da Amazônia e Fundação Universidade do Amazonas, Manaus, Brasil.

Harada, A. Y. and Benson, W. W. (1988). Espécies de *Azteca* (Hymenoptera, Formicidae) especializadas em *Cecropia* spp. (Moraceae): distribuição geográfica e consideraçõ ecológicas. *Revista brasiliera de Entomologia*, **32**, 423–35.

Heithaus, E. R., Culver, D. C., and Beattie, A. J. (1980). Models of some ant–plant mutualisms. *The American Naturalist*, **116**, 347–61.

Hespenheide, H. A. (1985). Insect visitors to extrafloral nectaries of *Byttneria aculeata* (Sterculiaceae): relative importance and roles. *Ecological Entomology*, **10**, 191–204.

Hocking, B. (1970). Insect associations with the swollen thorn acacias. *Transactions of the Royal Entomological Society London*, **122(7)**, 211–55.

Holthuijzen, A. M. A. and Boerboom, J. H. A. (1982). The *Cecropia* seedbank in the Surinam lowland rain forest. *Biotropica*, **14**, 62–8.

Holttum, R. E. (1954). *Plant life in Malaya*, 254 pp. Longmans, Green & Co, London.

Horich, C. K. (1977). Myrmecophilous orchids, aspects of a singular symbiosis. *Orquideologia*, **12**, 209–32.

Horvitz, C. C. and Schemske, D. W. (1984). Effects of ant-mutualists and an ant-sequestering herbivore on seed production of a tropical herb *Calathea ovandensis* (Marantaceae). *Ecology*, **65**, 1369–78.

Huxley, C. R. (1976). The ant-plants *Hydnophytum* and *Myrmecodia* (family: Rubiaceae) of Papua New Guinea and the ants which inhabit them. M.Sc. thesis, University of Papua New Guinea.

Huxley, C. R. (1978). The ant-plants *Myrmecodia* and *Hydnophytum* (Rubiaceae) and the relationships between their morphology, ant occupants, physiology and ecology. *New Phytologist*, **80**, 231–68.

Huxley, C. R. (1980). Symbiosis between ants and epiphytes. *Biological Reviews*, **55**, 321–40.

Huxley, C. R. (1981). Evolution and taxonomy of myrmecophytes with particular reference to *Myrmecodia* and *Hydnophytum* (Rubiaceae). D.Phil. thesis, University of Oxford.

Huxley, C. R. (1986). Evolution of benevolent ant–plant relationships. In *Insects and the plant surface* (ed. B. E. Juniper and T. R. E. Southwood). Edward Arnold, London.

Huxley, C. R. and Jebb, M. H. P. (1990). New taxa in the Rubiaceae Psychotriae. *Bulletin van de Nationale Plantentuin van Belgie*, **60**, 420–1.

Jack, W. (1823). Account of the *Lansium* and some other genera of Malayan Plants. *Transactions of the Linnean Society*, **14**, 122–5.

Jacob, F. (1977). Evolution and tinkering. *Science*, **196**, 1161–6.

Janet, C. (1907). *Anatomie du corselet et histolyse des muscles vibrateurs, apraas le vol nuptial, chez la reine de la fourmi* (Lasius niger). Ducourtieux et Gout, Limoges.

Janzen, D. H. (1966). Coevolution of mutualism between ants and acacias in Central America. *Evolution*, **20**, 249–75.

Janzen, D. H. (1967). Interaction of the bull's-horn acacia (*Acacia cornigera L.*) with an ant inhabitant (*Pseudomyrmex ferruginea* F. Smith) in eastern Mexico. *Kansas University Science Bulletin*, **47**, 315–558.

Janzen, D. H. (1969). Allelopathy by myrmecophytes: the ant *Azteca* as an allelopathic agent of *Cecropia. Ecology*, **50**, 147–53.

Janzen, D. H. (1972). Protection of *Barteria* (Passifloraceae) by *Pachysima* ants (Pseudomyrmecinae) in a Nigerian rain forest. *Ecology*, **53**, 885–92.

Janzen, D. H. (1973). Dissolution of mutualism between *Cecropia* and its *Azteca* ants. *Biotropica*, **5**, 15–28.

Janzen, D. H. (1974*a*). Swollen-thorn acacias of Central America. *Smithsonian Contributions to Botany*, **13**, 1–131.

Janzen, D. H. (1974*b*). Tropical blackwater rivers, animals, and mast fruiting by the Dipterocarpaceae. *Biotropica*, **6**, 69–103.

Janzen, D. H. (1974*c*). Epiphytic myrmecophytes in Sarawak: mutualism through feeding of plants by ants. *Biotropica*, **6**, 237.

Janzen, D. H. (1975). *Pseudomyrmex nigropilosa*: a parasite of a mutualism. *Science*, **188**, 936–7.

Janzen, D. H. (1979). New horizons in the biology of plant defenses. In *Herbivores: their interaction with secondary plant metabolites* (ed. G. A. Rosenthal and D. H. Janzen). Academic Press, New York.

Janzen, D. H. (1983). *Pseudomyrmex ferruginea* (Hormiga del Cornizuelo, *Acacia-Ant*). In *Costa Rican natural history* (ed. D. H. Janzen). University of Chicago Press, Chicago.

Janzen, D. H. (1985). The natural history of mutualisms. In *The biology of mutualisms* (ed. D. H. Boucher). Oxford University Press, New York.

Jebb, M. H. P. (1985). Taxonomy and tuber morphology of the Rubiaceous antplants. D.Phil. thesis, University of Oxford.

Jolivet, P. (1952). Quelques données sur la myrmécophilie des Clytrides (Col. Chrys.). *Bulletin Institut royal des Sciences Naturelles Belgique*, **28(8)**, 1–12.

Jolivet, P. (1973). Les Plantes Myrmécophiles du Sud-Est Asiatique. *Cahiers du Pacifique*, **17**, 41–69.

Jolivet, P. (1983). Un hémimyrmécophyte à Chrysomélides (Col.) du Sud-Est Asiatique, *Clerodendrum fragrans* (Vent.) Willd. (Verbenaceae). *Bulletin de la Société Linneienne Lyon*, **52(8)**, 242–61.

Jolivet, P. (1985). Un myrmécophyte hors de son pays d'origine: *Clerodendrum fallax* Lindley, 1844 (PVerbenaceae) aux îles du Cap Vert. *Bulletin de la Société Linneienne Lyon*, **54(5)**, 122–7.

Jolivet, P. (1986). *Les fourmis et les plantes. Un exemple de coévolution*, 254 pp. Boubée Editions, Paris.

Jolivet, P. (1987*a*). Nouvelles observations sur les plantes à fourmis. Réflexions sur la Myrmécophilie. *Entomologiste, Paris*, **43(1)**, 39–52.

Jolivet, P. (1987*b*). Remarques sur la biocénose des *Cecropia* (Cecropiaceae). Biologie des *Coelomera* Chevrolat avec la description d'une nouvelle espèce du Brésil. *Bulletin de la Société Linneienne Lyon*, **56(8)**, 255–76.

Jolivet, P. (1988*a*). Ants, leaf-beetles and plants. *Annals of International Symposium of Evolution and Ecology of Tropical Herbivores, Campinas*: p. 40.

Jolivet, P. (1988*b*). A case of parasitism of an ant–*Cecropia* mutualism in Brazil. *Advances in Myrmecology*, **1(20)**, 327–33.

Jolivet, P. (1989). The Chrysomelidae of *Cecropia* (Cecropiaceae). A strange cohabitation. *Entomography*, **6**, 391–5.

Jolivet, P. (1990). Relative protection of *Cecropia* trees against leaf-cutting ants in tropical America. *Applied myrmecology* (ed. Van der Meer *et al.*) **24**, 251–4.

Jolivet, P. (in press). *Host plants of the Chrysomelidae of the world*, 350 pp. Sandhill Crane Press, Books, Gainesville.

Jolivet, P., Petitpierre, E., and Hsiao, T. H. (ed.) (1988). *Biology of Chrysomelidae*, 613 pp. Kluwer Academic Publishers, Dordrecht.

Jordano, P. (1987). Patterns of mutualistic interactions in pollination and seed dispersal: connectance, dependence asymmetries, and coevolution. *American Naturalist*, **129**, 657–77.

Juniper, B. E., Robins, R. J., and Joel, D. M. (1989). *Carnivorous plants*, 353 pp. Academic Press, London.

Kabakov, O. N. (1967). The Coleoptera of the epiphytes in the tropical forests of Vietnam. *Entomological Review of Washington*, **46**, 410–14.

Keeler, K. H. (1981). Function of *Mentzelia nuda* (Loasaceae) postfloral nectaries in seed defense. *American Journal of Botany*, **68**, 295–9.

Kempf, W. W. (1958). Estudos sôbre *Pseudomyrmex*. II. (Hymenoptera: Formicidae). *Studia Entomologica (n.s.)*, **1**, 433—62.

Kleinfeldt, S. E. (1978). Ant-gardens: the interaction of *Codonanthe crassifolia* (Gesneriaceae) and *Crematogaster longispina* (Formicidae). *Ecology*, **59**, 449–56.

Kleinfeldt, S. E. (1986). Ant-gardens: mutual exploitation. In *Insects and the plant surface* (ed. B. E. Juniper and T. R. E. Southwood), pp. 282–94. Edward Arnold, London.

Kluge, A. G. and Farris, J. S. (1969). Quantitative phyletics and the evolution of anurans. *Systematic Zoology*, **18**, 1–32.

Koepcke, M. (1972). Uber die Resistenzformen der Vogelnestern in einem begrenzten Gebeit des tropischen Regenwaldes in Perú. *Journal of Ornithology*, **113**, 138–60.

Koptur, S. (1984). Experimental evidence for defense of *Inga* (Mimosoideae) sampling by ants. *Ecology*, **65**, 1787–93.

Koptur, S. and Lawton, J. H. (1988). Interactions among vetches bearing extrafloral nectaries, their biotic protective agents, and herbivores. *Ecology*, **69**, 278–83.

Kress, W. J. (1986). A symposium: the biology of tropical epiphytes. *Selbyana*, **9**, 1–22.

Lajtha, K. (1988). The use of ion-exchange resin bags for measuring nutrient availability in arid ecosystems. *Plant and Soil*, **105**, 105–12.

Larsson, S., Wiren, A., Ericsson, T., and Lundgren, L. (1986). Effects of light and nutrient stress on defensive chemistry and susceptibility to *Galerucella lineola* (Coleoptera) in two *Salix* species. *Oikos*, **47**, 205–10.

Lee, D. W. (1986). Unusual strategies of light adaptation observed in rainforest plants. In *On the economy of plant form and function* (ed. T. Givnish). Cambridge University Press, New York.

Léonard, J. (1951). Notulae Systematicae XI. Les Cynometra et les genres voisins en Afrique tropicale. *Bulletin du Jardin Botanique de l'Etat, Bruxelles*, **21**, 373–450.

Leston, D. (1973). The ant mosaic—tropical tree crops and the limiting of pests and disease. *Pest Articles News Summaries*, **19**, 311–36.

Letourneau, D. K. (1983). Passive aggression: an alternative hypothesis for the *Piper–Pheidole* association. *Oecologia*, **60**, 122–6.

Letourneau, D. K. (1990). Code of ant-plant mutualism broken by parasites. *Science*, **248**, 215–17.

Longino, J. T. (1986). Ants provide substrate for epiphytes. *Selbyana*, **9**, 100–3.

Longino, J. T. (1989a). Geographic variation and community structure in an ant–plant mutualism: *Azteca* and *Cecropia* in Costa Rica. *Biotropica*, **21**, 126–32.

Longino, J. T. (1989b). Taxonomy of the *Cecropia* inhabiting ants in the *Azteca alfari* species group (Hymenoptera: Formicidae): evidence for two broadly sympatric species. *Los Angeles County Museum, Contributions in Science*, **412**, 1–16.

Luizâo, F. J. and Fortunato de Carvalho, R. M. (1981). Estimativa da biomassa de raizes de duas espécies de *Cecropia* e sua relaçâo com a associaçâo ou nâo das plantas a formigas. *Acta Amazonica*, **11**, 93–6.

Madison, M. (1979). Additional observations on ant-gardens in Amazonas. *Selbyana*, **5**, 107–15.

Majer, J. D. (1976). The maintenance of the ant mosaic in Ghana cocoa farms. *Journal of Applied Ecology*, **13**, 123–44.

Maley, J. (1989). Late Quaternary climatic changes in the African rainforest: forest refugia and the major role of sea surface temperature variations. In *Paleoclimatology and paleometeorology: modern and past patterns of global atmospheric transport* (ed. M. Leinen and M. Sarnthein). Kluwer Academic Publishers, Dordrecht, Boston, and London.

Marquis, R. J. (1988). Phenological variation in the neotropical understorey shrub *Piper arieianum*: causes and consequences. *Ecology*, **69**, 1552–65.

Maschwitz, U. and Hölldobler, B. (1970). Der Nestkartonbau bei *Lasius fuliginosus*. *Zeitschrift für vergleichende Physiologie*, **66**, 176–89.

May, R. M. (1973). *Stability and complexity in model ecosystems*. Princeton University Press, Princeton.

May, R. M. (1974). *Stability and complexity in model ecosystems*, 2nd edn. Princeton University Press, Princeton.

May, R. M. (1976). Models for two interacting populations. In *Theoretical ecology, principles and applications* (ed. R. M. May). Saunders, Philadelphia.

McKey, D. (1984). Interaction of the ant-plant *Leonardoxa africana* (Caesalpiniaceae) with its obligate inhabitants in a rainforest in Cameroon. *Biotropica*, **16**, 81–99.

McKey, D. (1988). Promising new directions in the study of ant-plant mutualisms. In *Proceedings of the XIV International Botanical Congress* (ed. W. Greuter and B. Zimmer). Koeltz, Konigstein/Taunus.

McKey, D. (1989). Interactions between ants and leguminous plants. In *Advances in Legume Biology* (ed. C. H. Stirton and J. L. Zarruchi). Monographs in Systematic Botany from the Missouri Botanical Garden Number 29, St. Louis.

McKey, D. (in press). Comparative biology of ant–plant interactions in *Leonardoxa* (Leguminosae: Cesalpiniodeae) 1. A revision of the genus, with notes on the natural history. *Annals of the Missouri Botanical Garden*. St. Louis.

Miehe, H. (1911). Javanische Studien II. Untersuchungen über die javanische *Myrmecodia*. *Abhandlungen der kaiserische Sachsischen Gesellschaft de Wissenschaften. Mathematischen physischen Klasse*, **33(4)**, 312–61.

Mitter, C. and Brooks, D. R. (1983). Phylogenetic aspects of coevolution. In *Coevolution* (ed. D. J. Futuyma and M. Slatkin). Sinauer Associates, Sunderland, Massachusetts.

Moby Etia, P. (1980). Climate. In *Atlas of the United Republic of Cameroon*. Editions Jeune Afrique, Paris.

Mole, S., Ross, J. A. M., and Waterman, P. G. (1988). Light-induced variation in phenolic levels in foliage of rain-forest plants. *Journal of Chemical Ecology*, **14**, 1–21.

Mooney, H. A. and Gulmon, S. L. (1982). Constraints on leaf structure and function in reference to herbivory. *BioScience*, **32**, 198–206.

Müller, F. (1876). Üeber das Haarkissen am Blattstiel der Imbauba (*Cecropia*), das Gemüsebeet der Imbauba-Ameise. *Jenaische Zeitschrift für Naturwissenschaften*, **10**, 281–6.

Müller, F. (1880–1881). Die Imbauba und ihre Beschützer. *Kosmos*, **8**, 109–16.

Müller, R. N., Kalisz, P. J., and Kimmerer, T. W. (1987). Intraspecific variation in production of astringent phenolics over a vegetation-resource availability gradient. *Oecologia*, **72**, 211–15.

Nadkarni, N. (1986). The nutritional effects of epiphytes on host trees with special references to alteration of precipitation chemistry. *Selbyana*, **9**, 44–51.

Oliveira, P. S., de Silva, A. F., and Martins, A. B. (1987). Ant foraging on extrafloral nectaries of *Qualea grandiflora* (Vochysiaceae) in cerrado vegetation: ants as potential antiherbivore agents. *Oecologia*, **74**, 228–30.

Olmsted, I. C. and Dejean, A. (1987). Tree–epiphyte–ant relationships of the low inundated forest in Sain Ka'an Biosphere Reserve, Quintana Roo, Mexico. *Association for tropical biological abstracts*. Ohio State University, Columbus.

Ong, S. L. (1978). Ecology of the ant-association in *Macaranga triloba*. Ph.D. thesis, University of Malaya, Kuala Lumpur.

Pearce, E. A. and Smith, C. G. (1984). *The Times Books world weather guide*. New York Times Book Company, New York.

Penzig, O. (1894). Sopra una nuova pianta formicaria d'Africa (*Stereospermum dentatum* Rich.). *Malpighia*, **8**, 466–71.

Putz, F. W. and Holbrook, P. J. (1988). Further observations on the dissolution of mutualism between *Cecropia* and its ants: the Malaysian case. *Oikos*, **53**, 121–5.

Ramirez, B. W. (1974). Coevolution of *Ficus* and Agaonidae. *Annals of the Missouri Botanical Garden*, **61**, 770–80.

Raven, J. A. (1988). Acquisition of nitrogen by the shoots of land plants: its occurrence and implications for acid–base regulation. *New Phytologist*, **109**, 1–20.

Rehr, S. S., Feeny, P. P., and Janzen, D. H. (1973). Chemical defence in Central American non-ant-acacias. *Journal of Animal Ecology*, **42**, 405–16.

Rickson, F. R. (1971). Glycogen plastids in Müllerian body cells of *Cecropia peltata*—a higher green plant. *Science*, **173**, 344–7.

Rickson, F. R. (1973). Review of glycogen plastid differentiation in Müllerian body cells of *Cecropia peltata*. *Annals of the New York Academy of Science*, **210**, 104–14.

Rickson, F. R. (1976). Anatomical development of the leaf trichilium and Müllerian bodies of *Cecropia peltata* L. *American Journal of Botany*, **65**, 1266–71.

Rickson, F. R. (1977). Progressive loss of ant-related traits of *Cecropia peltata* on selected Caribbean islands. *American Journal of Botany*, **64**, 585–92.

Rickson, F. R. (1979). Absorption of animal breakdown products into a plant stem: the feeding of a plant by ants. *American Journal of Botany*, **66**, 87–90.

Rickson, F. R. (1980). Developmental anatomy of ultrastructure of the ant food bodies (Beccarian bodies) of *Macaranga triloba* and *M. hypoleuca* (Euphorbiaceae). *American Journal of Botany*, **67**, 285–92.

Rickson, F. R. and Rickson, M. M. (1986). Nutrient acquisition facilitated by litter collection and ant colonies on two Malaysian palms. *Biotropica*, **18**, 337–43.

Rico-Gray, V. and Thien, L. R. (1989). Ant–mealybug interaction decreases reproductive fitness of *Schomburgkia tibicinis* (Orchidaceae) in Mexico. *Journal of Tropical Ecology*, **5**, 109–12.

Ridley, M. (1983). *The explanation of organic diversity. The comparative method and adaptations of mating.* Clarendon Press, Oxford.

Risch, S. J. (1982). How *Pheidole* ants help *Piper* plants. *Brenesia*, **19/20**, 545–8.

Risch, S. J., McClure, M., Vandemeer, J., and Waltz, S. (1977). Mutualism between three species of tropical *Piper* (Piperaceae) and their ant inhabitants. *American Midland Naturalist*, **98**, 433–44.

Risch, S. J. and Rickson, F. R. (1981). Mutualism in which ants must be present before plants produce food bodies. *Nature*, **291**, 149–50.

Rohlf, F. J. (1982). Consensus indices for comparing classifications. *Mathematical Biosciences*, **59**, 131–44.

Rumphius, G. E. (1750). *Herbarium amboinense*, Vol. 6, pp. 119–20. Changnion and Hermann Utywerf, Amsterdam.

Santschi, F. (1936). Fourmis nouvelles ou intéressantes de la République Argentine. *Revista de Entomologia, Rio de Janeiro*, **6**, 402–21.

Schemske, D. W. and Horvitz, C. C. (1988). Plant–animal interactions and fruit production in a neotropical herb: a path analysis. *Ecology*, **69**, 1128–37.

Schnell, R. and Grout de Beaufort, F. (1966). Contribution à l'étude des plantes à myrmécodomaties de l'Afrique intertropicale. *Mémoires de l'institut Fondamental d'Afrique noire, Ifan-Dakar*, **75**, 1–66.

Schremmer, F. (1984). Untersuchungen und Beobachtungen zür Ökoethologie der Pflanzenameise *Pseudomyrmex triplarinus* welche die Ameisenbäume der Gattung *Triplaris* bewohnt. *Zoologisches Jahrbücher. Systematik, Ökologie und Geographie der Tiere*, **111**, 385–410.

Schupp, E. W. (1986). *Azteca* protection of *Cecropia*: ant occupation benefits juvenile trees. *Oecologia*, **70**, 379–85.

Seidel, J. L. (1988). The monoterpenes of *Gutierrezia sarothrae*: chemical interactions between ants and plants in neotropical ant-gardens. Ph.D. thesis, University of Utah.

Selman, B. J. (1962). Remarkable new chrysomeloids found in the nests of arboreal ants in Tanganyika (Col. Clytridae & Cryptocephalidae). *Annals and Magazine of Natural History*, **13(5)**, 295–9.

Siegler, D. S. and Ebinger, J. E. (1987). Cyanogenic glycosides in ant-acacias of Mexico and Central America. *Southwestern Naturalist*, **32**, 499–503.

Siegler, D. S., Saupe, S. G., Young, D. A., and Richardson, P. M. (1982). *Acacia rigidula*—a new ant-acacia. *Southwestern Naturalist*, **27**, 364–5.

Sillén-Tullberg, B. (1988). Evolution of gregariousness in aposematic butterfly larvae: a phylogenetic analysis. *Evolution*, **42**, 293–305.

Smith, W. (1903). *Macaranga triloba*, a new myrmecophilous plant. *New Phytologist*, **2**, 79–82.

Snelling, R. R. (1979*a*). *Aphomomyrmex* and a related new genus of arboreal African ants (Hymenoptera: Formicidae). *Contributions in Science from the Natural History Museum of Los Angeles County*, **316**, 1–8.

Snelling, R. R. (1979*b*). Three new species of the palaeotropical arboreal ant genus *Cataulacus* (Hymenoptera: Formicidae). *Contributions in Science from the Natural History Museum of Los Angeles County*, **315**, 1–8.

Spanner, L. (1938). Ein Beitrag zür Morphologie einiger Myrmecodien. *Biehefte zur Botanisches Centralblatt*, **58**, 267–94.

Spanner, L. (1939). Untersuchungen über dan Warme- und Wasser-haushalt von *Myrmecodia* und *Hydnophytum*, in Ein Beitrag zür Biologie der phanerogamen Epiphyten. *Jahrbuch für Botanik*, **88**, 243–83.

Spruce, R. (1908). *Notes of a botanist on the Amazon and Andes.* MacMillan & Co., London.

Stitz, H. (1910). Westafrikanische Ameisen. I. *Mitteilungen aus dem Zoologischen Museum in Berlin*, **5**, 125–51.

Stout, J. (1979). An association of an ant, a mealy bug, and an understorey tree from a Costa Rican rain forest. *Biotropica*, **11**, 309–11.

Sudd, J. H. and Franks, N. R. (1987). *The behavioural ecology of ants.* Chapman and Hall (in association with Methuen, Inc.), New York.

Swofford, D. L. (1985). PAUP (Phylogenetic analysis using parsimony) Version 2.4 User's Manual. Illinois Natural History Survey, Champaign.

Terborgh, J. (1983). *Five New World primates.* Princeton University Press, Princeton.

Tho, Y. P. (1978). Living in harmony. *Nature Malaysiana*, **3**, 34–9.

Thompson, J. N. (1981). Reversed animal–plant interactions: the evolution of insectivorous and ant-fed plants. *Biological Journal of the Linnean Society*, **16**, 147–55.

Treub, M. (1883). Sur le *Myrmecodia echinata* Gaudich. *Annales du Jardin Botanique de Buitenzorg*, **3**, 129–59.

Treub, M. (1888). Nouvelles recherches sur le *Myrmecodia* de Java. *Annales du Jardin Botanique de Buitenzorg*, **7**, 191–212.

Uhl, C., Clark, K., Clark, H., and Murphy, P. (1981). Early plant succession after cutting and burning in the upper Rio Negro region of the Amazon basin. *Journal of Ecology*, **69**, 631–49.

Ule, E. (1901). Ameisengarten im Amazonas-gebeit. *Engler's Botanische Jahrbücher*, **30**, 45–51.

Ule, E. (1905). Wechselbeziehungen zwischen Ameisen und Pflanzen. *Flora*, **94**, 491–7.

Ule, E. (1906). Ameisenpflanzen. *Engler's Botanische Jahrbücher*, **37**, 335–52.

Vandermeer, J. H. and Boucher, D. H. (1978). Varieties of mutualistic interaction in population models. *Journal of Theoretical Biology*, **74**, 549–58.

Vasconcellos Neto, J. and Jolivet, P. (1988). Une nouvelle stratégie de défense: la stratégie de défense annulaire (cycloalexie) chez quelques larves de Chrysomélides brésiliens. *Bulletin de la Société entomologique de France*, **92(9–10)**, 291–9.

Wallace, B. J. (1989). Vascular epiphytism in Australo-Asia. In *Ecosystems of the world* (ed. H. Lieth and M. J. A. Werger), Vol. 14B, pp. 261–82. Elsevier, New York.

Wanntorp, H. E. (1983). Historical constraints in adaptation theory: traits and non-traits. *Oikos*, **41**, 157–9.

Ward, P. S. (1989*a*). Genetic and social changes associated with ant speciation. In *The genetics of social evolution* (ed. M. D. Breed and R. E. Page). Westview Press, Boulder.

Ward, P. S. (1989*b*). Systematic studies on pseudomyrmecine ants: revision of the *Pseudomyrmex oculatus* and *P. subtilissimus* species groups, with taxonomic comments on other species. *Quaestiones Entomologicae*, **25**, 393–468.

Ward, P. S. (1990). The ant subfamily Pseudomyrmecinae: generic revision and relationship to other formicids. *Systematic Entomology*, **15**, 449–89.

Waterman, P. G., Ross, J. A. M., and McKey, D. B. (1984). Factors affecting levels of some phenolic compounds, digestibility, and nitrogen content of the mature leaves of *Barteria fistulosa* (Passifloraceae). *Journal of Chemical Ecology*, **10**, 387–401.

Weber, N. A. (1943). Parabiosis in neotropical 'ant-gardens'. *Ecology*, **24**, 400–4.

Weir, J. S. and Kiew, R. (1986). A reassessment of the relations in Malaysia between ants(*Crematogaster*) on trees (*Leptospermum* and *Dacrydium*) and epiphytes of the genus *Dischidia* (Asclepiadaceae) including 'ant-plants'. *Biological Journal of the Linnean Society*, **27**, 113–32.

Wheeler, W. M. (1910*a*). *Ants, their structure, development and behaviour.* Columbia University Press, New York.

Wheeler, W. M. (1910*b*). A new species of *Aphomomyrmex* from Borneo. *Psyche*, **17**, 131–5.

Wheeler, W. M. (1920). The subfamilies of formicidae, and other taxonomic notes. *Psyche*, **27**, 46–55.

Wheeler, W. M. (1921*a*). A new case of parabiosis and the 'ant gardens' of British Guiana. *Ecology*, **2**, 89–103.

Wheeler, W. M. (1921*b*). A study of some social beetles in British Guiana and of their relation to the ant-plant *Tachigalia. Zoologia New York*, **3**, 35–126.

Wheeler, W. M. (1922). Keys to the genera and subgenera of ants. *Bulletin of the American Museum of Natural History*, **45**, 631–710.

Wheeler, W. M. (1942). Studies of neotropical ant-plants and their ants. *Bulletin of the Museum of Comparative Zoology*, **90**, 3–262.

Wheeler, W. M. and Bequaert, J. C. (1929). Amazonian myrmecophytes and their ants. *Zoologischen Anzeiger*, **82**, 10–39.

White, P. S. (1983). Corner's rules in eastern deciduous trees: allometry and its implications for the adaptive architecture of the trees. *Bulletin of the Torrey Botanical Club*, **110**, 203–12.

Whitmore, T. C. (1967). Studies in *Macaranga*, an easy genus of Malayan wayside trees. *Malayan Nature Journal*, **20**, 89–99.

Wiley, E. O. (1981). *Phylogenetics. The theory and practice of phylogenetic systematics.* John Wiley, New York.

Williamson, M. (1972). *The analysis of biological populations.* Edward Arnold, London.

Wilson, E. O. (1971). *The insect societies.* Belknap Press, Cambridge, Massachusetts.

Wilson, E. O. (1987). The arboreal ant fauna of Peruvian Amazon forests: a first assessment. *Biotropica*, **19**, 245–82.

Wood, S. L. (1983). *Scolytodes atratus panamensis* (*Cecropia* petiole borer). In *Costa Rican natural history* (ed. D. H. Jansen), pp. 768–9. University of Chicago Press, Chicago.

Woodhead, S. (1981). Environmental and biotic factors affecting the phenolic content of different cultivars of *Sorghum bicolor. Journal of Chemical Ecology*, **7**, 1035–47.

Yapp, R. H. (1902). Two Malayan 'myrmecophilous' ferns, *Polypodium* (*Lecanopteris*) *carnosum* Blume and *Polypodium sinuosum* Wall. *Annals of Botany*, **62**, 185–231.

Part 5

Pollination, ant exclusion, and dispersal

27

The evidence for, and importance of, ant pollination

Rod Peakall, Steven N. Handel, and Andrew J. Beattie

Introduction

Despite the world abundance of ants, and the wide radiation of pollination vectors, very few plant species have been described as ant pollinated. Various authors have attempted to explain this lack of ant pollination but these explanations have never been supported by adequate data. However, a recent series of studies have suggested that antibiotic secretions may have contributed to the paucity of ant pollination systems and provide a new series of hypotheses which can be tested experimentally. One consequence of these studies is a call for a reappraisal of the published reports of ant pollination. In this chapter we briefly examine these reports and point to the need for critical experimental studies. It is hoped that this review will widen interest in ant pollination systems and encourage the discovery of other cases of ant pollination.

Taxonomic distribution of ant pollination

Anecdotal reports of ant pollination have been made for some 15 species of plants in a variety of families (Table 27.1). Ant pollination studies providing some supporting data are only available for ten other plant species. These species encompass an array of plant families, with the Orchidaceae being represented by three species (Table 27.2). However, even for these studies with some documentation three critical observations are often lacking:

1. observations of pollen transfer from anther to ant;
2. observations of pollen transfer from ant to stigma; and
3. experiments demonstrating that ant pollination leads to viable seed set.

Studies lacking observations of pollen transfer assumed ant pollination based on observations of ants carrying pollen and the frequency of ants at the flowers. The assumption that insects carrying pollen will be pollinators is frequently made in other non-ant pollinated cases. However, as will be seen later, this may be a dangerous assumption when applied to ants. Only three

Table 27.1. Plant species for which anecdotal reports of ant pollination have been made. None of these reports contain the data enumerated in Table 27.2.

Plant species	Plant family	Ant relationship[a]	Reference
Rohdea japonica[b]	Liliaceae	A	Migliorato 1910
Cocos nucifera	Arecaceae	AC	Patel 1938
Glaux maritima	Primulaceae	A	Dahl and Hadac 1940
Seseli libanotis	Apiaceae	A	Hagerup 1943
Theobroma leiocarpa	Sterculiaceae	B	Posnette 1942; Winder 1978
Orthocarpus pusillus	Scrophulariaceae	A	Kincaid 1963
Herniaria ciliolata	Caryophyllaceae	A	Proctor and Yeo 1973
Trinia glauca	Apiaceae	A	Proctor and Yeo 1973
Morinda royoc	Rubiaceae	B	Percival 1974
Cordia brownei	Boraginaceae	B	Percival 1974
Cordia globosa	Boraginaceae	B	Percival 1974
Euphorbia (3 species)	Euphorbiaceae	B	Ehrenfeld 1978
Coronopus didymus (= *Senebrera pinnatifida*)	Brassicaceae	AC	Chauhan 1979
Halogeton glomeratus	Chenopodiaceae	A	Blackwell and Powell 1981
Suaeda suffrutescens	Chenopodiaceae	AC	Blackwell and Powell 1981
Leptadenia reticulata	Asclepiadaceae	B	Pant *et al.* 1982

[a] A, ants were considered to be the primary pollinator; B, ants were not considered to be the primary pollinators; C, other insect species were observed on the flower and may have been secondary pollinators.
[b] Disputed by van der Pijl 1955.

studies have reported seed set results from exclusive ant pollination (Table 27.2).

With the exception of *Voandzeia subterranea* Thou., *Microtis parviflora* R.Br., and *Leporella fimbriata* (Lindl.) George, all other species in Table 27.2 were visited by an array of insects in addition to ants. For *Thlaspi alpina* L. syn.: *Hutchinsonia alpina* (L.) R.Br. *Diamorpha smallii* Britton *Scleranthus perennis* L., and *Epipactis palustris* Crantz., other insect groups were also reported as pollinators. Furthermore, in the case of *Thlaspi alpina* and *Epipactis palustris*, ants were considered by the authors to play only a secondary role as pollinators. An additional challenge to the importance of ants in reproduction is that *Voandzeia subterranea*, *Thlaspi alpina*, *Scleranthus perennis*, and *Eritrichium aretiodes* D.C. also set seed by autogamy (self fertilization) (Table 27.2).

In addition to the low number of ant-pollinated species, some of them may only be locally dependent on ants. In *Diamorpha smallii*, for example, ants were the predominant pollinators at Wyatt's (1981) study sites but honey-bees are known to be the primary pollinator at other sites. Evolution of specific floral traits for ants would be undermined by successful pollination by such different visitors. Similarly, Peakall and Beattie (1989) reported exclusive pollination of *Microtis parviflora* at only a single study site. The nectar of this species is hexose rich and may be preferred by short-tongued bees and flies as well as ants. It was concluded that small, winged pollinators may pollinate *M. parviflora* elsewhere. Consequently, *Leporella fimbriata* may be the only obligately ant-pollinated plant. This orchid is pollinated uniquely and exclusively by winged male ants of only one species, *Myrmecia urens* which pseudocopulate with the orchid flower in response to bio-chemical and morphological cues (Peakall *et al.* 1987; Peakall 1989).

Factors limiting pollination by ants

Ants have long been regarded as undesirable flower visitors that are only capable of nectar thieving. For example, Kerner von Marilaun (1895) discussed mechanisms such as extrafloral nectaries, sticky secretions, and glandular hairs which deterred or prevented ants from visiting flowers. Kerner van Marilaun rightly concluded that these devices also protected flowers from attack by animals such as caterpillars, and that extrafloral nectaries attracted ants which protected the plant from herbivores. More recently it was hypothesized that flowers repelled ant visitors by producing toxic floral nectars (Janzen, 1977). A series of subsequent studies, however, have found that toxic nectars are rare and even nectars with potentially toxic compounds do not inhibit nectar consumption by ants (Feinsinger and Swarm, 1978; Schubart and Anderson, 1978; Rico-Gray, 1980; Haber *et al.*, 1981).

It has been suggested that ants make poor pollinators for a variety of reasons. These include: the small size of ants which limits contact with the anthers and stigmas of many flowers; their smooth integuments; and frequent grooming which minimizes pollen carrying ability (van der Pijl, 1955; Faegri and van der Pijl 1971; Proctor and Yeo 1973; Buckley 1982). While these factors may prevent pollination by many ant species, other ants differ little from winged pollinators such as bees in size, hairiness, and grooming frequency (Beattie 1985).

Another reason suggested for the lack of ant pollination has been the assumption that, because worker ants are flightless they are unable to

Table 27.2. Plant species reported to be ant pollinated with some supportive data. Most species are not reported to be exclusively ant pollinated.

Plant species (family)	Insect visitors ranked by frequency (%)	Pollinator ranked by importance	Ant pollination observed	Ant seed se observe
Voandzeia subterranea (Fabaceae)	ants	autogamy ant	no	no[a]
Polygonum cascadense (Polygonaceae)	ants wasps syrphids	ant	yes	no[b]
Eritrichium aretiodes (Boraginaceae)	ants (58) other (42)	ant autogamy	no	no
Oreoxis alpina (Apiaceae)	ants (72) other (28)	ant	no	no
Hutchinsonia alpina (= *Thlaspi alpina*) (Brassicaceae)	beetles (70) ants (20) other (10)	beetle ant autogamy	no	no
Diamorpha smallii (Crassulaceae)	ants honeybee solitary bees	ant solitary bee honeybee	no	no
Scleranthus perennis (Caryophyllaceae)	ants beetles honeybees	ant honeybee autogamy	yes	no
Epipactis palustris (Orchidaceae)	solitary wasps ants digger wasps	solitary wasp digger wasp	yes	no
Epipactis palustris (Orchidaceae)	ants (50) honeybees (25) hoverflies (20) other (5)	ant honeybee hoverfly	yes	yes
Microtis parviflora (Orchidaceae)	ants	ant	yes	no
Microtis parviflora (Orchidaceae)	ants	ant	yes	yes
Leporella fimbriata (Orchidaceae)	male ants	male ants	yes	yes

[a] Doku (1968) reported reduced fruit formation for one of two cultivars when ants were excluded and suggested ants effected both self- and cross-pollination. Subsequently, Doku and Karikari (1971) concluded that while ants may be pollinators, autogamy also occurred in the cultivars studied. It was found that enhanced fruit formation in presence of ants was not due to increased fertilization but rather to the effect of ant activity on the ability of fertiliz

Inter-plant movement observed	Ants observed (% with pollen)	Total hours	Number of study sites (ants obs.)	Reference
yes	'occasionally'		1 (1)	Doku 1968 Doku and Karikari 1971
yes	(100)		1 (1)	Hickman 1974
yes	86	100		Petersen 1977a
no	306 (80)	100		Petersen 1977a
no	40 (100)	100		Petersen 1977a,b
yes	many (100)	'many'	'many'	Wyatt 1981 Wyatt and Stoneburner 1981
yes			1 (1)	Svensson 1985
no	67 (37)	93	12 (5)	Nilsson 1978
yes	78 (14)	9.5	1 (1)	Brantjes 1981
yes	484 (36)		2 (2)	Jones 1975
yes	100+ (80)	26	1 (1)	Peakall and Beattie 1989
yes	33 (82)	350	7 (7)	Peakall et al. 1987 Peakall 1989

flowers to penetrate the soil where normal fruit development occurs. The importance of ant pollination to fertilization of wild populations remains unknown.

 Greenhouse specimens from which ants but not flying insects were excluded, resulted in very reduced seed set compared with natural populations.

transport pollen far enough to effect cross pollination (Van der Pijl 1955; Proctor and Yeo 1973; Schubart and Anderson 1978). To test this idea we examined the three studies that have reported ant pollinator movements between flowers (Table 27.3). The mean distances moved by pollinating ants of *Diamorpha smallii* in five-minute intervals are reported to be 109 and 81 cm, with a maximum distance of 1000 cm. Estimates of ant-mediated pollen flow by fluorescent pollen analogues in *Scleranthus perennis* suggested a mean maximum pollen transfer distance between plants of 28.1 cm, with a maximum of 160 cm (Table 27.3). Finally, in *Microtis parviflora* a mean pollen transfer distance between inflorescences of 6.2 cm was observed with a maximum of 76 cm (Table 27.3). In all three studies the distributions are leptokurtic with short means similar to pollinator movement distributions observed for a wide array of winged pollen vectors (Levin and Kerster 1974; Levin 1981).

Table 27.3. Ant pollinator movements reported for three plant species.

Species	Mean (cm)	SD	Range	Kurtosis	n
Diamorpha smallii (Wyatt and Stoneburner 1981)					
Mean distance moved 1979[a]	109	107	1–500	8.4	20
Mean distance moved 1980[a]	81	208	1–1000	19.5	20
Scleranthus perennis (Svensson 1985)					
Interflower pollen transfers[b]	6.2	15.4	0–105		156
Interplant pollen transfers[b]	22.5	23.2	3–105		50
Mean maximum transfer distance	28.1	38.7	0–160		34
Microtis parviflora (Peakall and Beattie 1989)					
Interplant distances	6.2	10.1	1–76	25.7	79
Pollen transfer distances[c]	6.6	10.3	1–76	29.7	59

[a] Mean distance moved by pollen bearing ants in a 5 min interval.
[b] Pollen transfer distances estimated by fluorescent dye pollen analogues.
[c] Inter-plant distances by ants bearing pollen.

None of these studies estimated out-crossing directly. In *Diamorpha smallii* the ants tended to visit 2 or 3 of the 10 or so flowers per plant before moving to the next plant. Since the sticky pollen of this species is well attached over the entire pubescent body of the pollinating ant, pollen carry over to different plants is probably high and, indeed, essential for fruit set in this self-incompatible species (Wyatt and Stoneburner 1981). In *Scleranthus perennis*, Svensson (1985) estimated that 72 per cent of pollen transfers were within plants, leading to high levels of self-pollination. Secondary pollen carry-over did not change these results appreciably (Svensson 1986). In *Microtis parviflora* out-crossing estimates await elucidation of the clonal structure of orchid patches but some pollen tranfers are expected to represent out-crossing (Peakall and Beattie 1989). In all three systems the ant pollinators were capable of effecting some out-crossing. Inter-plant ant movements were also observed in *Voandzeia subterranea*, *Polygonum cascadense*, *Epipactis palustris*, and *Microtis parviflora* (Table 27.2). Consequently, the available studies do not support the contention that ants are unable to effect out-crossing. In fact, winged pollinators may also effect considerable pollination by different flowers on the same plant, and this may be the usual condition for many clonal and inflorescence arrangements (Wyatt 1982; Handel 1985).

In the orchid *Leporella fimbriata*, which is pollinated only by winged male ants, the distribution of pollinator flight movements was leptokurtic with a mean of 3.14 m. However, the orchid forms extensive clones and consequently 70 per cent of the pollinator moves between flowers were within clones. Out-crossing rates by these winged ants may rarely exceed 30 per cent (Peakall and James 1989). It is clear that in these ant pollinated systems it is not the magnitude of pollinator movements *per se*, but the magnitude of pollinator movements in relation to the distance of the nearest genotype which is critical for out-crossing. While there are large numbers of studies on pollinator flight distance and pollen flow in systems involving larger insect pollinators, such as bumblebees and honeybees, studies on small flying insects are only just beginning (Eichwort and Ginsberg 1980; Ginsberg 1984). These are important pollinators for many flowers, and their foraging patterns and visitation biology may be very different from those of the large social bees. We predict that patterns of movement and pollen dispersal by small winged insects will be similar to those of small ant pollinators. Careful methods will be needed to document behaviour of these smaller pollinators, to understand properly this key force in floral evolution.

It has been concluded that none of the above reasons for the paucity of ant pollination is satisfactory (Beattie 1985). Instead, a recent series of studies by Beattie and his co-workers have proposed a very different factor affecting the efficacy of ants as pollinators. The metapleural glands of ants have been found to be the source of antibiotic secretions (Maschwitz *et al.* 1970) which, in addition to killing potentially pathogenic micro-organisms, disrupt normal pollen function (Iwanami and Iwadare 1978; Beattie *et al.* 1984, 1985, 1986). Beattie *et al.* (1984) found that pollen exposed to species of Australian ants exhibited reduced pollen quality as assessed by the FCR procedure, reduced germinability, and shorter pollen tubes. More recently, Hull and Beattie (1988) reported significant reductions in pollen quality as assessed fluorochromatically for species of North American ants. All of the 20 species, representing six of the 11 subfamilies of ants, tested in these two studies were found to reduce pollen quality. In the ant *Myrmecia urens* all cases, including the male which lacks metapleural glands, have been found to have strong disruptive effects on pollen artificially applied to the integument (Peakall *et al* 1990). These findings suggest that ant pollination systems, when found, will be characterized by mechanisms for the avoidance of harmful metapleural secretions (see Chapter 37).

The evidence for ant pollination

The discovery that some ants possess antibiotic secretions on their surface which may seriously effect pollen viability calls for a critical reappraisal of the reported cases of ant pollination and its consequences. Reproduction in plants requires a sequence of events from the behaviour of the pollinating agent (pollen pick-up, movement, deposition), through pre-zygotic events (pollen germination, pollen tube growth, fertilization), to post-zygotic events (development of the embryo, seed maturation). All steps in this sequence must occur to legitimize the functional role of a pollen vector in reproduction. Since it is now known that pollen can be damaged by contact with many ants, the full biological sequence leading to seed set must be documented in order to demonstrate unequivocally that a plant species is ant pollinated. With this conservative viewpoint, we have summarized the data of the reported cases of ant pollination in Table 27.2. The evidence for ant pollination is thin, and rests on a shaky foundation. With the exceptions of studies in the orchid genera, *Leporella*, *Microtis*, and *Epipactis*, it has never been shown that ant-carried pollen will lead to viable seeds (Table 27.2). In the case of *Leporella* the pollinia were secured to the male ants, *Myrmecia urens*, by stigmatic secretions which insulated the pollen against antibiotic effects (Peakall *et al* 1990). In *Microtis*, pollen was found to be unaffected by contact with the primary ant pollinator, *Iridomyrmex gracilis*. Furthermore, the pollinia were separated from the ant integument by a short stalk (stipe) and carried on the head, remote from the metapleural glands (Peakall and Beattie 1989). Similar mechanisms may protect the pollen of *Epipactis*.

Despite the variety of factors which may have contributed to the rarity of ant pollination systems it can be expected that additional cases of ant pollination will be found. In the search for ant-pollinated species, where do we look? The syndrome of plant traits proposed by Hickman (1974) are consistent with the morphology of some species studied. However, the criteria offer little help in understanding the orchid species known to be associated with ants, and are not sufficient as rules to eliminate many other insect species that might pollinate plants with this syndrome (Beattie 1985, Chapter 7: Peakall *et al*. 1987). Reliance on imprecise criteria for an evolutionary relationship may mask other, more fundamental requirements (cf. Wheelwright and Orians 1982; Howe 1985). The first principles of ant pollination remain to be uncovered. We suspect that these rules will explain the paucity of ant pollinated species as well as characterize the few that will remain on the list.

We encourage critical and quantitative reappraisals of all reported ant pollination systems as well as searches for new cases. The following list details observations and experiments which are essential for unambiguous and valid assessments of ant pollination (points 1–3), and points 4–9 are desirable. Without these data, the syndrome and distribution of ant pollinated plants will remain ancecdotal. In deciding the case of whether ant pollination is a force in floral evolution, the jury is still out.

1. Observations of pollen transfer from the ant surface to the stigma.
2. Observations of pollen transfer between flowers, and between flowers on different plants.
3. Experiments demonstrating that exclusive ant pollination leads to normal seed set.
4. Assessment of pollen quality after being carried by ants.
5. Assessment of the presence or absence of metapleural glands on ants found transporting pollen.
6. Measurement of the foraging patterns and pollinator movement distances, emphasizing among-plant movements.
7. Assessments of out-crossing rates, pollen flow distances, and other population genetic parameters.
8. Where other pollinators are also servicing a plant species, what is the importance of the ant component of the pollinator guild?
9. Are there geographical differences in the importance of ant pollination?

Acknowledgements

We thank C. Gross for suggestions. R.P. was supported by an Australian Research Council grant, and S.N.H. by a Visiting Research Scholarship from Macquarie University and by the Bureau of Biological Research, Rutgers University.

28

The greasy pole syndrome

Ray Harley

While there are many examples of plant adaptations which attract ants to the mutual benefit of both plant and ant, there are very few known examples of mechanisms which actively deter them. Ants, while readily attracted to nectar, are unreliable pollinators and act as nectar thieves. Most of the cited examples of ant deterrence involve systems within the flower, either due to unpalatable nectar (van der Pijl 1955, Janzen 1977) or chemical or morphological modification of the floral tissues themselves (Guerrant and Fiedler 1981).

Other mechanical systems involving non-floral characters have been largely ignored, at least since the last century, when Kerner von Marilaun (1878) published observations on the European dioecious willow *Salix daphnoides* Vill. Kerner von Marilaun noted the presence of wax on the stems of this species and suggested that it effectively barred ants from visiting the nectar-rich flowers, thus ensuring that winged insects, such as bees, could more easily cross-pollinate them. Recently, however, it has been suggested that anti-robbing devices involving wax may be more frequent than previously thought (Juniper *et al*. 1989); two ferns, *Phlebodium aureum* and *Pityrogramma triangularis* being among the examples cited.

One very striking example, which has been under observation over the past ten years, occurs in two related genera of Lamiaceae, *Eriope* Kunth ex Benth. and *Hypenia* (Mart. ex Benth.) R. Harley which are found in the savannahs of tropical South America. These two genera, with about 30 and 20 species respectively, belong to the tribe Ocimeae, subtribe Hyptidinae with *Hypenia* having only recently been split off from the large and diverse neotropical genus *Hyptis* Jacq. (Harley 1988*a*).

The Ocimeae is characterized by having sternotribic flowers, the stamens being orientated towards the lower lip of the corolla, rather than towards the upper lip as in many familiar temperate genera of Lamiaceae, such as *Lamium* and *Salvia*. The pollen is thus typically deposited on the lower side of the pollinating insect.

In the subtribe Hyptidinae the flowers possess an explosive pollination mechanism, the lower lip of the corolla being more or less compressed and bearing a tensile hinge mechanism composed of thickened tissue at its base. The flowers are usually adapted to visits by solitary bees, foraging for nectar and pollen (Burkart 1937; Harley 1971, 1976). Although the *Hypenia macrantha* group, with large red flowers, is visited by humming-birds (Harley 1974; Brantjes and de Vos 1981).

At anthesis, the flowers are in the untriggered position, with the laterally compressed lower lip of the corolla retaining the four anthers within it, the stamen filaments being bent down under tension. The flower is now ready to be visited. The bee, perhaps in search of nectar, alights on the lower lip of the corolla, and due either to its weight or sometimes, perhaps when foraging for pollen, due to a certain amount of manipulation, the lip flicks back with great rapidity. As it does so, the tension on the stamens is released which flick up to deposit a mass of pollen on the underside of the bee's abdomen.

In *Hyptis*, however, where the mechanism is nevertheless usually fully functional, the flowers are much smaller and massed in compact heads, which are normally visited by a much wider range of insects, including Lepidoptera and Diptera as well as bees. Geitonogamy (fertilization by pollen from another flower on the same plant) is probably very common and can be achieved by unspecialized insects clambering among the flowers.

Where plants have larger, fewer flowers and laxer inflorescences, such as in *Eriope* and *Hypenia* however, marauding ants may interfere with normal pollination by triggering the flowers prematurely and thus reduce seed production.

All species of *Hypenia* and about ten species of *Eriope*, including many of the largest-flowered species in the tribe, possess a suite of characters which are designated here as 'the greasy pole syndrome' (Fig. 28.1). All are slender, wand-like shrubs from approximately 50 cm to 4 m in height, characterized by elongate, ± unbranched, erect stems. The lower internodes are densely covered by spreading, strigose hairs, while the upper ones are glabrous, the epidermis being covered with a conspicuous waxy layer. In many species the upper internodes are often strikingly fistulose, with the appearance of the swelling being characteristic of each species (Fig. 28.1). At one time it was believed that these swellings might be some kind of insect gall. However, several species of both *Eriope* and *Hypenia* have now been raised from seed at Kew and these all possessed the characteristic waxy stems and swellings, thus indicating the genetic nature of the phenomenon.

Scanning electron microscope studies reveal that the wax occurs as a series of minute irregular platelets. It would seem that these are easily dislodged by contact with foreign bodies, so that an insect attempting to scale the waxy stems would have great difficulty in gaining a foothold. Field observations indicate that ants of all sizes are unable to scale the stems and thus rob the flowers of nectar (Harley 1988*b*).

Experiments carried out at the Royal Botanic Garden, Kew, using cultivated *Hypenia vitifolia* and the common black ant, *Lasius niger* tend to confirm this. The ants introduced at ground level become hopelessly entangled in the hairs near the base of the stem, while those introduced onto

Fig. 28.1. Greasy pole syndrome in *Hypenia vitifolia*, Bahia: Morro do Chapeu. (Irwin, Harley & Smith 32425).

the leaves, (which would occur if they invaded from nearby vegetation) were unable to climb the waxy surface of the stems and fell to the ground.

One of the interesting aspects of the 'greasy pole syndrome' is the combined occurrence of three, or perhaps four, characters, all of which appear to achieve the same effect. It is suggested that each has been selected as a response to a different aspect of the same problem: to deter ant nectar thieves. The rigid hairs at the base of the stem will deter ground-foraging ants. However, these can climb neighbouring vegetation and reach the *Eriope* or *Hypenia* plant at a higher level, which is where the waxy stems come into play. The wand-like stems are extremely motile in windy conditions, but it is perhaps still possible for ants to move along horizontal waxy branches. In such conditions the fistulose swellings would provide the ultimate barrier, since, whichever way the stem was orientated, part of the swelling would be more or less vertical. Such extreme protective measures would probably be unnecessary if ants were not social insects. Where a rich nectar source was available, the path to it would soon become crowded with ants ready to exploit it.

Acknowledgements

I should like to thank Dr Paula Rudall, from the Royal Botanic Garden, Kew, for examining the stems and preparing scanning electron micrographs of the wax platelets, John Stone for design work, and Mrs Sandy Atkins for assistance in collating material. Finally, I wish to acknowledge the assistance of the Conselho Nacional de Desenvolvimento Cientifico e Tecnológico (CNPq) of Brazil.

29

Seed dispersal by ants; comparing infertile with fertile soils

M. Westoby, L. Hughes, and B. L. Rice

Introduction

Some plants, known as myrmecochores, have a food body (elaiosome) attached to their seeds. This food body acts as an incentive for ants to carry the diaspore (seed plus food body) back to their nest. There, the food body is chewed off and eaten, and the seed itself may be discarded intact, inside or outside the nest, where it may subsequently germinate. The seed has been moved from where it first fell, and thus dispersal has been achieved.

This chapter summarizes the work of our laboratory in seeking to explain why this dispersal syndrome should be found in a much larger percentage of plant species on infertile rather than on fertile soils in Australia. An explanation for this difference might also explain why the phenomenon is much more common in Australia and South Africa than elsewhere.

Abundance of plants adapted for dispersal by ants

Before 1975, adaptation for dispersal by ants was known from about 300 plant species, mostly herbs of mesic meadows or forest understorey in the northern hemisphere. Then Berg (1975) reported that about 1500 species, mostly sclerophyll shrubs of infertile soils, were adapted for dispersal by ants in Australia. Berg (1975) thought it 'unlikely . . . that future investigations [would] alter the fact that Australia is the world's main centre for ant-dispersed plants'. However, based on the association of dispersal by ants with sclerophyll vegetation on low-nutrient soils, one might expect that the syndrome should be common in South African fynbos also (Slingsby and Bond 1981; Westoby *et al.* 1982*b*). This has proved to be true: about 1300 species in that vegetation are adapted for dispersal by ants (Milewski 1982; Milewski and Bond 1982; Bond and Slingsby 1983).

Willson *et al.* (1990) have recently compiled dispersal spectra from a substantial number of plant communities. Outside Australia and South Africa, myrmecochores usually account for less than 10 per cent of the flora at a site, and rarely more than 20 per cent. (Fig 29.1). In contrast, in Australia it is common for 30–50 per cent of the flora to be ant-dispersed.

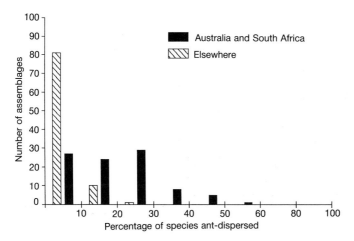

Fig. 29.1. Frequencies of species lists compiled by Willson *et al.* (1990) having different percentages of plant species adapted for seed dispersal by ants. Australia and South Africa are compared to all other lists.

The data set used by Willson *et al.* (1990) included mesic deciduous forests from the north temperate zone similar to those where Handel *et al.* (1981) and Beattie (1983) reported that approximately 30 per cent of the herb flora were myrmecochores. However, because such sites included substantial numbers of shrub and tree species, none of which were adapted for ant dispersal, the overall percentage of the flora adapted for ants in such sites proved to be modest. The data set included few tropical sites, so it remains conceivable that tropical vegetation types with many ant-dispersed species will be found in the future (Beattie 1983). However, it is known that most plant species in moist tropical forests are trees or shrubs, most of which are dispersed by vertebrates or by the wind (Gentry 1982; Tanner 1982; Foster 1986). Therefore, tropical forests are not likely to include vegetation with large percentages of species dispersed by ants. The only dispersal spectrum we know of from a tropical vegetation with many shrub species is Gottsberger and Silberbauer-Gottsberger's (1983) study of Brazilian cerrado and, here again, only a small percentage was ant-dispersed.

Within Australia, myrmecochores account for more of the flora in sclerophyll vegetation (Fig. 29.2). The sclerophyll vegetation is found on low-nutrient soils. Eucalypt forests with mesic fern and graminoid understorey are on more fertile soils, and rainforests on the most fertile (Beadle 1954, 1966). These vegetation types on more fertile soils have fewer myrmecochores. The higher level of adaptation for dispersal by ants on the infertile soils is at the expense of fleshy fruit for dispersal by vertebrates,

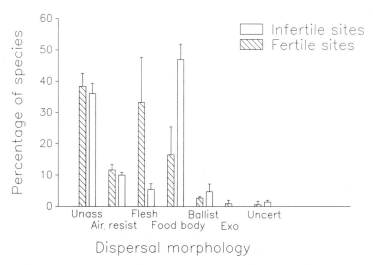

Fig. 29.2. Mean (± SD) percentage of plant species with different dispersal syndromes. Unass = unassisted; Air resist = having structures increasing air resistance and in that respect adapted for dispersal by wind; Flesh = fleshy fruits adapted for dispersal by vertebrates; Food body = having a food body or elaiosome and in that respect adapted for dispersal by ants; Ballist = having a structure propelling the seed ballistically; Exo = exozoochore, having hooks or spines adapted to attach the seed externally to animals; Uncert = dispersal structures of uncertain function. Four fertile-soil sites are compared to four infertile-soil sites near Sydney. (From Westoby *et al.* 1990.)

which accounts for a larger percentage of plant species on the fertile soils (Fig. 29.2).

Natural history background

Myrmecochores occur in a large number of families (Berg 1975), including some with a worldwide distribution (e.g. Apiaceae, Euphorbiaceae, Violaceae, Cyperaceae, and Fabaceae). Within a family some genera may include myrmecochores while others do not, and within many genera some species are myrmecochores while others are not. Myrmecochory has plainly evolved many times independently. It is not possible that the reason myrmecochory is especially common in Australia and South Africa is because of a single evolutionary development in the history of the Gondwanan flora.

The soil contrasts that we are discussing within Australia arise from differences in geomorphology. Total soil phosphorus is typically used as the best single indicator of soil fertility (Bowen 1981; Williams and Raupach 1983). The infertile sites we have worked at have total soil phosphorus of 70–120 p.p.m., compared to 200–800 p.p.m. at the fertile-soil sites. It is well established that sclerophylly in this region is a response to low soil nutrients, as indicated by phosphorus content (Beadle 1954, 1966). Vegetation at the infertile sites is known as dry sclerophyll woodland or forest, or as heath if there are no trees. Vegetation at the fertile sites is known as wet sclerophyll forest, if eucalypt-dominated, or as rainforest. The understorey in dry sclerophyll forest is dominated by shrubs, and in wet sclerophyll forest by graminoids and ferns. To use the terminology 'wet' versus 'dry' is confusing, since in many places one can move from 'wet' to 'dry' within ten metres, by crossing a soil boundary. Clearly, the vegetation differences are not due to climate. The vegetation is species rich, particularly on the infertile soils, with 70–100 spp in tenth of a hectare plots (Rice and Westoby 1983). Most species are shrubs, and there are virtually no annuals.

The ant species which remove most seeds have small nests, with entrances that range from simple holes to small craters of earth. From a randomly selected point on the ground, the average distance to the nearest nest ranges from less than a metre up to 3–5 m, depending on species.

Food bodies of Australian myrmecochores are usually waxy and persistent. Year-old diaspores are removed just as quickly as freshly collected ones (Hughes and Westoby 1990). Plants release diaspores singly rather than in aggregates, and diaspores are therefore usually encountered as single items by ants. Indeed, diaspores of many species are dispersed ballistically first (diplochory). Removal rates are high, with half-lives of a few hours in summer and a few days even in winter (Hughes and Westoby 1990). Removal by organisms other than ants appears to be insignificant in Australia.

Some hypotheses to account for dispersal by ants being more common on infertile than on fertile soils

The six hypotheses considered below do not exhaust the possibilities. The last four in the list are the ones on which we have data to contribute.

Mineral nutrient costs

If adaptations suited to dispersal by ants were cheap in mineral nutrients relative to their cost in energy, while other dispersal modes cost more mineral nutrients relative to energy, this could account for more plants having ant-dispersed seeds on low-nutrient soils.

We tried to test this hypothesis by comparing elaiosomes to wings and hairs for wind dispersal (Westoby et al. 1982b), before we realized that the comparison ought to be with fleshy fruits for vertebrate dispersal. We are currently retesting the hypothesis.

Fire avoidance

Within Australia, sclerophyll vegetation is more fire prone than vegetation on fertile soils. It could be argued that having seed that is buried by ants is more of an advantage in sclerophyll vegetation for this reason. The problem with this explanation is that fire prone sclerophyll shrub vegetation in

California and around the Mediterranean is not rich in myrmecochores. So, if the fire-avoidance hypothesis were accepted within Australia and South Africa some separate explanation would have to be sought for the scarcity of myrmecochory in fire-prone vegetation on other continents.

Benefits of reaching nutrient-enriched microsites

Ant nests can be chemically different from background soils (Davidson and Morton 1981b; Beattie and Culver 1983; Culver and Beattie 1983; Beattie 1985). Seedlings have been found to grow better in ant-nest soils by some worker (Culver and Beattie 1980; Beattie 1983) but not by others (Horvitz and Schemske 1986a). If dispersal by ants served to position seeds in nutrient-enriched microsites, this could plausibly be more of an advantage on infertile than on fertile soils.

Three lines of evidence indicate that nutrient-enriched microsites are not a factor in the Australian sclerophyll vegetation sites where we work. First, the ant species responsible for most seed removals move nest entrances frequently (Table 29.1) (Hughes (in press)). This must limit the scope for nutrients to accumulate. Secondly, ant-nests that are currently active do not appear to be enriched in mineral nutrients or in organic matter (Table 29.2). This is consistent with the rapid turnover of nest entrance sites. Thirdly, soil around the base of seedlings of ant-dispersed species is not nutrient enriched relative to soil around the base of seedlings dispersed by other means, or relative to soil at random places on the ground (Rice and Westoby 1986).

These tests did not measure all conceivable nutritional or physical properties of the microsites. However, a significant microsite effect resulting from some factor that we did not measure does not seem likely. Most seedling

Table 29.1. Turnover per year of nest entrance locations for some ant species responsible for removing seeds in sclerophyll vegetation near Sydney (Hughes (in press)).

Ant species	Turnover per year
Rhytidoponera metallica	4.3
Pheidole sp. 1	5.4
Pheidole sp. 4	10.3
Iridomyrmex darwinianus	8.0
Iridomyrmex sp. 8	1.6
Iridomyrmex sp. 13	5.9
Paratrechina vaga	14.3

Table 29.2. Summary of tests for differences between soil characteristics of nests of *Rhytidoponera metallica* (F. Smith), *Pheidole* sp. 1, *Iridomyrmex darwinianus* (Forel), and *Aphaenogaster longiceps* (F. Smith), and random places on the ground (Hughes (in press)). SNK tests after one-way ANOVA.

Characteristic	Nature of difference if any
Total phosphorus	n.s.
Total nitrogen	n.s.
% Organic matter	*Pheidole* < control = *Rhytidoponera* = *Iridomyrmex*
Exchangeable aluminium	*Rhytidoponera* < *Pheidole*
Exchangeable potassium	n.s.
Exchangeable magnesium	n.s.
Exchangeable sodium	n.s.
pH	n.s.

n.s. = not significant.

germination and successful establishment in sclerophyll vegetation is in the aftermath of fire, and intervals between fires can be greater than 30 years. Seeds coming to rest in ant-nests would often germinate eventually in microsites which had only been ant-nests for a few months, many years previously. Further, there is no evidence that the same locations tend to be used repeatedly for ant nests (Hughes (in press)). Most of the ground area will have been an ant-nest at some time during the between-fire interval.

Ant traffic

Might there be, for some reason, more traffic of ants carrying seeds in vegetation on infertile than on fertile soils? Mossop (1989) tested this hypothesis by measuring seed removal rates. Myrmecochore seeds were actually removed faster at fertile-soil sites, though the difference was not statistically significant.

Predation avoidance

Myrmecochore seeds, with their food bodies experimentally removed, are less likely to be carried away by ants and therefore stay on the ground surface for longer. In the northern hemisphere (O'Dowd and Hay 1980; Heithaus 1981) and in South Africa (Bond and Breytenbach 1985) it has been shown that, as a consequence, seeds without food bodies are more likely to be eaten by small mammals. However, in Australia, small mammal densities are low and mammals do not remove significant numbers of seeds from experimental depots (Fox *et al.* 1985; Auld 1986; Morton and Davidson 1988; Hughes and Westoby 1990). For this reason we discounted the predation avoidance hypothesis for some years.

Eventually, we appreciated that ants are also predators. It has proved difficult to classify ant species as dispersers versus predators. When nests are fed with seeds and then excavated, many seeds are not recovered, but this is due to imperfect excavation and to ants discarding seeds away from the nest, as well as to seeds being eaten. We have developed an assay in which nests are fed known numbers of seeds bearing food bodies and an equal number of plastic beads bearing food bodies. This assay has been applied to *Rhytidoponera metallica* and to *Pheidole* sp. 1. *R. metallica* is a ponerine which we had been treating as a disperser rather than a predator, based on unquantified earlier experience, and *Pheidole* sp. 1 belongs to a genus which includes many seed predators. For both species some seeds were recovered, but in a smaller percentage than for the beads (Table 29.3). This indicates that both the ants studied ate some, though not all, of the seeds they collected. Thus, *Rhytidoponera metallica* seems to be a predator as well as a disperser, and *Pheidole* sp. 1 seems to be a disperser as well as a predator. However, there was a quantitative difference between the species, with the seeds being exposed to less risk of being eaten if taken by *Rhytidoponera* than by *Pheidole*.

Table 29.3. Percentage of seeds and beads recovered from ant nests by excavation one week after they were fed to the nest (Hughes, unpublished data).

	Seeds	Beads
Rhytidoponera metallica	43	59
Pheidole sp. 1	7	33

When the food body was experimentally reduced to 10 per cent of its natural size, ants became less likely to carry the diaspore away, at any given encounter. The effect was greater in *Rhytidoponera* than in *Pheidole* (Fig. 29.3).

Taken together, these two pieces of evidence suggest that seeds bearing food bodies of natural size (the size to which natural selection has brought them) may have a better chance of being taken by an ant species which is less likely to eat the seeds themselves, than seeds with smaller food bodies would have. This is consistent with avoidance of predation by ants being a factor in myrmecochory in Australia.

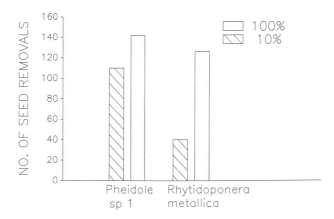

Fig. 29.3. When the food body of *Acacia myrtifolia* Willd. is reduced to 10 per cent of its natural size, the proportion of encounters with ants that lead to the seed being carried at least 5 cm is reduced. This reduction is relatively greater in *Rhytidoponera metallica* than in *Pheidole* sp. 1 (x^2 test, $P < 0.001$) (Tuffs 1988).

How might this relate to the contrast between fertile and infertile soils in Australia? Is predation by ants more likely for seeds on infertile than on fertile soils? Or are the opportunities for being dispersed without being eaten better on infertile soils? We do not yet know whether there are differences of these sorts between the ant assemblages on infertile compared to fertile soils. If such differences are proved to exist, the question of why such differences between ant assemblages should be associated with soil nutrients will be posed.

Secondary correlate of seed size or of plant stature

Diaspore size in sclerophyll vegetation might be concentrated, for reasons presumably to do with seedling establishment, in a size range which coincidentally is suited to dispersal by ants. The prospects of establishment by a seed are strongly influenced by the reserves of energy or mineral nutrient it contains (Janzen 1969; Harper *et al.* 1970; Harper 1977; Grime 1979; Willson 1983). Differences in mean seed mass found in different environments are thought to reflect the different problems posed for seedling establishment (Salisbury 1942, 1974; Harper *et al.* 1970; Baker 1972). Different seed sizes obviously tend to favour different seed dispersal methods. So, it might be that low-nutrient soils favour seed sizes in a range which, in turn, favours dispersal by ants.

Similarly, one could imagine an indirect correlation between soil nutrients and seed size via plant growth form. The growth form mix of vegetation is different depending on soil fertility (among other factors), and a variety of reasons could be suggested as to why plants of different growth forms might be pre-adapted to use different dispersal agents.

To test these hypotheses we listed species on replicate fertile-soil and infertile-soil sites, and classified them according to dispersal morphology, growth form, and diaspore size (Westoby *et al.* 1990). We dissected diaspores and weighed the embryo plus endosperm (hereafter called the 'reserve'), because this weight clearly represented part of the 'problem' which had to be solved by a dispersal mode; while flesh, wings, hairs, etc. could be regarded as part of the dispersal 'solution' adopted over evolutionary time. The result was a four-dimensional contingency table, site-type × diaspore mass × growth form × dispersal mode.

There was a strong relationship between reserve fresh weight and dispersal mode, with larger reserves more likely to be adapted for dispersal by vertebrates on both fertile and infertile soils (Fig. 29.4). There was also a strong effect of plant growth form, with herbaceous perennials and shrubs less than 2 m tall much more likely to be ant-dispersed than tall shrubs, trees, or climbers (Fig. 29.5). In both cases, however, there was also a difference between site-types, with species on fertile soils more likely to be vertebrate dispersed, even in a given band of reserve weights or within a given plant growth form (Figs 29.4 and 29.5).

Consider the association between soil fertility and dispersal mode. Indirect correlations via both reserve weight and plant stature were capable of

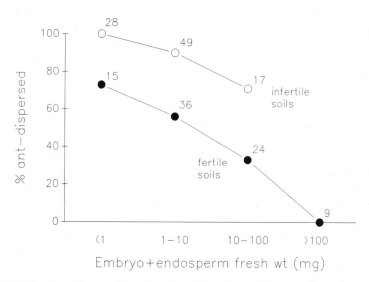

Fig. 29.4. Number of cases of species adapted for dispersal by ants as a percentage of those adapted for ants or for vertebrates, in relation to embryo-plus-endosperm fresh weight, on sites with fertile and infertile soils. Numbers by data points show total cases from which percentages were calculated. (From Westoby *et al.* 1990.)

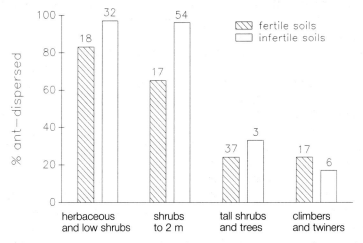

Fig. 29.5. Number of cases of species adapted for dispersal by ants as a percentage of those adapted for ants or for vertebrates, in relation to plant growth forms, on sites with fertile and infertile soils. Numbers over bars show total cases from which percentages were calculated. There are six total cases more than in Fig. 29.4, these being cases for which diaspore weights were not obtained. (From Westoby *et al.* 1990.)

accounting for as much as 85 per cent of this association (Table 29.4), but the remaining 15 per cent was still significant. The indirect correlation via plant stature alone was capable of accounting for as much as 84 per cent, and via reserve weight alone for as much as 23 per cent. The indirect relationship via reserve weight was comparatively weak, even though dispersal mode was strongly related to reserve weight (Fig. 29.4), because reserve weight proved

Table 29.4. Reduction in G^2 due to adding relevant terms to log–linear models of a 2^4 contingency table classified according to dispersal morphology (D), soil type (S), plant-growth form (F), and embryo-plus-endosperm fresh weight (W). (After Westoby *et al.* 1990).

Model	G^2	% Reduction in SD due to indirect effects
SD after S, F, W, D	40.4	100
SD after SF, FD	6.6	84
SD after SW, WD	31.2	23
SD after SF, FD, SW, WD	6.0	85

not to be strongly differentiated between these particular soil types. Thus, reserve weight was predominantly associated with variation within, rather than between, these soil types.

Why should dispersal mode have been strongly associated with plant stature in this comparison? One interpretation is that dispersal by ants achieves shorter distances but is cheaper. The biomass costs of procuring dispersal should be construed as including not only the dispersor (food body for ants or flesh for vertebrates) but also the seed coat; because seed coats are 2.3 times heavier, at a given reserve weight, for vertebrate-dispersed than for ant-dispersed species (Fig. 29.6; Westoby, Howell and Rice, unpublished work). Biomass costs counting both dispersor and seed coat are about 6.3 times greater for procuring dispersal by vertebrates than by ants, at any given reserve weight (Fig. 29.7; Westoby, Howell and Rice, unpublished work).

In sclerophyll vegetation near Sydney, most seeds are taken less than 2 m by ants, and distances over 4 m are rare (L. Hughes, unpublished work). Suppose it were the case that the benefits of dispersal accrue mainly at distances of 1–2 m for low shrubs and herbaceous perennials, but not for tall shrubs, trees, and climbers. Such taller plants might be selected to incur the extra biomass costs of procuring dispersal by vertebrates. We have no data on distances travelled by vertebrate-dispersed fruits in this particular system, but it seems likely that at least a proportion of diaspores taken by birds travel tens or hundreds of metres.

Discussion

The work described here has not been directed at finding out why any particular plant species has ant dispersal of its seeds, but at why more plant species have ant-dispersal on infertile than on fertile soils. Our evidence comes almost entirely from within Australia; indeed, almost entirely from the Sydney region. However, the distribution of myrmecochory between continents also has features which appear to be strongly associated with soil nutrients. So it is possible that conclusions reached from local comparisons will apply to differences between continents as well.

We believe some hypotheses can be definitely rejected. These include the idea that there might be more effective ant traffic on the ground under vegetation on infertile soils, and (surprisingly) the hypothesis that ant-dispersal targets seeds to nutrient-enriched microsites.

We do not yet have data to test the hypothesis that food bodies for ants might be especially cheap in mineral nutrients relative to other currencies, nor the predation avoidance hypothesis. We have some evidence that by having food bodies the size they are and not smaller, there are improved chances that diaspores will be removed by an ant which will disperse them without eating them. However, we do not yet know whether the ant assemblage on infertile soils makes this more worthwhile.

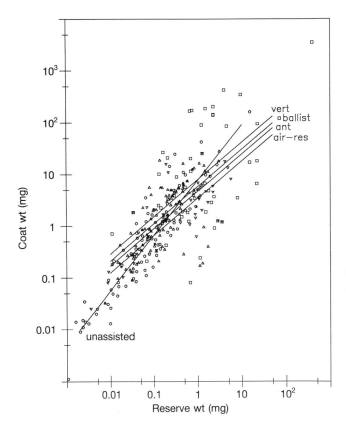

Fig. 29.6. Scaling of coat dry weight in relation to reserve dry weight for species belonging to dispersal syndromes 'vertebrate' (open squares, $n = 50$), 'air-resistant' (inverted triangles, $n = 32$), 'ballistic' (closed circles, $n = 18$), 'ants' (open triangles, $n = 109$), and 'unassisted' (open circles, $n = 85$). Diaspores were divided into embryo plus endosperm (reserve), structures serving dispersal such as wings or hairs, flesh, elaiosomes, or ballistic propulsors (dispersor), and the remainder (coat). Dispersal types other than unassisted had parallel slopes not significantly different from 2/3, indicating no size-related trend in coat weight per unit surface area within a dispersal type. The intercept for 'vertebrate' type lay significantly higher than for 'ant', by an amount equivalent to coats 2.3 times heavier at a given reserve weight.

The hypothesis that the association between soil fertility and dispersal mode arises because of an indirect correlation via reserve weight proved to have limited explanatory power, because the distribution of reserve weights was not very different between these soil types. However, dispersal mode was strongly associated with reserve weight, so indirect associations via reserve weight are worth investigating as possible explanations for differences in the spectrum of dispersal between other pairs of vegetation types.

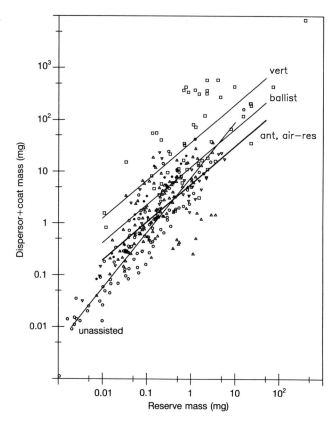

Fig. 29.7. Scaling of dispersor plus coat dry weight in relation to reserve dry weight (see Fig. 29.6). Slopes for dispersal types, other than unassisted, were parallel; mean slope 0.73. Dispersor plus coat weight at a given reserve weight was 6.3 times higher for vertebrate-syndrome species than for ant-syndrome species.

An indirect correlation via plant stature might account for up to 84 per cent of the difference in dispersal modes between soil types (including virtually all of that part of the difference which could alternatively be accounted for as an indirect correlation via reserve weight). Given that the biomass costs for the ant-dispersal syndrome are much less than for the vertebrate-dispersal syndrome, this suggests adaptations for ant-dispersal might be present in large part because they are cheap, and can achieve a distance which is effective for smaller plants. It would be nice to have some direct evidence as to the benefits obtained at different distances, however.

One problem for the cheap but adequate interpretation is that many vegetation types are dominated by plants of small stature but, nevertheless, do not have many species adapted for dispersal by ants. This emphasizes that just because an association via plant stature *can* account for up to 84 per cent of the difference in dispersal mode between fertile and infertile soils, this does not necessarily mean that it *does* account for it.

Finally, it should be repeated that there is a significant difference in dispersal mode between infertile and fertile soils, amounting to 15 per cent of the total difference, which remains unexplained by any of the hypotheses we have examined so far.

Acknowledgements

Jocelyn Howell, Kate Mossop, and Linda Tuffs have made important contributions to this research. It has been partly funded by the Australian Research Council and by a Commonwealth Postgraduate Research Award.

30

Myrmecochory in Cape fynbos

W. J. Bond, R. Yeaton, and W. D. Stock

Introduction

In 1975, Berg reported the widespread occurrence of myrmecochory in Australian heathlands. At the time, there was considerable interest in the hypothesis of convergent evolution among the mediterranean climate regions of the world (Cody and Mooney, 1978). Berg's discovery suggested a further test of the hypothesis in the arena of reproductive biology. Myrmecochory was virtually unknown in the Cape fynbos, the closest equivalent to the vegetation studied by Berg. Was fynbos also rich in myrmecochores? A survey to address this question indicated that perhaps one fifth of the Cape flora (>1000 species), mostly of little affinity with the Australian elements, were myrmecochorous (Slingsby and Bond 1981; Milewski and Bond 1982; Bond and Slingsby 1983). The discovery of this remarkable convergence in dispersal type, across diverse phylogenies, must rank as one of the more successful predictions of convergence theory. Ironically, the selective pressures which lead to the evolution of myrmecochory in diverse lineages within the Cape and Australian floras are still enigmatic.

In this chapter we review the studies of myrmecochory in Cape fynbos. The emphasis has generally been on the ecological significance of the phenomenon for the plants of the region. Predator evasion is important for larger seeded species, while recent work indicates that ant dispersal may also facilitate co-existence of competitors of Proteaceae. The broader question, of why so many plant lineages of diverse ancestry have evolved this trait, is of considerable interest and has stimulated a number of interesting and novel suggestions (Westoby *et al.* 1982*b*) but little consensus.

Distribution of myrmecochory in the Cape flora

Myrmecochory was first reported in five families of the Cape flora by Marloth (1915) who illustrated ants collecting seeds (see also Visser 1981 for *Mystropetalon* (Balanophoraceae) (all nomenclature follows Bond and Goldblatt 1984). Following Berg (1975), inventories of plant families, genera, and species have been compiled from numerous field cafeteria-type experiments, direct observation, and the occurrence of elaiosome-like structures on fruits and seeds (Bond and Slingsby 1983). Bond and Slingsby

(1983) reported probable myrmecochory in some 29 families, and 78 genera, in the Cape (the list is preliminary since only a fraction of species have been observed in the field). Myrmecochory occurs across diverse families in both monocotyledons and dicotyledons, including five of the seven families endemic to the Cape. Myrmecochory is found in tall tree-like shrubs, short shrubs, and perennial herbs, but is rare in geophytes (e.g. *Lachenalia* spp. (Liliaceae) and absent from annuals.

Within the Cape region myrmecochory is essentially a feature of fynbos shrublands on nutrient-poor soils and is rare in adjacent vegetation (Bond and Slingsby 1983). Very few genera are dispersed by ants in the Karoo, and genera that cross edaphic boundaries have myrmecochorous members in fynbos and wind-dispersed members in adjacent shrublands in arid climates or on nutrient-rich substrates (e.g. *Zygophyllum*, (Zygophyllaceae,) *Osteospermum* (Asteraceae).

Despite the diversity of plant myrmecochores, the ant side of the partnership is relatively depauperate. For most of the mid and lower mountain slopes and in coastal fynbos, *Anoplolepis* spp. (*A. custodiens* Smith and *A. steingroeveri* Forel) are the dominant seed dispersers (see below). They form large colonies of aggressive, ground foraging and nesting ants with a nest structure conducive to seed storage and germination. A number of other ant species also move seed; especially where *Anoplolepis* is absent such as in wetlands or mesic hygrophilous fynbos (Bond and Breytenbach 1985). In contrast to the plants, many of the ants have widespread distributions beyond the boundaries of fynbos; *Anoplolepis custodiens* for example, occurs throughout South Africa (Steyn 1954). The evolution of myrmecochory cannot, therefore, be attributed to special features of the ant fauna. Bond and Slingsby (1984) suggested that fynbos populations of *Anoplolepis* may respond more rapidly to elaiosomes than non-fynbos populations, implying local co-evolution. Breytenbech (1988), however, found that rates of seed discovery were as rapid in his Karoo study site, where no myrmecochores occur, as in the fynbos. There is, therefore, no evidence for co-evolutionary relationships between plants and their ant dispersers.

Ecological importance of myrmecochory

Most work on the ecological importance of myrmecochory has been on members of the Proteaceae, especially the genera *Leucospermum* (46 spp.) and *Mimetes* (14 spp.). These are all woody shrubs with relatively large achenes (up to 10 mm in length) with an elaiosome formed by the pericarp. Smaller seeded species have been less intensively studied. Their diaspores are not as attractive to ants and may be buried less readily, so the ecology of the interaction may differ from that of the Proteaceae (Bond and Slingsby 1983; Pierce 1987).

Ant behaviour

Anoplolepis appears to be very temperature sensitive, showing daily and seasonal variation in activity (Steyn 1954). It emerges from the nest at different times depending on the season. In subtropical regions, it emerges only at midday on the warmest days of winter, while in summer it is active in the early morning and late afternoon, retreating to the nest in the midday heat. There is some evidence that myrmecochorous plants exploit the period of peak ant activity in summer. In the Proteaceae, the myrmecochorous genera fruit in mid-summer, whereas non-myrmecochorous fruit ripens predominantly in autumn (Pierce 1984). The temperature sensitivity of *Anoplolepis* may also influence nest location; seeds are transported into nests in open areas unshaded by large Proteaceous shrubs in fynbos at Cape Point and never into shaded areas below the shrubs (Table 30.1). This has important ecological consequences for the plants (see below).

Table 30.1. The end point of dispersal of seeds of *Leucospermum conocarpodendron* found by three species of ants at Cape Point during December 1988.

Ant species	Number of seed piles found	Distance* (m)	Distance† (m)
Crematogaster peringueyi	14	< 0.1	< 0.1
Pheidole capensis	7	< 0.1	< 0.0
Anoplolepis custodiens	14	3.1 ± 0.4	2.1 ± 0.2
Unknown species	1	< 0.1	< 0.1
Not found	5		

* = Dispersal distance (± s.e.) from seed pile to nest entrance.
† = Average distance (± s.e.) from edge of *Leucospermum* canopy to nest entrance. (From Yeaton and Bond 1990.)

All seed-dispersing ant species recruit to Proteaceae seeds, but dispersal of seeds by solitary workers has been observed in smaller seeded species. In *Anoplolepis*, hundreds of workers may be recruited until the area is swarming with ants. Seeds are dragged into nest entrances and planted in small cavities 40–70 + mm below the soil surface. According to Stern (1954) the main part of the nest is well below ground level and the seeds are dispersed in a system of superficial horizontal channels. A fraction of the seed (25 per cent of *Mimetes*) may be ejected from the nest within a day or two of burial (Bond and Slingsby 1983).

The speed of ant reaction to Proteaceous seeds suggests that the ants are responding to some chemical signal released by the seed. Rodents are able to find intact seeds more readily than seeds with elaiosomes removed which also suggests the presence of a chemical attractant (Bond and Breytenbach 1985).

Response of plants

Several aspects of ant behaviour can influence the fate of the seeds they carry (Beattie 1985). First, the speed of discovery and seed burial in nests can reduce the loss of seeds to vertebrate predators (O'Dowd and Hay 1980; Heithaus 1981; Bond and Slingsby 1984; Bond and Breytenbach 1985). Secondly, the distance between seed source and ant-nests influences dispersal distance, whether seeds are planted in nests or not, and may influence seedling survival (Westoby and Rice 1981; Westoby et al. 1982a, b; Andersen 1988b). Thirdly, the depth at which seeds are buried may influence the micro-habitat, including the heat of intense fires (Berg 1981) and the availability of nutrients (Beattie and Culver 1983; Beattie 1985; Andersen 1989). Finally, the location of nests in special micro-sites may influence the spatial distribution and competitive interactions of myrme-cochorous plants (Handel 1978; Davidson and Morton 1981b).

Speed of discovery, seed burial, and predator evasion

Several studies have shown that myrmecochory in large-seeded Proteaceae is important for evading rodent predators. Experimental disruption of the dispersal process by removal of elaiosomes (Bond and Breytenbach 1985; Slingsby and Bond 1985) leads to failure of seedling establishment largely due to seed predation. Some studies have followed the fate of seedlings from dispersal in unburnt fynbos through to seedling establishment after subsequent fire (Bond and Slingsby 1984; Slingsby and Bond 1985). In both cases, seedling establishment failed without ant dispersal.

The dependence of plants on ant dispersal is of interest in the conservation of myrmecochorous Proteaceae. The Argentine ant, *Iridomyrmex humilis* has invaded small areas of fynbos (Kock and Giliomee 1989), mostly where vehicle access is possible, and displaced the dominant fynbos ants, particularly *Anoplolepis* and *Pheidole*. *Iridomyrmex* eats elaiosomes *in situ* and does not bury seeds. This has been shown to cause recruitment failure because of the prolonged exposure of seeds to rodents and to surface fires (Bond and Slingsby 1984). Although the areas invaded are small at present (Kock and Giliomee 1989) the threat may be large, as myrmecochorous Proteaceae include species with very small, local populations (Hall and Veldhuis 1985).

More recently, invasion of exotics has also been linked to myrmecochory. The Australian weeds, *Acacia saligna* Wendl. and *A. cyclops* A. Cunn. ex G. Don (Mimosaceae), depend on ants for their invasion of fynbos (Holmes

1990). Their seeds are heavily preyed on by rodents in the initial stages of invasion when the trees are only sparsely scattered. However, the seeds of the acacias have 'parasitized' the fynbos dispersal mutualism and are dispersed and planted by indigenous ants, thus evading vertebrate predators. Once an acacia stand has thickened (usually after a few fire cycles) the heavy seed rain satiates seed predators and ants are no longer needed for seedling regeneration.

Breytenbach (1986) has attributed poor recruitment of *Phylica arborea* Thou. (= *P. nitida* Lam.) (Rhamnaceae) on the subantarctic Marion Island to the recent introduction of rodents. *Phylica arborea*, like its fynbos relatives, has a caruncle which is attractive to mainland fynbos ants. Until recently neither ants nor rodents occurred on Marion Island. Breytenbach suggested that the current invasion of mice onto the island is leading to heavy seed predation in the absence of ants and is partly responsible for the poor seedling regeneration.

Dispersal distance

Dispersal distances of myrmecochores are typically small. The maximum distance recorded is 17 m for a *Leucospermum* seed. These short distances contrast with wind-dispersed serotinous Proteaceae. Serotinous plants retain seeds in woody cones on the plant until the above-ground parts are killed by fire. Then the seeds are released into the open, post-fire environment. Primary dispersal distances fall in the range of 20–40 m, but secondary dispersal by rolling across the soil surface can extend this by hundreds of metres (Bond 1988). The dispersal shadow of *Leucospermum conocarpo-dendron* (L.) St John, a myrmecochore, is contrasted with *Protea repens* Thunb, a serotinous species, in Fig. 30.1.

Since many fynbos species have non-overlapping generations (being killed by fire), it is unlikely that escape from parental competition or parent-centred concentrations of pathogens or predators is a major benefit of dispersal. Westoby *et al.* (1982*b*), however, showed that myrmecochores tended to be more widely spaced from conspecifics than non-myrmecochores, and Andersen (1988*b*) has suggested that the dispersal shadow of myrmeco-chores may confer unique, but unspecified, benefits.

Depth of burial and fire survival

Seeds of large-seeded myrmecochores are typically buried 40–70 mm below the soil surface, with some up to 120 mm deep (Bond and Slingsby 1983). As judged by the length of the hypocotyl, myrmecochores germinate from deeper in the soil than non-myrmecochores (Musil 1990). This may be significant for fire survival. Myrmecochorous seedlings are conspicuously good at surviving very hot fires which kill non-myrmecochorous seed banks (Richardson and van Wilgen 1986). The corollary is that some myrme-cochorous Proteaceae have very poor germination after the cool fires most

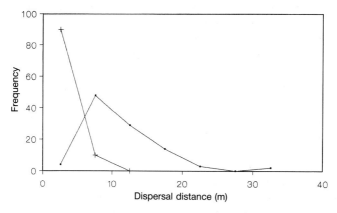

Fig. 30.1. Dispersal distances in *Protea repens* (—●—) (serotinous, wind dispersed) and *Leucospermum conocarpodendron* (+—+) (myrmecochorous). The *Protea* curve is for the free-fall phase of dispersal in a 6 m s^{-1} wind. (From Bond 1988.) The *Leucospermum* curve is derived from seedlings emerging from a known source after a fire. (From Slingsby and Bond 1985.)

favoured by management (W.J. Bond *et al.* 1990). Local variation in fire intensity may therefore be significant in controlling the distribution and density of myrmecochores.

The deep burial of seed implies unusual germination cues. Most large-seeded myrmecochores germinate *en masse* only after fire. Brits (1986 *a*, *b*) has shown that myrmecochorous Proteaceae seeds respond to changes in the range of soil temperature after fire and to increases in oxygen. He has found that temperature ranges experienced by seeds buried in the soil are close to the known optimum for germination in burnt, and not in unburnt, fynbos (Brits 1987).

Seed destination and nutrient concentrations

Beattie (1985) has argued that seed placement in ant-nest soils, which are enriched in nutrients, is of major benefit for myrmecochores. This would help explain the frequent occurrence of ant-dispersal in the nutrient-poor soils of Australia and the Cape. However, Rice and Westoby (1986) tested soils around myrmecochorous and non-myrmecochorous seedlings and found no evidence for higher nutrients in the former (Majer 1982; Westoby *et al.* Chapter 29, this volume).

Recently, Bond and Stock (1989) have explored the implications of ant behaviour for soil-nutrient supply to a myrmecochore, *Leucospermum conocarpodendron*. *Anoplolepis* carries seeds to nests in the open and buries the seeds at depth. We compared soil nutrients beneath the skeletons of parents in the first growing season after a burn, with soils in the open areas to which seeds had been transported by ants. We also compared nutrients immediately around the base of the hypocotyl of seedlings with soils 15 cm away, to test whether seeds are planted in micro-sites enriched with nutrients due to their deposition in ant-nests. We used the length of the hypocotyl as a measure of whether seeds had been dispersed by ants or had fallen passively, usually below parent canopies, and been buried under litter (see also Brits 1987).

The results (Table 30.2) are surprising and directly the reverse of those predicted by the nutrient-enrichment hypothesis. Ants carry seeds to sites that have less nutrients than sites to which seeds are dispersed passively. This is both because nutrients accumulate in the litter below parents and because more nutrients become available through mineralization by hotter fire temperatures under the bushes. In addition, nutrients are concentrated in the surface of fynbos soils so that passively dispersed seeds have access to higher nutrient concentrations. Differences in soil nutrients among the micro-sites were reflected in tissue nutrient concentrations of phosphorus, but not of nitrogen. Phosphorus is believed to be the major limiting nutrient in fynbos and Australian heathland (Specht 1979; Witkowski 1990).

The processes of litter accumulation which produce these results are not unusual and must be similar for most of the larger Proteaceae and other myrmecochorous taxa. Thus it seems unlikely that seedlings benefit from nutrient enrichment in ant-nest soils. On the contrary, myrmecochory has evolved despite the cost of seeds being placed in nutrient-poor sites. It is also

Table 30.2(a). Micro-site variation in total nitrogen (mg N g^{-1} soil) and total phosphorus (μg P g^{-1} soil) after a fynbos fire, Cape Peninsula. Soils were sampled from the vicinity of *Leucospermum conocarpodendron* seedlings germinating under the burnt canopies of parents or in open sites, and compared with soils sampled at the same depth but with no seedlings present. Data are means ($\pm$ s.d.), back-transformed after log transformation. n = Sample size. Depth of seed burial (measured by hypocotyl length) was used as a covariate. (From Bond and Stock 1989.).

	Under parents			Away from parents		
Seedlings	present	absent	mean	present	absent	mean
n	15	15	30	29	29	58
Total N	1.683	1.596	1.641	1.247	1.322	1.282
s.d. (mg N g^{-1} soil)	1.140	1.194	1.114	1.107	1.122	1.079
Total P	63.8	61.4	62.7	45.2	47.3	46.1
s.d. (mg P g^{-1} soil)	1.09	1.12	1.07	1.04	1.04	1.03

Table 30.2(b). Results of two-way analyses of covariance for total nitrogen and total soil phosphorus, data in Table 30.2(a). Values are F ratios. The covariate, hypocotyl length, indicates depth of sample. (d.f. = 1 hypocotyl length, 1 canopy, 1 seedling, 1 interaction, 55 residual).

	Hypocotyl length	Under parent vs. open	With seedling vs. without seedling	Interaction
Total N	7.2	0.3	0.01	0.1
P	<0.01	n.s.	n.s.	n.s.
Total P	12.3	6.0	0.00	0.2
P	<0.001	<0.1	n.s.	n.s.

n.s. = not significant.

Table 30.2(c). Nutrient content and biomass of above-ground tissues of *Leucospermum* seedlings under parents and dispersed into open sites. Statistical test = Mann–Whitney.

	n	Under canopy	n	In the open	P
Total N (mg)	7	4.51 ± 0.64	15	4.47 ± 0.84	n.s.
Total P (µg)	8	340 ± 27	15	260 ± 53	<0.01
Biomass (g)	15	0.12 ± 0.06	15	0.15 ± 0.08	n.s.

worth noting that seeds are not deposited in the heart of *Anoplolepis* nests but only in superficial galleries where, presumably, there is little other material that could act as a source of nutrients. The nests of other ant species have not yet been investigated.

Seed destination and competitor avoidance

Several recent theoretical papers have suggested that differences in seed dispersal may influence the outcome of interspecific competition between plants with similar vegetative niches, 'trophically equivalent species' (Shmida and Ellner 1984; Comins and Noble 1985). Shrubby members of the Proteaceae dominate many communities in Cape fynbos (Kruger 1979). Co-occurring species are typically broad-leaved, evergreen, and sclerophyllous, and share similar life histories; generations are non-overlapping with mature plants killed by fire and regeneration largely confined to the post-fire period. Despite their apparent 'trophic equivalence', co-occurring species differ strikingly in their dispersal biology. The myrmecochorous species are dispersed between fires, store seed in the soil, and use complex germination cues to emerge after fires (Brits 1986*a*, *b*).

In marked contrast are a serotinous group of species. Seeds are held in serotinous cones and released *en masse* after fire where, in the open post-fire conditions, they are dispersed by wind. The seeds are covered by long hairs and are blown over the burnt soil surface until caught by obstacles such as stones or burnt plant debris (Bond 1988). In this group seeds may be carried an order of magnitude further than by ants but the final location of seeds is subject to the vagaries of wind, surface wash, and barriers to abiotic seed movement (Manders 1986; Bond 1988).

Recently, we studied mechanisms of co-existence in two-occurring species on the Cape Peninsula; *Leucospermum conocarpodendron*, dispersed by ants, and *Protea lepidocarpodendron* L., a serotinous species (Yeaton and Bond 1990). Seedlings compete for space made available after periodic fires. By comparing the performance of *Leucospermum* individuals growing next to *Protea* with those growing in open sites, at a variety of post-fire ages, we found that *Leucospermum* is strongly suppressed if it grows within the canopy radius of a *Protea* (Table 30.3). Fewer *Leucospermum* individuals reach maturity and those that do have fewer branches to bear inflorescences at their apices, and flowering is negligible. *Leucospermum* has thick bark and can survive cool fires by regrowth from the canopy. However, if *Leucospermum* canopies are adjacent to *Protea* the heavier crown fuel loads lead to higher mortality of *Leucospermum* after fire (Table 30.3). Thus, in growth, reproduction, and mortality *Leucospermum* is adversely affected by neighbouring *Protea*.

The co-existence of the two species depends, in part, on non-uniform seed dispersal between patches and the contrasting destinations of the two seed types. *Protea* seeds accumulate under the burnt skeletons of their parents but also under the skeletons of *Leucospermum* bushes. Few *Protea* seeds accumulate in open areas between shrubs because of the absence of barriers to wind dispersal. *Leucospermum*, in contrast, has some seedlings beneath the canopy radius of the burnt parents from passive dispersal, almost none under *Protea* bushes, and many in open areas where seeds were carried to nests by sun-seeking ants (Table 30.4). The seeds of the myrmecochore are thus placed in the most favourable micro-site to escape their *Protea* competitor as a result of the nest choice of the ants.

The importance of the ants can be gauged by modelling the system as a three-state Markov process (Horn 1975; McAuliffe 1988). The probabilities of changing from one state to another after a fire can be assessed by counting seedlings in open areas and beneath the burnt canopies of *Protea* and *Leucospermum*. Seedling counts can be translated into transition probabilities

Table 30.3. Competitive effects of *Protea lepidocarpodendron* on *Leucospermum conocarpodendron*. (From Yeaton and Bond 1990).

(1) Seedling growth
Height of neighbouring plants (within 50 cm) 3 years after a fire at Cape Point. ($P < 0.001$)

	n	Height (cm)
Protea	50	29.5 ± 1.0
Leucospermum	50	15.6 ± 0.6

(2) Shrub size at maturity
The effect of *Protea* canopies on stem diameter of *Leucospermum* at Steenberg Plateau. Vegetation was 15–16 years old. ($P < 0.001$)

	n	stem diameter (cm)
In the open	25	6.9 ± 0.2
Under *Protea*	25	3.3 ± 0.2

(3) Reproductive performance
The effect of *Protea* canopies on flowering of *Leucospermum* at Steenberg Plateau during spring 1988. ($P < 0.001$)

	n	Mean no. of inflorescences
In the open	25	8.9 ± 1.7
Under *Protea*	25	0.2 ± 0.2

(4) Fire survival
Neighbour effects on survival of *Leucospermum* after a light burn at Rooihoogte, Cape Point (x^2 10.6, $P < 0.001$).

	n	% survival
In the open	62	56.5
Next to *Protea*	38	21.1

by assuming that the probability of a *Protea* being replaced by a *Leucospermum*, for example, depends on the sum of the relative frequency of the seedlings of the two species for a large sample of individuals (Culver 1981). Alternatively, one can include knowledge of the biology of the system and count patches in which both *Protea* and *Leucospermum* seedlings occur as moving to *Protea* before the next fire cycle. The matrices and projected population growth, assuming similar post-fire regeneration, are shown in Table 30.5 and Fig. 30.2. According to the model, *Leucospermum* will persist in

Table 30.4. The mean density ($\pm$ se) of seedlings of *Protea lepidocarpodendron* and *Leucospermum conocarpodendron* in three sub-habitats in fynbos at Miller's Point, Cape Peninsula. (From Yeaton and Bond 1990).

Subhabitats	n	Seedlings m^{-2}	
		Protea	*Leucospermum*
Open	50	0.3 ($\pm$ 0.1)	1.7 ($\pm$ 0.4)
Under *Leucospermum*	50	3.5 ($\pm$ 0.7)	2.6 ($\pm$ 0.3)
Under *Protea*	50	12.8 ($\pm$ 7.1)	0.6 ($\pm$ 0.1)

Table 30.5. Transition matrices derived from seedling distributions of ant-dispersed *Leucospermum conocarpodendron* and serotinous *Protea lepidocarpodendron*. Values of the matrices are derived under the assumption that if both *Protea* and *Leucospermum* occur in the same patch, *Protea* will dominate by the next fire. Assume, in addition, no dispersal of *Leucospermum* seeds beyond its canopy radius. The predicted stationary states are the proportions of the two species (and open space) in the community after many fire cycles with the same transition characteristics. (From Yeaton and Bond 1990).

	Probability of subhabitat becoming			
	Open	*Protea*	*Leucospermum*	Stationary state
With ants				
Open	0.30	0.16	0.54	0.08
Under *Protea*	0.06	0.92	0.02	0.84
Under *Leucospermum*	0.08	0.66	0.26	0.08
No ants				
Open	0.84	0.16	0.00	0.33
Under *Protea*	0.08	0.92	0.00	0.67
Under *Leucospermum*	0.08	0.66	0.26	0.00

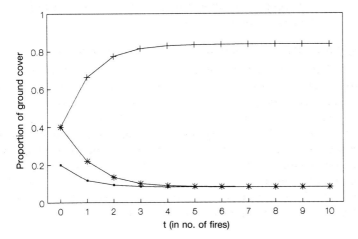

Fig. 30.2. Projected system dynamics at Miller's Point, Cape Peninsula. Each time unit represents the interval from one fire to the next. Proportions of *Protea* (—+—), *Leucospermum* (—*—), and open space (—•—) are calculated from $N_t = N_{t-1} \times T$, where $N_0 = 0.2, 0.4, 0.4$ and T is the matrix of transition probabilities given in Table 30.5. (From Yeaton and Bond 1990.)

the community, though at low levels. However, should ant dispersal fail (as it has where *Iridomyrmex humilis* invades, see Bond and Slingsby 1984, Fig. 3), *Leucospermum conocarpodendron* would be eliminated from the community. Hamilton and May (1977) presented a hypothetical model of a similar situation to argue that some dispersal will *always* be selectively advantageous.

Cowling (1987), using models developed by Chesson and Warner (1981), has suggested that the variability of fire regimes is important in maintaining diversity in fynbos. In this case, *Leucospermum* can apparently persist even if the fire regime remains identical (i.e. transition probabilities remain unchanged). Nevertheless, population projections suggest a rapid rate of decline of *Leucospermum* in the population we studied. This is of interest for the management of fynbos. There has been some debate as to whether prescribed burning would lead to loss of species diversity if the same fire regime is applied consistently in an area. In the case of *Protea* and *Leucospermum*, *Leucospermum* would reach low population levels within only three fire cycles. It would be possible to manipulate fire frequencies or fire seasons to favour vegetative *Leucospermum* survival at the expense of *Protea lepidocarpodendron* seedling recruitment. Indeed, such an event probably created the pre-burn community we studied. The Markov model projections indicate that short interval or unseasonal fires, which reduce *Protea* but not

Leucospermum populations, should recur every third or fourth fire cycle to ensure persistence of *Leucospermum*.

Ecological benefits of myrmecochory in small-seeded species

Small-seeded myrmecochores may not have the same ecological benefits as the larger seeded (Proteaceae) species that have received most research attention. Pierce (1987) found that seeds of small-seeded species with elaiosomes were removed at the same rate as small seeds without elaiosomes. Westoby *et al.* (1982*b*) similarly noted that elaiosomes of many Australian species are not soft and collapsing and therefore are not dependent on rapid discovery and removal. The inference is that the seeds of such species will not escape predators through the agency of ants. Westoby *et al.* (1982*b*) have argued that enhanced dispersal distance is the main benefit for such species. As yet we have no information on this topic in the Cape.

Evolution of myrmecochory

The ecological benefits of myrmecochory described above may be important for the persistence of species in communities, but they are not necessarily reasons for the evolution of myrmecochory in the Cape fynbos. Many genera are uniformly myrmecochorous. All species of *Leucospermum*, for example, are myrmecochorous yet differ greatly in other features, including growth form (Rourke 1972). The competitive interaction between *Leucospermum conocarpodendron* and *Potea lepidocarpodendron* is not general for these genera. Some *Leucospermum* species outgrow their serotinous neighbours, others evade them by resprouting from root stocks after fire. It is difficult to argue from the particular case of avoidance of competition by a single species to a general selective advantage for myrmecochory, leading to the evolution of the syndrome in an entire genus.

There are two major problems that require explanation. First, there is massive convergence in dispersal type in fynbos. Why is ant dispersal so common in fynbos and so rare in adjacent vegetation types? This is the problem that has attracted the most attention (e.g. Berg 1981; Milewski and Bond 1982; Westoby *et al.* 1982*b*; Beattie 1985; Rice and Westoby 1986). We are little further in our understanding despite considerable research on the ecological implications of myrmecochory. For example, predator evasion is known to be important for larger seeded species but there are several objections to the argument that this is the major advantage of myrmecochory (Westoby *et al.* 1982*b*; Bond and Slingsby 1983; Beattie 1985), and predator pressures seem to be no less in adjacent vegetation types. The hypothesis that ant-nest soils are richer in nutrients and provide more fertile seedling beds has received no support from studies in Australia (Majer 1982; Rice and Westoby 1986; Andersen 1989) or South Africa (Bond and Stock 1989). The argument that seed burial is important in fire environments is also not persuasive, since fires also occur in vegetation adjacent to fynbos with few myr-mecochores (renosterveld) and in other parts of the world (chaparral) where myrmecochory is very rare.

Westoby *et al.* (1982*b*) argued that there are no special benefits of myr-mecochory but rather a saving on costs, especially in nutrients expended on the elaiosome. However, there have been no studies on the Cape flora com-paring, for example, reproductive costs of wind and myrmecochorous mem-bers of related taxa on nutrient-rich and nutrient-poor soils.

Finally, one may ask what alternative dispersal agents exist in the fynbos environment and why they are not used more widely. Siegfried (1983) has shown that bird dispersal is particularly rare in fynbos compared to adjacent vegetation on soils richer in nutrients (Table 30.6). S. M. Pierce (personal

Table 30.6. Distribution of myrmecochory and ornithochory in the south-western Cape, South Africa. Data are plant species numbers relativized to the total of bird dispersed species. (From Siegfried 1983, Appendix 3).

	Myrmecochory	Ornithochory
Nutrient poor		
Mountain fynbos	28	8
Coastal fynbos	5	1
Nutrient rich		
Strandveld	6	19
Renosterveld	3	14

communication) has suggested that bird dispersal may be excluded for mechanical reasons from fynbos vegetation. First, vertebrate-dispersed fruits might be expected to have a larger minimum size than other kinds of fruit to provide a sufficient reward for vertebrates. Larger fruits require stouter twigs to support them and the birds that consume the fruits. Large twigs, however, are allometrically related to large leaf sizes (Midgley and Bond 1989; McKey Chapter 21, this volume) Thus, vertebrate dispersal will generally be associated with large leaves. However, nutrient-poor soils are associated with a high frequency of species with small leaves (Cowling and Campbell 1980; Givnish 1987).

Since small leaves are borne on thin shoots, vertebrate dispersal is unlikely to develop without special architectural modification of the plants. There is some support for this hypothesis (S. M. Pierce, unpublished work) which would lend weight to the argument that myrmecochory in nutrient-poor soils in the Cape and Australia may have developed by default as one of the few feasible dispersal mechanisms in the absence of dispersal by birds.

The second evolutionary problem is why some taxa, entirely confined to fynbos, have both myrmecochorous and non-myrmecochorous members. The genus *Leucadendron* in the Proteaceae is particularly interesting. Midgley (1987) has shown that myrmecochory is a highly derived trait in this genus and has evolved from serotinous ancestors. He was unable to find differences in growth form, height, or geographical distribution which consistently separated myrmecochorous species from serotinous species of *Leucadendron*, or any selective advantage of myrmecochory over serotiny. In addition, some species in the genus have seeds neither dispersed by ants nor stored in the canopy. Their biology is still poorly understood but they provide an interesting control group of large seeded Proteaceae with soil-stored seeds that survive without the assistance of ants (Midgley 1987).

Conclusions

Myrmecochory in Cape fynbos has provided several fascinating insights into how mechanisms maintaining fynbos populations and communities function in this extreme environment. It is likely that future progress will depend on more intensive study of the neglected ant half of the interaction and comparative studies of related plant species pairs with different dispersal mechanisms. In particular, the notion that dispersal costs are less for myrmecochory than for alternatives needs testing in the context of nutrient-poor fynbos versus adjacent environments.

Acknowledgements

W. B. thanks Peter Slingsby for many pleasant hours of ant watching and G. J. Breytenbach, J. Vlok, J. J. Midgley, and S. M. Pierce for critical discussion. Several workers have generously made unpublished results available: A. de Kock, P. M. Holmes, C. Musil, and S. M. Pierce. We thank the Research Committee of the University of Cape Town for funding part of this research and the Directorate of Forestry for encouragement and support.

31

Light environments, stage structure, and dispersal syndromes of Costa Rican Marantaceae

Carol C. Horvitz

Introduction

Even though morphological traits of fruits, seeds, and infructescences are widely assumed to be adaptive to different kinds of dispersal (van der Pijl 1972; Howe and Smallwood 1982; Augspurger 1984), we still understand very little about the factors that select for one type of dispersal morphology instead of another (Janson 1983; Charles-Dominique 1986; Howe 1986; Janson *et al*. 1986). The usual paradigm is that plants that inhabit early seres of successional environments (newly disturbed sites) are characterized by small, widely dispersed, frequently dormant seeds that are packaged in relatively low-energy fruits adapted for either wind dispersal or a wide array of animal foragers. In contrast, late successional plants are characterized by larger, more locally dispersed, usually non-dormant (especially in tropical species) seeds that are packaged in energy-rich fruits which, because of their large size, are available to a more restricted group of animal foragers. This paradigm (McKey 1975; Platt 1975; Werner 1977; Bazzaz and Pickett 1980; Gross and Werner 1982; Vazquez-Yanes and Orozco-Segovia 1984; Wheelwright 1985; Platt and Hermann 1986) was born out of the general intuition that optimal seed dispersal characteristics must depend upon the interaction between spatio-temporal heterogeneity of the environment and plant life histories; but it is insufficient to account for the diversity of dispersal syndromes observed in nature.

A comparative study of closely related taxa with divergent dispersal syndromes may provide deeper insights into the nature of evolution of seed dispersal characters and plant life history characters, as well as selection for specialization of plants in relation to animals that have differing attributes as seed dispersers. Here, I address the question of the adaptive significance of divergent dispersal syndromes by comparing population attributes of ant-dispersed and bird-dispersed species in the Marantaceae, a family of tropical forest herbs. A model of forest and population dynamics (briefly described below in the section on theoretical background to the model) predicted that stage-specific growth, survival, and reproduction would decline much more rapidly for bird-dispersed than for ant-dispersed herbs as a gap fills in following a tree fall (Horvitz and Schemske 1986a). According to this hypothesis, demographic success, as measured by stage-specific growth, survival, and reproduction, would be enhanced in light gaps and inhibited in the deep shade for all understorey herbs, but ant-dispersed species are expected to do better in intermediate habitats and to have a wider regeneration niche (*sensu* Denslow 1980, 1987) than bird-dispersed species.

The seed shadows created by these two animal groups are likely to be on very different scales, 1–2 m for ants (Horvitz and Beattie 1980; Horvitz and Schemske 1986b, unpublished work) and much farther for birds (O'Daniel 1987). Seed shadows created by understorey birds may be quite large (many seeds may be moved 100–200 m), as indicated by Murray's (1986a,b) radio tracking study of avian understorey frugivores in a cloud forest. Consequently, seeds carried by these different animals differ in their probability of encountering forest gaps. Seeds carried by birds are more likely to encounter gaps simply because of scale and chance alone. It is not necessary to assume that understorey birds preferentially visit gaps, although there is evidence that they do in some forests (Thompson and Willson 1978; Levey 1988a) but not in others (Schemske and Brokaw 1981). Seeds carried by ants are unlikely to reach new large-scale gaps, but they may reach micro-sites that enhance germination and seedling success either through soil nutrients (Beattie and Culver 1983; Beattie 1985), escape from predators (O'Dowd and Hay 1980), or sunflecks (Smallwood 1982).

Predictions of the model

The model predicted that the evolution of dispersal characters is related to the evolution of characters that determine how a plant utilizes the open space created by disturbances, particularly those characters that determine how long a plant may persist during the years following the disturbance. In this sense the evolution of the dispersal syndrome and the evolution of the plant's response to disturbance may be said to be correlated or interdependent. As shadiness increases during gap-phase regeneration after tree fall, demographic parameters of ant-dispersed species are expected to decline slowly, while demographic parameters of bird-dispersed species are expected to decline rapidly. This prediction concerns how demography is expected to change with time, and will require a long-term study of demographic change from year to year in sites undergoing gap-phase regeneration for both ant- and bird-dispersed species. Testing this hypothesis will be the subject of future research.

To determine whether the hypothesis about the divergent dynamic demographics of bird- and ant-dispersed species predicted by the model was worth pursuing, I formulated a hypothesis about the *static* stage habitat

population structure based on insights from the *dynamic* model. I hypothesized that seedlings of bird-dispersed species will be restricted to the lightest habitats, while seedlings of ant-dispersed species will be common in
darker habitats. Testing this hypothesis is the subject of this chapter.

Theoretical background to the model

In Horvitz and Schemske's (1986*a*) matrix model of transition probabilities,
each entry represented the probability that a plant in a particular stage-
habitat category would become a plant (or contribute to the number of
plants) in another stage habitat category after one year. This linear
Markovian model included five stage classes (from seeds to large reproductives) and ten habitats (from new tree fall to closed canopy), resulting in a
50×50 matrix summarizing the dynamics of the entire metapopulation in
the dynamic landscape (different approaches to metapopulations are found
in Addicott 1978 and Olivieri and Gouyon 1985). The dominant eigenvalue
of this matrix corresponds to the population growth rate (Caswell 1982) and
thus to a measure of the average fitness of the population (Fisher 1930;
Charlesworth 1980); in this case, the global average fitness over all ten
dynamic habitats. When the dominant eigenvalue is equal to one the
metapopulation will be stable; if it is less than one the metapopulation will be
becoming extinct. Any character that causes an increase in the value of this
parameter will be selected for. Three dispersal characters were investigated:
long-distance dispersal, local dispersal to safe micro-sites, and seed
dormancy.

Forest patch dynamics were modelled by a process in which:

1. the probability of gap formation in each patch of forest depended upon
   the time since the last gap in that patch; and
2. patches that did not return to gap were assumed to become progressively
   shadier with age.

These assumptions led to hypothetical forests with structures and turnover
rates similar to forests that have been studied empirically.

Each habitat in the sequence had its own complete set of plant demographic parameters, probabilities of growth, survival, and reproduction for
each life history stage of the understorey herb. These parameters declined in
value as patches aged; juveniles being more sensitive than adults to the
changing habitat. Since actual rates of demographic change with patch age
were unknown, the model emphasized an investigation of three possible rates
of demographic change (non-linear slow, linear, and non-linear fast) in the
sequence of intermediate habitats of gap-phase regeneration following a treefall.

Utilizing real demographic data for an ant-dispersed species (*Calathea
ovandensis*) living in a new tree fall, a very interesting result was found in that

(a)

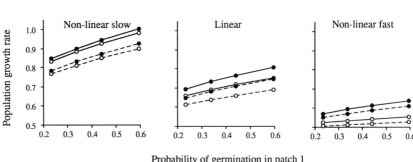

Probability of germination in patch 1

(b)

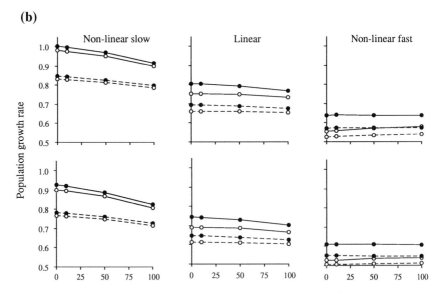

Percentage of seeds dispersed out of patch

Fig. 31.1. (a) Effects of within-patch dispersal to safe sites on overall population growth rate for three models of demographic change for an understorey herb in a dynamic environment. The effects of local dispersal to safe sites were modelled by improvement in germination. Along the horizontal axis, the germination rate in a new tree fall (patch 1) is given. Germination rates in the other patches (patch 2–patch 10) were calculated relative to this rate according to the models (see Horvitz and Schemske (1986a) for details). Solid lines and dashed lines represent two different disturbance regimes: $k = 0.02$ and $k = 0.01$, respectively, where k is the probability of a tree falling in the closed canopy patch. Solid circles and open circles represent two different degrees to which germination may be inhibited in the closed canopy; 50 and 100 per cent decrease, respectively, relative to germination in a new tree fall. Results are presented for four different germination rates. Note that positive slopes throughout indicate selection in favour of local dispersal to safe sites, most strongly for shade-tolerant (non-linear slow) models. (Reprinted by permission from Horvitz and

the *non-linear slow* rate of change of demography through time (shade-tolerant) was the only pattern that predicted a stable metapopulation, rather than a metapopulation in the process of becoming rapidly extinct (Fig. 31.1). Associated with this pattern were (a) selection *against* long-distance dispersal (Fig. 31.1*b*); (b) strong selection for local dispersal to safe-sites (Fig. 31.1 (*a*)), and (c) weak selection for seed dormancy. In contrast, the *non-linear fast* rate of change of demography through time (shade-intolerant) predicted weak selection *in favour* of long distance dispersal (Fig. 31.1*b*). Also predicted were (a) weak selection for local dispersal to safe micro-sites (Fig. 31.1*a*) and (b) strong selection for seed dormancy (Horvitz and Schemske 1986*a*). In intuitive terms, demographic success in intermediate habitats would be associated with selection against long-distance dispersal and strong selection for local dispersal to safe sites, while only a pattern of poor demographic performance in intermediate habitats should result in selection in favour of long-distance dispersal.

Note that these simulations did not make an assumption that some birds will preferentially disperse into gaps. There was also no inclusion of a diplochorous system in which bird-dispersed seeds are secondarily dispersed by ants. The simulations only required that seeds that were dispersed away from the patch of the mother plant encountered habitats at random, that is in proportion to their occurrence in the overall environment. By contrast, seeds that were dispersed within the same patch of forest as the mother plant experienced the same sequential change of environments as the mother plant.

For *Calathea*, selection for long-distance or short-distance dispersal, means specialization in relation to birds or ants, respectively. Under the only regime that showed selection for long-distance dispersal, the *non-linear fast* (shade-intolerant) model, the metapopulation growth rate was « 1.0, indicating rapid extinction of the metapopulation (Fig. 31.1*b*). Thus, the demography of bird-dispersed species in a new tree-fall must differ from the demography that was observed for an ant-dispersed species. Demographic differences must exist to predict stable populations of bird-dispersed species,

---

Schemske 1986*a*, Fig. 7.) (b) Effects of long-distance dispersal on overall population growth rate for three models of demographic change of an understorey herb in a dynamic environment. Long-distance dispersal was modelled by movement away from the maternal patch of the forest. For each model the upper and lower figures represent two different disturbance regimes: $k = 0.02$ and $k = 0.01$, respectively. Solid lines and dashed lines represent the results for the highest and lowest germination rates, respectively, used in the simulations presented in (a). Solid circles and open circles are as in part (a). Results are presented for four different levels of long-distance dispersal. Note that negative slopes indicate selection against long-distance dispersal; the only conditions that favour long-distance dispersal at all are some of the shade-intolerant (non-linear fast) models. (Reprinted by permission from Horvitz and Schemske, 1986*a*, Fig. 8.)

for example possibly higher seed production. I recently found that a stable metapopulation could be simulated by including in the model a ten-fold increase in seed production compared to the ant-dispersed species of the original model (Fig. 31.2). Many of the bird-dispersed species have very large and highly branched infructescences and are much larger plants (see Appendices 31.1, 31.2 and 31.3), suggesting that a ten-fold difference in seed production may be quite possible.

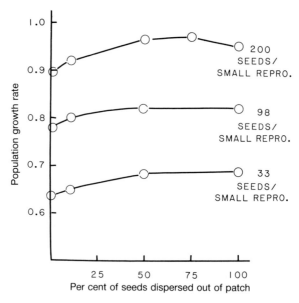

Fig. 31.2. Effects of long-distance dispersal and increased seed production on population growth rate, utilizing the model of population dynamics and tree fall gap dynamics of Horvitz and Schemske (1986a) for a relatively shade-intolerant ('non-linear fast') life history. Parameter values utilized: disturbance regime of $k = 0.02$; germination in the closed canopy was decreased by 100 per cent relative to a new tree fall; germination in a new tree fall was 0.75; seed production of small reproductives was increased from two times to 10 times that of the earlier model. All other parameter values were as specified in the original model.

Study species and site

The tropical forest understorey herbs in the family Marantaceae provide an excellent opportunity for such a study. Some species show clear morpho-logical adaptation for seed dispersal by ants: infructescence of low stature, fresh seeds presented on the forest floor, dark seeds bearing white, lipid-rich appendages (Horvitz and Beattie 1980; Horvitz and Schemske 1986b, c). Others exhibit a typical syndrome for bird dispersal (Stiles 1982; Willson and Melampy 1983; Wheelwright and Janson 1985): infructescences taller (up to 3 m), mature seeds retained on the plant and displayed conspicuously, utilizing contrasting colours in a bi-coloured display. I chose two wet forest field sites in Costa Rica, Corcovado National Park (Osa Peninsula, Pacific slopes) and La Selva Biological Reserve (Heredia Province, Atlantic slopes) (Hartshorn 1983) because they were reputed to be rich in Marantaceae (Hammel 1984; H. Kennedy, personal communication; L. E. Gilbert, personal communication). In September 1987, I located populations of fourteen species of Marantaceae (mostly *Calathea*) in these forests.

Descriptions of displays

Previous descriptive accounts of Costa Rican Marantaceae did not explicitly include dispersal display (Kennedy 1977, 1978, 1983; Hammel 1984). The first task for each species was to describe the dispersal syndrome. The position of fruits and seeds at maturity was observed, noting which structures were involved in the display. Seeds and seed arils were also collected, oven-dried and weighed.

Four dispersal syndromes could be distinguished as relevant to the current study, based upon retention of seeds on infructescences at maturity, height and branching of infructescences, structures which were actually involved in the display of the mature seeds, and dry masses of seeds and arils (see Appendix 31.1 for details and Table 31.1 for a summary). There were one ant-dispersal syndrome and three distinct bird-dispersal syndromes.

Two of the bird syndromes were associated with life history characters likely to influence the demography of juveniles. The 'Bird II' plants regularly reproduce by forming clonal propagules (about the size of large seedlings) at the nodes of the branches bearing the infructescences (this was also noted by Kennedy (1978) for one of these species, *C. warscewiczii* Koern.) (Table 31.1). The 'Bird III' plants have very large seeds, three to five times larger than the seeds of the other species (Table 31.1). Large seed size (Foster and Janson 1985), as well as reproduction by clonal propagules (Cook 1979; Caswell 1985), may well be associated with increased shade tolerance of juveniles.

Stage structure and light environment

For each study population, a rough measure of the population stage structure and of the light environment was obtained. The location of each population with respect to trail markers was also recorded. Stage structure was estimated by noting reproduction and by measuring the size of all the individuals, using the length of the longest leaf blade as a quick index of plant size (Horvitz and Schemske 1988). Plants were classified as seedlings, intermediates, or reproductives, using different criteria for each species (see Appendix 31.3). The light environment of each site was crudely estimated by taking a wide-angle (28 mm) lens image of the canopy on 35 mm slide film, at most sites. (At

Table 31.1. Summary of dispersal syndromes of some Costa Rican Marantaceae.

Type	Infructescence		Display	Seed mass (mg)	Reward % dry mass (aril/aril + seed) × 100
	height (m)	branched?			
Bird I	2–3	yes	seeds sticking out of flattened bracts	500–1000	7–17
Bird II (with clonal propagules)	1–1.5	no	seeds embedded in infructescence	500–1000	5–6
Bird III (with large seeds)	1–2.5	no	seeds showing in waxy fruit capsule	2500–3000	8
Ant	0.1–0.3	no	white arils against dark seeds, forest floor	<500	5

some sites, a qualitative estimate of the relative light environment was recorded and later assigned a category approximately comparable to one of the categories in the classification scheme derived from the quantitative estimates.) The slide was later projected onto a piece of paper to trace the image. Percentage open sky was calculated using a DT 110 Houston Instruments digitizing pad. Light environments were classified into six categories, in increments of 15 per cent open sky (0–15, 16–30, etc. up to 90 per cent).

The main result was that dispersal syndrome, light environment, and stage structure were not independent $(G_{Williams}) = 562$, d.f. $= 61$, $P < 0.001$, test for the complete independence of three factors; all the tests of conditional independence and the three-way interaction term were all highly significant) (Sokal and Rohlf 1981). In particular, plants with each of the four dispersal syndromes were distributed significantly differently across light environments, and the seedlings were the most heterogeneous of the three stage classes (Table 31.2). The seedlings of the 'typical' (Bird I) bird-dispersed species were almost entirely restricted to the two lightest habitats, while most of the seedlings of the ant-dispersed species were found in the darkest habitat (Fig. 31.3). Small plants, the size of seedlings, of both the

Table 31.2. Tests of independence of light environment from dispersal syndrome, by stage class. Within a column (stage class), dispersal syndromes sharing the same letter did not differ significantly $(P > 0.05)$ in distribution of individuals across light environments (pair-wise tests of the STP procedure for R × C contingency tables as given by Sokal and Rohlf 1981). Tests for independence of light and dispersal syndromes for each stage class across all dispersal syndromes are given by G values; all G values are adjusted by the Williams correction procedure (Sokal and Rohlf 1981).

Dispersal syndrome	Plant stage class			
	Seedlings	Intermediates	Reproductives	Total plants (Sum all stages)
Ant	A	M	X	D
Bird III	AB	N	Y	E
Bird II	B	MN	XZ	F
Bird I	C	P	Z	G
G	233*	114*	79*	367*
d.f.	15	15	15	15

* $P < 0.001$.

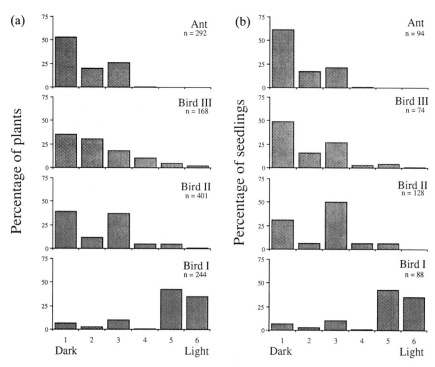

Fig. 31.3. Relative abundance of Marantaceae with different dispersal syndromes in different light environments. (a) Percentage of plants with each syndrome found in each environment. (b) Percentage of seedlings for each dispersal syndrome found in each light environment.

clonally reproducing and large-seeded (Bird II and Bird III) bird-dispersed species were found across a wide range of light environments, but they were mostly in the three darkest habitats (Fig. 31.3). Furthermore, the stage structure of populations differed significantly among the light environments for all the bird-dispersed species, but not for the ant-dispersed species (Table 31.3, Fig. 31.4).

Preliminary checklist of birds and ants

Finally, preliminary observations of birds and ants interacting with seeds of Marantaceae were conducted. Birds were observed visiting naturally fruiting plants by using binoculars and standing quietly 10–20 m from plants (a total of 10 person-hours). The bird species and the proximate fruit-handling behaviour were recorded. Additional data from other more extensive studies (O'Daniel 1987; D. Levey, personal communication) are also shown (Table 31.4). Ants were observed by watching their interactions with fresh seeds (a

Table 31.3. Tests of independence of stage structure from light environment, by dispersal syndromes. Within a column (dispersal syndrome), light environments sharing the same letter did not differ significantly ($P > 0.05$) in stage structure (pair-wise tests of the STP procedure for R × C contingency tables as given by Sokal and Rohlf 1981). Dashes indicate fewer than ten individuals for a given combination; in such cases stage structure was not analysed. Tests for independence of stage and light across all light environment are given by G values; all G values are adjusted by the Williams correction procedure (Sokal and Rohlf 1981).

Light environment	Dispersal syndrome			
	Ant	Bird III	Bird II	Bird I
1	A	M	U	F
2	A	N	V	G
3	A	M	W	H
4	—	N	X	IJ
5	—	—	Y	I
6	—	—	—	J
G	7.9 n.s.	34.2*	62.2*	78.5*
d.f.	4	6	8	10

* $P < 0.001$.

total of 14 person-hours). Seeds from newly opened fruit capsules were obtained and placed on the forest floor.

At least 12 understorey bird species feed on the seed arils of Marantaceae (Table 31.4). Twelve species of ants were attracted to seeds, although only the ponerines and a large species of *Aphaenogaster* acted as dispersers (Table 31.5), a finding consistent with previous work on Mexican ant-dispersed *Calathea* (Horvitz and Beattie 1980; Horvitz and Schemske 1986*b*). Seeds of both the ant- and bird-dispersed Marantaceae have lipid-rich arils that are highly attractive to ants if fresh seeds reach the forest floor (Table 31.5).

Discussion

The distribution and stage structures of the 'typical' (Bird I) bird-dispersed and the ant-dispersed populations were consistent with the static hypothesis, while the bird-dispersed plants that had clonal propagules or large seeds

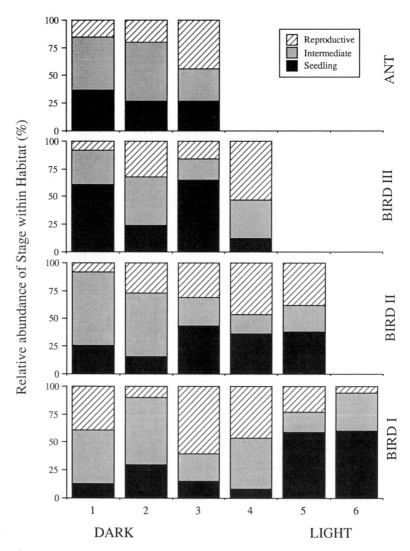

Fig. 31.4. Relative abundance (percentage of plants) of stages within each light environment for each dispersal syndrome; the length of each pattern bar represents the relative abundance of a stage within a light environment, for each dispersal syndrome. Only habitat dispersal categories with at least ten individuals were included. Sample sizes for each habitat dispersal category, listed for each dispersal syndrome in order from darkest (1) to lightest (up to 6) were, respectively, for 'ant': 154, 60, 75; for 'Bird III': 59, 50, 31, 17; for 'Bird II': 156, 49, 150, 22, 21; for 'Bird I': 44, 10, 60, 13, 65, 52.

showed a more complex interaction with the light environment. If the 'typical' (Bird I) syndrome is ancestral, one may argue that the other syndromes are all derived adaptations that allow the plants to inhabit shadier habitats. It is plausible that the ant-dispersed species are derived from the clonal species group. There are two characters which make this a reasonable argument. First, the seeds of the clonal species are closest in size, colour and morphology to those of the ant-dispersed species. Secondly, the peculiar growth habit, in which the stem that bears both the infructescence and clonal propagules grows actively towards the ground, may be a pre-adaptation that gave rise to seed dispersal by litter-foraging ants. These results suggest an interaction between dispersal syndrome and habitat in their effects on demography and indicate the need for a detailed comparative study of both the changing light environments and the changing demography of plants with each syndrome during the gap-phase regeneration process.

Ant-dispersal is widespread throughout the angiosperms and the diverse habitats of the world (Beattie 1983, 1985) and is predominant in the spring-flowering understorey herbs of temperate woodlands (Culver and Beattie 1978; Handel *et al.* 1981; Thompson 1980, 1981). Directed short-distance dispersal to local safe micro-sites has been proposed as the major adaptive significance of ant dispersal (Culver and Beattie 1978, 1980; Heithaus *et al.* 1980; Davidson and Morton 1981*a*, *b*; Beattie and Culver 1982, 1983; Beattie 1985; Levey 1988*b*; but see also Horvitz and Schemske 1986*c*).

Ant dispersal is generally considered to be a derived syndrome that has arisen independently from many different ancestral origins (van der Pijl 1972; Beattie 1985). Of interest here is that ant-dispersal has probably been derived from bird-dispersal in several groups (Nordhagen 1932; Berg 1958; Davidson and Morton 1984; O'Dowd and Gill 1986; also discussed in Thompson 1981; Higashi and Ito Chapter 32, this volume). In the Marantaceae, ant-dispersal is probably derived, since the highly branched inflorescences characteristic of many of the bird-dispersed species are considered relatively primitive (Andersson 1976). My preliminary observations and O'Daniel's study (1987) suggest that after *Calathea* seeds are dropped by birds they may be secondarily moved by ants, as occurs in Australian acacias (Davidson and Morton 1984), *Ficus* (Roberts and Heithaus 1986; Kaufmann *et al.* (in press)), and *Miconia* (Levey 1988*b*).

Previous comparisons of closely related ant- and bird-dispersed species have focused on morphological, nutritional, and display characters. O'Dowd and Gill (1986) presented a multivariate analysis that quantitatively separated Australian acacias into bird- and ant-dispersed species. They argued that ant dispersal may have arisen when bird abundance fell during the Miocene due to low rainfall.

Davidson and Morton (1984), working with Australian acacias of the arid zone community, compared the nutritional value of the rewards, rates of seed parasitization, and habitat associations of bird- and ant-dispersed species.

Table 31.4. Behaviour of birds interacting with seeds of Costa Rican *Calathea*.

Bird species	Species of *Calathea*				
	C. insignis	*C. inocephala*	*C. lutea*	*C. marantifolia*	
Arremon aurantirostris Orange-billed sparrow	e	carry 0–20 m eat aril/ discard seed[a]	crush[a]	e	
Cyanocompsa cyanoides Blue-black grosbeak	e	e	crush[a]	crush[b]	
Habia atrimaxillaris Black-cheeked ant tanager	e	e	swallow[c] crush[a]	e	
Manacus aurantiacus Orange-collared manakin	e	e	swallow/ regurgitate[a]	e	
M. candei White-collared manakin	seeds found in regurgitate[d]	e	e	e	
Mionectes oleagineus Ochre-bellied flycatcher	swallow[c] seeds found in regurgitate[d]	e	swallow/ regurgitate[a] seeds found in regurgitate[d]	swallow[c]	

Species				
Pipra coronata Blue-crowned manakin	e	swallow[a]	swallow/ regurgitate[a]	e
P. mentalis Red-capped manakin	e	e	swallow[a] seeds found in regurgitate[d]	e
Ramphocelus passerini Scarlet-rumped tanager	e	e	eat aril/ discard seed[a]	e
Saltator maximus Bull-throated saltator	seed found in regurgitate[d]	e	seed found in regurgitate[d]	e
Schiffornis turdinus Thrush-like manakin	e	swallow[a]	e	e
Sporophilla aurita Variable seed-eater	e	e	eat aril/ discard seed[a]	e

Data source:
[a] O'Daniel (1987), observation of birds on plants.
[b] Doug Levey (personal communication), bird on plants.
[c] birds on plants.
[d] Doug Levey (personal communication), birds caught, kept in pail $\frac{1}{2}$ hr.
[e] interaction not observed.

Table 31.5. Behaviour of ants observed interacting with seeds of Costa Rican Marantaceae. Fresh seeds removed from plants and placed on forest floor.

Ant species	Plant species						
	Calathea insignis	C. inocephala	C. lutea	C. marantifolia	C. micans	Pleiostachya pruinosa	Ischnosiphon inflatus
Myrmicinae							
Aphaenogaster arenoides			carry*	carry (41 cm)		carry (187 cm)	
Solenopsis geminata		rob aril	rob aril			rob aril	rob aril
Pheidole radoszkowskii		rob aril	rob aril	drag (19–41 cm)		rob aril	
P. hirsuta						antennate	
P. sp.	antennate						
Wasmannia auropunctata						antennate	
Ponerinae							
Ectatomma sp.	carry				carry (151 cm)		
Pachycondyla apicalis	antennate		carry (73 cm)	carry	carry		antennate
P. harpax			carry*	carry			
Odontomachus bauri					carry		
O. laticeps							
O. erythrocephalus							drag (117 cm)

* Seeds picked up by ants following bird-handling (O'Daniel 1987)
No entry in the table indicates interaction not observed.

While bird-dispersed species invested more in dispersal rewards, they also suffered less from seed parasitization. This result emphasized a balanced trade off between the two syndromes. Ants sometimes carried seeds of 'bird' plants, but the reverse did not occur. Juveniles of a bird-dispersed species were significantly associated with habitats beneath typical bird-perching locations in trees; interestingly, in one treeless habitat, this species was significantly associated with ant mounds. In contrast, the juveniles of an ant-dispersed species were not associated with ant-mounds or with bird perches, but they tended to be found more often in the vicinity of conspecific, mature individuals. Davidson and Morton (1984) emphasized the similarity in the chemistry of the micro-sites to which both birds and ants carried seeds, and they generally considered the two syndromes as more or less equally good alternatives for at least some acacias. However, they tended to view ant dispersal as a sort of back-up mechanism that would ensure plants at least some germination success 'despite spatial and temporal patchiness in the availability of avian frugivores', in light of the more ubiquitous availability of the ants.

Different comparisons between ant and bird dispersal were made by Thompson (1981) for plants in temperate deciduous forests. He proposed that early fruiting and low stature constrained the spring understorey herbs to utilize ants as dispersal agents due to a seasonal paucity of birds and inability to make an attractive display so low in the forest as compared with plants that fruit later in the summer or in the autumn. In an earlier paper, Thompson (1980) noted that the edges of rotting logs were colonized by locally occurring ant-dispersed plants, while bird-dispersed plants reaching the logs were from relatively far away. These earlier observations are particularly relevant to this chapter.

The predictability of gaps may have a profound effect on the availability of niches for understorey herbs. Smith (1987) proposed that the seasonally deciduous canopy of temperate, broad-leaved deciduous forests provides predictable canopy gaps every spring and autumn resulting in numerous niches for understorey herbs. These seasonal gaps are predictable in time and ubiquitous in space as compared to the gaps due to tree falls that are utilized by understorey herbs living in tropical evergreen forests; these gaps are both rare and relatively unpredictable in both time and space. (With respect to Smith's observations about 'gaps' over spring-flowering herbs in temperate deciduous forests, it is interesting to note that these are the very species noted for ant-dispersal.) Perhaps these ant-dispersed species achieve an apparent 'shade-tolerance' (ability to recruit juveniles in the mature forest) like that predicted by the model of Horvitz and Schemske (1986a), by living in a temporally predictable seasonal gap.

Light gaps created by annual leaf fall in tropical forests may differ substantially in their effects on herbs from those in temperate forests. First, in many tropical forests, not all species are deciduous. Thus, annual gaps may not be ubiquitous in space. Secondly, tropical leaf fall is associated with the dry season; if moisture is limiting, herbs may not be able to take advantage of increased light levels. Preliminary data from the seasonally deciduous forest of Barro Colorado Island, Panama indicate that many understorey herbs continue growing well through the dry season, while others are dormant (A. Smith, personal communication).

Previous studies have implied that ant-dispersal is a 'poor relation' to bird dispersal in that it arises when birds are less available. Although the data of both Davidson and Morton (1984) and Thompson (1981) suggest that juvenile recruitment close to adults is successful in ant-dispersed species but not in bird-dispersed species, this result was not emphasized by these authors. I suggest that for some kinds of dynamic demography (patterns of demographic change in heterogeneous environments), long-distance (i.e. bird) dispersal is actually undesirable. Depending upon suites of demographic traits as they interact with spatio-temporal heterogeneity of the environment, the 'best' dispersal syndrome may be local dispersal or, alternatively long-distance dispersal. Dispersal some distance away from the maternal patch may be a very risky business (Olivieri and Gouyon 1985) and is only to be expected under special circumstances (but see, Gadgil 1971; Hamilton and May 1977). Density-dependent mortality due to biotic interactions (Janzen 1970, 1971; Dirzo and Dominguez 1986) or competition may lead to dispersal away from the parent plant canopy but otherwise close to 'home'. In contrast, predictable deterioration of the abiotic environment near the mother due to successional changes (e.g. increasing shadiness) may favour long-distance dispersal. Ant dispersal may be favoured in species that have populations that are able to persist locally beyond an initial colonization of a disturbance. Any character that enables persistence in intermediate habitats would be likely to result in selection against long-distance dispersal and a concomitant strengthening of selection in favour of adaptations to find favourable micro-sites locally. One could imagine many morphological scenarios that would co-occur with both dispersal and shade-tolerance. For example, one way plants may evolve shade-tolerance is to be slower growing and of smaller stature; such plants may also produce smaller seeds which would be more easily dispersed by ants than by birds (C. R. Huxley, personal communication). Such a scenario is not incompatible with the main predictions of the model, although the proximal mechanism of selection is different.

In summary, I have presented both a theoretical framework and some comparative empirical data on habitat utilization by closely related ant- and bird-dispersed species in support of the idea that the evolution of dispersal characters is correlated with the evolution of plant characters that mediate the plants' response to environmental changes brought about by disturbance/recovery cycles. Thus, the best way to understand the evolution of ant dispersal may be to examine community level dynamics that dictate the disturbance/recovery regime for the environment that the plants inhabit, as well as the micro-site characteristics associated with ant behaviour and ant-nests. Understanding how plants differ in their ability to exploit both large and small patches in their heterogeneous environments should help us in understanding why plants differ in dispersal syndromes and how selection may lead to specialization on ants as seed dispersal agents instead of animals that regularly move longer distances.

Acknowledgements

I thank Linda DeLay and Pamela Phillips for ornithological, logistical, and demographical assistance in the field; Linda DeLay for data entry; Don Feener and Jack Longino for help in identifying the ants; Donna O'Daniel and Douglas Levey for sharing observations, data, and insights about the birds; and a University of Miami General Research Support Award for funding the field work. This is Contribution No. 354 from the Program in Tropical Biology, Ecology and Behavior of the Department of Biology, University of Miami.

Appendix 31.1.

Dispersal syndromes of some Costa Rican Marantaceae, characters.

Species (Infructescence height (m))	Colour				Seed mass (mg)		Aril mass (mg)		Reward aril/ seed + aril %
	bracts	seed	aril	capsule	x̄ (s.d.)	n	x̄ (s.d.)	n	
Bird I (see below) *Calathea insignis* (= *crotalifera*) Petersen (2–3)									
C. lasiostachya Donnell Smith (2–2.5)	yellow	blue	white	yellow	404 (126)	5	76 (16)	5	15.8
C. lutea (Aublet) Schultes (2–3)	beige	blue	white	?	675 (94)	3	83 (12)	3	11.0
C. similis Kennedy (2–3)	brown	orange	orange	pink	658 (31)	9	138 (51)	9	17.4
Pleiostachya pruinosa (Reg.) K. Schumann (2–3)	orange	blue	white	?	673 (116)	4	?		
Bird II (clonal propagules) (see below) *C. donnell-smithii* K. Schumann (2–2.5)	beige	blue	white	?	972 (129)	2	71 (—)	1	6.8
(1–1.7)	purple	grey	cream	?	?				

Species									
C. leucostachys Hooker f. (1–1.2)	purple	grey	cream	?	986 (23)	2	67 (7)	2	6.4
C. marantifolia Standley (1–1.5)	yellow	grey	cream	yellow	568 (101)	8	31 (15)	8	5.2
C. warscewiczii (Mathieu) Koernicke (1–1.2)	pink	grey	cream	?	?				
Bird III (large seeds) (see below)									
C. gymnocarpa Kennedy (1–1.5)	decid.	blue	white	orange	?				
C. inocephala (O. Kuntze) H. Kennedy and Nicholson (1–2.5)	rotten orangish	blue	white	orange	2811 (361)	7	217 (63)	7	8.4
Ant (see below)									
C. cleistantha Standley (0.1)	purple	black	white	purple	?				
C. cuneata Kennedy (0.3)	?								
C. micans (Mathieu) Koernicke (0.1)	purple	black	white	purple	209 (33)	5	7 (3)	5	3.2
Unknown (see below)									
Ischnosiphon inflatus Andersson (2–2.5)	beige	brown	cream	beige	3056 (231)	3	271 (101)	3	9.0

Appendix 31.2.

Descriptions of dispersal syndromes of Costa Rican Marantaceae

Bird I: Tall, highly branched infructescence; seeds pushed out of bracts and displayed against contrastingly coloured bracts. Aril and capsule not part of display. Aril 'wrap' style.

Bird II: Solitary infructescence; buds, unopened flowers, persistent calyces and bracts contribute to display; at maturity, calyx dehisces and seeds are exposed, but embedded within infructescences. Arils and capsules not part of display. Arils 'arm' style. Following flowering, a new vegetative clonal shoot frequently develops in the axil of the leaf subtending the infructescence and the structure grows towards the ground and roots.

Bird III: Solitary infructescence; bright blue seeds poking out of orange, waxy capsules constitute display. Arils not part of display. Arils 'arm' style. Seeds much larger than other *Calathea* spp.

Ant: Very short infructescence; contrasting light aril against dark seed displayed on forest floor; no 'on-plant' display. At maturity, capsule splits apart completely and seeds are scattered onto the forest floor. Aril 'arm' style.

Unknown: Tall, branched infuctescence, very slender and beige, inconspicuous. At maturity, linear seed displayed at right angles to infructescence. Pungent varnishy musky odour. Aril and capsule not part of display. Aril 'wrap' style.

Appendix 31.3.

Stage classification by largest leaf size for some Costa Rican Marantaceae.

Species	Leaf sizes for each stage (cm)			Number of plants measured		
	seedling	intermediate	reproductive	Corcovado	La Selva	Total
Calathea cleistantha	$\leqslant 10$	11–34	$\geqslant 35$	0	168	168
C. cuneata	$\leqslant 10$	11–34	$\geqslant 35$	0	24	24
C. donnell-smithii	$\leqslant 10$	11–28	$\geqslant 29$	182	1	183
C. gymnocarpa	$\leqslant 15$	16–44	$\geqslant 45$	8	17	25
C. inocephala	$\leqslant 20$	21–74	$\geqslant 75$	122	21	143
C. insignis	$\leqslant 20$	21–64	$\geqslant 65$	0	59	59
C. lasiostachya	$\leqslant 20$	21–47	$\geqslant 48$	0	14	14
C. leucostachys	$\leqslant 15$	16–31	$\geqslant 32$	0	28	28
C. lutea	$\leqslant 20$	21–64	$\geqslant 65$	89	14	103
C. marantifolia	$\leqslant 15$	16–27	$\geqslant 28$	82	66	148
C. micans	$\leqslant 5$	6–7	$\geqslant 8$	0	100	100
C. similis	$\leqslant 20$	21–64	$\geqslant 65$	0	4	4
C. warscewiczii	$\leqslant 15$	16–23	$\geqslant 24$	0	42	42
Pleiostachya pruinosa	$\leqslant 15$	16–54	$\geqslant 55$	40	24	64
Total for all species				523	582	1105

32

Ground beetles and seed dispersal of the myrmecochorous plant *Trillium tschonoskii* (Trilliacae)

Seigo Higashi and Fuminori Ito

Introduction

Trillium species are myrmecochorous, perennial herbs occuring principally in woodlands of North America and eastern Asia, including Japan. The fruits of this genus contain several scores of seeds which are transported by such ants as *Formica*, *Lasius*, *Myrmica*, *Aphaenogaster*, and *Camponotus*, which are attracted to their juicy elaiosomes (Gates 1940, 1941; Berg 1958; Beattie and Culver 1981; Mesler and Lu 1983; Nesom and La Duke 1985). The present study involves *Trillium tschonoskii* Maxim., which grows in the cool temperate broad-leaved deciduous woodland of Hokkaido, northern Japan. This species is in full bloom in May, and in late July produces fruits containing about 80 seeds each. Each seed bears a juicy elaiosome about twice the size of the seed.

As shown by Higashi *et al.* (1989), the fruits initially fall close to the parent plants; over 50 per cent within 20 cm. Ants, principally *Myrmica ruginodis* Nylander (Myrmicinae) and *Aphaenogaster japonica* Forel (Myrmicinae), transport some seeds to their nests, a mean distance of 64 cm (Fig. 32.1). The nests are very short lived, more than 70 per cent being abandoned within a year and, consequently, nest soils are no richer in nitrogen or phosphorus than surrounding soils. There are proportionally more older than younger juvenile plants at distances more than 60 cm from the nearest parent plant than at distances less than 60 cm (Table 32.1). This indicates that the relatively short-distance dispersal of seeds by ants is sufficient to reduce seedling mortality, by reducing competition between seedlings. This is probably the primary advantage of myrmecochory to this species of *Trillium*. This contrasts with studies involving ant species with longer lived nests, where the main advantage to the plant is nutrient-enhanced micro-sites for the seedlings (Beattie 1985).

However, Higashi *et al.* (1989) also found that seed removal per fruit was less than 25 per cent in over 75 per cent of fruits. Many seeds were left near parent plants without being dispersed by ants. This is consistent with

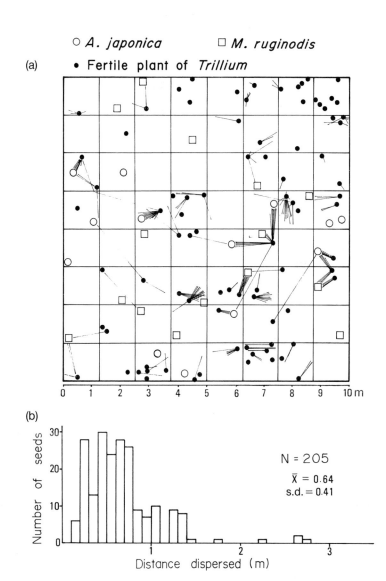

Fig. 32.1. Seed dispersal of *Trillium tschonoskii* within a 10 × 10 m quadrat in deciduous forest in Hokkaido. (a) Spatial distribution of ant nests and *Trillium* plants, and dispersal of marked seeds. Each line connects a recovered seed to its parent plant. (b) Frequency distribution of dispersal distance for the seeds which were moved and recovered. (From Higashi *et al*. 1989.)

Table 32.1. A 3 × 3 contingency table showing the association between age class structure of *Trillium tschonoskii* plants and distance from the nearest fertile plant within a 2 × 2 m quadrat (Higashi *et al*. 1989). Expected values are given in parentheses. The association is significant at $P < 0.005$ ($x^2 = 30.9$).

Age class	Distance class			
	< 30 cm	30–60 cm	> 60 cm	Total
Seedling	108 (89.6)	12 (27.2)	10 (13.2)	130
One-leaved	176 (188.9)	77 (57.2)	21 (27.9)	274
Three-leaved	89 (94.5)	24 (28.6)	24 (13.9)	137
Total	373	113	55	541

observations by Ohara and Kawano (1986) who analysed the population structure of *Trillium* spp. and showed that seedlings were highly aggregated, forming patchily distributed clumps, while fertile plants showed nearly random or slightly over dispersed distributions. The clumps of seedlings were not found near ant nests, but near parental plants, suggesting that a high proportion of the seed had not been dispersed.

The role of ground beetles

In order to determine the reason for the patchy seed distribution, we counted all animal visitors attracted to fresh *T. tschonoskii* fruits until 100 individuals were recorded both in the daytime (11.00–16.00) and at night (20.00–04.00 h) during the fruiting season (Table 32.2). The fruits were frequently visited by ants such as *Myrmica ruginodis*, *Aphaenogaster japonica*, *Lasius niger* L., *Camponotus obscuripes* Mayr, and *Paratrechina flavipes* (F. Smith) in the daytime (Fig. 32.2a), but by ground beetles such as *Apotomopterus japonicus* Motschulsky, *Carabus conciliator* Fischer von Waldheim, *C. opaculus* Putzeys, *Pterostichus thunbergi* Morawitz, and *P. subovatus* Motschulsky at night (Fig. 32.2b). Mammals were not attracted to the fruits even at night, even though we avoided potential disturbance by remaining distant from the baits. The secondary forests in which Japanese *Trillium* spp. grow have high populations of mice, e.g. *Clethrionomys rufocanus bedfordiiae* Thomas, *Apodemus argenteus* Temininck, and *A. speciosus ainu* Thomas, which may occasionally forage on *Trillium* fruits. However, the effect of mammals on the dispersal of *Trillium* is probably slight in Hokkaido.

Table 32.2. Relative abundance of animals attracted to the elaiosomes of *Trillium tschonoskii* in deciduous forest in Hokkaido. The first 100 individuals noted by day and by night were counted.

	Day (11.00–16.00)	Night (20.00–04.00)
Ants		
Myrmica ruginodis	31	12
Aphaenogaster japonica	27	9
Lasius niger	13	2
Camponotus obscuripes	0	4
Paratrechina flavipes	3	0
Ground beetles		
Apotomopterus japonicus	1	17
Carabus conciliator	0	5
C. opaculus	0	8
Pterostichus thunbergi	2	20
P. subovatus	0	8
P. orientalis	0	1
Synuchus melantho	1	2
Silpha perforata venatoria	3	4
Eusilpha japonica	0	1
Others	19	7
Total	100	100

Close study of the behaviour of nocturnal ground beetles was made by Ohara and Higashi (1987). Beetles of the Carabinae often ate the whole elaiosome of a seed within 20 seconds; beetles of the Harpalidae and Silphidae took several minutes to eat a whole elaiosome and they occasionally left it partly eaten. The extent of elaiosome damage by ground beetles was observed by putting 168 intact seeds on the forest floor during one night. About half (49 per cent) of the seeds remained intact, but the elaiosomes of the other half (51 per cent) were wholly or partly damaged: 25 per cent were partly eaten and 26 per cent were wholly eaten. Subsequently, a field experiment was conducted to observe the reaction of ants to differently damaged elaiosomes. Seeds were grouped into three classes: intact seeds; seeds with half the elaiosome removed with forceps; and seeds with all the elaiosome removed. Fifty seeds from each class were put together on plates which were placed near the nests of *Aphaenogaster japonica* and *Myrmica ruginodis* at 10.00. Both ant species were indifferent to the seeds for the first two to three

Fig. 32.2. (a) *Aphaenogaster japonica* drawing out seeds from a *Trillium tschonskii* fruit and (b) *Apotomopterus japonicus* devouring the elaiosome of a seed.

hours. Thereafter most of the intact seeds were rapidly transported to their nests within three to four hours. In contrast, seeds without elaiosomes were not carried by either ant species, except rarely when the seed surface was wet with elaiosome juice. Most of the seeds with half an elaiosome were also ignored by *M. ruginodis*. Although *A. japonica* was a little more interested than *M. ruginodis*, 33 of the 50 seeds were ignored. These findings indicate that elaiosome damage by ground beetles seriously interrupts the dispersal of *Trillium* seeds by ants.

These observations suggest that the seed removal frequency is considerably dependent on the activity of ants and ground beetles. To compare ant and beetle activity, 50, 1 × 1 m white sheets, and 50 pit fall traps, 6.5 cm wide and 9 cm deep, were set up on the forest floor. The ants walking on the white sheets and the trapped beetles were counted every four hours for two days during the fruiting season (Fig. 32.3). The ants were active day and night with little fluctuation in numbers. The effective agents, *Myrmica ruginodis* and

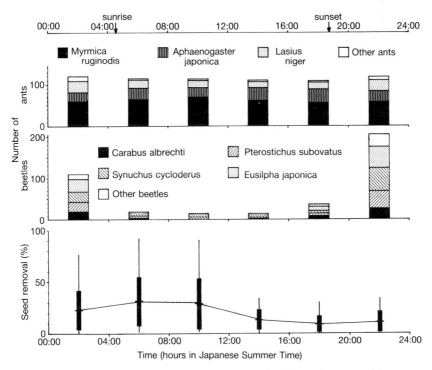

Fig. 32.3. Effects of daily activity of ants and ground beetles on the removal frequency of *Trillium tschonoskii* seeds in deciduous forest in Hokkaido. The numbers of ants and beetles are the totals for two days. The seed removal frequency is given by the average, standard deviation (thick line) and range (thin line) for 20 fruits, each containing 50 seeds.

Aphaenogaster japonica, constituted 68 per cent (at 02.00) to 84 per cent (at 14.00) of all ants recorded. Unlike the ants, ground beetles showed marked fluctuation in activity; the number trapped around midnight (20.00–24.00) was about 17 times the number trapped around midday (12.00–16.00).

Concurrently, the effect of the time of day of fruit dispersal on the frequency of seed removal was studied as follows. Every four hours, 20 plates, each with a fruit containing 50 seeds, were put on the forest floor, i.e. $20 \times 6 = 120$ plates in all. One week later when all seeds had lost their elaiosomes, the seed removal frequency per fruit was counted. The frequency of seed removal was greatest for fruits placed on the forest floor in the morning, and least for those dispersed in the afternoon (Fig. 32.3); i.e. of the fruits dispersed at 6.00, 31 per cent $\pm$ sd 24 of the seeds were removed on average, suggesting a high probability that the intact fruit would be encountered by ants during the daytime; while for the fruits dispersed at 18.00, the seed removal frequency was only 9 per cent $\pm$ 8, suggesting frequent interference by nocturnal ground beetles before the fruits were found by ants.

Conclusions

Trillium tschonoskii is myrmecochorous, but seed removal frequency is controlled less by ants than by ground beetles. Unlike specialized mutualisms, where mutualistic partners might co-adapt to avoid interference by outsiders, these ants and plants are in loose mutualism where the ants are not specialized elaiosome feeders and are hardly induced to expel the outsiders. Moreover, many species of *Trillium* may be in the process of evolving from endozoochory to myrmecochory (Berg 1958). According to our recent studies (S. Higashi and F. Ito, unpublished work), the frequency of seed removal depends on the system of seed separation from the mother plants; the most efficient myrmecochorous species do not drop all their seeds at once, but sporadically over many days. About 80 seeds are contained in *T. tschonoskii* fruits; this appears to be too many for ants to transport to their nests before interference by nocturnal ground beetles.

Most ground beetles are voracious omnivores that can potentially eat elaiosomes of any plant. Hence, interference by beetles in seed dispersal is also likely in other myrmecochorous plants; although no one has reported this phenomenon, probably due to lack of night observation. The interference by ground beetles in seed dispersal could have considerable significance for the ecological and evolutionary study of myrmecochorous plants.

33

Seed harvesting by ants in Australia

Alan N. Andersen

Introduction

Seed harvesting by ants refers to the removal and subsequent consumption of seeds (i.e. seed predation), as distinct from the dispersal of seeds possessing an ant-attracting appendage (i.e. myrmecochory). Outside Australia, seed harvesting is primarily a phenomenon of arid regions. The seeds of ephemeral plants represent a valuable food source in desert environments, which typically support large populations of granivores, mostly rodents, birds, and ants (Mares and Rosenzweig 1978; Brown *et al.* 1979; Abramsky 1983). The ants are primarily species of the myrmicine genera *Pogonomyrmex, Messor, Veromessor, Monomorium* and *Pheidole*, and, although strictly they are not specialist granivores (Bohart and Knowlton 1953; Brown *et al.* 1979), seeds nevertheless constitute a major portion of their diets. Foraging by these ants can have a marked impact on seed densities and distributions (Reichman 1979) and consequently on the community structure of desert plants (Inouye *et al.* 1980). Most of these ant species are confined to deserts, and harvester ants are often absent altogether from other habitats (e.g. Brown *et al.* 1975). Harvester ants can be prominent in seasonally arid habitats, such as in mediterranean California (Hobbs 1985), tropical Mexico (Carrol and Risch 1984), and African savannahs (Levieux 1983), but there are no records of them being important in temperate habitats anywhere in America, Europe, or Africa.

The Australian scene is remarkably different. In the first place, seed harvesting by ants occurs throughout the continent, and is prominent in all major habitat types (with the possible exception of tropical rainforests, where seed harvesting has not yet been investigated). Secondly, harvester ants are the major post-dispersal seed predators in most plant communities. Australia possesses a depauperate fauna of granivorous mammals (Morton 1979), and in many cases mammalian seed predation is negligible (O'Dowd and Gill 1984; Andersen and Ashton 1985). Birds can be important granivores in arid environments (Morton and Davies 1983), but even then they play second fiddle to ants (Morton 1985). Thirdly, only a few groups of Australian harvester ants, notably those characteristic of arid regions, can even loosely be considered as specialist granivores. All harvesting species in mesic regions, and many from the arid zone, are omnivorous, feeding on seeds opportunistically. The seeds eaten by these omnivorous species are often from woody plants, rather than primarily from grasses or other herbs. Finally, whereas granivory outside Australia is almost entirely restricted to myrmicine ants, in Australia ponerine and formicine as well as myrmicine species are prominent harvesters. Indeed, in mesic southern Australia, species of *Rhytidoponera* (Ponerinae) and *Prolasius* (Formicinae) are often the dominant seed-eating ants (Ashton 1979; Andersen and Ashton 1985).

The ubiquity of harvester ants in Australia means that post-dispersal seed predation by ants is virtually a universal process in Australian plant communities. It is a key factor in Australia's ecological landscape. This means that seed-harvesting by ants demands consideration when managing Australia's natural ecosystems. It also means that the ants sometimes come into conflict with human interests, particularly in agricultural and forestry seeding operations (Butler 1971; Campbell 1982; Abbot and van Heurck 1985), but also with rehabilitation of degraded ecosystems (Major 1990).

The ants

The biology of most Australian ant species is extremely poorly known, so what follows is very incomplete. Ten Australian genera are known to contain seed-eating species (Table 33.1); three of these (*Heteroponera*, *Mayriella* and *Solenopsis*) are represented by single, uncommon or localized, harvesting species (they are therefore of minor importance, and will not be considered further).

The distribution of granivory within a genus varies enormously, ranging from *Rhytidoponera*, *Tetramorium*, and *Melophorus*, where only a small proportion of total species eat seeds, to *Pheidole* where seed harvesting is virtually universal (Table 33.1). Within *Meranoplus* seed harvesting is characteristic of at least two major species groups, and appears to be scattered in others. Almost all species formerly referred to as *Chelaner* eat seeds (with the notable exception of the *rubriceps* group, which forage predominantly on vegetation), whereas the vast majority of mainstream Australian *Monomorium* appear not to. Within *Prolasius* seed harvesting occurs in the *flavicornis*, *bruneus* and *pallidus* groups, but apparently not in the *niger* and allied groups.

The major seed harvesting genera can be divided into three classes according to their distribution:

1. *Melophorus*, *Meranoplus*, and *Tetramorium*, with most harvesting species confined to arid, semi-arid, and seasonally arid regions;
2. *Rhytidoponera* and *Prolasius*, with harvesting species mostly confined to mesic southern, particularly south-eastern, Australia (the only exception is *Rhytidoponera metallica*, which extends into the central arid zone); and

3. *Pheidole* and *Monomorium*, which have harvesting species throughout Australia.

Within *Pheidole* many of the major species groups are themselves widespread, whereas the harvesting groups of *Monomorium* typically have either arid-zone, cool-temperate, or tropical distributions.

Seed selection

Australian seed-eating ants span the full spectrum from generalist omnivores to specialist granivores (Fig. 33.1). Most of the species restricted to mesic southern Australia (i.e. species of *Rhytidoponera* and *Prolasius*, and cool temperate species of *Monomorium*) are generalist omnivores. Seeds are not a particularly important food resource in this region, and, although quantitative data are lacking, the diets of these seed-eating species probably consist mostly of insect material. Many harvester ants of the arid zone, on the other hand, have much more specialized diets. Species of the *Monomorium rothsteini* group represent the most abundant and widespread group of harvesters in arid regions, and seeds comprise more than 90 per cent of their diets (Briese and Macauley 1981; Davison 1982). Species of the *whitei* group, also prominent arid zone harvesters, subsist entirely on seeds (Briese and Macauley 1981; Davison 1982), as apparently do species of the *Meranoplus diversus* group (Andrew 1986; A. N. Andersen, unpublished work), which is widely distributed in the northern arid zone and in the seasonal tropics.

Most harvester ants eat a wide variety of seeds, with their choices determined largely by seed size, morphology, and availability (Smith and Atherton 1944; Briese and Macauley 1981; Andersen 1982; Davison 1982). The most notable exceptions are species of the *Meranoplus diversus* group, which appear to feed exclusively on the seeds of only one or a few grass species (Andrew 1986; Andersen, unpublished work). They therefore rate among the most specialized of all known post-dispersal granivores. Two or three species often occur together, each specializing on different grass species, and each immediately recognizable by their distinctive nest middens consisting of discarded seed husks (Andersen and Lonsdale 1990).

The degree of dietary specialization is often reflected in the degree of modification of the clypeus, the lower margin of which forms the anterior margin of a typical ant's head (Fig. 33.2). Most broadly adapted omnivores (e.g. species of *Rhytidoponera* and *Prolasius*) have an unspecialized clypeus, whereas in many of the more specialized harvesters, the clypeus is produced into a pair of teeth, presumably to aid seed collection and transport. A toothed clypeus occurs independently in harvesting species of *Mayriella*, *Solenopsis*, and many groups of *Monomorium* and *Meranoplus*. The greatest clypeal modification occurs in species of the *Meranoplus diversus*

Table 33.1. Summary of records of seed-eating ants in Australia.

Genus	Species	Distribution	Reference(s)
subfamily Ponerinae			
Heteroponera	*?leae*	semi-arid Victoria	Andersen and Yen 1985
Rhytidoponera	*metallica* group, including *metallica, tasmaniensis,* and *inornata*	throughout southern Australia	Andersen 1983, 1988a; Majer 1982; Andersen and Ashton 1985
	victoriae	mesic south-eastern Australia	Andersen and Ashton 1985; Andersen 1988
	violacea	south-western Australia	Majer 1982
subfamily Myrmicinae			
Mayriella	*abstinens*	mesic south-eastern Australia	A. N. Andersen, unpublished work
Meranoplus	*diversus* group	northern Australia	Andrew 1986; Watkinson *et al.* 1989
	many other species	throughout Australia	Andersen 1982; Greenslade 1982; Morton and Davidson 1988

Monomorium	many species, mostly those previously known as *Chelaner*	throughout Australia	Greenslade 1982; Briese 1982; Davison 1982; Andersen and Ashton 1985
Pheidole	many species	throughout Australia	Campbell 1982; Briese 1982; Andersen 1982; Andersen and Ashton 1985
Solenopsis	*geminata* (introduced)	Darwin region	Risch and Carroll 1986
	possibly some native species	throughout mesic Australia	A. N. Andersen, unpublished work
Tetramorium	probably only a few species	arid zone	Briese and Macauley 1981; Morton and Davidson 1988; P. J. M. Greenslade, personal communication
subfamily Formicinae			
Melophorus	at least several species	southern arid zone	Briese and Macauley 1981; Greenslade 1982; Morton and Davidson 1988
Prolasius	*bruneus*, *flavicornis*, and *pallidus* groups	mesic south-eastern and south-western Australia	Ashton 1979

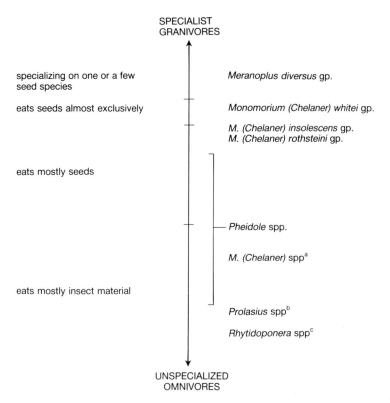

Fig. 33.1. Dietary specialization in the major Australian taxa of seed harvesting ants (see Table 33.1 for references). (a) Species of the *flavigaster, flavipes, kiliani, leae,* and *sculpturatus* groups, (b) Species of the *brunneus, flavicornis,* and *pallidus* groups, (c) *victoriae,* and species of the *metallica* group.

group which, as previously stated, have extremely specialized diets. The combination of a specialist diet and a highly modified clypeus also occurs in certain harvesting species of *Melophorus* (P. J. M. Greenslade, personal communication).

Diversity

Australian harvester ant communities appear to be unusually rich in species by world standards. A study of the harvester ants at 19 sites (each approximately 0.2 hectares) located throughout the arid zone, by Morton and Davidson (1988), found that species richness ranged from 6–13 (mean = 8.8), compared with 2–8 (5.2) in North American deserts. Species richness in North America was strongly correlated with annual rainfall, but

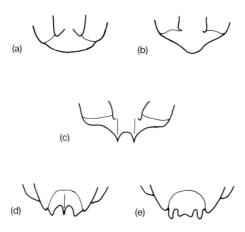

Fig. 33.2. Clypeal specialization in some Australian harvester ants: (a) *Rhytidoponera tasmaniensis*; (b) *Monomorium (Chelaner)* sp. *kiliani* gp; (c) *M. (Chelaner)* sp. *whitei* gp; (d), (e) *Meranoplus* spp. *diversus* gp. Omnivorous species of *Rhytidoponera* have an unspecialized clypeus, with greatest clypeal modification occurring in species of the *Meranoplus diversus* group, which are specialist granivores. Drawings not to scale.

no such relationship was found in Australia. Between-habitat (beta) diversity of harvester ants was also found to be far higher in Australia: a total of only about 20 species was recorded in North American deserts, whereas the Australian arid zone supports hundreds (P. J. M. Greenslade, unpublished work). The entire central Namib desert in southern Africa supports only about 10 harvester species (Marsh 1986), with seven recorded from a single (16 hectare) site (Marsh 1985). Abramsky (1983) recorded just one harvester species at an Israeli desert site.

The species richness of harvester ants at other Australian sites is comparable with that in the arid zone. For example, ten species were recorded at saltbush and grassland sites in semi-arid New South Wales (Briese 1982), and nine species at a woodland site (approximately 0.25 hectares) in mesic south-eastern Australia (Andersen and Ashton 1985). In a savannah woodland of Kakadu National Park in monsoonal northern Australia, where ant diversity is exceptionally high, a total of 19 harvester species have been recorded within 0.5 hectares (Andersen and Lonsdale (1990). As stated previously, elsewhere in the world harvester ants are typically far less diverse outside the arid zone, and often absent altogether.

The local diversity and abundance of harvester ants is clearly influenced by seed availability (Davison 1982). However, seeds are often not a limiting resource (Briese 1982; Westoby *et al.* 1982*a*), so other factors may be more important. This is particularly true for the less specialized ant species, where seeds are not such a critical part of the diet. The unspecialized habits of many

Australian seed harvesters make them especially susceptible to competition from species of *Iridomyrmex*, which have a dominant influence on most Australian ant communities (Greenslade 1976; Greenslade and Halliday 1983; Andersen 1986*a,b*; Andersen, Chapter 36, this volume) Greenslade and Mott (1979) found that the incidence of seed harvesters was inversely proportional to total ant diversity, perhaps reflecting their lack of competitive ability. Greenslade (1982) found that harvesters tend to have more specialized diets in the more diverse ant communities.

The influence of competition is illustrated by the impact of habitat disturbance, which often breaks down the dominance exerted by *Iridomyrmex*, leading to marked increases in foraging populations of unspecialized species (Yeatman and Greenslade 1980; Andersen and Yen 1985; Andersen and McKaige 1987; Andersen 1988*a*). This has important implications for conservation management, as disturbance arising from land management practices can lead to markedly increased activity by seed-eating species (particularly *Rhytidoponera* spp. (Andersen 1990)) and, therefore, to rates of harvesting. It also has important implications for seed harvesting in agricultural situations, where the replacement of native vegetation with exotic pastures and crops often results in the proliferation of seed-eating species of *Pheidole* (Greenslade and Mott 1979; Greenslade 1982).

Rates of harvesting

Wherever they occur, harvester ants are capable of inflicting severe seed losses. Briese (1982) estimated that ants collected nearly 4000 seeds m^{-2} during an autumn bout of harvesting in semi-arid New South Wales, and up to 5000 seeds h^{-1} have been recorded passing into a single nest in North Queensland (Smith and Atherton 1944). Ants commonly remove more than 50 per cent of the seeds from experimental baits within 24 hours (Andersen 1982; Andersen and Ashton 1985; Wellington and Noble 1985). High rates of removal have been recorded in almost all habitats studied, including arid shrublands and hummock grasslands (Davison 1982; Greenslade 1982; Morton 1985; Morton and Davidson 1988); semi-arid heathlands and shrublands (Briese and Macauley 1981; Andersen 1982); seasonally arid woodlands in northern Australia (Greenslade and Mott 1979); mediterranean sclerophyll habitats in south-western Australia (Majer 1982); and heathlands, woodlands, open forests, and tall open forests in south-eastern Australia (Ashton 1979; Drake 1981; Andersen and Ashton 1985; Andersen 1987, 1988*a*). Harvesting occurs in the tropical rainforests of Queensland (P. J. M. Greenslade, personal communication), but quantitative data are lacking. The only habitats where harvesting is not likely to be important are those supporting few ants in general, such as in alpine regions and at sites that are seasonally water-logged or inundated. The more specialized harvesting species cache seeds in extensive underground granaries (Briese 1982; Davison 1982), whereas seed storage often does not occur in broadly omnivorous species (Ashton 1979; Andersen 1982).

Rates of harvesting vary markedly on both daily and seasonal scales, following closely the patterns of foraging activity of harvester ants (Johns and Greenup 1976; Twigg *et al*. 1982; Andersen and Ashton 1985; Wellington and Noble 1985; Panetta 1988). Foraging populations of harvester ants are often largest when seed fall is greatest, suggesting an evolutionary or ecological response to changes in seed supplies. This is probably true for specialized harvesters of the arid zone (e.g. Briese 1982; Davison 1982), and is certainly true for the *Meranoplus diversus* group, where colony activity is finely tuned to seed availability (Andrew 1986). However, in mesic southern Australia, where both seed fall and the abundance of seed-eating (omnivorous) ants are highest during summer months (Twigg *et al*. 1982; Andersen and Ashton 1985), it is more likely that ants and plants respond independently to the season. Rates of harvesting are also notoriously patchy, both between nearby sites and on very small ($<$ 1 m) spatial scales (Johns and Greenup 1976; Andersen 1982; Wellington and Noble 1985).

It has been mentioned that rates of seed harvesting often increase following habitat disturbance. Fire is the most important agent of disturbance throughout most of Australia (Gill 1981), and from a conservation perspective is a major land management tool (Gill 1977; Morton and Andrew 1987; Press 1987). A common immediate effect of fire is to satiate harvester ants by inducing a mass relase of seeds from woody fruit (O'Dowd and Gill 1984; Andersen 1988*a*). This may play an important role in the reproductive biology of many serotinous species (Ashton 1979). In the longer term fire can markedly increase rates of seed-harvesting due to its positive effects on populations of seed-eating ants (Andersen 1988*a*). These effects can persist for two years or more, and might have an important impact on seedling establishment. Furthermore, the lower fire intensities typical of management burns might reduce the effectiveness of predator satiation following fire, as seed fall is not as synchronous as it is after high-intensity fires.

As also mentioned previously, agricultural land is often colonized by unspecialized species of *Pheidole*, therefore increasing harvesting rates. Particularly high rates of seed removal by *Pheidole* have been reported in agricultural land throughout the country, including Queensland (Smith and Atherton 1944; Russell *et al* 1967), New South Wales (Campbell 1966; Campbell and Swain 1973; Johns and Greenup 1976), Victoria (McGowan 1969), Western Australia (Twigg *et al*. 1982; Panetta 1988), and the Northern Territory (Mott and Mckeon 1977).

Impact of seed losses

The sheer volume of seeds removed by ants is often impressive. However, the impact of these losses on plant populations has been poorly studied. From a plant's perspective, the impact of seed losses is likely to depend more on how many seeds are remaining, and where they remain, rather than on how many are lost. Moreover, seed losses must be placed in the context of the overall dynamics of the population. For example, seedling recruitment is negligible in the absence of disturbance in many stable populations of long-lived perennials. In these cases recruitment is limited by a rarity of safe sites rather than by seed supply, so that seed losses to predators are possibly irrelevant for much of the time (Andersen 1989b). This is illustrated by a study of *Eucalyptus baxteri* Maiden & Nakely in south-eastern Australia (Andersen 1987). The elimination of harvester ants increased seedling densities 15-fold, but all these seedlings perished within a year, so in this instance losses to ants had no impact on recruitment.

Whereas the impact of harvester ants on populations of long-lived perennials can be difficult to determine, their effects on ephemerals are straightforward. Populations of ephemerals are frequently limited by seed supply, in which case seed losses to predators have a direct influence on them (e.g. Watkinson *et al.* 1989). Losses to predators can obviously have a direct bearing on the success of agricultural and forestry seeding operations (Cremer 1966; McGowan 1969; Mott and McKeon; 1977; Campbell 1982). Harvester ants can also affect adversely the persistence of improved pastures (McGowan 1969).

Conclusions

It is clear that seed-harvesting by ants has special significance in Australia compared with elsewhere in the world. Interestingly, this is also the case for myrmecochory, which is extraordinarily prevalent in Australia (Berg 1975, 1981; Rice and Westoby 1981; Westoby *et al.* Chapter 29, this volume). It is worth noting that the highest concentrations of myrmecochorous species occur in sclerophyllous vegetation of southern Australia, where dispersal is predominantly by unspecialized species of *Pheidole* and *Rhytidoponera* (Auld 1986; Andersen 1988b). These same ants double as the major seed predators (of other species) in the region. Has the availability of unspecialized seed-collecting ants contributed to the prevalence of myrmecochory?

Why is seed harvesting by ants so different in Australia? One of the notable Australian features is the diversity of harvesting species. However, this might just be a reflection of the diversity of Australian ants in general, which is exceptional (Andersen and Yen 1985; Greenslade and Greenslade 1989), rather than anything specifically to do with harvesting. Seed harvesting by ants in Australia is really distinctive in its prevalence in temperate habitats, where the major harvesting species are unspecialized omnivores. It is not clear why this is so, but a possible explanation lies in the dominance of the vegetation by sclerophyllous Myrtaceae (notably species of *Eucalyptus*, *Leptospermum* and *Melaleuca*) which produce large numbers of small, unprotected, thin-coated seeds. Little specialization is therefore required to make these seeds available to ants. The sclerophyllous Myrtaceae are best developed in southern Australia, where seed-eating omnivores are most prevalent, and these Myrtaceae appear to constitute the most favoured seed species (Andersen and Ashton 1985).

Acknowledgements

I thank Drs R. W. Braithwaite, R. C. Buckley, P. J. M. Greenslade, R. J. Hobbs, J. D. Majer, and S. R. Morton for their helpful comments on the manuscript. This is TERC library contribution no. 676.

References to Part 5

Abbot, I. and van Heurck, P. (1985). Comparison of insects and vertebrates as removers of seed and fruit in a Western Australian forest. *Australian Journal of Ecology*, **10**, 165–8.

Abramsky, Z. (1983). Experiments on seed predation by rodents and ants in the Israeli desert. *Oecologia*, **57**, 328–32.

Addicott, J. F. (1978). The population dynamics of aphids on fireweed: a comparison of local population and metapopulations. *Canadian Journal of Zoology*, **56**, 254–64.

Andersen, A. N. (1982). Seed removal by ants in the mallee of northwestern Victoria. In *Ant–plant interactions in Australia*, (ed. R. C. Buckley), pp. 31–44. W. Junk, The Hague.

Andersen, A. N. (1983). A brief survey of ants in Glenaladale National Park, with particular reference to seed-harvesting. *Victorian Naturalist* **100**, 233–7.

Andersen, A. N. (1986*a*). Patterns of ant community organisation in mesic south-eastern Australia. *Australian Journal of Ecology*, **11**, 87–97.

Andersen, A. N. (1986*b*) Diversity, seasonality and community organisation of ants at adjacent heath and woodland sites in south-eastern Australia. *Australian Journal of Zoology*, **34**, 53–64.

Andersen, A. N. (1987). Effects of seed predation by ants on seedling densities at a woodland site in SE Australia. *Oikos*, **48**, 171–4.

Andersen, A. N. (1988*a*). Immediate and longer-term effects of fire on seed predation by ants in sclerophyllous vegetation in south-eastern Australia. *Australian Journal of Ecology*, **31**, 285–93.

Andersen, A. N. (1988*b*). Dispersal distance as a benefit of myrmecochory. *Oecologia*, **75**, 507–11.

Andersen, A. N. (1989*a*). Soil of the nest-mound of the seed dispersing ant, *Aphaenogaster longiceps*, enhances seedling growth. Australian Journal of Ecology **13**, 469–471.

Andersen, A. N. (1989*b*). How important is seed predation to recruitment in stable populations of long-lived perennials? *Oecologia*, **81**, 310–5.

Andersen, A. N. (1990). The use of ant communities to evaluate change in Australian terrestrial ecosystems: a review and a recipe. Proceedings of the Ecological Society of Australia, **16**, 347–57.

Andersen, A. N. and Ashton, D. H. (1985). Rates of seed removal by ants at heath and woodland sites in southeastern Australia. *Australian Journal of Ecology*, **10**, 381–90.

Andersen, A. N. and Lonsdale, W. M. (1990). Herbivory by insects in Australian tropical savannas: a review. *Journal of Biogeography*, **17**, 433–44.

Andersen, A. N. and McKaige, M. E. (1987). Ant communities at Rotamah Island, Victoria, with particular reference to disturbance and *Rhytidoponera tasmaniensis*. *Proceedings of the Royal Society of Victoria*, **99**, 141–6.

Andersen, A. N. and Yen, A. L. (1985). Immediate effects of fire on ants in the semi-arid mallee region of northwestern Victoria. *Australian Journal of Ecology*, **19**, 25–30.

Andersson, L. (1976). The synflorescence of the Marantaceae. Organization and descriptive terminology. *Botanischer Notiser*, **129**, 39–48.

Andrew, M. H. (1986) Granivory of the annual grass *Sorghum intrans* by the harvester and *Meranoplus* sp. in tropical Australia. *Biotropica*, **18**, 334–9.

Ashton, D. H. (1979). Seed harvesting by ants in forests of *Eucalyptus regnans* F. Muell. in central Victoria. *Australian Journal of Ecology*, **4**, 265–77.

Augspurger, C. K. (1984). Seedling survival of tropical tree species: interactions of dispersal distance, light-gaps, and pathogens. *Ecology*, **65**, 1705–12.

Auld, T. D. (1986) Population dynamics of the shrub *Acacia suaveolens* (Sm.) Willd.: dispersal and the dynamics of the seed-bank. *Australian Journal of Ecology*, **11**, 235–54.

Baker, H.G. (1972). Seed weight in relation to environmental conditions in California. *Ecology*, **53**, 997–1010.

Bazzaz, F. A. and Pickett, S. D. A. (1980). The physiological ecology of tropical succession: a comparative review. *Annual Review of Ecology and Systematics*, **11**, 287–310.

Beadle, N. C. W. (1954). Soil phosphate and the delimitation of plant communities in eastern Australia. *Ecology*, **35**, 370–5.

Beadle, N. C. W. (1966). Soil phosphate and its role in molding segments of the Australian flora and vegetation with special reference to xeromorphy and sclerophylly. *Ecology*, **47**, 991–1007.

Beattie, A. J. (1982). Ants and gene dispersal in flowering plants. In *Pollination and evolution*. (ed. J. A. Armstrong). Royal Botanic Gardens Press, Sydney.

Beattie, A. J. (1983). The distribution of ant-dispersed plants. In *Dispersal and distribution*, (ed. K. Kubitzki), pp. 249–70. Parey, Hamburg.

Beattie, A. J. (1985). *The evolutionary ecology of ant–plant mutualisms*, 182 pp. Cambridge University Press, Cambridge.

Beattie, A. J. and Culver, D. C. (1981). The guild of myrmecochores in the herbaceous flora of West Virginia forests. *Ecology*, **62**, 107–15.

Beattie, A. J. and Culver, D. C. (1982). Inhumation: how ants and other invertebrates help seeds. *Nature*, **297**, 627.

Beattie, A. J. and Culver, D. C. (1983). The nest chemistry of two seed dispersing ant species. *Oecologia* (Berlin); **56**, 475–508.

Beattie, A. J., Turnbull, C. L. Knox, R. B. and Williams, E. (1984). Ant inhibition of pollen function: a possible reason why ant pollination is rare. *American Journal of Botany*, **71**, 421–6.

Beattie, A. J., Turnbull, C. L., Hough, T., Jobson, S., and Knox, B. (1985). The vulnerability of pollen and fungal spores to ant secretions: evidence and some evolutionary implications. *American Journal of Botany*, **72**, 606–14.

Beattie, A. J., Turnbull, C. L., Hough, T., and Knox, B. (1986). Antibiotic production: a possible function for the metapleural gland of ants (Hymenoptera: Formicidae). *Annals of the Entomological Society of America*, **79**, 448–50.

Berg, R. Y. (1958). Seed dispersal, morphology and phylogeny of *Trillium*. *Skrifer Utgitt av Det Norske Videnskaps-Akademi*, **1958**, 1,1–36.

Berg, R. Y. (1975). Myrmecochorous plants in Australia and their dispersal by ants. *Australian Journal of Botany*, **23**, 475–508.

Berg, R. Y. (1981). The role of ants in seed dispersal in Australian lowland heathland. In *Heathlands and related shrublands: analytical studies*. (ed. R. L. Specht), pp. 41–50. Elsevier, Amsterdam.

Blackwell, W. H. and Powell, M. H. (1981). A preliminary note on pollination in the Chenopodiaceae. *Annals of the Missouri Botanic Gardens*, **68**, 524–6.

Bohart, G. E. and Knowlton, G. F. (1953). Notes on food habits of the western harvester ant. *Proceedings of the Entomological Society of Washington*, **55**, 151–3.

Bond, P. and Goldblatt, P. (1984). Plants of the Cape flora: a descriptive catalogue. *Journal of South African Botany*, **13**.

Bond, W. J. (1988) Proteas as 'tumbleseeds': wind-dispersal through the air and over soil. *South African Journal of Botany*, **54**, 455–60.

Bond, W. J. and Breytenbach, G. J. (1985). Ants, rodents, and seed predation in Proteaceae. *South African Journal of Zoology*, **20**, 150–54.

Bond, W. J., Le Roux, D. and Emtzen, R. (1990). Fire intensity and regeneration of myrmecochorous proteaceae. *South African Journal of Botany*, **56**, 326–30.

Bond, W. J. and Slingsby, P. (1983). Seed dispersal by ants in shrublands of the Cape Province and its evolutionary implications. *South African Journal of Science*, **79**, 231–3.

Bond, W. J. and Slingsby, P. (1984). Collapse of an ant–plant mutualism: the Argentine ant (*Iridomyrmex humilis*) and myrmecochorous Proteaceae. *Ecology*, **65**, 1031–7.

Bond, W. J. and Stock, W. D. (1989). The costs of leaving home: ants disperse seeds to low nutrient sites. *Oecologia*, **81**, 412–17.

Bowen, G. D. (1981). Coping with low nutrients. In *The biology of Australian plants* (ed. J. S. Pate and A. J. McComb), pp. 33–64. University of Western Australia Press, Nedlands.

Brantjes, N. B. M. (1981). Ant, bee and fly pollination in *Epipactis palustris* (L.) Crantz (Orchidaceae). *Acta Botanica Neerlandica*, **30**, 59–68.

Brantjes, N. B. M. and de Vos, O. C. (1981). The explosive release of pollen in the flowers of *Hyptis* (Lamiaceae). *New Phytologist*, **87**, 425–30.

Breytenbach, G. J. (1986) Dispersal: the case of the missing ant and the introduced mouse. *South African Journal of Botany*, **52**, 463–66.

Breytenbach, G. J. (1988). Why are myrmecochorous plants limited to fynbos (Macchia) vegetation types? *South African Forestry Journal*, **144**, 3–5.

Briese, D. T. (1982). Relationship between the seed-harvesting ants and the plant community in a semi-arid environment. In *Ant–plant interactions in Australia*, (ed. R. C. Buckley), pp. 11–24. W. Junk, The Hague.

Briese, D. T. and Macauley, B. J. (1981). Food collection within an ant community in semi-arid Australia, with special reference to seed harvesters. *Australian Journal of Ecology* **6**, 1–19.

Brits, G. (1986*a*). Influence of fluctuating temperatures and H_2O_2 treatment on germination of *Leucospermum cordifolium* and *Serruria florida* (Proteaceae) seeds. *South African Journal of Botany*, **52**, 286–90.

Brits, G. (1986*b*). The effect of hydrogen peroxide treatment on germination in Proteaceae species with serotinous and nut-like achenes. *South African Journal of Botany*, **52**, 291–93.

Brits, G. (1987). Germination depth vs. temperature requirements in naturally dispersed seeds of *Leucospermum cordifolium* and *L. cuneiforme* Proteaceae. South African Journal of Botany, **53**, 119–24.

Brown, J. H., Grover, J. J., Davidson, D. W., and Lieberman, G. A. (1975). A preliminary study of seed predation in desert and montane habitats. *Ecology*, **56**, 987–92.

Brown, J. H., Reichman, O. J., and Davidson, D. W. (1979). Granivory in desert ecosystems. *Annual Review of Ecology and Systematics*, **10**, 201–27.

Buckley, R. C. (1982). Ant–plant interactions: a world review. In *Ant–plant interactions in Australia*. (ed. R. C. Buckley), pp. 111–41. Dr W. Junk, The Hague.

Burkart, A. (1937). El mecanismo floral de la labiada *Hyptis mutabilis. Darwiniana*, **3**, 425–7.

Butler, S. C. (1971). The eucalypt ash resource of West Gippsland. *Proceedings of the Royal Society of Victoria*, **84**, 53–9.

Campbell, M. H. (1966). Theft by harvesting ants of pasture seed broadcast on unploughed land. *Australian Journal of Experimental Agriculture and Animal Husbandry*, **6**, 334–8.

Campbell, M. H. (1982). Restricting losses of aerially sown seed due to seed-harvesting ants. In *Ant–plant interactions in Australia*. (ed. R. C. Buckley), pp. 25–30. W. Junk, The Hague.

Campbell. M. H. and Swain, F. G. (1973). Factors causing losses during the establishment of surface-sown pastures. *Journal of Range Management*, **26**, 355–9.

Carroll, C. R. and Risch, S. J. (1984). The dynamics of seed harvesting in early successional communities by a tropical ant *Solenopsis germinata Oecologia*, **61**, 338–92.

Caswell, H. (1982). Stable population structure and reproductive value for populations with complex life cycles. *Ecology*, **63**, 1223–31.

Caswell, H. (1985) The evolutionary demography of clonal reproduction. In *Population biology and evolution of clonal organisms*. (ed. Jackson, Buss, and Cook). pp. 187–224. Yale University Press. New Haven, CT.

Charles-Dominique, P. (1986). Inter-relations between frugivorous vertebrates and pioneer plants: *Cecropia*, birds and bats in French Guyana. In *Frugivores and seed dispersal*. (ed. A. E. Estrada and T. H. Fleming), pp. 119–36. Dr W. Junk, The Hague.

Charlesworth, B. (1980). *Evolution in age-structured populations.* Cambridge Studies in Mathematical Biology. Cambridge University Press, Cambridge.

Chauhan, E. (1979). Pollination by ants in *Coronopus didymus* (L) Sm. *New Botanist*, **6**, 39–40.

Chesson, P. L. and Warner, R. R. (1981). Environmental variability promotes coexistence in lottery competitive systems. *American Naturalist*, **117**, 923–43.

Cody, M. L. and Mooney, H. A. (1978). Convergence versus nonconvergence in Mediterranean-climate ecosystems. *Annual Review of Ecology and Systematics*, **9**, 265–321.

Comins, H. N. and Noble, I. R. (1985). Dispersal, variability and transient niches: species coexistence in a uniformly variable environment. *American Naturalist*, **126**, 706–23.

Cook, R. E. (1979). Patterns of juvenile mortality and recruitment in plants. In *Topics in plant population biology*, (ed. O. T. Solbrig, S. Jain, G. Johnson, and P. H. Raven), pp. 207–31. Columbia University Press, New York.

Cowling, R. M. (1987). Fire and its role in coexistence and speciation in Gondwana shrublands. *South African Journal of Science*, **83**, 106–11.

Cowling, R. M. and Campbell, B. M. (1980). Convergence in vegetation structure in the mediterranean communities of California, Chile and South Africa. *Vegetatio*, **43**, 191–97.

Cremer, K. W. (1966). Treatment of *Eucalyptus regnans* seed to reduce losses to insects after sowing. *Australian Forestry*, **30**, 162–74.

Culver, D. C. (1981). On using Horn's Markov succession model. *American Naturalist*, **117**, 572–574.

Culver, D. C. and Beattie, A. J. (1978). Myrmecochory in *Viola*: dynamics of seed–ant interactions in some West Virginia species. *Journal of Ecology*, **66**, 53–72.

Culver, D. C. and Beattie, A. J. (1980). The fate of *Viola*: seeds dispersed by ants. *American Journal of Botany*, **67**, 710–14.

Culver, D. C. and Beattie, A. J. (1983). Effects of ant mounds on soil chemistry and vegetation patterns in a Colorado montane meadow. *Ecology*, **64**, 485–92.

Dahl, E. and Hadac, E. (1940). Maur som blomsterbestovere. *Nytt Magasin Naturvidensk*, **81**, 46–8.

Davidson, D. W. and Morton, S. R. (1981*a*). Myrmecochory in some plants (F. Chenopodiaceae) of the Australian arid zone. *Oecologia*, **50**, 357–66.

Davidson, D. W. and Morton, S. R. (1981*b*). Competition for dispersal in ant-dispersed plants. *Science*, **213**, 1259–61.

Davidson, D.W. and Morton, S. R. (1984). Dispersal adaptations of some *Acacia* species in the Australian arid zone. *Ecology*, **65**, 1038–51.

Davison, E. A. (1982). Seed utilization by harvester ants. In *Ant–plant interactions in Australia*. (ed. R. C. Buckley), pp. 1–6. W. Junk, The Hague.

Denslow, J. S. (1980). Gap partitioning among tropical rainforest trees. *Biotropica*, **12** (Suppl.), 47–55.

Denslow, J. S. (1987). Tropical rainforest gaps and tree species diversity. *Annual Review of Ecology and Systematics*, **18**, 431–51.

Dirzo, R. and Dominguez, C. A. (1986). Seed shadows, seed predation and the advantages of dispersal. In *Frugivores and seed dispersal*, (ed. A. E. Estrada and T. H. Fleming), pp. 237–50.Dr. W. Junk, Dordrecht.

Doku, E. V. (1968). Flowering, pollination and pod formation in Bambara Groundnut (*Volandzeia subterranea*) in Ghana. *Experimental Agriculture*, **4**, 41–8.

Doku, E. V. and Karikari, S. K. (1971). The role of ants in pollination and pod formation of Bambarra Groundnut. *Economic Botany*, **25**, 357–62.

Drake, W. E. (1981). Ant–seed interaction in dry sclerophyll forest on North Stradbroke island, Queensland. *Australian Journal of Botany*, **29**, 293–309.

Ehrenfeld, J. (1978). Pollination of three species of *Euphorbia* subgenus Chamesyce, with special reference to bees. *American Midland Naturalist*, **101**, 87–9.

Eichwort, G. C. and Ginsberg, H. S. (1980). Foraging and mating behaviour in Apoidea. *Annual Review of Ecology and Systematics*, **25**, 421–46.

Faegri, K. and van der Pijl, L. (1971). *The principles of pollination ecology*, 2nd edn. Pergamon Press, Oxford.

Feinsinger, P. and Swarm, L. A. (1978). How common are ant-repellent nectars? *Biotropica*, **10**, 238–9.

Fisher, R. A. (1930). *The genetical theory of natural selection*. Oxford University Press, Oxford.

Foster, S. A. (1986). On the adaptive value of large seeds for tropical moist forest trees: a review and synthesis. *Botanical Review*, **52**, 260–99.

Foster, S. A. and Janson, C. H. (1985). The relationship between seed size and establishment conditions in tropical woody plants. *Ecology*, **66**, 773–80.

Fox, B. J., Quinn, R. D., and Breytenbach, G. J. (1985). A comparison of small-mammal succession following fire in shrublands of Australia, California and South Africa. *Proceedings of the Ecological Society of Australia*, **14**, 179–97.

Gadgil, M. (1971). Dispersal: population consequences and evolution. *Ecology*, **52**, 253–61.

Gates, B. N. (1940). Dissemination by ants of the seeds of *Trillium grandiflorum*. *Rhodora*, **42**, 194–6.

Gates, B. N. (1941). Observation in 1940 on the dissemination by ants of the seeds of *Trillium grandiflorum*. *Rhodora*, **43**, 206–7.

Gentry, A. L. (1982). Patterns of neotropical plant species diversity. *Evolutionary Biology*, **15**, 1–84.

Gill, A. M. (1977). Management of fire-prone vegetation for plant species conservation in Australia. *Search*, **8**, 20–6.

Gill, A. M. (1981). Adaptive responses of Australian vascular plant species to fire. In *Fire and the Australian biota*, (eds. A. M. Gill, R. H. Groves, and I. R. Noble), pp. 234–71. Australian Academy of Science, Canberra.

Ginsberg, H. S. (1984). Foraging behaviour of the bees *Halictus ligatus* (Hymenoptera: Halictidae) and *Ceratina calcarata* (Hymenoptera: Anthophoridae). *Journal of the New York Entomological Society*, **92**, 162–8.

Givnish, T. J. (1987). Comparative studies of leaf form: assessing the relative roles of selective pressures and phylogenetic constraints. *New Phytologist*, **106**, 31–60.

Gottsberger, G. and Silberbauer-Gottsberger, I. (1983). Dispersal and distribution in the cerrado vegetation of Brazil. In *Dispersal and distribution*. (ed. K. Kubitzki), pp. 315–52. Parey, Hamburg.

Greenslade, P. J. M. (1976). The meat ant *Iridomyrmex purpureus* (Hymenoptera: Formicidae) as a dominant member of ant communities. *Journal of the Australian Entomological Society*, **15**, 237–40.

Greenslade, P. J. M. (1982). Diversity and food specificity of seed-harvesting ants in relation to habitat and community structure. In *Proceedings of the 3rd Australasian conference on grassland invertebrate ecology*. (ed. K. Lee), pp. 227–33. South Australian Government Printer, Adelaide.

Greenslade, P. J. M. and Greenslade, P. (1989). Ground layer invertebrate fauna. In *Mediterranean landscapes in Australia: Mallee ecosystems and their management*. (ed. J. C. Noble and R. A. Bradstock), pp. 266–84. CSIRO Press, Melbourne.

Greenslade, P. J. M. and Halliday, R. B. (1983). Colony dispersion and relationships of meat ants *Iridomyrmex purpureus* and allies in an arid locality in South Australia. *Insectes Sociaux*, **30**, 82–99.

Greenslade, P. J. M. and Mott, J. J. (1979). Ants of native and sown pastures in the Katherine area, Northern Territory, Australia (Hymenoptera: Formicidae). In *Proceedings of the 2nd Australasian Conference on grassland invertebrate ecology*, (ed. T. K. Crosby and R. P. Pontinger), pp. 153–6. Government Printer, Wellington.

Grime, J. P. (1979). *Plant strategies and vegetation processes*. Wiley, New York.

Gross, K. L. and Werner, P. A. (1982). Colonizing abilities of 'biennial' plant species in relation to ground cover: implications for their distributions in a successional sere. *Ecology*, **63**, 921–31.

Guerrant, E. O. and Fiedler, P. L. (1981). Flower defenses against nectar-pilferage by ants. *Biotropica* (Suppl. *Reproductive Botany*), 25–33.

Haber, W. A., Frankie, G. W., Baker, H. G., Baker, I., and Koptur, S. (1981). Ants like flower nectar. *Biotropica*, **13**, 211–4.

Hagerup, O. (1943). Myre-bestovning. *Botanisk Tiddsskrift*, **46**, 116–23.

Hall, A. V. and Veldhuis, H. A. (1985). *South African red data book: plants—fynbos and Karoo Biomes*. South African National Scientific Programmes Report No. 117, CSIR, Pretoria.

Hamilton, W. D. and May, R. M. (1977). Dispersal in stable habitats. *Nature*, **269**, 578–81.

Hammel, B. (1984). Systematic treatment of the Cyclanthaceae, Marantaceae, Cecropiaceae, Clusiaceae, Lauraceae and Moraceae for the flora of a wet lowland tropical forest, Finca La Selva, Costa Rica, Doctoral Dissertation. Duke University. North Carolina.

Handel, S. N. (1978). The competitive relationship of three woodland sedges and its bearing on the evolution of ant dispersal of *Carex pedunculata*. *Evolution*, **32**, 151–63.

Handel, S. N. (1985). The intrusion of clonal growth patterns on plant breeding systems. *American Naturalist*, **125**, 367–84.

Handel, S. N., Fisch, S. B., and Schatz, G. E. 1981). Ants disperse a majority of herbs in a mesic forest community in New York State. *Bulletin of the Torrey Botanical Club*, **108**, 430–7.

Hanzawa, F. M., Beattie, A. J., and Culver, D. C. (1988). Directed dispersal: demographic analysis of an ant–seed mutualism. *American Naturalist*, **131**, 1013.

Harley, R. M. (1971). An explosive pollination mechanism in *Eriope crassipes*, a Brazilian labiate. *Biological Journal of the Linnean Society*, **3**, 159–64.

Harley, R. M. (1974). New collections of Labiatae from Brazil. Notes on New World Labiatae, III. *Kew Bulletin*, **29**, 125–40.

Harley, R. M. (1976). A Review of *Eriope* and *Eriopidion* (Labiatae). *Hooker's Icones Plantarum*, **8**, 1–107.

Harley, R. M. (1988*a*). Revision of generic limits in *Hyptis* Jacq. (Labiatae) and its allies. *Botanical Journal of the Linnean Society*, **98**, 87–95.

Harley, R. M. (1988*b*). Evolution and distribution of *Eriope* (Labiatae) and its relatives in Brazil. In *Proceedings of a workshop on neotropical distributions*, pp. 71–120. Academia Brasileira de Ciências, Rio de Janeiro.

Harper, J. L. (1977). *Population biology of plants*. Academic Press, London.

Harper, J. L., Lovell, P. H., and Moore, K. G. (1970). The shapes and sizes of seeds. *Annual Review of Ecology and Systematics*, **1**, 327–57.

Hartshorn, G. S. (1983). Plants. In *Costa Rican natural history*, (ed. D. H. Janzen), pp. 118–57. University of Chicago Press, Chicago.

Heithaus, E. R. (1981). Seed predation by rodents on three ant-dispersed plants. *Ecology*, **62**, 136–45.

Heithaus, E. R., Culver, D. C., and Beattie, A. J. (1980). Models of some ant–plant mutualisms. *American Naturalist*, **116**, 347–16.

Hickman, J. C. (1974). Pollination by ants: a low energy system. *Science*, **184**, 1290–2.

Higashi, S., Tsuyuzaki, Sh., Ohara, M., and Ito, F. (1989). Adaptive advantages of ant-dispersed seeds in the myrmecochorous plant *Trillium tschonoskii* (Liliaceae). *Oikos*, **54**, 389–94.

Hobbs, R. J. (1985). Harvester ant foraging and plant species distribution in annual grassland. *Oecologia*, **67**, 519–23.

Holmes, P. M. (1990). Dispersal and predation of alien *Acacia* seeds: effects of season and invading stand density. (in press, *Oecologia*).

Horn, H. S. (1975). Markovian properties of forest succession. In *Ecology and evolution of communities*, (ed. M. L. Cody and J. M. Diamond, pp. 196–211. Belknap Press of Harvard University Press, Cambridge.

Horvitz, C. C. and Beattie, A. J. (1980). Ant dispersal of *Calathea* (Marantaceae) seeds by carnivorous ponerines (Formicidae) in a tropical rain forest. *American Journal of Botany*, **67**, 321–6.

Horvitz, C. C. and Schemske, D. W. (1986a). Seed dispersal and environmental heterogeneity in a neotropical herb: a model of population and patch dynamics. In *Frugivores and seed dispersal*, (eds. A. E. Estrada and T. H. Fleming), pp. 169–86. Dr. W. Junk, The Hague.

Horvitz, C. C. and Schemske, D. W. (1986b). Seed dispersal of a neotropical myrmecochore: variation in removal rates and dispersal distance. *Biotropica*, **8**, 319–23.

Horvitz, C. C. and Schemske, D. W. (1986c). Ant-nest soil and seedling-growth in a neotropical ant-dispersed herb. *Oecologia*, **70** 318–20.

Horvitz, C. C. and Schemske, D. W. (1988). An experimental analysis of the cost of reproduction in a neotropical herb. *Ecology*, **69**, 1741–5.

Howe, H. F. (1985). Gomphothere fruits: a critique. *American Naturalist*, **125**, 853–65.

Howe, H. F. (1986). Seed dispersal by fruit-eating birds and mammals. In *Seed dispersal*, (ed. D. R. Murray), pp. 123–90. Academic Press, Australia.

Howe, H. .F. and Smallwood, J. (1982). Ecology of seed dispersal. *Annual Review of Ecology and Systematics*, **13**, 201–28.

Hughes, L. (in press). The relocation of ant nest entrances: potential consequences for ant-dispersed seeds. *Australian Journal of Ecology*.

Hughes, L. and Westoby, M. (1990). Removal rates of seeds adapted for dispersal by ants. *Ecology*, **71**, 138–48.

Hull, D. A. and Beattie, A. J. (1988). Adverse effects on pollen exposed to *Atta texana* and other North American ants: implications for ant pollination. *Oecologia*, **75**, 153–5.

Inouye, R. S., Byers, G. S., and Brown, J. H. (1980). Effects of predation and competition on survivorship, fecundity and community structure of desert annuals. *Ecology*, **61**, 1344–51.

Iwanami, Y. and Iwadare, T. (1978). Inhibiting effects of Myrmicacin on pollen growth and pollen tube mitosis. *Botanical Gazette*, **139**, 42–5.

Janson, C. H. (1983). Adaptation of fruit morphology to dispersal agents in a neotropical rainforest. *Science*, **219**, 187–9.

Janson, C. H., Stiles, E. W., and White, D. W. (1986). Selection on plant fruiting traits by brown capuchin monkeys: a multivariate approach. In *Frugivores and seed dispersal*. (ed. A. E. Estrada and T. H. Fleming). pp. 83–92. Dr. W. Junk, The Hague.

Janzen, D. H. (1969). Seed eaters versus seed size, number, toxicity and dispersal. *Evolution*, **23**, 1–27.

Janzen, D. H. (1970). Herbivores and the number of tree species in tropical forests. *American Naturalist*, **104**, 501–28.

Janzen, D. H. (1971). Seed predation by animals. *Annual Review of Ecology and Systematics*, **2**, 465–92.

Janzen, D. H. (1977). Why don't ants visit flowers? *Biotropica*, **9**, **4**, 252.

Johns, G. G. and Greenup, L. R. (1976). Pasture seed theft by ants in northern N.S.W. *Australian Journal of Experimental Agriculture and Animal Husbandry*, **16**, 257–64.

Jones, D. L. (1975). The pollination of *Microtis parviflora* R. Br. *Annals of Botany*, **39**, 585–9.

Juniper, B. E., Robins, R. J., and Joel, D. M. (1989). *The carnivorous plants*: 298 *et seq.*, London.

Kaufmann, S., McKey, D. B., Hossaer-McKey, M., and Horvitz, C. C. (in press).

Adaptations for a two phase seed dispersal system involving vertebrates and ants in a hemiepiphytic fig, *Ficus microcarpa*, Moraceae. *American Journal of Botany*.

Kennedy, H. (1977). An unusual flowering strategy and new species in *Calathea*. *Botanischer Notiser*, **130**, 333–9.

Kennedy, H. (1978). Systematics and pollination of the 'closed-flowered' species of *Calathea* (Marantaceae). *University of California Publications in Botany*, **71**, 1–90.

Kennedy, H. (1983). *Calathea insignis* (Hoja Negra, hoja de sal, Bijagua, Rattlesnake Plant). Plant species accounts. In *Costa Rican natural history*, (ed. D. H. Janzen), pp. 204–6. University of Chicago Press, Chicago.

Kerner von Marilaun, A. (1878). *Flowers and their unbidden guests*, (translated by W. Ogle). C. Kegan Paul & Co., London.

Kerner von Marilaun, A. (1895). *The natural history of plants*, Vol II, (translated and edited by F. W. Oliver). Blackie & Son Ltd, London.

Kincaid, T. (1963). The ant–plant, *Orthocarpus pusillus* Bentham. *Transactions of the American Microscopical Society*, **82**, 101–5.

Kock, A. E. de and Giliomee, J. (1989). A survey of the Argentine ant *Iridomyrmex humilis* (Mayr.) (Hymenoptera: Formicidae) in South African fynbos. *Journal of the Entomological Society of Southern Africa*, **52**, 157–64.

Kruger, F. J. (1979). South African heathlands. In *Ecosystems of the world. Heathlands and related shrublands. Descriptive studies*, (ed. R. L. Specht), pp. 19–80. Elsevier, Amsterdam.

Levey, D. J. (1988*a*). Tropical wet forest treefall gaps and distributions of understorey birds and plants. *Ecology*, **69**, 1076–89.

Levey, D. J. (1988*b*). Evidence for a nested dispersal system in *Miconia*. *Bulletin of the Ecological Society of America*, **69**, 207.

Levieux, J. (1983). The soil fauna of tropical savannas. *IV*: The ants. In *Tropical savannas*. (ed. F. Bourliere), pp. 525–40. Elsevier, Amsterdam.

Levin, D. A. (1981). Dispersal versus gene flow in plants. *Annals of the Missouri Botanic Gardens*, **68**, 233–53.

Levin, D. A. and Kerster, H. W. (1974). Gene flow in seed plants. In *Evolutionary biology* (ed. T. Dobzhansky, M. K. Hecht, and W. C. Steere), Vol. 7, pp. 139–220. Plenum Press, New York.

McAuliffe, J. R. (1988). Markovian dynamics of simple and complex desert plant communities. *American Naturalist*, **131**, 459–90.

McGowan, A. A. (1969). Effect of seed harvesting ants on the persistence of Wimmura rye grass in pastures of north east Victoria. *Australian Journal of Experimental Agriculture and Animal Husbandry*, **9**, 37–40.

McKey, D. (1975). The ecology of coevolved seed dispersal systems. In *Coevolution of animals and plants*, (ed. L. E. Gilbert and P. H. Raven), pp. 159–91. University of Texas Press, Austin.

Majer, J. D. (1982). Ant–plant interactions in the Darling botanical district of Western Australia. In *Ant–plant interactions in Australia*, (ed. R. C. Buckley), pp. 45–61. W. Junk, The Hague.

Major, J. D. (1990). The role of ants in Australian land reclamation seeding operations. In *Applied myrmecology: A world perspective*, (ed. R. K. Vander Meer, K. Jaffe and A. Cedeno), pp. 544–54. Westview, Boulder, Colorado.

Manders, P. T. (1986). Seed dispersal and seedling recruitment in *Protea laurifolia*. *South African Journal of Botany*. **52**, 421–4.

Mares, M. A. and Rosenzweig, M. L. (1978). Granivory in North and South American deserts: rodents, birds and ants. *Ecology*, **59**, 235–41.

Marloth, R. (1915). *The flora of South Africa*, Vols I–IV. Wesley, London.

Marsh, A. C. (1985). Forager abundance and dietary relationships in a Namib Desert ant community. *South African Journal of Zoology*, **20**, 197–203.

Marsh, A. C. (1986). Ant species richness along a climatic gradient in the Namib Desert. *Journal of Arid Environments*, **11**, 235–41.

Maschwitz, U., Koob, K., and Schildknecht, H. (1970). Ein Beitrag zür Funktion der Metathorakaldruse der Ameisen. *Journal of Insect Physiology*, **16**, 387–404.

Mesler, M. R. and Lu, K. L. (1983). Seed dispersal of *Trillium ovatum* (Liliaceae) in second-growth redwood forests. *American Journal of Botany*, **70**, 1460–7.

Midgley, J. J. (1987). Aspects of the evolutionary biology of the Proteaceae, with emphasis on the genus *Leucadendron* and its phylogeny. Ph.D. thesis, University of Cape Town.

Midgley, J. J. and Bond, W. (1989). Leaf size and inflorescence size may be allometrically related traits. *Oecologia*, **78**, 427–9.

Migliorato, E. (1910). Sull̦impollinazsione di *Rohdea japonica* Roth per mezzo dell formiche. *Annali di Botanica*, **8**, 241–2.

Milewski, A. V. (1982). The occurrence of seeds and fruits taken by ants versus birds in mediterranean Australia and southern Africa, in relation to the availability of soil potassium. *Journal of Biogeography*, **9**, 505–16.

Milewski, A. V. and Bond, W. J. (1982). Convergence of myrmecochory in mediterranean Australia and South Africa. In *Ant–plant interactions in Australia*, (ed. R. C. Buckley), pp. 89–98. Dr W. Junk, The Hague.

Morton, S. R. (1979). Diversity of desert-dwelling mammals: a comparison of Australia and North America. *Journal of Mammalogy*, **60**, 253–64.

Morton, S. R. (1985). Granivory in arid regions: comparison of Australia with North and South America. *Ecology*, **66**, 1859–66.

Morton, S. R. and Andrew, M. H. (1987). Ecological impact and management of fire in northern Australia. *Search*, **18**, 77–82.

Morton, S. R. and Davidson, D. W. (1988). Comparative structure of harvester ant communities in arid Australia and North America. *Ecological Monographs*, **58**, 19–38.

Morton, S. R. and Davies, P. H. (1983). Food of the zebra finch (*Poephila guttata*), and an examination of granivory in birds of the Australian arid zone. *Australian Journal of Ecology*, **8**, 235–43.

Mossop, M. K. (1989). Comparison of seed removal by ants in vegetation on fertile and infertile soils. *Australian Journal of Ecology*, **14**, 367–73.

Mott, J. J. and McKeon, G. M. (1977). A note on the selection of seed types by harvester ants in northern Australia. *Australian Journal of Ecology*, **2**, 231–5.

Murray, K. G. (1986*a*). Consequences of seed dispersal for gap-dependent plants: relationships between seed shadows, germination requirements, and forest dynamic processes. In *Frugivores and seed dispersal*, (ed. A. E. Estrada and T. H. Fleming). pp. 187–98. Dr. W. Junk, Dordrecht.

Murray, K. G. (1986*b*). Avian seed dispersal of neotropical gap-dependent plants. Dissertation. University of Florida, Gainesville.

Musil, C. (1990). An analysis of post-fire regeneration in a sand-plain lowland fynbos community. *South African Journal of Botany*, (in press).

Nesom, G. L. and La Duke, J. C. (1985). Biology of *Trillium nivale* (Liliaceae). *Canadian Journal of Botany*, **63**, 7–14.

Nilsson, L. A. (1978). Pollination ecology of *Epipactis palustris* (Orchidaceae). *Botanischer Notiser*, **131**, 355–68.

Nordhhagen, R. (1932). Zür Morphologie und Verbreitungsbiologie der Gattung *Roscoea. Sm. Bergen Mus. Arbok Naturv*, **5**, 5–57.

O'Daniel, D. (1987). Seed dispersal and seed predation in two species of *Calathea* in a Costa Rican rainforest. Master's Dissertation. University of Texas at Austin.

O'Dowd, D. J. and Gill, A. M. (1984). Predator satiation and site alteration: mass reproduction of alpine ash (*Eucalyptus delegatensis*) following fire in southeastern Australia. *Ecology*, **65**, 1052–66.

O'Dowd, D. J. and Gill, A. M. (1986). Seed dispersal syndromes in Australian Acacia. In *Seed dispersal*, (ed. D. R. Murray), pp. 87–121. Academic Press.

O'Dowd, D. J. and Hay, M. E. (1980). Mutualism between harvester ants and a desert ephemeral: seed escape from rodents. *Ecology*, **61**, 531–40.

Ohara, M. and Higashi, S. (1987). Interference by ground beetles with the dispersal by ants of seeds of *Trillium* spp. (Liliaceae). *Journal of Ecology*, **75**, 1091–8.

Ohara, M. and Kawano, S. (1986). Life history studies on the genus *Trillium* (Liliaceae). IV. Population structures and spatial distribution of four Japanese species. *Plant Species Biology*, **1**, 147–61.

Olivieri, I. and Gouyon, P. H. (1985). Seed dimorphism and dispersal: theory and implications. In *Structure and function in plant populations*, (ed. J. Haeck and J. W. Woldendorp). North Holland Publishing Company, Amsterdam.

Panetta, F. D. (1988). Factors determining seed persistence of *Chondrilla juncea* L. (skeleton weed) in southern Western Australia. *Australian Journal of Ecology*, **13**, 211–24.

Pant, D. D. Nautiyal, D. D. and Chaturvedi, S. K. (1982). Pollination ecology of some Indian ascelepiads. *Phytomorphology*, **32**, 302–13.

Patel, J. S. (1938). *The coconut, a monograph*. Government Press, Madras.

Peakall, R. (1989). The unique pollination of *Leporella fimbriata* (Orchidaceae) by pseudocopulating winged male ants *Myrmecia urens* (Formicidae). *Plant Systematics and Evolution*, **167**, 137–48.

Peakall, R. and Beattie, A. J. (1989). Pollination of the orchid *Microtis parviflora* by flightless worker ants. *Functional Ecology*, **3**, 515–22.

Peakall, R. and James, S. H. (1989). Outcrossing in an ant pollinated clonal orchid. *Heredity*, **62**, 161–7.

Peakall, R., Beattie, A. J., and James, S. H. (1987). Pseudocopulation of an orchid by male ants: a test of two hypotheses accounting for the rarity of ant pollination. *Oecologia*, **73**, 552–4.

Peakall, R., Angus, C. J., and Beattie, A. J. (1990). The significance of ant and plant traits for ant pollination in *Leporella fimbriata*. *Oecologia*, **84**, 457–60.

Percival, M. (1974). Floral ecology of coastal scrub in Southeast Jamaica. *Biotropica*, **6**, 104–29.

Petersen, B. (1977a). Pollination by ants in the alpine tundra of Colarado, USA. *Transactions Illinois State Academy of Science*, **70**, 349–55.

Petersen, B. (1977b). Pollination of *Thlaspi alpestre* by selfing and by insects in the alpine zone of Colorado. *Arctic and Alpine Research*, **9**, 211–15.

Pierce, S. M. (1984). *A synthesis of plant phenology in the fynbos biome*. South African National Scientific Programmes Report No. 88, 57 pp. CSIR, Pretoria.

Pierce, S. M. (1987). Dynamics of soil-stored seed banks in relation to disturbance. In *Disturbance and the dynamics of fynbos biome communities*, (ed. R. M. Cowling, D. C. Le Maitre, B. McKenzie, R. P. Prys Jones, and B. W. van Wilgen, pp. 46–55. South African National Scientific Programmes Report No. 135. CSIR, Pretoria.

Platt, W. J. (1975). The colonization and formation of equilibrium plant-associations on badger disturbances in a tall-grass prairie. *Ecological Monographs*, **45**, 285–305.

Platt, W. J. and Hermann, S. M. (1986). Relationship between dispersal syndrome and characteristics of populations of trees in a mixed-species forest. In *Frugivores and seed dispersal*, (ed. A. E. Estrada and T. H. Fleming), pp. 309–22. Dr. W. Junk, The Netherlands.

Posnette, A. F. (1942). Natural pollination of Cocoa. *Theobroma leiocarpa*, on the Gold Coast. *Tropical Agriculture*, **19**, 12–16.

Press, A. J. (1987). Fire management in Kakdu National Park: the ecological basis for the active use of fire. *Search*, **18**, 244–8.

Proctor, M. and Yeo, P. (1973). *The pollination of flowers*. William Collins, Glasgow.

Reichman, O. J. (1979). Desert granivore foraging and its impact on seed densities and distributions. *Ecology*, **60**, 1085–92.

Rice, B. L. and Westoby, M. (1981). Myrmecochory in sclerophyll vegetation of the West Head, NSW. *Australian Journal of Ecology*, **6**, 291–8.

Rice, B. L. and Westoby, M. (1983). Species richness at tenth-hectare scale in Australian vegetation compared to other continents. *Vegetatio*, **52**, 129–40.

Rice, B. L. and Westoby, M. (1986). Evidence against the hypothesis that ant-dispersed seeds reach nutrient-enriched microsites. *Ecology*, **67**, 1270–4.

Richardson, D. M. and van Wilgen, B. W. (1986). The effects of fire on felled *Hakea sericea* and natural fynbos and implications for weed control in mountain catchments. *South African Forestry Journal*, **139**, 4–14.

Rico-Gray, V. (1980). Ants and tropical flowers. *Biotropica*, **12**, 223–4.

Risch, S. J. and Caroll, C. R. (1986). Effects of seed predaton by a tropical ant on competition among weeds. *Ecology*, **67**, 1319–27.

Roberts, J. T. and Heithaus, E. R. (1986). Ants rearrange the vertebrate-generated seed shadow of a neotropical fig tree. *Ecology*, **67**, 1046–51.

Rourke, J. P. (1972). Taxonomic studies on *Leucospermum* R.Br. *Journal of South African Botany*, **8** (Suppl.)

Russell, M. J. Coaldrake, J. E., and Sanders, A. M. (1967). Comparative effectiveness of some insecticides, repellents and seed-pelleting devices in the prevention of ant removal of pasture seeds. *Tropical Grasslands*, **1**, 153–66.

Salisbury, E. J. (1942). *The reproductive capacity of plants*. Bell, London.

Salisbury, E. J. (1974). Seed size, and mass in relation to environment. *Proceedings of the Royal Society, Series B*, **186**, 83–8.

Schemske, D. W. and Brokaw, N. V. L. (1981). Treefalls and the distribution of understorey birds in a tropical forest. *Ecology*, **62**, 938–45.

Schubart, H. O. R. and Anderson, A. B. (1978). Why don't ants visit flowers? A reply to D. H. Janzen. *Biotropica*, **10**, 310–11.

Shmida, A. and Ellner, S. (1984). Coexistence of plant species with similar niches. *Vegetatio*, **58**, 29–55.

Siegfried, W. R. (1983). Trophic structure of some communities of fynbos birds. *Journal of South African Botany*, **49**, 1–43.

Slingsby, P. and Bond, W. J. (1981). Ants, friends of the fynbos. *Veld and Flora*, **67**, 39–45.

Slingsby, P. and Bond, W. J. (1985). The influence of ants on the dispersal distance and recruitment of *Leucospermum conocarpodendron* (L.) Buek (Proteaceae). *South African Journal of Botany*, **51**, 30–4.

Smallwood, J. (1982). The effect of shade and competition on emigration rate in the ant *Aphaenogaster rudis*. *Ecology*, **63**, 124–34.

Smith, A. P. (1987). Respuestas de hierbas del sotobosque tropical a claros ocasionados por la caida de arboles. *Revista de Biologia Tropical* **1**, (Suppl.) 111–18.

Smith, J. H. and Atherton, D. O. (1944). Seed-harvesting and other ants in the tobacco-growing districts of North Queensland. *Queensland Journal of Agricultural Science*, **1**, 33–61.

Sokal, R. R. and Rohlf, F. J. (1981). *Biometry*, 2nd edn. W. H. Freeman and Company, New York.

Specht, R. L. (1979). Heathland and related shrublands of the world. In *Ecosystems of the world. Heathlands and related shrublands*. (ed. R. L. Specht), Vol. 9. Elsevier, Amsterdam.

Steyn, J. J. (1954). The pugnacious ant (*Anoplolepis custodiens*) and its relation to the control of citrus scales at Letaba. *Memoirs of the Entomological Society of Southern Africa*, **3**, 1–96.

Stiles, E. W. (1982). Fruit flags: two hypotheses. *American Naturalist*, **120**, 500–9.

Svensson, K. (1985). An estimate of pollen carryover by ants in a natural population of *Scleranthus perennis* L. (Caryophyllaceae). *Oecologia* **56**, 373–7.

Svensson, K. (1986). Secondary pollen carryover by ants in a natural population of *Scleranthus perennis* L. (Caryophyllaceae). *Oceologia*, **70**, 631–2.

Tanner, E. V. R. (1982). Species diversity and reproductive mechanisms in Jamaican trees. *Biological Journal of the Linnean Society*, **18**, 263–78.

Thompson, J. N. (1980). Treefalls and colonization patterns of temperate forest herbs. *American Midland Naturalist*, **104**, 176–84.

Thompson, J. N. (1981). Elaiosomes and fleshy fruits: phenology and selection pressures for ant-dispersed seeds. *American Naturalist*, **117**, 104–8.

Thompson, J. N. (1982). *Interaction and coevolution*. John Wiley, New York.

Thompson, J. N. and Willson, M. F. (1978). Disturbance and dispersal of fleshy fruits. *Science*, **200**, 1161–3.

Tuffs, L. C. (1988). The importance of elaiosome size to seed dispersal in *Acacia myrtifolia*. Honours thesis, Macquarie University.

Twigg, L., Majer, J. D., and Stynes, B. A. (1982). Influence of seed-harvesting ants in annual ryegrass and their possible effects on the epidemiology of ryegrass toxity. In *Proceedings of the 3rd Australian conference on grassland invertebrate ecology*, (ed. K. E. Lee), pp. 235–43. South Australian Government Printer, Adelaide.

van der Pijl, L. (1955). Some remarks on myrmecophytes. *Phytomorphology*, **5**, 190–200.

van der Pijl, L. (1972). *Principles of dispersal in higher plants*. Springer Verlag, New York.

Vazquez-Yanes, C. and Orozco-Segovia, A. (1984). Ecophysiology of seed germination in tropical humid forests of the world: a review. In *Physiological ecology of plants in the wet tropics*. (ed. E. Medina, H. A. Mooney, and C. Vazquez-Yanes, pp. 37–50. W. Junk, The Hague.

Visser, J. (1981). *South African parasitic flowering plants*. Juta, Cape Town.

Watkinson, A. R., Lonsdale, W. M., and Andrew, M. H. (1989). Modelling the population dynamics of an annual plant: *Sorghum intrans* in the Wet-Dry tropics. *Journal of Ecology*, **17**, 162–81.

Wellington, A. B. and Noble, I. R. (1985). Seed dynamics and factors limiting recruitment of the mallee *Eucalyptus incrassata* Labill, in semi-arid, south-eastern Australia. *Journal of Ecology*, **73**, 657–66.

Werner, P. A. (1977). Ecology of plant populations in successional environments. *Systematic Botany*, **1**, 246–68.

Westoby, M. and Rice, B. (1981). A note on combining two methods of dispersal for distance. *Australian Journal of Ecology* **6**, 189–92.

Westoby, M., Cousins, J. M., and Grice, A. C. (1982*a*). Rate of decline of some soil seed populations during drought in western New South Wales. In *Ant–plant interactions in Australia*, (ed. R. C. Buckley), pp. 7–10. Junk, The Hague.

Westoby, M., Rice, B., Shelley, J. M., Haig, D., and Kohen, J. L. (1982*b*). Plants' use of ants for dispersal at West Head, New South Wales. In *Ant–plant interactions in Australia*, (ed. R. C. Buckley), pp. 75–87. Junk, The Hague.

Westoby, M., Rice, B., and Howell, J. (1990). Seed size and plant growth form as factors in dispersal spectra. *Ecology*, **71**, 1307–15.

Wheelwright, N. T. (1985). Fruit size, gape width and the diets of fruit-eating birds. *Ecology*, **66**, 808–18.

Wheelwright, N. T. and Janson, C. H. (1985). Colors of fruit displays of bird-dispersed plants in two tropical forests. *American Naturalist*, **126**, 777–99.

Wheelwright, N. T. and Orians, G. (1982). Seed dispersal by animals: contrasts with pollen dispersal, problems of terminology, and constraints on coevolution. *American Naturalist*, **119**, 402–13.

Williams, C. H. and Raupach, M. (1983). Plant nutrients in Australian soils. In *Soils: an Australian viewpoint*, pp. 777–93. CSIRO Division of Soils, CSIRO/Academic Press, Melbourne.

Willson, M. F. (1983). *Plant reproductive ecology*. Wiley, New York.

Willson, M. F. and Melampy, M. N. (1983). The effect of bi-coloured fruit displays on fruit removal by avian frugivores. *Oikos*, **41**, 27–31.

Willson, M. F., Rice, B. L., and Westoby, M. (1990). Seed dispersal spectra: a comparison of temperate plant communities. *Journal of Vegetation Science*, **1**, 547–62.

Wilson, R. G. (1956). Tyre rollers on the Darling Downs. *Queensland Agricultural Journal*, **82**, 249–53.

Winder, J. A. (1978). The role of nondipterous insects in the pollination of Cocoa in Brazil. *Bulletin of Entomological Research*, **68**, 559–74.

Witkowski, E. (1990). Response to nutrient additions by the plant growth forms of sand-plain lowland fynbos, South Africa. *Vegetatio*, (in press).

Wyatt, R. (1981). Ant-pollination of the granite outcrop endemic *Diamorpha smallii* (Crassulaceae). *American Journal of Botany*, **68**, 1212–7.

Wyatt, R. (1982). Inflorescence architecture: how flower number, arrangement, and phenology affect pollination and fruit-set. *American Journal of Botany*, **69**, 845–51.

Wyatt, R. and Stoneburner, A. (1981). Patterns of ant-mediated pollen dispersal in *Diamorpha smallii* (Crassulaceae). *Systematic Botany*, **6**, 1–7.

Yeatman, E. M. and Greenslade, P. J. M. (1980). Ants as indicators of habitat in three conservation parks in South Australia. *South Australian Naturalist*, **55**, 20–6, 30.

Yeaton, R. I. and Bond, W. J. (1990). Competition between two shrub species: dispersal differences and fire promote coexistence. *American Naturalist*, (in press).

Part 6

Ants, vegetation ecology, and the future of ant–plant research

34

The influence of mound-building ants on British lowland vegetation

Stanley R. J. Woodell and Timothy J. King

Introduction

There are forty-seven species of ant in Britain, most of which occur in the lowlands, but only nine are relatively common. The majority nest in the soil and feed mainly on honeydew and aphids. None of these ants habitually eats much vegetation or harvest many seeds. Although ants are numerous, and important consumers, obtaining over 200 KJ m^{-1} year^{-1} in British grasslands (King, 1981a), most colonies are inconspicuous, merely having an entrance hole, perhaps with parts of flowers, seeds, or fruits scattered nearby. It may be for this reason that British research on the effects of ants on plants has concentrated upon the yellow ant, *Lasius flavus* F., which constructs prominent mounds in grasslands, sand dunes, and saltmarshes, and on the red wood ant, *Formica rufa* L., a large species which builds mounds of pine needles and other elongated vegetable matter and forages along trails across the soil surface and into trees. We shall concentrate upon *Lasius flavus*, which feeds on root aphid honeydew; Whittaker deals with *Formica rufa* in Chapter 6.

There are many different ways in which ants might affect species composition and growth rates in plant communities, either directly or indirectly. Acting directly, ants may bite off leaves or roots, transport seeds to their nests, or heap soil over the surfaces of plants. Acting more subtly, ants may accumulate nutrients near the colony, which could alter the competitive balance between plant species. Alternatively, predation of aphids and other herbivores by ants may increase the growth rates of infested plants (see Chapter 6). Promotion of honeydew aphids may adversely affect plants.

The effects on plants are best understood in the case of *Lasius flavus*, whose mounds often carry a markedly different vegetation from that of the surrounding soil. The effects on plant distribution patterns have been studied quantitatively (Thomas 1962; Elton 1966; Grubb *et al.* 1969; King 1972, 1977a,b, 1981b; Woodell 1974; Kay and Woodell 1976) and experimentally (King 1972, 1976, 1977c, 1981a). The main conclusions are that *Lasius flavus* affects plant patterns by smothering plants with soil, providing permanently open 'gaps', and producing micro-topographical heterogeneity.

On the other hand, *Formica rufa* builds enormous mounds of plant material in pine forests and these mounds, like their foraging trails, tend to be bare of plants. Although there are reports that certain plant species are associated with *F. rufa* mounds (Sernander 1906; Ulbricht 1939), the ants often remove the herbaceous seedlings nearby.

The ecology of *Lasius flavus* mounds

Ant-hills built by *Lasius flavus*, and occasionally by *Lasius niger* L., occur abundantly in saltmarshes, railway embankments, and permanent grass-lands, particularly below altitudes of 250 m. It seems likely that mounds can be occupied for hundreds of years, although the longest continuous occupa-tion observed so far is 22 years (T. C. E. Wells personal communication).

The ecology of *Lasius flavus* itself is well known (Pontin 1957, 1960*a*, 1963, 1978; Haarløv 1960; Odum and Pontin 1961; Waloff and Blackith 1962). Each active yellow ant mound is the breeding centre of a colony which may contain several thousand workers (Haarlov 1960; Odum and Pontin 1961). These forage below ground along tunnels which extend up to a metre around the mound in all directions. The ants' food consists mainly of root-feeding aphids and their honeydew. Pontin (1960*b*) found that in limestone grassland in Oxfordshire, 16 species of aphid and one coccid occurred on roots in an area exploited by both *L. flavus* and *L. niger*.

Newly fertilized queens burrow into the soil and may establish new colonies. The first sign is the heaping of soil on to the surface. Usually the workers construct an earth mound, supported by grass stems, but sometimes they colonize molehills. The selective advantage of the mound to an ant colony probably arises from the increased temperatures for brood rearing. The workers move the brood around inside the ant-hill, in relation to the position of the sun. Workers select soil particles predominantly below 600 μm in diameter and as a result the soil is immature, and lacks organic matter, a crumb structure, and stones. As a mound becomes larger, it is stabilized by plants. Some of the plants buried originally grow up through the heaped soil, others invade vegetatively or by seed. Apart from erosion by rainfall and trampling, ant-hills are often excavated by badgers (*Meles meles* L.) or rabbits (*Oryctolagus cuniculus* L.), and attacked by pheasants (*Phasianus colchicus* L.), grey partridge (*Perdix perdix* L.), and green woodpeckers (*Picus viridis* Hartest), for which the ants may be an important food source (Potts 1971).

Rabbits use ant-hills preferentially as latrines (Thomas 1962; Elton 1966), urinating and defecating on the mounds. They disturb the soil, sometimes killing seedlings, and may graze ant-hill vegetation in preference to that nearby (Pigott 1955; King 1977*a,c*). In tall grasslands, rabbit grazing maintains a short vegetation on the ant-hills, allowing ant colonies to survive. In short grasslands, however, growth may be limited by shortage of water, nutrients (Wells *et al.* 1976), or even by microbial immobilization of inorganic nitrogen (Prescott 1988). Rabbit grazing in such circumstances does not seem to influence the composition of the ant-hill vegetation.

The vegetation on the mounds

Several comparisons between the vegetation on the mounds and the surrounding vegetation have shown that ant-hills frequently have a different relative abundance of plant species. The response of each plant species to mounds seems to be consistent from site to site. In general, the vegetation on the tops of occupied ant-hills in calcareous and acid grasslands resembles that of sand dunes of similar base status (King 1977*a*).

Some of the species with particularly marked responses to ant-hills in lowland Britain are listed in Table 34.1. At least 28 winter annuals, such as *Arenaria serpyllifolia* L. *sensu lato*, *Veronica arvensis* L., and *Aira praecox* L. are frequently restricted in grassland to ant-hills (King 1977*a*). Perennials which can grow up through heaped soil, such as *Thymus praecox* Opiz ssp. *arcticus* (Durrand) Jalas, *Helianthemum nummularium* (L.) Miller, and the mosses, *Polytrichum piliferum* Hedw. and *P. juniperinum* Hedw., are often much more abundant on mounds than around them. On the other hand, several rosette perennials, such as *Cirsium acaule* Scop. and *Sanguisorba minor* Scop., hardly occur on ant-hills.

The extent to which ant activity, rather than the topography or soil of the mound, produces the distinctive vegetation is shown by vegetation changes when an ant-hill is abandoned. The colony of *Lasius flavus* in a mound may disappear, sometimes because it is ousted by *L. niger* or a *Myrmica* species, or when the hill is shaded by tall grasses, shrubs, or trees. The ant-hill vegetation changes when the mound is abandoned. For instance, King (1974) and Wells *et al.* (1976) found that in calcareous grassland in Wiltshire, abandoned hills had been colonized by *Avenula pratensis* (L.) Dumort, *A. pubescens* (Hudson) Dumort, and *Bromus erectus* Hudson. They postulated that the onset of myxomatosis had led to the mounds being shaded by taller vegetation in the absence of rabbit grazing, and abandoned because they became too cool. Subsequently tall grasses seeded onto the abandoned mounds, and developed tussocks from which the normal ant-mound species were eliminated. Later in succession, ant-hills, devoid of vegetation, may persist under scrub for decades (Grubb *et al.* 1969). We know of sites where the mounds are still distinguishable 50 years after scrub invasion or tree plantation.

Detailed comparisons of the vegetation on occupied and abandoned ant-hills have not been published. In calcareous grassland, however, species such as *Hieracium pilosella* L. and *Carex flacca* Schreber, which are almost confined to the edges of active mounds, colonize the whole surface of deserted ant-hills (King 1977*b*). Woodell (1974) noted a similar invasion of

Table 34.1. Affinity of plants for ant-hills in calcareous and acidic
grassland. The scale runs from 0 (does not occur on mounds) to 100 (only
occurs on mounds). Only those plant species which show a distinct
preference are listed (King 1977*a*, 1981*a*). The affinity index is the total
percentage cover of a species on ant-hills in a large sample from several
different sites, divided by the sum of the percentage cover of the same
species on the same ant-hills and in an equal number of surrounding
vegetation quadrats. Nomenclature follows *Flora Europea*, Cambridge
University Press, 1964–1980.

Species	Family	Affinity index	Growth form
Calcareous grassland—ant-hills			
Arenaria serpyllifolia	Caryophyllaceae	100	Winter annual
Veronica arvensis	Scrophulariaceae	100	Winter annual
Cerastium fontanum	Caryophyllaceae	88	Perennial herb
Thymus praecox	Lamiaceae	87	Dwarf shrub
Bryum sp.		80	Moss
Helianthemum nummularium	Cistaceae	73	Dwarf shrub
Calcareous grassland—pasture			
Cirsium acaule	Asteraceae	1	Rosette perennial
Sanguisorba minor	Rosaceae	4	Rosette perennial
Succisa pratensis	Dipsacaceae	5	Rosette perennial
Carex flacca	Cyperaceae	6	Tufted perennial
Filipendula vulgaris	Rosaceae	11	Perennial herb
Hypochaeris radicata	Asteraceae	12	Rosette perennial
Leontodon hispidus	Asteraceae	14	Rosette perennial
Polygala vulgaris	Polygalaceae	14	Perennial herb
Hippocrepis comosa	Fabaceae	15	Perennial herb
Bromus erectus	Poaceae	16	Tufted grass
Plantago media	Plantaginaceae	17	Rosette perennial
Trifolium pratense	Fabaceae	17	Perennial herb
Cynosurus cristatus	Poaceae	18	Tufted perennial
Dactylis glomerata	Poaceae	18	Tufted perennial
Acidic grassland—ant-hills			
Polytrichum piliferum		99	Moss
Polytrichum juniperinum		98	Moss
Acidic grassland—pasture			
Nardus stricta	Poaceae	3	Rhizomatous grass
Potentilla erecta	Rosaceae	9	Perennial herb
Luzula campestris	Juncaceae	10	Tufted perennial
Anthoxanthum odoratum	Poaceae	18	Tufted perennial
Danthonia decumbens	Poaceae	18	Tufted perennial

abandoned mounds on a saltmarsh by *Plantago maritima* L. and *Limonium vulgare* Miller, both absent from occupied ant-hills.

Evidence that the vegetation on ant-hills alters significantly after abandonment was collected by King (1972) at Aston Rowant National Nature Reserve, Oxfordshire. By using the 'affinity index' for each species (Table 34.1), a mean affinity index for each ant-hill was calculated in the following way. The percentage cover of each species was multiplied by its affinity index for ant-hills and the resulting figure was added to the products for all the other species on the mound. The result was divided by the total number of species on the ant-hill. A group of 24 small and medium occupied ant-hills had a 'mean affinity index' of 57.1 (s.e. 1.2), whilst the surrounding vegetation scored 44.8 (s.e. 0.62). Small and medium abandoned ant-hills scored 52.1 (s.e. 0.69), significantly different from both groups at the 5 per cent level. This suggests both that ant-hill vegetation becomes more like the surrounding sward after being abandoned, and that it retains some ant-hill character. The legacy left by the ants, and the 'squatters' rights' which existing plants have established, may ensure that the vegetation remains distinctive.

The direct influence of ants

The workers of *Lasius flavus* heap soil on to the ant-hill surface in spring and summer, at night, after rain. Several estimates suggest that the largest, continually occupied mounds increase in net volume by about 11 yr^{-1} (King 1981*a*). This soil may kill some plants, especially rosette species susceptible to burial. In a survey of 206 mounds at Aston Rowant NNR, for example, 20 per cent of the *Carex flacca* and 9 per cent of the *Plantago lanceolata* L. rosettes were at least partly buried, as were 17 per cent of *Cirsium acaule* seedlings and 21 per cent of *Sanguisorba minor* plants. These data may well underestimate the lethal effects of burial; some plants were probably buried too deeply to be recorded, and only a small proportion of the ant-hills sampled were likely to be in a peak building year. The factors which determine the position of heaped soil on the ant-hill surface are poorly understood. The lowered soil temperatures caused by shading by plant leaves might stimulate an increased rate of soil deposition. In contrast, soil heaping favours some plant species. *Thymus praecox* ssp. *arcticus* can grow up through 2–6 cm of heaped soil a year, and experiments on *Helianthemum nummularium* show that 3–4 cm of heaped soil promotes the growth of axillary shoots which reach the soil surface; burial causes branching. The many shoots of *T. praecox* or *H. nummularium* on the surface of a mound are often parts of a single plant, the stem and branches of which provide a skeleton supporting the ant-hill structure. It seems most likely that the original plant was in the grassland when it was covered by ant activity, and it responded by growing up through the soil. On acidic soils the mosses *Polytrichum piliferum* and *P. juniperinum* branch and grow up through ant-heaped soil in a similar manner (Leach 1931). These species may all dominate ant-hill vegetation.

The bare soil which results from heaping constitutes a 'regeneration niche' (Grubb 1977) ripe for invasion by seeds which are able to reach such elevated 'gaps'. For the seeds of winter annuals, which have an after-ripening requirement and do not germinate in mid-summer, such sites could be called 'superior gaps', because they are kept open by the ants when they deposit excavated soil onto the surface each year. The heaped soil provides micro-sites which are more suitable for the germination of some species than others. For instance, sown seeds of the winter annual *Arenaria serpyllifolia* exhibited 41 per cent germination on ant-hills but only 5 per cent in the swards. For the rosette perennial *Leontodon hispidus* L., on the other hand, germination was 5 per cent and 8 per cent respectively (King 1977c; Newman 1982). If, despite the intense seedling mortality on this bare soil (King 1977c), short-lived species flower and fruit there, they may establish a seed bank and take up permanent residence. Given volumes of ant-hill soils yielded 140 times as many seedlings of *Arenaria serpyllifolia* and 8.7 times as many of *Aira praecox* as the same volumes of pasture soil (King 1972, 1981a).

Initially, young mounds seem to be devoid of seeds. Some plant species invade mounds by vigorous vegetative reproduction. Others, particularly the forbs whose rosettes are killed by heaped soil, must invade by seed. Their seeds, however, are frequently heavy and poorly dispersed and are produced at low densities, especially in grazed grassland (King 1977c). In this category are species such as *Sanguisorba minor* and *Cirsium acaule*. It seems likely that dispersal of their seeds onto the mounds is a limiting factor in recruit-ment because, for example, seeds of *S. minor* sown onto ant-hills increase its population size. The projection of the mound above the surface of the grassland exerts a major effect, by limiting invasion by seed. These species were never seen flowering on ant-hills. The factors which may influence the abundance of plant species on ant-hills (King 1977c) are summarized in Fig. 34.1. The time which elapses between seed germination and fruit production, and the ability of a plant to withstand or avoid burial by ant-heaped soil, are crucial in determining the abundance of a species on ant-hills.

Lasius flavus workers might eliminate some plants by biting their roots. Since these ants forage underground, their behaviour cannot easily be observed. Pickles (1940) found the workers chewing a rhizome of bracken, *Pteridium aquilinum* (L.) Kuhn. However, there seems to be no need to postulate a plant food source for *L. flavus*. Pontin (1978) found about 3000 aphids associated with each ant colony at Staines Moor, Surrey. More than 3000 first instars were lost from the aphid population per ant nest per day; these were probably eaten by the ants, in addition to some older aphids and the honeydew produced. The energy flow through this ant population was approximately $310 \text{ kJ m}^{-2} \text{ yr}^{-1}$, 68 per cent from aphid honeydew and the rest from live aphids. This is comparable to the energy flow through other ant

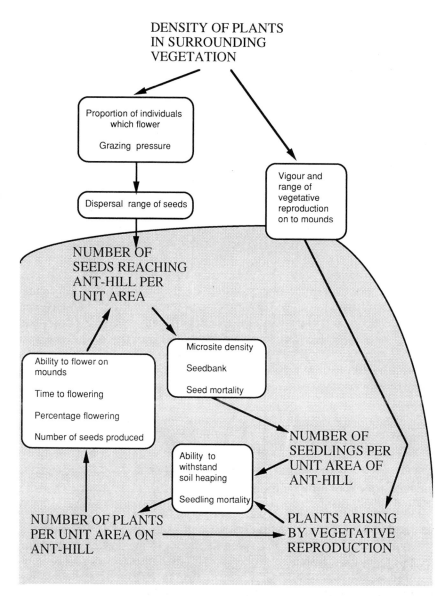

Fig. 34.1. Stages in the life cycle of an angiosperm showing (in white rectangles) the main factors which influence the population size of plant species on ant mounds.

species in grasslands. For example, energy flow through *Pogonomyrmex badius* Latreille in old fields in America was 50–201 kJ m^{-2} yr^{-1} (Golley and Gentry 1964) and through *Lasius alienus* Forst. in a corynephoretum in Denmark 272 kJ m^{-2} yr^{-1} (Neilson 1972). Thus *Lasius flavus* need not bite roots for food. Although the aphids may reduce plant growth away from the mounds, it seems unlikely that they affect the relative abundance of species.

Seed harvesting of a type frequent in deserts and dry grasslands of tropical, subtropical, and warm temperate areas (Beattie 1985; Andersen Chapter 33, this volume) is not so common in Europe, but there are several reports of significant percentages of a local flora being ant-dispersed (Sernander 1906). Among the better known British herbs with ant-dispersed seeds are *Viola* spp. (Beattie and Lyons 1975; Culver and Beattie 1980), *Primula vulgaris* Hudson (Valentine 1948), *Veronica hederifolia* L., *Lamium purpureum* L., and *Luzula* spp. There are no reports that *L. flavus* carries seeds, and seeds placed into their burrows are ignored (King 1977*a*). It is possible that small seeds might be carried to the tops of mounds instead of soil particles; but the seeds of most grasses and rosette perennials are too large to be carried by workers.

Kay and Woodell (1976) suggested that ant-hills in saltmarshes might have acted as refuges for some plants during the post-glacial forest maximum. Presumably *Lasius flavus* was abundant in unforested coastal habitats such as saltmarshes, and especially sand dunes. Recent evidence (Bush and Flenley 1987) suggests that limestone grassland survived continuously in Yorkshire from post-glacial times and there is no reason why ants, and their associated short-lived plant species, should not have persisted in such places. Early forest clearance would have created grassland and provided suitable conditions for ant colonization. In addition, natural gaps in the forest cover may have been frequent and persistent. Ants may have colonized such gaps for long enough for seed banks of ant-hill tolerant species to have been established.

Indirect effects of ants on mound vegetation

Ant-hill soils differ in nutrient content from surrounding soils and are more susceptible to drought, at least on the south sides. In fact, the markedly different micro-climates on the different aspects of the ant-hills determine the asymmetrical distribution of plant species over their surfaces.

Ant-hill soils tend to have a much higher content of exchangeable potassium ions than their surroundings (Thomas 1962; Lambley 1967; Czerwinski *et al*. 1971; Jakubczyk *et al*. 1972; Wells *et al*.1976; King 1977*a*; Wells 1986). This is the case even where rabbits are uncommon. The reason for this is probably the high potassium content of phloem sap (approx. 950 mg l^{-1} (Bidwell 1974)) which, as aphid honeydew, is the main component of an ant's diet. At some sites, however, there will be a major additional input from rabbit urine. The nutrient requirements of plant species are not well enough known to be able to judge whether soil potassium levels influence plant distribution and/or longevity. Culver and Beattie (1983) similarly found that mounds of *Formica canadensis* were higher in potassium and phosphorus than surrounding soils. They also concluded that such differences in soil nutrients were unlikely to be responsible for the differential distribution of plants on and off the mounds.

In former times ant-hill soils were sometimes spread over pastures by farmers. They may have been considered a nuisance in hay meadows; several nineteenth century farming books recommend their destruction by cutting them up and spreading or even burning them. There may also have been a belief that they had a fertilizing effect. Certainly, the lush green growth which appears first on ant-hills in the spring suggests that the dynamics of ammonification and mineralization of nitrogenous compounds would repay further investigation. Even though mound soils may have a greater nutrient content per unit mass, they would not necessarily contain more nutrients per unit volume, because ant-hill soils have a significantly ($P\langle0.05$) lower bulk density than surrounding soils (King 1977*a*). On the other hand, the ants would accumulate nutrients from the surrounding pasture in their nests, if dead bodies and egesta were concentrated there.

Compared with surrounding soils, ant-hills often have significantly higher pH and cation contents, which suggests that ant activities counteract leaching (King 1981*a*). Soils near the mounds may exhibit various degrees of podsolization, with some bleached horizons and iron pan formation, but the soils beneath occupied ant-mounds show no evidence of this (King 1981*a*).

Soils of ant-hills are significantly ($P\langle0.001$) drier than those beneath the surrounding grassland, at least in the upper layers (King 1972). Persistent droughts frequently kill the vegetation on mounds, long before the vegetation around suffers damage. The effect of this on plant distribution patterns has not been investigated, but droughting may contribute to the slow growth rates, and hence burial, of many rosette plants on ant-hills (King 1977*c*). The dryness of ant-hill soils may also result in relative advantage to winter annuals which survive the drought period as seed.

The creation of a mound, however small, will produce micro-climatic differences akin to those which result from larger topographic heterogeneity (Perring 1959). King (1977*a*) found the evaporation rate on the south side of a mound to be 2.7 times that from the north side, on a summer's day. The soil water content in the top 50 mm was significantly lower ($P < 0.01$) on the south side of ant-hills than on the north (King 1972). S. R. J. Woodell (unpublished work.) found that on a sunny January day the soil on the south side of ant-hills could reach 15 °C at 10 mm depth, while the north side was still frozen solid. Plant species occur differentially on the northern or southern aspects. These patterns may also reflect the frequency of heaped soil (Fig. 34.2 (a)), and perhaps the competition from

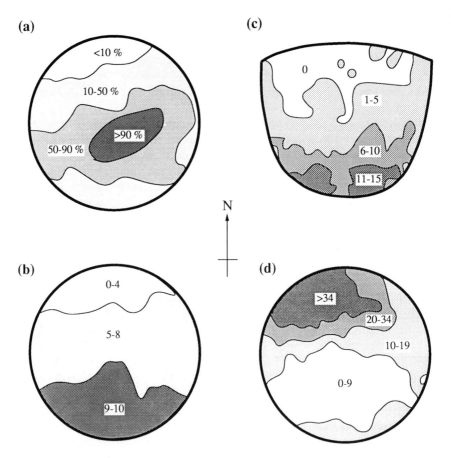

Fig. 34.2. Percentage frequency of (a) soil heaping; (b) the dwarf shrub *Helianthemum nummularium*; (c) *Thymus praecox* spp. *arcticus*; and (d) *Pseudoscleropodium purum* over the surfaces of *Lasius flavus* mounds on chalk. (a) and (b) on ten randomly selected mounds in Oxfordshire, 1971; (c) on 20 randomly selected mounds in Wiltshire, 1969; (d) on 250 systematically selected mounds in Oxfordshire, 1969. (From King 1972.)

grasses and bryophytes in the moister conditions on the north side. Plants predominantly on southerly slopes include *Thymus praecox* ssp. *articus*, *Helianthemum nummularium*, *Frankenia laevis* L., and the moss *Polytrichum piliferum* (Fig. 34.2 (b), (c)). Proctor (1956) commented that *H. nummularium* exhibits a preference for south to south-west facing slopes in Britain. A more extreme example is *Frankenia laevis*, a Mediterranean plant which reaches the northern limits of its range on the east coast of Britain. Woodell (1974) showed that it occurs preferentially on the southern sides of ant-hills at Scolt Head Island, Norfolk, and on the sheltered southern sides of bushes of *Suaeda vera* J. R. Gmelin.

The one ant-hill bryophyte with a southern preference is the endohydric acrocarp *Polytrichum piliferum* (King 1981a). *Polytrichum juniperinum* occurs on the same ant-hills as *P. piliferum*, in the Gower peninsula, South Wales, but nearer the summits. The latter species prefers drier and more unstable micro-habitats. Penny and Bayfield (1982) dried both species for nine days. Only 13 per cent of *P. piliferum* apices died and photosynthesis rapidly resumed on wetting; 93 per cent of *P. juniperinum* apices died and photosynthesis was still depressed 28 days later.

Species more frequent on northerly aspects include the mosses *Pleurozium schreberi* (Brid.) Mitt. and *Pseudoscleropodium purum* (Hedw.) Fleisch (Fig. 34.2 (d)). The distribution of these ectohydric, pleurocarpous mosses has been a useful aid to teaching the effects of micro-topography on plant distribution (King and Woodell 1973). It is likely that these mosses are limited by desiccation. King (1972) rotated ant-hills through 180° and, after a week in which the maximum air temperature exceeded 21 °C each day, counted the leaf cells which replasmolysed in the shoot tips of *P. purum*. He found a significantly higher percentage ($P < 0.001$) of dead cells in the moss growing on the south side than in that on the north side.

Modification of the physical soil structure by ants

In some grasslands, both acidic and calcareous, the frequency of ant-hills can be so great that a substantial percentage of the sward is composed of mounds. Brian (1977) reported that the mounds of *Lasius flavus* in grassland on sandy gravel in Berkshire covered about 10 per cent of the total area. In an acidic grassland in the Gower peninsula, South Wales, we observed one area of 1000 m² which contained over 900 mounds. If the largest mounds increase in volume by about 1 l yr⁻¹, probably about 0.5 l m⁻² yr⁻¹ of soil is deposited by the ants at this site, equivalent to 0.5 mm yr⁻¹ over the whole area. Darwin (1881) calculated from observations in several different grasslands that earthworms accumulated 2.1–5.6 mm yr⁻¹ on the soil surface. On this basis worms in worm-rich grassland may deposit soil only four to eleven times faster than ants. Significantly, Darwin commented, 'Nor must we overlook other agencies which in all ordinary cases add to the amount of mould, and which would not be included in the castings that were collected, namely the fine earth brought up to the surface by burrowing larvae and insects, *especially by ants*.' (our italics). In acidic grasslands, containing few or no worms (Edwards and Lofty 1977), ants may be far more important than earthworms as soil movers. Brian (1983) stated that in some soils clay-rich minerals are taken selectively from the subsoil and produce steeper and firmer mounds than could be constructed with sandy material. For *Lasius flavus* in the first two or three years of an ant-hill, the growth rate is high and the ants may excavate a vast system of galleries and chambers down to 2 m on deep, well-drained soils. Rarely, the upper layers of the soil seem to have been created by ants. At Roche Court Down on the Porton ranges, Wiltshire (Wells *et al*. 1976, transect 13), the top 10 cm of soil beneath *Bromus erectus* grassland appears to be ant-created (King 1974). The site was last ploughed in about 1884 suggesting a deposition rate of 1 mm yr^{-1}. The upper soil on the site resembled that of an ant-hill: it was porous, friable, loose, fluffy, and immature, with a low humus content, few stones, and very small structural aggregates. In contrast, the soil of abandoned ant-hills in the area, beneath *B. erectus*, frequently had an A humus horizon 2–3 cm deep, with structural aggregates 2–5 mm across, stained with humus beneath. A feature of the site is the high density (1.7 m^{-2}) of small (< 6 l) active and abandoned mounds beneath the *Bromus erectus*. The colonies in these mounds are probably shaded out within a year or two, and the mounds are eroded down to the level of the surrounding soil by heavy rain, rabbits, or voles.

The rapidity with which ants can build such temporary mounds is obvious on lawns, where *Lasius niger* is the species most often involved. Several other species move quantities of soil (Brian 1977). For instance, *Tetramorium caespitum* L. makes a large system of shafts and chambers which may extend more than a metre deep in heathland. Here the soil cast out is quickly blown away. Brian found that *Formica fusca* L. can spread about 5 kg of subsoil on a square metre of soil per year. He suggested that beneficial effects of ants on soil structure may be important in gardens and in hedgerows which are often the last refuge for ants in areas of arable farming.

Conclusion

Many of the papers in this volume describe interactions between ants and plants in which both partners are morphologically, behaviourally, and/or biochemically specialized to participate in the interaction. The examples described here, however, illustrate the subtlety and complexity of inter-actions between unspecialized, temperate species of ants and plants. The impact of ants and aphids on plant distribution and growth rate, is probably greater than plant ecologists and soil scientists have hitherto realized. The direct effects of ants on soil profile, structure, and homogeneity may be very important in some habitats.

The presence of ants enhances the floral and faunal diversity of British lowland plant communities. The mounds built by *Lasius flavus* maintain a range of short-lived plant species in the community which would otherwise probably become extinct. Ant-hills should therefore be taken into account in conservation strategy. Areas with ant-hills, other things being equal, should be selected as reserves in preference to areas without them, and the com-munities should be managed with the long-term persistence of ant-hills in mind. *Formica rufa* has much more precise habitat requirements than *Lasius flavus*, and is declining (Barrett 1979). Reserves which contain this species must also maintain a range of mature trees on which the ants can forage. In view of the ecological importance and intrinsic interest of both species, experimental introductions into new sites might be considered. Such introductions would enable us to develop techniques of preserving these species and also allow the ecological impacts of the ant species to be assessed more precisely.

In recent years, effort has been devoted to the creation of species-rich grasslands, using seed from existing meadows. The invertebrate fauna of such grasslands is depauperate compared with real ancient grassland (J. Thomas, personal communication), and the flora lacks many of the species favoured by the presence of ant-hills. Those concerned with attempts to recreate species-rich grasslands could consider introducing *Lasius flavus*. In fact, it may colonize quite rapidly; T. C. E. Wells (personal communication) found *L. flavus* building mounds in a newly created grassland in his garden, after two years. A mobile species like this may indeed rapidly invade new meadows. There is much to be said for artificially introducing the ant early, and as soon as mounds are obvious, introducing seed of some of the characteristic plants, which are unlikely to colonize unaided. The resulting communities will be more interesting and diverse, and certain declining plants will be given a greater chance of survival.

References

Barrett, K. J. (1979). *Provisional atlas of the insects of the British Isles. 5. Hymenoptera Formicidae, Ants*, (2nd edn). NERC, ITE, Monks Wood.

Beattie, A. J. and Lyons, N. (1975). Seed dispersal in *Viola*: adaptations and strategies. *American Journal of Botany*, **62**, 714–22.

Beattie, A. J. (1985). The Evolutionary Ecology of ant–plant mutualisms. Cambridge University Press, Cambridge.

Bidwell, R. G. S. (1974). *Plant Physiology*. Macmillan, New York.

Brian, M. V. (1977). *Ants*. Collins, London.

Brian, M. V. (1983). *Social insects, ecology and behavioural biology*. Chapman and Hall, London.

Bush, M. B. and Flenley, J. R. (1987). The age of the British chalk grassland. *Nature*, **329**, 434–8.

Culver, D. C. and Beattie, A. J. (1980). The fate of *Viola* seeds dispersed by ants. *American Journal of Botany*, **67**, 710–14.

Culver, D. C. and Beattie, A. J. (1983). Effects of ant mounds on soil chemistry and vegetation patterns in a Colorado montane meadow. *Ecology*, **64**, 458–92.

Czerwinski, Z., Jakubczyk, H., and Petal, J. (1971). Influence of the ant-hills on the meadow soils. *Pedobiologia*, **11**, 227–85.

Darwin, C. (1881). *The formation of vegetable mould by the action of worms*. John Murray, London.

Edwards, C. A. and Lofty, J. R. (1977). *Biology of earthworms*, (2nd edn). Chapman and Hall, London.

Elton, C. S. (1966). *The pattern of animal communities*. Methuen, London.

Gaspar, C. (1972). Actions des fourmis du genre *Lasius* dans l'ecosysteme prairie. *Ecologia Polslka*, **20**, 145–52.

Golley, F. B. and Gentry, J. B. (1964). Bioenergetics of the southern harvester ant, *Pogonomyrmex badius*. *Ecology*, **45**, 217–25.

Grubb, P. J. (1977). The maintenance of species-richness in plant communities: the importance of the regeneration niche. *Biological reviews*, **52**, 107–45.

Grubb, P. J., Green, H. E., and Merrifield, R. C. J. (1969). The ecology of chalk heath: its relevance to the calcicole-calcifuge and soil acidification problems. *Journal of Ecology*, **57**, 175–212.

Haarløv, N. (1960). Microarthropods from Danish soils. *Oikos Supplementum*, **3**, 1–176.

Jakubczyk, H., Czerwinski, Z., and Petal, J. (1972). Ants as agents of the soil habitat changes. *Ecologia Polska*, **20**, 153–61.

Kay, Q. O. N. and Woodell, S. R. J. (1976). The vegetation of ant-hills in West Glamorgan saltmarshes. *Nature in Wales*, **15**, 81–7.

King, T. J. (1972). *The plant ecology of ant-hills in grasslands*. D. Phil. thesis, University of Oxford.

King, T. J. (1974). *Notes on Porton Down ant-hills*. Unpublished report, ITE, Monk's Wood.

King, T. J. (1976). The viable seed contents of ant-hill and pasture soil. *New Phytologist*, **77**, 143–7.

King, T. J. (1977*a*). The plant ecology of ant-hills in calcareous grasslands. I. Patterns of species in relation to ant-hills in southern England. *Journal of Ecology*, **65**, 235–56.

King, T. J. (1977*b*). The plant ecology of ant-hills in calcareous grasslands. II. Succession on the mounds. *Journal of Ecology*, **65**, 257–78.

King, T. J. (1977*c*). The plant ecology of ant-hills in calcareous grasslands. III. Factors affecting the population sizes of selected species. *Journal of Ecology*, **65**, 279–315.

King, T. J. (1981*a*). Ant-hill vegetation in acidic grasslands in the Gower peninsula, South Wales. *New Phytologist*, **88**, 559–71.

King, T. J. (1981*b*). Ant-hills and grassland history. *Journal of Biogeography*, **8**, 329–334.

King, T. J. and Woodell, S. R. J. (1973). The use of the mounds of *Lasius flavus* in teaching some principles of ecological investigation. *Journal of Biological Education*, **9**, 109–13.

Lambley, P. W. (1967). *Special investigation into ant-hills*. B.Sc. thesis, University of Leicester.

Leach, W. (1931). The importance of some mosses as pioneers on unstable soils. *Journal of Ecology*, **19**, 98–102.

Neilson, M. G. (1972). An attempt to estimate energy flow through a population of workers of *Lasius alienus* (Forst) (Hymenoptera, Formicidae). *Natura Jutlandica*, **16**, 99–107.

Newman, E. I. (1982). Niche separation and species diversity in terrestrial vegetation. In *The plant community as a working mechanism*, (ed. E. I. Newman), pp. 61–77. British Ecological Society Special Publication Series 1.

Odum, E. P. and Pontin, A. J. (1961). Population density of the underground ant, *Lasius flavus*, as determined by tagging with P32. *Ecology*, **42**, 186–8.

Penny, M. G., and Bayfield, N. G. (1982). Photosynthesis in desiccated shoots of *Polytrichum*. *New Phytologist*, **91**, 637–45.

Perring, F. H. (1959). Topographical gradients of chalk grassland. *Journal of Ecology*, **47**, 447–81.

Pickles, W. (1940). Fluctuations in the populations, weights and biomasses of ants at Thornhill, Yorkshire from 1935–9, *Transactions of the Royal Entomological Society of London*, **90**, 467–85.

Pigott, C. D. (1955). Biological Flora of the British Isles: *Thymus* L. *Journal of Ecology*, **43**, 365–87.

Pontin, A. J. (1957). An investigation of the interaction between *Lasius flavus* (F.) and *L. niger* (L.). *Proceedings of the Royal Entomological Society pf London (C)*, **22**, 49.

Pontin, A. J. (1960a). Observations on the keeping of aphid eggs by ants of the genus *Lasius* (Hym., Formicidae). *Entomologists' Monthly Magazine*, **96**, 198–9.

Pontin, A. J. (1960b). Field experiments on colony foundation by *Lasius niger* (L.) and *Lasius flavus* (F.) (Hym., Formicidae). *Insectes Sociaux*, **7**, 227–30.

Pontin, A. J. (1963). Further considerations of competition and the ecology of the ants *Lasius flavus* (F.) and *L. niger* (L.). *Journal of Animal Ecology*, **32**, 565–74.

Pontin, A. J. (1978). The numbers and distribution of subterranean aphids and their exploitation by the ant *Lasius flavus* (Fabr.). *Ecological Entomology*, **3**, 203–7.

Potts, G. R. (1971). Recent changes in the farmland fauna with special reference to the decline in the grey partridge. *Bird Study*, **17**, 145–66.

Prescott, C. V. (1988). *Factors responsible for the maintenance of the chalk grassland plagioclimax at Shorehill Down, Kemsing, Kent.* CNAA Ph.D. thesis, Thames Polytechnic.

Proctor, M. C. F. (1956). Biological flora of the British Isles: *Helianthemum* Mill. *Journal of Ecology* **44**, 675–92.

Sernander, R. (1906). Entwurf einer Monographie der europaischen Myrmekochoren. *Kunglica Svenska Vetenskapakademien Handlingar*, **41 (7)**, 1–410.

Thomas, A. S. (1962). Ant-hills and termite mounds in pastures. *Journal of the British Grassland Society*, **17**, 103–8.

Ulbricht, E. (1939). Deutsche Myrmekochoren. *Repertorium novarum specierum regni vegetabilis, Berlin*, **67**, 1–56.

Valentine, D. H. (1948). Studies in British Primulas. II. Ecology and Taxonomy of Primrose and Oxlip (*Primula vulgaris* Huds. and *P. elatior* Schreb.) *New Phytologist*, **47**, 113–30.

Waloff, N. and Blackith, R. E. (1962). The growth and distribution of the mounds of *Lasius flavus* (Fabricus) (Hym., Formicidae) in Silwood Park, Berkshire. *Journal of Animal Ecology*, **31**, 421–37.

Wells, S. (1986). *Ant-hill vegetation at Upwood Meadows*. B.Sc. thesis, University College, London.

Wells, T. C. E., Sheail, J., Ball, D.F., and Ward, L. K. (1976). Ecological studies on the Porton Ranges: relationships between vegetation, soils and land-use history. *Journal of Ecology*, **64**, 589–626.

Woodell, S. R. J. (1974). Ant-hill vegetation in a Norfolk salt marsh. *Oecologia (Berlin)*, **16**, 221–5.

35

A neotropical rainforest canopy, ant community: some ecological considerations

John E. Tobin

Introduction

A Smithsonian project to characterize the entire arthropod fauna of a small patch of tropical rainforest canopy from Manu National Park, in Amazonian Peru, is currently underway. Though several studies have addressed the questions of species richness, biomass, and guild composition (e.g. Moran and Southwood 1982; Southwood *et al.* 1982; Adis *et al.* 1984), the Smithsonian project is an attempt to determine the number of species as well as the abundance of each species. This chapter presents a preliminary description and analysis of the ant fauna collected as part of this project.

Materials and methods

Two adjacent canopy trees, *Matisia cordata* (Bombacaceae) and *Hirtella triandra* (Chrysobalanaceae), and eleven associated vines were treated with insecticidal fog (Erwin 1983). These trees formed a small patch of canopy that was incontiguous with any nearby trees. An area of 93.6m² (1008 square feet) of forest floor was covered with plastic sheets to collect the specimens as they fell from the study trees. The specimens were then placed in containers with alcohol and shipped to the lab. At the present, just over half of the material has been processed and sorted to order or family. Specialists in various arthropod groups are now in the process of identifying specimens to the specific level.

Results and discussion

Of the 28 279 specimens examined so far, 69.7 per cent (19 702 specimens) are ants. The rest of the arthropods are primarily beetles (9.2 per cent), psocopterans (4.0 per cent), dipterans (2.5 per cent), collembolans (2.2 per cent), and spiders (2.0 per cent). Twenty-four orders of arthropods are represented in the sample (Erwin 1989).

So far I have identified 52 species belonging to 28 genera and 5 subfamilies of ants. The dry weight of the six most abundant species has been determined: *Monacis bispinosa*, 21.24 g; *Dolichoderus* sp.1, 6.76 g; *Paraponera clavata*,

1.02 g; *Azteca* sp. 1, 0.97 g; *Azteca* sp. 2, 0.11 g; and *Camponotus* sp. 1, 0.08 g. The biomass of each of the other species amounts to less than 0.05 g. It is worth noting that four of these species, accounting for over 90 per cent of the biomass, belong to the subfamily Dolichoderinae. The importance of dolichoderines in the rainforest canopy, a fact also noted by Wilson (1987), stands in contrast to their relative scarcity in most other major habitat types. I shall argue that diet may partly explain their unusually large biomass in the rainforest canopy.

Biomass data for arthropods other than ants are not yet available, but the ants are expected to contribute more than half of the total dry weight (T. L. Erwin, personal communication). Other studies show a comparably high frequency of ants (e.g. Adis *et al.* 1984). This appears to run counter to the common view that ants are generalized predators. For example, Sudd and Franks (1987) stated that 'ants are in general carnivores'. If animal prey were the ants' principal source of energy, one might predict that they should be dramatically less important (in terms of biomass) than they appear to be in the canopy. However, the large biomass of canopy ants suggests the possibility that they obtain most of their energy close to the base of the trophic pyramid, from nectar or other plant sources. The dominant canopy ant species may effectively function as primary consumers, only complementing their diets with animal protein, rather than the other way round. This points to a more intimate association between numerous arboreal ants and their host trees than has been previously recognized.

Four different explanations for the ants' unusually large biomass were considered:

1. there is an extremely high turnover rate at the level of the ants' prey;
2. there is a large influx of ant biomass into the canopy from outside;
3. the fogging method is grossly underestimating the herbivore biomass, on which the ants are feeding; and
4. the ants (at least the most abundant species, such as the dolichoderines) may be obtaining a large proportion of their energy from plant matter, such as nectar and so effectively functioning as primary consumers.

The last explanation appears, at this stage, to be the most consistent with the available information. Dietary studies of the dominant canopy ant species will be conducted to test this hypothesis.

Acknowledgements

I am indebted to Terry Erwin and the BIOLAT Program of the Smithsonian Institution for financial and logistical support. Edward O. Wilson, Bert Hölldobler, Stefan Cover, Catherine Lindell, and Mark Moffett provided invaluable advice and encouragement. This work was supported in part by a National Science Foundation Graduate Fellowship.

References

Adis J., Lubin, Y. D., and Montgomery, G. G. (1984). Arthropods from the canopy of inundated and terra firme forests near Manaus, Brazil, with critical considerations on the pyrethrum-fogging technique. *Studies on Neotropical Fauna and Environment*, **19** (4), 233–46.

Erwin, T. L. (1983). Tropical forest canopies: the last biotic frontier. *Bulletin of the Entomological Society of America*, **29** (1), 14–19.

Erwin, T. L. (1989). Sorting tropical forest canopy samples. *Insect Collection News*, **2** (1), 8.

Moran, V. C. and Southwood, T. R. E. (1982). The guild composition of arthropod communities in trees. *Journal of Animal Ecology*, **51**, 289–306.

Southwood, T. R. E., Moran, V. C., and Kennedy, C. E. J. (1982). The richness, abundance and biomass of the arthropod communities on trees. *Journal of Animal Ecology* **51**, 635–49.

Sudd, J. H., and Franks, N. R. (1987). *The behavioural ecology of ants*. Chapman and Hall, New York.

Wilson, E. O. (1987). The arboreal ant fauna of Peruvian Amazon forests: a first assessment. *Biotropica*, **19** (3), 245–51.

36

Parallels between ants and plants: implications for community ecology

Alan N. Andersen

Botanists and zoologists often take little notice of each other. This is especially pronounced in the field (or is it fields?) of community ecology, where plant and animal ecologists traditionally tread separate paths. For example, plant ecologists talk of patterns and processes, animal ecologists of structure and dynamics. Zoologists hotly dispute the importance of competition as a factor structuring (?patterning) communities (Strong *et al.* 1984; den Boer 1986), whereas botanists accept it as fundamental (Harper 1977; Grime 1979; Tilman 1982).

The problem of the isolation of botanical and zoological schools of thought has been recognized at the population level. As Begon and Mortimer (1981) put it, 'plant and animal populations have had their own, independent ecologists for too long, and, since the same fundamental principles apply to both, there is most to be gained at present from a concentration on similarities rather than differences.' This recognition has led to new perspectives on population ecology (see de Stasio (1989) for a plant ecologists' approach to animal population dynamics), and has resulted in substantial conceptual advances in the field, as illustrated by the recent spate of texts on animal–plant, and particularly ant–plant, interactions (e.g. Buckley 1982; Thompson 1982; Crawley 1983; Beattie 1985; Howe and Westley 1988). It can be claimed that animal–plant, and particularly ant–plant, mutualisms have inspired valuable mutualisms between plant and animal ecologists. Similarly, the recognition that evolutionary forces operate similarly on plants and animals has now made plant ecologists keen students of sociobiology, sexual selection and optimality theory, thereby providing novel perspectives on plant reproduction (e.g. Lloyd 1979; Charnov 1982; Willson and Burley 1983; Wiens 1984; Lovett Doust and Lovett Doust 1988; Willson 1990).

The point I wish to make here is this: just as the field of ant–plant interactions has acted as something of a melting pot for botanists and zoologists interested in population ecology, ant communities provide a unique opportunity for plant and animal ecologists to find common ground and increase our understanding of processes operating at the community level. It has long been recognized (e.g. Brian 1965) that in many ways ant colonies behave more like plants than like animals, and that the methods of studying them are often those of the plant ecologist (Waloff and Blackith 1962). Ant colonies may compete with plants for resources as if they were themselves plants (Rissing 1988) and botanical nomenclature, such as 'regeneration niche' (Grubb 1977), has been applied to ant colonies (Ryti and Case 1988). I will start by outlining some similarities between ant colonies and plants, and then illustrate how these have important consequences for community organization.

Parallels between ant colonies and plants

Modularity

Ant colonies, like plants, are modular organisms, consisting of an indeterminate number of repeated units of multicellular structure ('modules'; Harper 1977). Examples of plant modules include roots, stems, branches, leaves, and flowers; the modules of an ant colony comprise individuals of the worker and reproductive castes. Co-ordination between modules is facilitated by chemical communication: hormones in plants and pheromones in ants. Modularity is also widely recognized in colonial, sessile animals such as corals, bryozoans, and hydroids (van Valen 1978; Rosen 1979; Harper 1981, 1985; Vuorisalo and Tuomi 1986), but, other than a brief reference by Harper (1977), the relevance of modularity to social insects seems to have been ignored.

Modularity has a profound impact on an organism's demography and evolution (Begon and Mortimer 1981; Harper 1981). Modular organisms grow by the addition of modules, and factors causing mortality in unitary organisms, such as predation, often only affect individual modules of a modular organism, thereby merely slowing its growth rate. This gives a modular organism an 'internal' population structure. Plants, like ants, often have polymorphic modules (Harper 1981).

Module architecture has an important influence on the way a plant exploits resources and interacts with other plants (Harper 1981). Clegg (1978) has distinguished two types of module architecture in herbs (see also Sackville Hamilton *et al.* 1987). 'Phalanx' species, including most trussock grasses, have short internodes and clumped modules. They are better able to monopolize local resources then are 'guerilla' species, which have long internodes and well-spaced modules. The parallels with recruit versus solitary foraging in ants are compelling.

As in unitary organisms, natural selection operates on the transition from zygote to zygote (Harper 1981), and not directly on the behaviour of individual modules (although zygote to zygote transition is of course strongly influenced by module behaviour; see Tuomi and Vuorisalo 1989). Conversely, the taxonomy of modular organisms is based primarily on the modules themselves (e.g. plant leaves and flowers, worker ants), rather than on module architecture or behaviour. This is because module morphology is relatively fixed within a species, whereas module architecture and behaviour are typically plastic (Harper 1981).

Fixedness, foraging, and resource requirements

Most ant and most plant species 'nest' in or on the ground and use their modules to forage in the surrounding environment. Most plants have very similar resource requirements (Grime and Hodgson 1987), and so too do most ants. Plants use roots and stolons to forage in the soil for water and nutrients, and use shoots and leaves to forage above ground for sunlight. Worker ants forage in the soil, on the soil surface, and on vegetation, as scavengers, predators, and collectors of plant exudates. In plants, foraging involves the growth and addition of modules, in contrast to module mobility in ants.

There are, of course, both ants and plants that do not fit the above description, but even these exceptions occur in parallel. For example, many species of both taxa are epiphytic, nesting on vegetation. Most of these forage exclusively above the ground, with others send foraging modules to the ground. For both taxa, the greatest diversity and abundance of epiphytic species occurs in tropical rainforests. Still other plant and ant species are parasites, exploiting the foraging modules of other species and Dumpert (1981) for parasitism in ants.

The fact that ant colonies and plants are spatially fixed means that the competitive interactions of both taxa are confined to well-defined zones (Harper 1981). Their foraging modules interact only with those of their neighbours, so that it is the spatial distribution of individuals, more than their density, that determines their competitive interactions. An individual's neighbour can have profound effects on the number and architecture of that individual's modules (see Mack and Harper 1977, and Ryti and Case 1986, for examples for plants and ants respectively).

Parallels between ant and plant communities

Interspecific competition

Three of the attributes discussed above explain why competition is such a ubiquitous and potent force structuring plant communities. These are:

1. occupation of a fixed position, combined with,
2. an ability to exploit surrounding resources intensively using foraging modules, and
3. a set of resource requirements (sunlight, water, and mineral nutrients) that are shared by virtually all species.

Given that these attributes also apply to ants, it is not surprising that ants hold a somewhat special position among animals in that the importance of interspecific competition in their communities has not been seriously questioned. Indeed, the evidence is incontrovertible that competition is a major, if not primary, factor structuring ant communities (Greenslade 1971; Levings 1982; Torres 1984; Fox *et al*. 1985; Savolainen and Vepsä“läinen 1988). Competitive interactions can be direct, involving pairs of ecologically similar species (Brian 1956; Pontin 1963; Fox *et al*. 1985), indirect, involving a third species, or diffuse, involving several species simultaneously (Davidson 1980).

Ants possess two other attributes, not shared by plants, which have an important bearing on competitive interactions. First, whereas competition among plants is almost entirely exploitative (allelopathy is one possible exception), interference competition is of utmost importance in ants. This mostly takes the form of aggression between foraging workers (DeVita 1979; Adams and Traniello 1981; Hölldobler 1982) and, to a lesser extent, prey-robbing (Hölldobler 1986). However, it also includes specialized behaviour patterns such as stone dropping (Moglich and Alpert 1979) and nest plugging (Gordon 1988).

Secondly, ant species with similar ecological requirements are able to reduce competition with each other either by foraging at different times of day, or by specializing in different sized food items. In plants, seasonal differences in growth may be considered as niche differentiation (Grime *et al*. 1985), and some temporal partitioning appears to have occurred in response to competition for pollinators (Pleasants 1980; Gross and Werner 1983) and possibly also for seed dispersers. However, temporal niche differentiation generally appears to play a limited role in community organization. In ant communities, on the other hand, it is a widespread and apparently important mechanism facilitating species co-existence (Whitford 1978; Briese and Macauley 1980; Andersen 1983; Fellers 1989). Similarly, food resources for ants come in different sized packets, and these are not equally available to all ants. There is often a correlation between worker size and the size of the food items they take, with the implication that differences in worker size reduce competition between species (Chew 1977; Davidson 1977; Hansen 1978; but see Traniello 1989). Such partitioning of food resources is not available to plants, with large trees competing directly with small herbs both for sunlight and for water and nutrients in the soil. Competition in ant communities is therefore more complex than in plant communities: ants posses a type of competition (interference), and also mechanisms for reducing competition (food and time partitioning), that are not well-developed in plant communities.

Functional groups

Plant and animal ecologists all classify species into functional groups that transcend taxonomic barriers, thereby reducing the apparent complexity of ecological systems and allowing for comparative studies of communities that are taxonomically remote. In animal communities, functional groups are typically 'guilds', sets of species exploiting a common pool of resources (Terborgh and Robinson 1986). In practice, such guilds are almost always trophically based; for example, guilds of insectivorous birds (Root 1967), desert granivores (Brown and Davidson 1977), and herbivorous insects (Moran and Southwood 1982).

Clearly, a functional group classification based solely on resource utilization has limited use for plant communities. Functional group classifications of plants are therefore based on a broader range of ecological characters (see Cody 1986) including plant life form (Raunkiaer 1934), morphology (Halloy 1990), use of pollinators (Pleasants 1980) and seed dispersal agents (Beattie and Culver 1981), growth and flowering phenology (Sarmiento and Monasterio 1983), 'vital attributes' (Noble and Slatyer 1980), and dispersal and colonization ability (Platt 1975; Platt and Weis 1977; Wilson 1989). Similarly, some guilds of ants can be recognized (Torres 1984; Serrano *et al.* 1987), such as seed harvesters (e.g. Briese and Macauley 1981; Marsh 1985; Morton and Davidson 1988), but generally the guild concept in its limited sense is inappropriate for detailed studies of ant communities (Greenslade and Greenslade 1989). This is compounded by the fact that interference competition, which cuts across guild boundaries (Greenslade 1982), is so prominent in ants.

Greenslade has developed a functional group classification of Australian ants based largely on their habitat requirements and competitive interactions. Each group can be considered to represent a broad ecological strategy. The scheme was derived from studies of the southern arid zone (e.g. Greenslade 1978; Greenslade and Halliday 1983), but has been widely applied in Australia, including the semi-arid south (Greenslade and Greenslade 1989), temperate south-east (Andersen 1986*a*, *b*; Andersen and McKaige 1987), sub-tropical east (Greenslade and Thompson 1981), monsoonal north (Greenslade 1985; Andersen (in press); Andersen and Majer (1991)), and mediterranean south-west (Majer and Brown 1986). There are seven groups (following Andersen 1990), most of which have strong parallels with functional groups of Australian plants (Table 36.1).

Dominant species

A dominant species is one that is both abundant and capable of exerting a strong influence on other species (see Greenslade (1976) for ants, and Grime (1987) for plants). Plant ecologists often use the term 'dominance' synonymously with 'predominance', based on the reasonable assumption that a highly abundant species (or one with high foliage cover) will have a significant impact on the resources (at least for sunlight, if not for water or nutrients) available to other species. However, predominance in animal communities does not necessarily imply influence, so that a highly abundant species is not necessarily a functionally dominant one.

Table 36.1. Functional groups of Australian ants and their plant parallels
(see text for details).

Functional group	Major taxa	Plant parallel
1. Dominant species	*Iridomyrmex*	*Eucalyptus* (canopy trees in general)
2. Subordinate species	Camponotinae, especially *Camponotus*	Mid-storey trees and shrubs
3. (a) Hot climate specialists	*Melophorus, Meranoplus*	No obvious parallels
(b) Cold climate specialists	*Prolasius, Notoncus*	*Nothofagus*
4. Cryptic species	Tiny Ponerinae (e.g. *Hypoponera*) and Myrmicinae (e.g. *Solenopsis*)	Cryptogams
5. Opportunists	*Rhytidoponera, Paratrechina, Tetramorium*	Ruderal species (colonizing 'weeds')
6. Generalized Myrmicinae	*Monomorium, Pheidole, Crematogaster*	*Acacia*
7. Large, solitary foragers	*Myrmecia, Leptogenys, Pachycondyla*	No obvious parallels

Canopy trees are dominant components of plant communities. In Australian ant communities, species of *Iridomyrmex* are dominant ants (Greenslade 1976). The hundreds of species are distributed throughout the continent. They are extremely aggressive ants, especially abundant and diverse in arid, semi-arid, and seasonally arid regions, where high temperatures and open vegetation provide a particularly suitable environment for their high rates of activity and rapid recruitment. At a woodland site (18×24 m) in the seasonally arid tropics, for example, the nine species of *Iridomyrmex* represented 54 per cent of total ants captured in pitfall traps, and more than 80 per cent of ants at tuna baits; although total species numbered about 100 (A. N. Andersen unpublished work).

The dominance of Australian ant communities by *Iridomyrmex* parallels the well-known dominance of Australia's woody vegetation by *Eucalyptus* (see Groves 1981). Such continental dominance by a single genus of either an ant fauna or a woody flora is unique to Australia. Although *Eucalyptus* is virtually confined to Australia (Barlow, 1981), *Iridomyrmex* is widely distributed throughout south-east Asia and the New World. Some species achieve local dominance elsewhere (e.g. Hölldobler 1982; Bond and Slingsby 1984), but *Iridomyrmex* is generally a minor component of ant faunas outside Australia. *Iridomyrmex* and *Eucalyptus* are therefore both distinctively Australian.

The level of ground insolation is a critical factor controlling the local distribution and abundance of *Iridomyrmex*, with the genus often being absent altogether from areas of dense vegetation (Greenslade and Thompson 1981; Greenslade 1985; Andersen 1986*b*). For *Eucalyptus*, fire plays a critical role in maintaining its dominance (Mount 1964). Throughout the wetter parts of its range, for example, *Eucalyptus* is replaced by broad-leaved (rainforest) species in the long-term absence of fire (Webb and Tracey 1981). Species of *Eucalyptus* often possess a suite of fire-promoting characteristics such as highly flammable leaves and decorticating bark, and most regenerate vigorously following fire, either from seed (e.g. Ashton 1976), from epicormic buds or lignotubers (Gill 1981), or from root suckers or rhizomes (Lacey 1974). Interestingly, the dominance of *Iridomyrmex* is also promoted by fire in the wetter parts of its range. This is because fire simplifies vegetation structure, thereby increasing the level of ground insolation and favouring *Iridomyrmex* at the expense of shade-tolerant species (Andersen (in press)). The dominance of the Australian landscape by *Eucalyptus* and *Iridomyrmex* is therefore linked by fire.

Subordinate species

Subordinate species are taxa that co-occur successfully with dominant species but are inferior competitors: they are often widespread and reasonably abundant, but never attain dominant status (see Grime (1987) for plants). In a forest, mid-storey trees and shrubs are subordinate to canopy trees. In Australian ant communities, genera of the tribe Camponotini (especially *Camponotus*) are subordinate to *Iridomyrmex*. The distribution of *Camponotus* parallels that of *Iridomyrmex*: they both favour well-insolated habitats and are most abundant and diverse in arid regions. Features which enable *Camponotus* species to co-exist successfully with *Iridomyrmex* include large body size, polymorphism, nocturnal activity, foraging on vegetation, and submissive behaviour. They can recruit rapidly to food resources, but are easily displaced by *Iridomyrmex*.

Climate specialists

This group includes a wide range of taxa that are restricted to hot and cold climates respectively. The latter comprise mostly species of *Prolasius*, *Notoncus* and cool-temperate groups of *Monomorium* (ex *Chelaner*), which occur predominantly in south-eastern, and to a lesser extent south-western, Australia, where the influence of *Iridomyrmex* is reduced. They are often among the most abundant ants in these regions, particularly in shaded habitats (Andersen 1986*a*) or during winter months (Andersen 1983, 1986*b*).

Hot climate specialists, on the other hand, occur where *Iridomyrmex* is most abundant, and show evidence of having had a long evolutionary history of association with *Iridomyrmex*. The exceptionally diverse endemic genus *Melophorus*, which has a similar distribution to that of *Iridomyrmex* within Australia, is perhaps the best example. All known species have extremely restricted foraging schedules, only being active during periods of hot weather. Their foraging is therefore virtually confined to the middle of hot days (Greenslade 1979; Andersen 1983). In the Kakadu region of northern Australia, for example, species of *Melophorus* only become active when surface soil temperatures exceed 30 °C, and they continue activity beyond 60 °C (A. N. Andersen unpublished work). The mean surface temperature experienced by foraging workers was calculated at 48.2 °C, and at this temperature virtually no other ants are active.

The specializations exhibited by hot climate species appear to be in response to interference competition, so they do not have close parallels with any functional group of Australian plants. However, cold climate specialists have parallels with Gondwanan taxa such as *Nothofagus* (Barlow 1981), which is also locally dominant at the wettest sites of south-eastern Australia and has close relatives in New Zealand and South America.

Cryptic species

Cryptic species of ants forage predominantly within soil and litter rather than on open ground, and parallel cryptogams (especially bryophytes) in plant communities. Both functional groups are best developed in wet forests, where they are important but inconspicuous components of the ecosystem and have little or no competitive interaction with dominant species.

Opportunists

These are ruderal species (Grime 1979), characteristic of 'waste' or disturbed habitats. They have obvious parallels with 'weedy' plants that colonize new habitats (e.g. the 'colonizers', of Bazzaz (1986). The major taxa are species of *Rhytidoponera*, *Paratrechina*, and *Tetramorium* (Yeatman and Greenslade 1980; Greenslade 1985; Andersen and McKaige 1987; Andersen and Majer (1990).

Opportunists are taxonomically diverse (the three major genera represent three different subfamilies), but share the following attributes.

1. They have generalized habits, with omnivorous diets, flexible foraging times, and an ability to tolerate a wide range of physical conditions.
2. They are poor competitors. They are timid ants, and even when recruit-foraging occurs, as in *Paratrechina*, they are easily displaced from food resources.
3. They are most abundant at sites where *Iridomyrmex* is least abundant, such as cool-temperate regions (e.g. Andersen 1986*a,b*) and in heavily

shaded habitats elsewhere (e.g. Greenslade 1985; Andersen and Majer 1991).
4. They increase in abundance following habitat disturbance (e.g. Yeatman and Greenslade 1980; Andersen and McKaige 1987; Andersen 1988; Andersen and Burbidge 1991).

Generalized habits, poor competitive ability and proliferation following disturbance are attributes shared by ruderal plants (Baker 1974; Hill 1977; Grime 1979; Bazzaz 1986; Groves 1986). Ruderal plants also possess other traits that are potentially shared by opportunist ants, including effective dispersal, rapid growth rates, early production of propagules, and a flexible breeding system (e.g. propensity for both in- and out-crossing). Unfortunately, the autecology of Australian ant species is too poorly known to pursue these parallels, although it is worth noting that the peculiar method of reproduction (through mated workers rather than queens) in *Rhytidoponera* may contribute to its success as an opportunist (Andersen and McKaige 1987).

Generalized Myrmicinae

The three cosmopolitan myrmicine genera *Monomorium*, *Pheidole*, and *Crematogaster* are ubiquitous in Australia. Unlike members of other functional groups, they appear to be poorly integrated into Australian ant communities, possibly because they represent recent arrivals in evolutionary time (Greenslade and Thompson 1981; Greenslade 1985). They have extremely broad niches but, unlike opportunists, are often good competitors and are commonly among the most abundant ants. Species of *Monomorium* in particular are successful at sites where *Iridomyrmex* is highly abundant (e.g. Andersen 1983; Greenslade 1985; Andersen (in press)). This can be attributed to their rapid recruitment and ability to defend resources using topically applied venom alkaloids (see Jones *et al*. 1988).

The behaviour of generalized myrmicines in Australian ant communities has parallels with that of acacia in Australian plant communities. Both taxa are represented by very many (> 1000) species distributed throughout the world, and are virtually ubiquitous within Australia. Together *Eucalyptus* and *Acacia* dominate almost every tall woody vegetation type in the continent (Barlow 1981), except at the very wettest sites (Groves 1981). Together *Iridomyrmex* and generalized myrmicines are the predominant ants almost everywhere in Australia, except at the very coolest and shadiest sites. Generalized myrmicines almost always co-occur with dominant *Iridomyrmex* and are often the most abundant ants at less favourable (i.e. cooler or shadier) sites. Acacia almost always co-occurs with dominant *Eucalyptus* (as a mid-storey tree or understorey shrub), and often dominates the less favourable (i.e. more arid) sites (Barlow 1981; Johnson and Burrows 1981). Finally, generalized myrmicines and *Acacia* both include species that are extremely successful colonists. Species of acacia commonly dominate the woody vegetation following major disturbance such as severe fire or clearing, and are widely used as colonists in revegetation programs. Similarly, severely disturbed sites are often colonized by generalized myrmicines; in particular, the clearing of land for agriculture commonly results in a proliferation of *Pheidole* species, which then become major seed-harvesting pests (Andersen Chapter 33, this volume).

Large, solitary foragers

This final group contains taxa whose large body size (typically 8–20 mm) and low population densities are more characteristic of other predacious arthropods (e.g. carabid beetles, scorpions, and spiders) than of ants. The most common of these are species of *Myrmecia* in southern Australia, and large ponerines (e.g. *Pachycondyla*, *Leptogenys* spp.) in northern Australia. They have no obvious parallels with any functional group of plants.

Functional group composition

From the above discussion it is clear that climate and vegetation structure are key factors determining the relative abundance of different functional groups in Australian ant communities. The two physical factors limiting plant production, stress and disturbance (Grime 1979), can be applied to ants within this framework. As far as ants are concerned, the major stresses are climatic (low temperature) and micro-climatic (poor insolation), and the major manifestations of disturbance are destruction of vegetation and litter, and disruption of soil.

Grime (1979) recognizes three fundamental plant strategies, which respectively predominate in communities for which competition, stress, and disturbance are the prime ecological factors. Similarly, three fundamental types of Australian ant community can be recognized, each paralleling one of Grime's primary strategies in plants (Figs 36.1 and 36.2). 'Competitive' communities are those occurring at favourable sites, namely open habitats in

Fig. 36.1. The three primary types of Australian ant communities following Grime's (1979) triangular model of plant strategies (see text for details). 1 = Competitive, 2 = stress-tolerant, 3 = ruderal.

hot regions (i.e. throughout the arid zone and in monsoonal northern Australia). Dominant species of *Iridomyrmex* are highly abundant and, typically, so too are species of *Camponotus* and *Melophorus*. These represent well-integrated 'core' taxa of a tightly structured community (Greenslade 1979) and together usually contribute well over half the total ants and total species. Some highly competitive generalized myrmicines, particularly species of *Monomorium*, are also often abundant. Opportunists and cryptic species are poorly represented; the former because of competition, the latter because of lack of suitable habitat. Examples of competitive communities are described in Greenslade (1978), Briese and Macauley (1977, 1980, 1981), Greenslade and Halliday (1983), Andersen (1983, 1984), and Andersen and Yen (1985).

'Stress-tolerant' communities occur in the wetter forests of cool-temperate south-eastern and south-western Australia. *Iridomyrmex* and *Camponotus* are poorly represented, while *Melophorus* is absent altogether. The overwhelming majority of ants are either cold-climate specialists (mostly species of *Prolasius*, *Notoncus*, and certain *Monomorium* (ex *Chelaner*)) or cryptic species (Fig. 36.2). Examples of stress-tolerant communities are those of the tall open forest and closed-forest sites of Andersen (1986*a*).

'Ruderal' communities are characteristic of highly disturbed sites. The most abundant ants are typically opportunists (especially small species of *Rhytidoponera*; see Andersen (1990)) and colonizing species of *Pheidole*. Members of all other functional groups can be present, but almost always in low numbers. An example of a ruderal community is that of the grazed site of Andersen and McKaige (1987).

The above three primary communities represent extremes of the full spectrum of Australian ant communities, and various intermediate or secondary types (Grime 1979) can also be recognized. Two of these appear to be particularly important (Fig. 36.2): 'competitive ruderal communities', occupying low-stress habitats subject to moderate disturbance; and 'stress-tolerant competitive communities', occupying undisturbed habitats subject to moderate stress. Examples of competitive ruderal communities occur throughout the frequently burnt savannah woodlands and open forests of northern Australia. Frequent fires cause a structural simplification of the vegetation, thereby increasing ground insolation. This favours dominant *Iridomyrmex* at the expense of generalized myrmicines (Fig. 36.2).

Stress-tolerant competitive communities, characterized by *Iridomyrmex* occurring at moderate densities, occupy two types of habitat, depending on whether the stress is climatic (i.e. low temperatures) or micro-climatic (i.e. poor insolation). The former occur in open habitats of cool-temperate regions ((b) in Fig. 36.2), where *Iridomyrmex* is often abundant in well-insolated patches. The small-scale distribution of *Iridomyrmex* in these communities has a marked influence on the distribution of taxa predominating in poorly insolated patches, particularly species of *Rhytidoponera* and *Pheidole*.

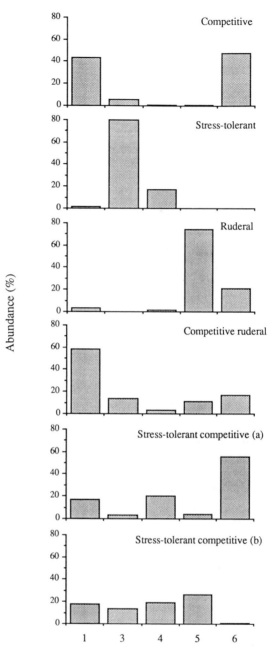

Fig. 36.2. Functional group composition (1 = dominant species; 3 = climate specialists; 4 = cryptic species; 5 = opportunists; 6 = generalized myrmicinae) of some Australian ant communities following Grime's (1979) plant strategy concepts and nomenclature (see text and Fig. 36.1). Examples are given of each of the three types of primary communities (see Fig. 36.1), and also of some secondary communities. The habitat of each community is as follows: competitive—semi-arid wood-

Examples of the second type of stress-tolerant competitive community occur in open forests of northern Australia that are burnt infrequently ((a) in Fig. 36.2). The relatively poor insolation and thick litter of these habitats act as a constraint on *Iridomyrmex*; generalized myrmicines are often the most abundant ants, and cryptic species are an important component of the fauna (Andersen (in press)).

Dominant species and diversity

Local diversity within plant communities often follows a distinctive pattern: it initially increases with environmental favourability, but then decreases as conditions allow highly competitive species to become so dominant that they exclude other species. Therefore, there is often a humped relationship between local diversity and gradients of resource availability, productivity, and disturbance (Grime 1973; Connell 1978; Lubchenco 1978; Huston 1979; Tilman 1982). A similar relationship can also occur in sessile marine invertebrates (Paine 1974; Connell 1978), but has not been observed in terrestrial animals and is not likely to be in most cases (Fuentes and Jaksic 1988).

Given the many parallels between ants and plants, humped diversity patterns might also be expected in ant communities. My unpublished observations on exceptionally rich ant communities in Kakadu National Park indicate that this is indeed the case, with a humped relationship existing between the abundance of dominant species and total species richness at tuna baits at any point in time (Fig. 36.3). When the number of dominant ants at baits was low, so too was total species richness. This suggests that at these times environmental conditions (probably related to temperature) were generally unfavourable for ants. As the abundance of dominant species (and presumably environmental favourability) increased, so did total species richness. However, as the number of dominant ants continued to increase, species richness began decreasing, presumably because of competitive exclusion, and reached its lowest levels when dominant species were most abundant. A humped temporal relationship between diversity and the abundance of dominant species, although less pronounced, was also observed in counts of ants inside quadrats. Humped spatial patterns of ant diversity have also been documented in Australia (Majer 1985; Fox *et al.* 1988).

---

land and heathland (Andersen 1984); stress-tolerant—cool-temperate closed forest (Andersen 1986*a*); ruderal—cool-temperate grazing paddock (Andersen and McKaige 1987); competitive ruderal—annually burnt monsoonal open forest (Andersen (in press)); stress-tolerant competitive (a)—long-unburnt monsoonal open forest (Andersen (in press)); stress-tolerant competitive (b)—cool-temperate woodland (Andersen 1986*b*). All data are from pitfall traps.

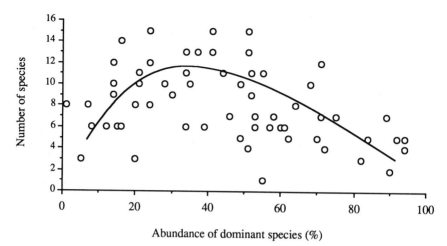

Fig. 36.3. Relationship between the abundance of dominant species and total species richness at tuna baits at woodland and open forest sites in monsoonal northern Australia (Andersen, unpublished work). The data are for the same sites at different points in time, and suggest that dominant species regulate the number of ant species foraging at any time in a manner similar to the regulation of local diversity by dominant species in plant communities.

Conclusion

Three attributes shared by plants and ants have particularly important consequences for community organization: their modularity, their fixedness of position, and their common resource requirements. The sharing of these attributes appears to have led to parallels in competitive interactions within ant and plant communities, in the organization of communities into functional groups, and in the evolution of ecological strategies.

Recognition of the parallels between ant and plant communities provides fertile ground for a broader application of narrowly based ecological principles. For example, according to Grime (1987), important characteristics of a dominant plant include a robust perennial life form, a rapid commitment of resources to the production of new modules, high morphological plasticity during module differentiation, and short lifespans of individual modules. Are these attributes also shared by dominant ants? Similarly, can plant ecologists learn something from the subordinate behaviour of *Camponotus*, or the opportunistic behaviour of *Rhytidoponera*? Is the dual dominance of *Eucalyptus* and *Iridomyrmex* merely a coincidence, or does it say something about the Australian landscape, such as the role of fire? Ant communities have traditionally been considered extremely diverse relative to those of other animals, but is it more realistic to compare their diversity with that of plant communities? Since ant communities unite many features of plant communities with those of animals, they provide an opportunity for zoologist and botanists to develop a more comprehensive understanding of biological communities.

References

Adams, E. A. and Traniello, J. F. A. (1981). Chemical interference competition by *Monomorium minimum* (Hymenoptera: Formicidae). *Oecologia*, **51**, 265–70.

Andersen, A. N. (1983). Species diversity and temporal distribution of ants in the semi-arid mallee region of northwestern Victoria. *Australian Journal of Ecology*, **8**, 127–37.

Andersen, A. N. (1986*a*). Patterns of ant community organization in mesic south-eastern Australia. *Australian Journal of Ecology*, **11**, 87–97.

Andersen, A. N. (1986*b*). Diversity, seasonality and community organization of ants at adjacent heath and woodland sites in south-eastern Australia. *Australian Journal of Zoology*, **34**, 53–64.

Andersen, A. N. (1988). Immediate and longer-term effects of fire or seed predation by ants in sclerophyllous vegetation in south-eastern Australia. *Australian Journal of Ecology*, **13**, 285–93.

Andersen, A. N. (1990). The use of ant communities to evaluate change in Australian terrestrial ecosystems: a review and a recipe. *Proceedings of the Ecological Society of Australia*, **16**, 347–57.

Andersen, A. N. (in press). Responses of ground foraging ant communities to three experimental fire regimes in a savanna forest of tropical Australia. *Biotropica*.

Andersen, A. N. and Burbidge, A. H. (1991). The ants of a vine thicket near Broome: a comparison with the northwest Kimberley. *Journal of the Royal Society of Western Australia*, **73**, 79–82.

Andersen, A. N. and McKaige, M. E. (1987). Ant communities at Rotamah Island, Victoria, with particular reference to disturbance and *Rhytidoponera tasmaniensis*. *Proceedings of the Royal Society of Victoria*, **99**, 141–6.

Andersen, A. N. and Majer, J. D. (in press). The structure and biogeography of rainforest ant communities in the Kimberley region of northwestern Australia. In: *Rainforests*, Kimberley (eds N. L. McKenzie, R. B. Johnston, and P. J. Kendrick). Surrey Beatty & Sons, Sydney.

Andersen A. N. and Yen, A. L. (1985). Immediate effects of fire on ants in the mallee of northwestern Victoria. *Australian Journal of Ecology*, **10**, 25–30.

Ashton, D. H. (1976). The development of even-aged stands of *Eucalyptus regnans* F. Muell. in Central Victoria. *Australian Journal of Botany*, **24**, 397–414.

Baker, H. G. (1974). The evolution of weeds. *Annual Review of Ecology and Systematics*, **5**, 1–24.

Barlow, B. A. (1981). The Australian flora: its origin and evolution. In *Flora of Australia*, vol. 1. Introduction (ed. A. S. George), pp. 25–75. Australian Government Publishing Service, Canberra.

Bazzaz, F. A. (1986). Life history of colonizing plants: some demographic, genetic, and physiological features. In *Ecology of biological invasions of North America and Hawaii* (ed. H. A. Mooney and J. A. Drake), pp. 96–110. Springer-Verlag, New York.

Beattie, A. J. (1985). *The evolutionary ecology of ant-plant mutualisms*. Cambridge University Press, Cambridge.

Beattie, A. J. and Culver, D. C. (1981). The guild of myrmecochores in the herbaceous flora of West Virginia forests. *Ecology*, **62**, 107–15.

Begon, M. and Mortimer, M. (1981). *Population ecology: a unified study of animals and plants*. Blackwell Scientific Publications, Oxford.

Bond, W. J. and Slingsby, P. (1984). Collapse of an ant-plant mutualism: the argentine ant (*Iridomyrmex humilis*) and myrmecochorous Proteaceae. *Ecology*, **65**, 1031–7.

Brian, M. V. (1956). Segregation of species of the ant genus *Myrmica*. *Journal of Animal Ecology*, **25**, 319–37.

Brian, M. V. (1965). *Social insect populations*. Academic Press, London.

Briese, D. T. and Macauley, B. J. (1977). Physical structure of an ant community in semi-arid Australia. *Australian Journal of Ecology*, **2**, 107–20.

Briese, D. T. and Macauley, B. J. (1980). Temporal structure of an ant community in semi-arid Australia. *Australian Journal of Ecology*, **5**, 121–34.

Briese, D. T. and Macauley, B. J. (1981). Food collection within an ant community in semi-arid Australia, with special reference to seed harvesters. *Australian Journal of Ecology*, **6**, 1–19.

Brown, J. H. and Davidson, D. W. (1977). Competition between seed-eating rodents and ants in desert ecosystems. *Science*, **196**, 880–2.

Buckley, R. C. (ed.) (1982). *Ant-plant interactions in Australia*. Dr. W. Junk, The Hague.

Charnov, E. L. (1982). *The theory of sex allocation*. Princeton University Press.

Chew, R. M. (1977). Some ecological characteristics of the ants of a desert-scrub community in southeastern Arizona. *American Midland Naturalist*, **98**, 33–49.

Clegg, L. M. (1978). The morphology of clonal growth and its relevance to the population dynamics of perennial plants. Ph.D. thesis, University College of North Wales, Bangor.

Cody, M. L. (1986). Structural niches in plant communities. In *Community ecology* (eds J. Diamond and T. J. Case), pp 381–405. Harper and Row, New York.

Connell, J. H. (1978). Diversity in tropical rainforests and coral reefs. *Science*, **199**, 1302–10.

Crawley, M. J. (1983). *Herbivory: the dynamics of animal-plant interactions*. Blackwell Scientific Publications, Oxford.

Davidson, D. W. (1977). Species diversity and community organization in desert seed-eating ants. *Ecology*, **58**, 711–24.

Davidson, D. W. (1980). Some consequences of diffuse competition in a desert ant community. *American Naturalist*, **116**, 92–105.

den Boer, P. J. (1986). The present status of the competitive exclusion principle. *Trends in Ecology and Evolution*, **1**, 25–8.

de Stasio, B. T. (1989). The seed bank of a freshwater crustacean: copepodology for the plant ecologist. *Ecology*, **70**, 1377–89.

deVita, J. (1979). Mechanisms of interference and foraging among colonies of the harvester ant *Pogonomyrmex californicus* in the Mojave desert. *Ecology*, **60**, 729–37.

Dumpert, K. (1981). *The social biology of ants*. Pitman, Boston.

Fellers, J. H. (1987). Interference and exploitation in a guild of woodland ants. *Ecology*, **68**, 1466–78.

Fellers, J. H. (1989). Daily and seasonal activity in woodland ants. *Oecologia*, **78**, 69–76.

Fox, B. J., Archer, E., and Fox, M. D. (1988). Ant communities along a moisture gradient. In *Time scales and water stress* (ed. F. diCastri, C. Floret, S. Rambal and J. Roy), pp. 661–7. IUBS, Paris.

Fox, B. J., Fox, M. D., and Archer, E. (1985). Experimental confirmation of competition between two dominant species of *Iridomyrmex* (Hymenoptera: Formicidae). *Australian Journal of Ecology*, **10**, 105–10.

Fuentes, E. R. and Jaksic, F. M. (1988). The hump-backed species diversity curve: why has it not been found among land animals? *Oikos*, **53**, 139–43.

Gill, A. M. (1981). Adaptive responses of Australian vascular plant species to fires. In *Fire and the Australian biota* (eds A. M. Gill, R. H. Groves, and I. R. Noble), pp. 243–71. Australian Academy of Science, Canberra.

Gordon, D. M. (1988). Nest-plugging: interference competition in desert ants (*Novomessor cockerelli* and *Pogonomyrmex babatus*). *Oecologia*, **75**, 114–18.

Greenslade, P. J. M. (1971). Interspecific competition and frequency changes among ants in Solomon Islands coconut plantations. *Journal of Applied Ecology*, **8**, 323–52.

Greenslade, P. J. M. (1976). The meat ant *Iridomyrmex purpureus* (Hymenoptera: Formicidae) as a dominant member of ant communities. *Journal of the Australian Entomological Society*, **15**, 237–40.

Greenslade, P. J. M. (1978). Ants. In *The physical and biological features of Kunnoth Paddock in Central Australia* (ed. W. A. Low). CSIRO Division of Land Resources Management Technical Paper No. 4.

Greenslade, P. J. M. (1979). *A guide to ants of South Australia*. South Australian Museum, Adelaide.

Greenslade, P. J. M. (1982). Diversity and food specificity of seed-harvesting ants in relation to habitat and community structure. In *Proceedings of the 3rd Australasian Conference on Grassland Invertebrate Ecology*, (ed. K. E. Lee), pp. 227–33. South Australian Government Printer, Adelaide.

Greenslade, P. J. M. (1985). Preliminary observations on ants (Hymenoptera: Formicidae) of forest and woodland in the Alligator Rivers Region, N. T. *Proceedings of the Ecological Society of Australia*, **13**, 153–60.

Greenslade, P. J. M. and Greenslade, P. (1989). Ground layer invertebrate fauna. In *Mediterranean landscapes in Australia: mallee ecosystems and their management*. (ed. J. C. Noble and R. A. Bradstock), pp. 266–284. CSIRO, Melbourne.

Greenslade, P. J. M. and Halliday, R. B. (1983). Colony dispersion and relationships of meat ants *Iridomyrmex purpureus* and allies in an arid locality in South Australia. *Insectes Sociaux*, **30**, 82–99.

Greenslade, P. J. M. and Thompson, C. H. (1981). Ant distribution, vegetation, and soil relationships in the Cooloola-Noosa River area, Queensland. In *Vegetation classification in Australia* (eds. A. N. Gillison and D. J. Anderson), pp. 192–207. CSIRO and ANU Press, Canberra.

Grime, J. P. (1973). Competitive exclusion in herbaceous vegetation. *Nature*, **242**, 344–47.

Grime, J. P. (1979). *Plant strategies and vegetation processes*. John Wiley, Chichester.

Grime, J. P. (1987). Dominant and subordinate components of plant communities: implications for succession, stability and diversity. In *Colonization, succession and stability* (eds A. J. Gray, M. J. Crawley, and P. J. Edwards), pp. 413–28. Blackwell Scientific Publications, Oxford.

Grime, J. P. and Hodgson, J. G. (1987). Botanical contributions to contemporary ecological theory. *New Phytologist*, **88** (Suppl.), 283–95.

Grime, J. P., Shacklock, J. M. L. and Bond, S. R. (1985). Nuclear DNA contents, shoot phenology and species coexistance in a limestone grassland community. *New Phytologist*, **100**, 435–45.

Gross, R. S. and Werner, P. A. (1983). Relationships among flowering phenology, insect visitors, and seed-set of individuals: experimental studies on four co-occurring species of goldenrod (*Solidago*: Compositae). *Ecological Monographs*, **53**, 95–117.

Groves, R. H. (ed.) (1981). *Australian vegetation*. Cambridge University Press.

Groves, R. H. (1986). Plant invasions of Australia: an overview. In *Ecology of Biological invasions: an Australian perspective* (eds R. H. Groves and J. J. Burdon), pp. 138–49.

Grubb, P. J. (1977). The maintenance of species-richness in plant communities: the importance of the regeneration niche. *Biological Reviews*, **52**, 107–45.

Halloy, S. (1990). A morphological classification of plants and animals with special reference to the New Zealand alpine flora. *Journal of Vegetation Science*, **1**, 291–304.

Hansen, S. R. (1978). Resource utilization and coexistence of three species of *Pogonomyrmex* ants in an upper Sonoran grassland community. *Oecologia*, **35**, 109–17.

Harper, J. L. (1977). *Population biology of plants*. Academic Press, New York.

Harper, J. L. (1981). The concept of population in modular organisms. In *Theoretical ecology* (second edition) (ed. R. H. May), pp.53–77. Blackwell Scientific Publications, Oxford.

Harper, J. L. (1985). Modules, branches, and the capture of resources. In *Population biology and evolution of clonal organisms*, (eds J. B. C. Jackson, L. W. Buss, and R. E. Cook), pp. 1–33. Yale University Press, New Haven.

Hill, T. A. (1977). *The biology of weeds*. Arnold, London.

Hölldobler, B. (1982). Interference strategy of *Iridomyrmex pruinosum* (Hymenoptera: Formicidae) during foraging. *Oecologia*, **52**, 208–13.

Hölldobler, B. (1986). Food robbing in ants, a form of interference competition. *Oecologia*, **69**, 12–15.

Hölldobler, B., Standon, R. C., and Markl, H. (1978). Recruitment and food-retrieving behaviour in *Novomessor* (Formicidae, Hymenoptera) I. Chemical signals. *Behavioural Ecology and Sociobiology*, **4**, 163–81.

Howe, H. F. and Westley, L. C. (1988). *Ecological relationships of plants and animals*. Oxford University Press, New York.

Huston, M. (1979). A general hypothesis of species diversity. *American Naturalist* **113**, 81–101.

Johnson, R. W. and Burrows, W. H. (1981). *Acacia* open-forests, woodlands and shrublands. In *Australian vegetation* (ed. R. H. Groves), pp. 198–226.

Jones, T. H., Blum, M. S., Andersen, A. N., Fales, H. M., and Escoubas, P. (1988). Novel 2-ethyl-5-alkylpyrrolidines in the venom of an Australian ant of the genus *Monomorium*. *Journal of Chemical Ecology*, **14**, 35–45.

Lacey, C. J. (1974). Rhizomes in tropical eucalypts and their role in recovery from fire damage. *Australian Journal of Botany*, **22**, 29–38.

Levings, S. C. (1982). Patterns of nest dispersion in a tropical ground ant community. *Ecology*, **63**, 338–44.

Lloyd, D. G. (1979). Parental strategies of angiosperms. *New Zealand Journal of Botany*, **17**, 595–606.

Lovett Doust, J. and Lovett Doust, L. (eds) (1988). *Plant reproductive ecology: patterns and strategies*. Oxford University Press, Oxford.

Lubchenco, J. (1978). Plant species diversity in a marine inter-tidal community: importance of herbivore food preference and algal competitive abilities. *American Naturalist*, **112**, 23–29.

Mack, R. N. and Harper, J. L. (1977). Interference in dune annuals: spatial pattern and neighbourhood effects. *Journal of Ecology*, **65**, 345–63.

Majer, J. D. (1985). Recolonization by ants of rehabilitated mineral sand mines on Northstradbrook Island, Queensland, with particular reference to seed removal. Australian Journal of Ecology, **10**, 31–48.

Majer, J. D. and Brown, K. R. (1986). The effects of urbanization on the ant fauna of the Swan Coastal Plain near Perth, Western Australia. *Journal of the Royal Society of Western Australia*, **60**, 13–17.

Marsh, A. C. (1985). Forager abundance and dietary relationships in a Namib Desert ant community. *South African Journal of Zoology*, **20**, 197–203.

Moglich, M. H. J. and Alpert, G. D. (1979). Stone dropping by *Conomyrma bicolor*: a new technique of interference competition. *Behavioural Ecology and Sociobiology*, **6**, 105–13.

Moran, V. C. and Southwood, T. R. E. (1982). The guild composition of arthropod communities in trees. *Journal of Animal Ecology*, **51**, 289–306.

Morton, S. R. and Davidson, D. W. (1988). Comparative structure of harvester ant communities in arid Australia and North America. *Ecological Monographs*, **58**, 19–38.

Mount, A. B. (1964). The interdependence of the eucalypts and forest fires in southern Australia. *Australian Forestry*, **28**, 166–72.

Noble, I. R. and Slatyer, R. O. (1980). The use of vital attributes to predict successional changes in plant communities subject to recurrent disturbances. *Vegetatio*, **43**, 5–21.

Paine, R. T. (1974). Intertidal community structure: experimental studies on the relationship between a dominant competitor and its principal predator. *Oecologia*, **15**, 93–120.

Platt, W. J. (1975). The colonization and formation of equilibrium plant species associations on badger disturbances in a tall-grass prairie. *Ecological Monographs*, **45**, 285–305.

Platt, W. J. and Weis, I. M. (1977). Resource partitioning and competition within a guild of fugitive prairie plants. *American Naturalist*, **111**, 479–513.

Pleasants, J. M. (1980). Competition for bumblebee pollinators in Rocky Mountain plant communities. *Ecology*, **61**, 1446–60.

Pontin, A. J. (1963). Further considerations of competition and the ecology of the ants *Lasius flavus (F.)* and *L. niger (L.)*. *Journal of Animal Ecology*, **32**, 565–74.

Raunkiaer, C. (1934). *The life forms of plants and statistical plant geography*. Clarendon Press, Oxford.

Rissing, S. W. (1988). Seed-harvester ant association with shrubs: competition for water in the Mohave Desert? *Ecology*, **69**, 809–13.

Root, R. B. (1967). The niche exploitation pattern of the blue-grey gnatcatcher. *Ecological Monographs*, **37**, 317–50.

Rosen, B. R. (1979). Modules members and communes: a postscript introduction to social organisms. In *Biology and systematics of colonial organisms* (ed. J. Larwood and B. R. Rosen) pp. 13–35. Academic Press, London.

Ryti, R. T. and Case, T. J. (1986). Overdispersion of ant colonies: a test of hypotheses. *Oecologia*, **69**, 446–53.

Ryti, R. T. and Case, T. J. (1988). The regeneration niche of desert ants: effects of established colonies. *Oecologia*, **75**, 303–6.

Sackville Hamilton, N. R., Schmid, B., and Harper, J. L. (1987). Life history concepts and the population biology of clonal organisms. *Proceedings of the Royal Society of London* B, **232**, 35–57.

Sarmiento, G. and Monasterio. M. (1983). Life forms and phenology. *Tropical savannas*, (ed. F. Bourlier), pp. 79–108. Elsevier, Amsterdam.

Savolainen, R. and Vepsäläinen, K. (1988). A competition hierarchy among boreal ants: impact on resource partitioning and community structure. *Oikos*, **51**, 135–55.

Serrano, J. M., Acosta, F. J., and Alvarez, M. (1987). Estructura de las comunidades de hormigas en eriales mediterräneos según criterios funcionales. *Graellsia*, **43** 211–23.

Strong, D. R., Simberloff, D., Abele, L. G., and Thistle, A. B. (eds) (1984). *Ecological communities: conceptual issues and the evidence*. Princeton University Press, New Jersey.

Terborgh, J. and Robinson, S. (1986). Guilds and their utility in ecology. In *Community ecology: patterns and processes*, (ed. J. Kikkawa and D. J. Anderson), pp. 65–90. Blackwell Scientific Publications, Melbourne.

Thompson, J. N. (1982). *Interaction and coevolution*. Wiley, New York.

Tilman, D. (1982). *Resource competition and community structure*. Princeton University Press, New Jersey.

Torres, J. A. (1984). Niches and coexistence of ant communities in Puerto Rico: repeated patterns. *Biotropica*, **16**, 284–95.

Traniello, J. F. A. (1989). Foraging strategies of ants. *Annual Review of Entomology*, **34**, 191–210.

Tuomi, J. and Vuorisalo, T. (1989). What are the units of selection in modular organisms? *Oikos*, **54**, 227–33.

Van Valen, L. (1978). Aborescent animals and other colonoids. *Nature*, **276**, 318.

Vuorisalo, T. and Tuomi, J. (1986). Unitary and modular organisms: criteria for ecological division. *Oikos*, **47**, 382–5.

Waloff, N. and Blackith, R. E. (1962). Growth and distribution of the mounds of *Lasius flavus* (Fabricius) (Hym: Formicidae) in Silwood Park, Berkshire. *Journal of Animal Ecology*, **31**, 421–37.

Webb, L. J. and Tracey, J. G. (1981). Australian rainforests: patterns and change. In *Ecological biogeography of Australia*, (ed. A. Keast) pp. 605–94. Dr. W. Junk, The Hague.

Whitford, W. G. (1978). Structure and seasonal activity of Chihuahua Desert ant communities. *Insectes Sociaux*, **25**, 79–88.

Wiens, D. (1984). Ovule survivorship, brood size, life history, breeding systems, and reproductive success in plants. *Oecologia*, **64**, 47–53.

Willson, M. F. (1990). Sexual selection in plants and animals. *Trends in Ecology and Evolution*, **5**, 210–14.

Willson, M. F. and Burley, N. (1983). *Mate choice in plants: tactics, mechanisms, and consequences*. Princeton University Press, New Jersey.

Wilson, J. B. (1989). Relations between native and exotic plant guilds in the upper Clutha, New Zealand. *Journal of Ecology*, **77**, 223–35.

Yeatman, E. M. and Greenslade, P. J. M. (1980). Ants as indicators of habitats in three conservation parks in South Australia. *South Australian Naturalist*, **55**, 20–6,30.

37

Problems outstanding in ant–plant interaction research

Andrew J. Beattie

Introduction

The last few decades have witnessed a phenomenal growth in our understanding of ant–plant interactions. Nevertheless, many important questions remain to be answered. The five topics discussed here have been selected to show that these questions are both fundamental and diverse and that ant–plant research may be seminal in advancing our understanding of species interactions, their genetics, and their role in population dynamics and community organization. My remarks will be confined almost wholly to those interactions that are regarded as mutualistic but some of the major points apply equally to other types of interaction such as leaf-cutting and seed-eating.

Demography and mutualism

After many years of neglect mutualisms are now the subject of much research but there is a need for more (Boucher 1985). It appears that they are abundant, that most species are involved in at least one, and that community organization is, in part, a function of their number and diversity. And yet there are doubts. They arise because of the difficulty in identifying the benefits to all the participants. And even when the benefits are known we generally do not understand how they translate into a measurable selective advantage such as survivorship and reproductive rate. Plant-borne rewards, including food, may have little quantitative impact on ant demography. Even among the best understood mutualisms, such as insect pollination, there are few data on the effects of floral rewards on the demography of the insect populations or colonies.

Benefits to plants involved with ants have been shown in a variety of ways (e.g. Inouye and Taylor 1979; Huxley 1980), but demonstrable benefits to the ants are almost entirely lacking. We have strong reasons for thinking we are dealing with mutualisms, but the data for the ants do not exist.

This concern is aggravated by the increasing number of studies showing extreme variation in the outcome of ant–plant interactions for the better known partners, the plants (e.g. Horvitz and Schemske 1986*a*; Koptur and Lawton 1988), and the demonstrations that apparent mutualisms such as ant-guard systems do not necessarily benefit the plants (e.g. O'Dowd and Catchpole, 1983; Tempel 1983; Rashbrook *et al*. Chapter 16, this volume; Mackay and Whalen Chapter 17, this volume). Studies show:

1. that simple proximity of potentially mutualistic species is often insufficient for a mutualism to occur;
2. the importance of what Cushman and Addicott (Chapter 8) and Cushman and Whitham (1989) have correctly called the conditional nature of the interactions; and
3. that habitat patchiness is vital to the outcome (Thompson 1988).

May (1976) and Schaffer and Kot (1985) put it another way in their studies of dynamic systems, by showing that the outcome of a particular interaction will depend on where the system happens to be in phase space.

Demographic studies have identified selective advantages to the plants in terms of life history parameters (Horvitz and Schemske 1986*b*; Hanzawa *et al*. 1988). Similar studies in ant populations will identify benefits of evolutionary significance to the ants. Ant–plant interactions offer us a huge array of systems from which a few can be selected for manipulation in the laboratory and in the field. In addition to identifying the true mutualists among ant–plant interactions, such studies will generate useful models for understanding changes in the sign of interaction terms in time and space.

The crucial factor for the measurement of colony fitness in ants is the net rate of reproductive formation (Wilson 1971; Brian 1983). However, large numbers of colony parameters are available for monitoring, and their measurement may indicate exactly where plant rewards have their greatest demographic impact. Such parameters include the ratio of workers to reproductives; the ratio of larvae to workers; period of production of reproductives; biomass; growth rates and survivorship of forager, nurse, defender, and nanitic worker castes; ratios of these castes; rate of trophic egg production; rate of fertile egg production; number of males and queens produced; and colony survivorship. Life table and matrix analysis are appropriate for the data generated by laboratory and field experiments of this kind (Varley *et al*. 1973; Southwood 1978).

Several models of ant colony demography have been proposed and all of them emphasize the importance of the timing of the switch from worker to reproductive formation (Lovgren 1958; Oster and Wilson 1978; Brian *et al*. 1981). During the ergonomic phase of colony growth, selection acts to increase the numbers of the worker caste. If a colony initiates reproductives too early there may be an extended period for production of alates but insufficient workers to nurse them to maturity. Delay in switching may generate a wasteful excess of workers relative to the numbers of reproductives to be reared.

The switching time is an essential component of colony fitness and is strongly influenced by the efficiency with which workers harvest resources. Oster and Wilson (1978) showed that the resource return function $R(t)$ is an integral part of equations governing the growth of worker and reproductive populations which, in turn, determine switching time. For the workers,

$$\frac{dW(t)}{dt} = bu(t) R (t) W(t) - mW(t) \qquad (37.1)$$

where W = workers, b = a conversion constant, $u(t)$ = fraction of colony energy for worker production, $R(t)$ = the net rate at which resources are returned per forager, and m = worker mortality rate; and

$$\frac{dQ(t)}{dt} = bc(1 - u(t))R(t) W(t) - vQ(t) \qquad (37.2)$$

where Q = reproductives, bc is the conversion constant for reproductives, $1 - u(t)$ = fraction of colony energy for reproductive production, and v = reproductive mortality rate.

The resource allocation function for the reproductives $(1 - u(t))R(t)$ is determined by the distribution and the quality of resources, as shown by Brian et al (1981). Therefore, plant resources such as ant rewards can play a crucial role in colony demography. Their precise importance will be discovered by manipulative demographic experiments, varying the availability of rewards such as elaiosomes, food bodies, and extrafloral nectar, and assessing the effects on parameters such as switching time.

Figure 37.1 illustrates how colony growth may appear in experimental colonies with and without access to plant food rewards. Starting conditions for each are assumed to be as similar as possible; for example, a single fertile queen. Colonies provided with plant rewards undergo sigmoid growth (Wilson 1971) and initiate reproductives at t_1, while colonies on a basic diet exhibit a much lower growth rate with a delayed switching time at t_2. Studies by D. R. Nash and A. Bennett (unpublished) appear to be the first and only attempts to discover the benefits of plant rewards to ants.

My remarks so far have been confined to interactions assumed to be mutualistic. However, there remain large gaps in our knowledge of other types of interaction such as leaf cutting and seed harvesting. What do we know about the demographic effects of leaf-cutters or seed-harvesters on the populations of plant species upon which they feed? How do they affect growth rates and survivorship? Similarly, cut leaves and harvested seeds are the primary food of many kinds of ants, but what are the effects of individual prey species on the demography and distribution of colonies?

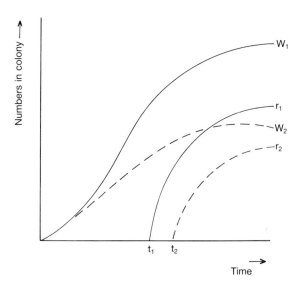

Fig. 37.1. Possible results of ant colony manipulation in the laboratory, W_1 and W_2 show the numbers of workers in colonies with and without plant rewards respectively. r_1 and r_2 show the numbers of reproductives produced by colonies with and without plant rewards respectively. t_1 and t_2 are the switching times for the two types of colony.

Genetics of ant–plant interactions

Very little is known about this aspect of ant–plant interactions and I shall focus on just two areas. First, the ways in which ants might affect the distribution of gene frequencies and gene flow in plant populations in space and time, and secondly, the genetic traits specific to the function of mutualistic interactions.

Central to the first question is the concept of the neighbourhood, which is a useful indicator of gene flow and micro-evolutionary potential (Rai and Jain 1982). Ants may affect these through their contributions to gene flow by transporting either pollen (gametes) or seeds (individuals).

Wingless ants that pollinate the orchid *Microtis parviflora* R.Br. do not forage very far (see below). Between-plant movements range from 1 to 76 cm with a mean of 6 ± 10 cm, and the distribution of forage movements is strongly leptokurtic, so that pollen flow mediated by ants is much the same as has been described for many other insect pollination systems (Levin and Kerster 1974; Peakall and Beattie 1989) promoting small neighbourhoods and spatial differentiation of gene frequencies.

These data emphasize that the physical distance moved by pollinators is not necessarily important; many pollinators besides ants are either small or territorial. The crucial factor for the plants is the array of genotypes within the foraging areas. Thus, ants are likely to have similar effects on plant populations as larger pollinators, albeit on a smaller scale.

While distance itself may not be crucial, the foraging pattern described above is based on floral food rewards, and the majority of between plant movements were very short as the ants, in common with many animal pollinators, maximized food intake with minimal expenditure of energy. By contrast, ant pollinators in the pseudocopulation system of the orchid *Leporella* were seeking mates (see below). The location of attractive flowers varied widely in time and space and consecutive pollinator visits not only involved widely separated plants but also intervening visits to perches for post-coital rests. The pattern of pollen flow promoted by this system, (we might call it optimal mate-searching behaviour), may differ from that promoted by the more frequent optimal foraging behaviour (Peakall 1989; Peakall and James 1989). The importance of seed dispersal to neighbour-hood size has been emphasized by Crawford (1984). When self-pollination is frequent, for example, gene flow by seeds may greatly influence neighbour-hood size and area.

$$N_a = 4\pi \left(\frac{t\delta^2 p}{2} + \delta^2 s \right) \tag{37.3}$$

Where N_a = neighbourhood area, t = the proportion of progeny from cross-pollination, ^2p = the variance of pollen dispersal distances, and ^2s = the variance of seed dispersal distances. When there is total selfing this expression reduces to

$$N_a = 4\pi\delta^2 s \tag{37.4}$$

showing that seed dispersal is the crucial term. In general, the role of seed dispersal in the determination of neighbourhood size is little known and study of ant dispersal systems may have much to contribute to this question.

The potential genetic effects of ant dispersal of seeds may be illustrated by the system studied by Davidson and Morton (1981 *a,b*). Seeds of *Sclerolaena diacantha* Benth. are dispersed to the mounds of the ant *Rhytidoponera* sp.B with the result that the adults are over dispersed and crowded on the nests. Investigations into the relationship between movements of ants dispersing seeds and the genetic structure of the populations have begun with the hypo-thesis that ant activity generates nest-oriented population subdivision.

Figure 37.2 is a diagram of nests within arrays of plants of varying density, to illustrate the ways in which seed dispersal by ants may contribute to population subdivision. An isolation-by-distance model would apply in the case of a continuous plant array. Alternatives are suggested by (i) the top set of arrows indicating highly localized seed dispersal that may conform to a stepping stone model, and (ii) the lower arrows suggesting random seed

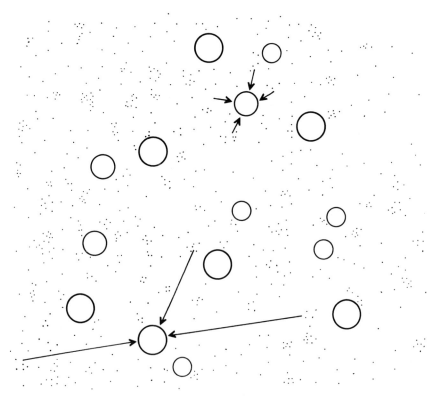

Fig. 37.2. Diagram to show possible patterns of gene flow among ant-dispersed plants (dots) growing in the presence of the nests (circles) of their dispersal agent. See text for full explanation.

accumulation from arrays of plants at varying distances that may conform to an island model.

The identity of the appropriate model depends heavily on the behaviour of ant seed dispersers. The frequency distribution of seed dispersal distances may be platykurtic (Beattie and Culver 1979; Higashi and Ito (Chapter 32, this volume) and this may reflect the importance of ant territories. The movements of ants as individuals, recruits, or columns are immensely variable (Wilson 1971; Brian 1983) and their effects on plant genetic structure will reflect this. Small or territorial ants will restrict seed movement but large, aggressive species can forage widely with impunity.

Studies are needed which will identify the effects of ants on the distribution of gene frequencies, but will also shed light on the wider question of the significance to population genetic structure of the dispersal of entire individuals (seeds). i.e. the movements of genets as opposed to gametes. Migrant animals frequently fail to contribute genes to the populations they enter for various behavioural reasons (Wilson 1975). In plants, numerous pollination studies show the contribution of pollen to population genetic structure while the role of seeds is less well known. We can hypothesize that ant dispersal results in bursts of new genotypes arriving on nests, creating high genet diversity, and that subsequent selection leads to a steady decline in this diversity until the next period of ant seed activity. Ant–plant systems such as this will contribute much to our understanding of seed dispersal, local selective forces, and the role of spatial heterogeneity in maintaining genetic polymorphism.

This brief discussion has again been confined to presumed mutualisms, and, in addition to myrmecochory, we have very little insight as to how ants rearrange gene frequencies in the plants that produce domatia, food bodies, or extrafloral nectar. The same can be said for other types of interactions such as leaf-cutting and seed-eating. Conversely, how does the distribution, density, physiology, or genotype of the plant populations effect ant genetics?

Patterns of changes in gene frequencies are relevant to the question of co-evolution. In many ant–plant interactions we are looking for mutualistic co-evolution, involving reciprocal selection on heritable traits. Gene frequencies for these traits will vary in space, oscillate through time, and even show chaotic trajectories (Schaffer and Kot 1986), in a mosaic of changing neighbourhoods. In this context co-evolution involving reciprocity appears improbable. Some traits such as extrafloral nectaries may show relatively simple inheritance (Adjei-Maafo et al. 1983; Gottlieb 1984), but many will show complex polygenic inheritance (Beattie 1985). However, strong linkage disequilibrium or pleiotropic effects may facilitate the cohesive transmission of vital traits from one generation to the next (Thompson 1989).

The evolution of many ant–plant mutualisms does not appear to require reciprocity. Interactions can arise as a result of strong directional, selective regimes on ant and plant genomes, acting independently at first but converging through time. Many plant traits involved in mutualisms may have evolved in response to directional selection unrelated to ants. For example, in Chapter 24 Jebb describes the evolution of tuber domatia from water-trapping structures and in Chapter 21 Mckey suggests that some petiole domatia arise through selection for larger leaves in new light environments. Reciprocity is not required but traits that benefit one partner because of the presence of the other may increase in frequency. Selection on ant traits may be of little initial relevance to the plants and may only produce a set of general application to the ant–plant interaction. This scenario may produce interactions that are functional (e.g. mutualistic) only under certain, perhaps infrequent, conditions but the fully operational interaction confers a selective advantage massive enough to maintain the crucial traits.

Ant–plant mutualisms may provide insights into the murk of what we call co-evolution. The key problem is to distinguish between a process of reciprocating selection on specific mutualistic traits and the *ad hoc* utilization of resources among ants and plants; what D. H. Janzen has called 'ecological fitting'. The first process is co-evolution while the second is co-ecology. Demographic studies of plant populations with and without particular traits such as extrafloral nectaries, and in different habitats with different ant communities, will begin to reveal the selective advantage of those traits.

The chemical basis of plant-borne ant rewards

Analyses of plant–borne rewards that are ingested by ants, including food bodies, extrafloral nectar, elaiosomes, and homopteran honeydew, has revealed that they contain a variety of sugars, amino acids, and lipids (Auclair 1963; Rickson 1971; Bentley 1977; Baker *et al.* 1978; Marshall *et al.* 1979; Lanza 1988). They are all required by the colony, by one caste or another, at some stage during development or reproduction (Wilson 1971). However, many ant species involved in ant–plant mutualisms are omnivores and obtain these nutrients from a variety of sources. The plant–borne rewards are only a fraction of the total diet and alternative sources of nutrition undoubtedly attract ants away from them. Competition with other food sources may explain much of the variation in the outcome of ant–plant mutualisms. In these circumstances the evolution of chemical mechanisms that increase the probability with which ants will exhibit fidelity to rewarding plant species may occur.

There are several ways in which plants might increase ant fidelity, including the following:

1. The supply of a complete diet. Some ant plants such as central American *Acacia* appear to do this (Janzen 1966) but the majority do not.
2. The utilization of a behavioural releaser. Wilson *et al.* (1958) described specific chemical releasers for necrophoric behaviour in ants and Brew *et al.* (1989) have suggested that the same releasers stimulate seed-carrying behaviour in myrmecochorous interactions.
3. A third mechanism for the manipulation of ants may be the provision of chemicals that are in short supply or unavailable elsewhere.

Spatial or temporal limitation of even a common commodity such as sugar may make a reward-bearing plant attractive and selection may result in reward production for a limited season. In the event of more specific ant requirements, plants may produce specialized chemicals. Davidson and Epstein (1989) have made some intriguing suggestions on this subject.

One aspect of this mechanism may be the provision by the plants of the dietary requirements of ants. Insects have an absolute dietary requirement for sterols (Clayton 1964; Svoboda and Thompson 1985), which are essential components of insect membranes, moulting hormones, cuticular waxes, and lipoprotein carrier molecules (Chippendale 1972; Dadd 1985). Cholesterol satisfies most insect dietary requirements and is commonly synthesized by plants. Plants also produce a variety of phytosterols which insects either use directly or by metabolizing them to cholesterol (Robbins *et al.* 1971).

The lipid content of ant rewards has been little studied but there are a few tantalizing leads. Marshall *et al.* (1979) identified the diglyceride 1,2-diolein as the attractant in the elaiosome of the European *Viola odorata*. Similar lipids are the ant attractants in other elaiosomes from North America and Australia (Brew *et al.* 1989). Lipids often constitute a major fraction of food bodies (Rickson 1971; Risch *et al.* 1977; Huxley 1986) and O'Dowd (1980) identified sterols in the pearl bodies of *Ochroma pyramidale*. Lipids have been found in some extrafloral nectars (H. and I. Baker, personal communication) and sterols are abundant in honeydew (Forrest and Knights 1972).

Intriguing indirect evidence for the possible importance of lipids as ant attractants comes from the life cycles of nematode and cestode parasites of a variety of vertebrates including lizards, rabbits, and chickens. The intermediate hosts of these parasites are ants and it is essential that the eggs or dispersal stages are taken by them (Horsefall 1938; Lee 1957; Bartel 1965; Holmes 1976). In his review of the biochemistry of these parasites Barrett (1981) pointed out that the eggs contain much lipid, but that its distribution is curious being extra-embryonic, just beneath the eggshell, and therefore of no use to the embryo. As these lipid reserves are discarded when the egg hatches, their primary function may be the attraction and manipulation of ant intermediate hosts. In the context of tapeworm dispersal Barrett speculated that 'in some cases it is conceivable that the lipid in the (ripe) proglottides might attract the intermediate host to eat them'.

A recent series of experiments by S. Berry (unpublished) explored the basic proposition that artificial nectars containing sterols were more attractive to ants than those without sterols. Several sterols and phytosterols were used but only the trials with cholesterol are reported here. Solvent-washed filter paper discs treated with 0.8 per cent cholesterol in a 10 per cent solution of sucrose, fructose, and glucose (2:1:1) were placed in the foraging areas of laboratory cultures of four ant species. Control discs contained the sugars but no sterol. The number of ants at each disc at one-minute intervals for each of 30 minutes quantified the ant responses. Table 37.1 shows that there was a preference for nectars containing cholesterol. This was especially interesting as the ant species tested had a range of natural diets. All were omnivores but *Meranoplus* and *Rhytidoponera* exhibit a preference for seeds, while *Dolichoderus* and *Iridomyrmex* prefer animal prey, extrafloral nectar, and honeydew.

Table 37.1. Ant responses to discs treated with 0.8 per cent cholesterol and sugars, or with sugars only (control). Each trial had four replica discs and the data are pooled for each ant species. The data are in 'ant minutes' (see text) and treatment and controls are compared by the G-test (Sokal and Rohlfe 1981). The data have been kindly provided by S. Berry.

Ant species	Cholesterol + sugar	Sugar only	G	P
Meranoplus sp.	709	444	61.46	< 0.001
Rhytidoponera metallica	412	323	10.80	< 0.005
Iridomyrmex sp.	236	82	77.81	< 0.001
Dolichoderus sp.	299	117	82.38	< 0.001

Preferences for other phytosterols, such as stigmasterol and ergosterol, were evident but varied according to ant species (S. Berry, unpublished work). Responses to phytosterols may depend on physiological adaptations to particular natural diets. For example, carnivores may respond most strongly to cholesterol, while omnivorous and fungivorous ants may metabolize other phytosterols more readily. The fungi cultivated by leaf-cutting ants contain high levels of ergosterol (Martin *et al.* 1969) and while this sterol elicited a mixed response from the ants in these trials, it may be of special significance to leaf-cutters. Perhaps fungus gardening evolved, in part, for its value in supplying sterol dietary requirements.

The data suggest that ants are attracted by sterols in rewards and raise the possibility that sterols may be one mechanism whereby plants maintain ant fidelity. There are a great many sterols involved in ant metabolism (Ritter *et al.* 1982) and there could be specificity, or chemical co-evolution, between plant sterol providers and particular ant species. There are also other lipid dietary requirements, such as linoleic and linolenic acids, that may be important.

Ant–plant biotechnology

There are a couple of ways in which ant–plant interactions may be useful: for pest control and for conservation.

Pest control

There have been many attempts to use ants for the control of pests on plants in orchards, plantations, forests, and row crops, with mixed success (reviewed by Majer 1982; Beattie 1985). Problems encountered have included the ants' propensity for culturing homopterans, ant predation of beneficial insects, and the effectiveness of ant predation being limited to pest outbreaks. Majer listed many properties required of ant species to be considered as candidates for biological control and Bentley (1983) suggested particular plant traits, such as extrafloral nectaries, which may increase the usefulness of ants. As the population dynamics of ant–plant interactions, the structure of ant communities, the predator–prey relationships, and the ant–homopteran relationship become better understood, it is likely that experimental manipulation will lead to greater and more long-term successes. Studies by Whittaker (Chapter 6), Rosengren and Sundotröm (Chapter 7), and Higashi and Ito (Chapter 32) are examples of recent research that advance our knowledge of potential applications for ant–plant interactions. Indeed, most studies contribute indirectly and, as data accumulate and concepts clarify, some of the problems encountered may be solved. A symposium on this subject would be immensely worthwhile.

Conservation

Majer (1984) suggested that ants, being abundant, small, and often very species rich, could be useful indicators of environmental conditions. A comparative study of seed dispersal by ants (Pudlo *et al*. 1980) in a disturbed forest in Illinois and an undisturbed forest in West Virginia, USA, suggested a similar role for ant–plant interactions. Disturbance led to the reduction or elimination of seed dispersing ant species, and this radically altered the demography of the plant populations involved. The most marked effect was that, in the absence of seed dispersal in the disturbed sites, there was no seedling recruitment and the adult population, although apparently healthy, appeared to be living off borrowed time.

The manifold effects of habitat disturbance generally result in the disruption of species interactions and conservationists now take pains to identify and preserve key interactions as a way to conserve individual or sets of species. Ant–plant systems may be advantageous for conservation research as the ants, and often the plants, can be relatively easily monitored. Identical or similar systems often occur in a variety of habitats as well as across large geographical areas, so that both local and more global changes can be measured with a reasonable amount of control.

Ant pollination

The literature on ant pollination is reviewed by Peakall *et al*. in Chapter 27. Until 1982 it appeared that ant-pollinated plant species would be readily predicted on the basis of a suite of characters governing the visits of ants to flowers within and between plants. These characters included small stature and growth close to the ground, multiple branching, inconspicuous flowers with easily accessible rewards borne externally to the foliage, and close proximity of many individual plants (Hickman 1974). However, an explanation for the small number of plants shown to be ant pollinated and the large number of flowers with, what appear to be, ant-exclusion devices remains elusive.

A substance called myrmicacin, isolated from the metapleural glands of four kinds of ants, has antibiotic properties (Maschwitz *et al.* 1970). The same substance has been shown to inhibit pollen germination and growth *in vitro* (Iwanami and Iwadare 1978). Beattie (1982) and Beattie *et al.* (1984, 1985, 1986) showed that pollen exposed to the integument of ants and to metapleural secretions inhibited pollen function. The active components of the secretions analysed were not myrmicacins but different molecules named metapleurins (A. J. Beattie and M. Lacey, unpublished work). Whatever the chemical identity of the secretions they appear to have militated against the evolution of ant pollination systems.

Two ant-pollinated orchids shed some light on this hypothesis. The first, *Leporella fimbriata* (Lindl.) George, has one or two flowers borne on stalks up to 30 cm high, and is pollinated by pseudocopulation with a male ant, *Myrmecia urens*. The workers and queen of this ant species bear metapleural glands but the males do not (Peakall *et al.* 1987). The absence of metapleural glands in the male was thought at first to explain the ability of the ant to pollinate. However, subsequent work has shown that the integuments of all three castes, males, females, and workers, disrupt pollen function and that insulation from this effect is provided by stigmatic secretions that form a layer between insect and pollen (Peakall, Angus, and Beattie 1990).

The second orchid, *Microtis parviflora*, is pollinated by flightless foragers of the ant *Iridomyrmex gracilis*. Field and laboratory experiments showed that workers ascended flowering stalks repeatedly in their search for nectar and that their visits resulted in full fruits containing fertile seeds (Peakall and Beattie 1989). The ants did possess metapleural glands but electron microscopy suggested that the secretory epithelium was not active at the time of pollination. Certainly, pollen contact with the ant integument was not harmful (Peakall and Beattie 1989). The pollen was always positioned on the face (frons) of the ant and held away from the integument by a short stalk (stipe).

These data, together with others (Hull and Beattie 1988), suggest a variety of traits that may facilitate the evolution of ant pollination.

Ant traits are:

1. Absence of metapleural glands. A preliminary survey showed that some male ants and four ant genera do not possess metapleural glands (Hölldobler and Engel-Siegal 1984). However, our work suggests that generalizations will be difficult. Within *Iridomyrmex* the males of some species possess the glands but others do not, while in *Myrmecia* the males of some species possess apparent glandular openings but there is no internal secretory tissue (C. J. Angus, unpublished work). In addition, the secre-

tion of antibiotics for distribution over the ant integument may not be limited to the metapleural gland. For example, in *Calomyrmex* it is the mandibular gland secretions that are antibiotic (Brough 1983).

2. Small or seasonally inactive glands. There is great variation in the size of metapleural glands (Hölldobler and Engel-Siegal 1984) but little is known of the timing of secretion. The formation of brood in the colony may stimulate activity in the metapleural glands.

3. Variation in the antibiotic activity of secretions. Variation in the potency of secretions relative to pollen function has been shown by Hull and Beattie (1988). Some ants may encounter fewer pathogenic micro-organisms than others and their antibiotics may be less active. Others may be unable to penetrate or damage pollen at all.

4. Ants are hairy or bristly so that pollen does not contact the secretions. The taxonomic literature shows that integuments of this type are very common.

5. Ant species that construct nests out of carton may impregnate the fabric with antibiotics from salivary glands.

Plant traits are:

1. Pollen is held away from the ant integument. This is the case in many orchids, including the ant-pollinated ones, and in the Asclepiadaceae.

2. The pollen grain is impervious to ant secretions. Some types have thick exines, no germinal pores (acolpate), or germinal pores hidden in deep furrows (Wodehouse 1935) and these traits may prevent penetration by harmful secretions.

3. Pollen is embedded in materials that minimize the effects of secretions. Kerner and Oliver (1895) noted that pollen may be invested by oils, viscin, and waxy or glutinous materials.

4. Pollen is produced in such great amounts that many grains escape harmful secretions.

5. Pollen is deposited on body parts remote from the metapleural glands. Beattie *et al.* (1985) showed that the detrimental effects on pollen increased with proximity to the metapleural gland. In *Microtis parviflora* pollen is deposited on the head, well away from the glandular openings.

The presence of these traits in the ants and angiosperms suggests that more cases of ant pollination will be found. However, the 'antibiotic hypothesis' is not the only basis for believing this. Ant pollination may be most frequent in those habitats where the traits elaborated by Hickman (1974) are adaptations to the physical environment. For example, diminutive 'cushion' plants are frequent in many arid and alpine habitats. This growth form may be advantageous to ants seeking floral nectar and plants requiring pollination where high winds and extreme temperatures mean that flying insects are infrequent. In addition, pathogenic micro-organisms may be of less significance as sources of mortality to ant colonies in these habitats, and metapleural secretions may be less potent. On the other hand, in alpine areas at least, ant diversity is very low and the array of potential ant pollinators small. By contrast, some arid and semi-arid habitats, including grasslands, exhibit extraordinarily high ant species richness and apparently low numbers of potential winged pollinators (Hagerup 1943; Anderson 1983; A. J. Beattie and R. Peakall, personal observation). These may be places where ant pollination is more common.

The search for conditions under which ant pollination has evolved is just one aspect of the very rich and rewarding field of ant–plant interactions. The future of the field depends on the will of the human race to preserve the environments in which the interactions occur.

Acknowledgements

I thank John Richards, Department of Botany, University of Newcastle-upon-Tyne for his hospitality, and Macquarie University for financial support while on study leave.

References

Adjei-Maafo, I. K., Wilson, L. T., Thomson, N. J., and Blood, P. R. B. (1983). Effect of pest damage, intensity on the growth, maturation and yield of nectaried and nectariless cotton. *Environmental Entomology*, **12**, 353–8.

Anderson, A. (1983). Species diversity and temporal distribution of ants in the semi-arid mallee region of northwestern Victoria. *Australian Journal of Ecology*, **8**, 127–37.

Auclair, J. L. (1963). Aphid feeding and nutrition. *Annual Review of Entomology*, **8**, 439–90.

Baker, H. G., Opler, P. A., and Baker, I. (1978). A comparison of the amino acid complements of floral and extrafloral nectars. *Botanical Gazette*, **139**, 322–32.

Barrett, J. (1981). *Biochemistry of parasitic helminths*. Macmillan, London.

Bartel, M. H. (1965). The life cycle of *Raillietina loeweni* Bartel and Hansen (Cestoda) from the Black-tailed Jackrabbit, *Lepus californicus melanotis* Mearns. *Journal of Parasitology*, **51**, 800–6.

Beattie, A. J. (1982). Ants and gene dispersal in flowering plants. In *Pollination and evolution* (ed. J. A. Armstrong, J. M. Powell, and A. J. Richards) pp. 1–8. Royal Botanic Gardens, Sydney.

Beattie, A. J. (1985). *The evolutionary ecology of ant plant mutualisms*. Cambridge University Press, New York.

Beattie, A. J. and Culver, D. C. (1979). Neighbourhood size in *Viola. Evolution*, **33**, 1226–9.

Beattie, A. J., Turnbull, C. L., Knox, R. B., and Williams, E. G. (1984). Ant inhibition of pollen function: A possible reason why ant pollination is rare. *American Journal of Botany*, **71**, 421–6.

Beattie, A. J., Turnbull, C. L., Hough, T., Jobson, S., and Knox, R. B. (1985). The vulnerability of pollen and fungal spores to ant secretions: Evidence and some evolutionary implications. *American Journal of Botany*, **74**, 606–14.

Beattie, A. J., Turnbull, C. L., Hough, T., and Knox, R. B. (1986). Antibiotic production: A possible function for the metapleural glands of ants. *Annals of the Entomological Society of America*, **79**, 448–50.

Bentley, B. L. (1977). Extrafloral nectaries and protection by pugnacious bodyguards. *Annual Review of Ecology and Systematics*, **8**, 407–27.

Bentley, B. L. (1983). Nectaries in agriculture, with an emphasis on the tropics. In *The biology of nectaries* (ed. B. L. Bentley and T. Elias), pp. 204–22. Columbia University Press, New York.

Boucher, D. H. (1985). *The biology of mutualism*. Croom Helm, London.

Brew, C. R., O'Dowd, D. J., and Rae, I. D. (1989). Seed dispersal by ants: Behaviour-releasing compounds in elaiosomes. *Oecologia* (in press).

Brian, M. V. (1983). *Social insects: ecology and behavioural biology*. Chapman and Hall, London.

Brian, M. V., Clarke, R. T., and Jones, R. M. (1981). A numerical model of an ant society. *Journal of Animal Ecology*, **50**, 387–405.

Brough, E. J. (1983). The antimicrobial activity of the mandibular gland of a formicine ant *Calomyrmex* sp. *Journal of Invertebrate Pathology*, **42**, 306–11.

Chippendale, G. M. (1972). Insect metabolism of dietary sterols and essential fatty acids. In *Insect and mite nutrition* (ed. J. G. Rodriguez. North Holland Publishing Company, Amsterdam.

Clayton, R. B. (1964). The utilization of sterols by insects. *Journal of Lipid Research*, **5**, 3–19.

Crawford, T. J. (1984). What is a population? In *Evolutionary ecology*, 23rd Symposium of the British Ecological Society (ed. B. Shorrocks) pp. 135–73. Blackwell, Oxford.

Cushman, J. H. and Whitham, T. G. (1989). Conditional mutualism in a membracidant association: temporal, age-specific, and density-dependent effects. *Ecology* (in press).

Dadd, R. H. (1985). Nutrition. In *Comprehensive insect physiology, biochemistry and pharmacology* (ed. G. A. Kerkut and L. I. Gilbert). pp. 313–90. Pergamon Press, Oxford.

Davidson, D. W. and Epstein, W. W. (1989). Epiphytic associations with ants. *Ecological Studies*, **76**, 200–33.

Davidson, D. W. and Morton, S. R. (1981*a*). Competition for dispersal in ant dispersed plants. *Science*, **213**, 1259–61.

Davidson, D. W. and Morton, S. R. (1981*b*). Myrmecochory in chenopodiaceous plants of the Australian arid zone. *Oecologia*, **50**, 357–66.

Forrest, J. M. S. and Knights, B. A. (1972). Presence of phytosterols in the food of *Myzus persicae*. *Journal of Insect Physiology*, **18**, 723–8.

Gottlieb, L. D. (1984). Genetics and morphological evolution in plants. *American Naturalist*, **123**, 681–709.

Hagerup, O. (1932). Myre-bestovning. *Botanisk Tidsskrift*, **46**, 116–23.

Hanzawa, F., Beattie, A. J., and Culver, D. C. (1988)). Directed dispersal: A demographic analysis of an ant-seed mutualism. *American Naturalist*, **103**, 1–13.

Hickman, J. C. (1974). Pollination by ants: A low energy system. *Science*, **184**, 1290–2.

Hölldobler, B. and Engel-Siegal, H. (1984). On the metapleural gland of ants. *Psyche*, **91**, 201–24.

Holmes, J. C. (1976). Host selection and its consequences. In *Ecological aspects of parasitology*, (ed. C. R. Kennedy) pp. 21–39. North Holland Publishing Company, Amsterdam.

Horsefall, M. W. (1938). Observations on the life history of *Raillietina echinobothrida* (Cestoda). *Journal of Parasitology*, **24**, 409–22.

Horvitz, C. C. and Schemske, D. W. (1986*a*). Seed dispersal of a neotropical myrmecochore: variation in removal rates and dispersal distance. *Biotropica*, **18**, 319–23.

Horvitz, C. C. and Schemske, D. W. (1986*b*). Seed dispersal and environmental heterogeneity in a neotropical herb: a model of population and patch dynamics. In *Frugivores and seed dispersal* (ed. A. Estrada and T. Fleming), pp. 169–86. Junk, Dordrecht.

Hull, D. and Beattie, A. J. (1988). Adverse effects on pollen exposed to *Atta texana* and other North American ants. *Oecologia*, **75**, 153–5.

Huxley, C. R. (1980). Symbiosis between ants and epiphytes. *Biological Reviews*, **55**, 321–40.

Huxley, C. R. (1986). Evolution of benevolent ant-plant relationships. In *Insects and the plant surface* (ed. B. E. Juniper and T. R. E. Southwood) pp. 257–94. Edward Arnold, London.

Inouye, D. W. and Taylor, O. R. (1979). A temperate region plant-ant-seed predator system: Consequences of extrafloral nectar secretion by *Helianthella quinqinervis*. *Ecology*, **60**, 1–7.

Iwanami, Y. and Iwadare, T. (1978). Inhibiting effects of myrmicacin on pollen growth and pollen tube mitosis. *Botanical Gazette*, **139**, 42–5.

Janzen, D. H. (1966). Coevolution of mutualism between ants and acacias in central America. *Evolution*, **20**, 249–75.

Kerner von Marilaun, A. and Oliver, F. W. (1895). *The natural history of plants*. 2 vols. Blackie & Son, London.

Koptur, S. and Lawton, J. H. (1988). Interactions among vetches bearing extrafloral nectaries, their biotic protective agents, and herbivores. *Ecology*, **69**, 278–83.

Lanza, J. (1988). Ant preferences for *Passiflora* nectar mimics that contain amino acids. *Biotropica*, **20**, 341–4.

Lee, S. H. (1957). The life cycle of *Skrjabinoptera phrynosoma* (Ortlepp) Schultz (Nematoda): A gastric nematode of texas horned toads. *Journal of Parasitology*, **43**, 66–73.

Levin, D. A. and Kerster, H. W. (1964). Gene flow in seeds plants. *Evolutionary Biology*, **7**, 139–220.

Lovgren, B. (1958). A mathematical treatment of the development of colonies of different kinds of social wasps. *Bulletin of Mathematical Biophysics*, **20**, 119–48.

Maschwitz, U., Koob, K., and Schildnecht, H. (1970). Ein Beitrag zur Funktion der Metathoracaldrüse der Ameisen. *Journal of Insect Physiology*, **16**, 387–404.

Majer, J. D. (1982). Ant manipulation in agro- and forest ecosystems. In *The biology of social insects* (ed. M. D. Breed, C. D. Michener, and H. E. Evans), pp. 90–7. Westview Press, Boulder.

Majer, J. D. (1984). Ant return in rehabilitated mines—an indicator of ecosystem resilience. In *Proceedings of the 4th International Conference on Mediterranean Ecosystems* (ed. B. Bell), pp. 105–6. University of Western Australia, Nedlands.

Marshall, D. L., Beattie, A. J., and Bollenbacher, W. E. (1979). Evidence for diglycerides as attractants in an ant-seed interaction. *Journal of Chemical Ecology*, **5**, 335–44.

Martin, M. M., Carman, R. M., and MacConnell, J. G. (1969). Nutrients derived from the fungus cultured by the fungus-growing ant *Atta colombica tonsipes*. *Annals of the Entomological Society*, **62**, 11–13.

May, R. M. (1976). Simple mathematical models with very complicated dynamics. *Nature*, **261**, 459–67.

O'Dowd, D. J. (1980). Pearl bodies of a neotropical tree. *Ochroma pyramidale*: ecological implications. *American Journal of Botany*, **67**, 543–9.

O'Dowd, D. J. and Catchpole, E. Q. (1983). Ants and extrafloral nectaries: No evidence for plant protection in *Helichrysum* sp. -ant interactions. *Oecologia*, **59**, 191–200.

Oster, G. F. and Wilson, E. O. (1978). *Caste and ecology in the social insects*. Princeton University Press, Princeton, New Jersey.

Peakall, R. (1989). The unique pollination of *Leporella fimbriata* by pseudocopulating winged male ants *Myrmecia urens*. *Plant Systematics and Evolution*, **167**, 137–48.

Peakall, R., Beattie, A. J., and James, S. H. (1987). Pseudocopulation of an orchid by male ants: A test of two hypotheses accounting for the rarity of ant pollination. *Oecologia*, **73**, 522–4.

Peakall, R., Angus, C. J., and Beattie, A. J. (1990). The significance of ant and plant traits for ant pollination in *Leporella fimbriata*. *Oecologia*, **84**, 457–60.

Peakall, R. and Beattie, A. J. (1989). Pollination of the orchid *Microtis parviflora* by flightless worker ants. *Functional Ecology*, **3**, 515–22.

Peakall, R. and James, S. H. (1989). Outcrossing in an ant pollinated clonal orchid. *Heredity*, **62**, 161–7.

Pudlo, R. J., Beattie, A. J., and Culver, D. C. (1980). Population consequences of changes in an ant-seed mutualism in *Sanguinaria canadensis*. *Oecologia*, **146**, 32–7.

Rai, K. N. and Jain, S. K. (1982). Population biology of *Avena* (ix). Gene flow and neighbourhood size in relation to microgeographic variation in *Avena barbata*. *Oecologia*, **53**, 399–405.

Rickson, F. R. (1971). Glycogen plastids in Mullerian body cells of *Cecropia peltata*—a higher green plant. *Science*, **173**, 344–7.

Risch, S., McClure, M., Vandermeer, J., and Waltz, S. (1977). Mutualism between three species of tropical piper (Piperaceae) and their ant inhabitants. *The American Midland Naturalist*, **98**, 433–44.

Ritter, K. S., Weiss, B. A., Norrbom, A. L., and Nes, W. R. (1982). Identification of delta 5, 7-24-methylene-methylsterols in the brain and whole body of *Atta cephalotes isthmicola*. *Comparative Biochemistry and Physiology* **71B**, 345–9.

Robbins, W. E., Kaplanis, J. N., Svoboda, J. A., and Thompson, M. J. (1971). Steroid metabolism in insects. *Annual Review of Entomology*, **16**, 53–72.

Schaffer, W. M. and Kot, M. (1985). Do strange attractors govern ecological systems? *Bioscience*, **35**, 342–50.

Schaffer, W. M. and Kot, M. (1986). Chaos in ecological systems: The coals that Newcastle forgot. *Trends in Ecology and Evolution*, **1**, 58–63.

Sokal, R. R. and Rohlf, F. J. (1981). *Biometry*. (2nd edn). Freeman, San Francisco.

Southwood, T. R. E. (1978). *Ecological methods*. (2nd edn). Chapman and Hall, London.

Svoboda, J. A. and Thompson, M. J. (1985). Steroids. In *Comprehensive insect physiology, biochemistry and pharmacology*, (ed. G. A. Kerkut and L. I. Gilbert). Vol. 10, pp. 137–75. Pergamon Press, Oxford.

Tempel, A. S. (1983). Bracken fern (*Pteridium aquilinum*) and nectar-feeding ants: A non-mutualistic interaction. *Ecology*, **64**, 1411–22.

Thompson, J. N. (1988). Variation in interspecific interactions. *Annual Review of Ecology and Systematics*, **19**, 65–87.

Thompson, J. N. (1989). Concepts of coevolution. *Trends in Ecology and Evolution*, **4**, 179–83.

Varley, C., Gradwell, G. R., and Hassell, M. P. (1973). *Insect population ecology: an analytical approach*. University of California Press, Berkeley.

Wilson, E. O. (1971). *The insect societies*. Belknap Press, Harvard University, Cambridge, Massachusetts.

Wilson, E. O. (1975). *Sociobiology: the new synthesis*. Belknap Press of Harvard University Press, Cambridge, Massachusetts.

Wilson, E. O., Durlach, N. I., and Roth, L. M. (1958). Chemical releasers of necrophoric behaviour in ants. *Psyche*, **65**, 108–14.

Wodehouse, R. P. (1935). *Pollen grains*. McGraw-Hill, London.

Index